LADY MARGARET HALL LIBRARY
OXFORD

302210190I

Plate tectonics exposed: ancient ocean floor (ophiolite) thrust almost horizontally over rocks of a former continental margin in western Newfoundland. Courtesy of H. Williams and the Geological Survey of Canada.

Second Edition

INTRODUCTION TO GEOPHYSICS

(Mantle, Core and Crust)

GEORGE D. GARLAND

University of Toronto
Toronto, Canada

W. B. SAUNDERS COMPANY
Philadelphia London Toronto

W. B. Saunders Company: West Washington Square
Philadelphia, PA 19105

1 St. Anne's Road
Eastbourne, East Sussex BN21 3UN, England

1 Goldthorne Avenue
Toronto, M8Z 5T9, Canada

Introduction to Geophysics—Mantle, Core and Crust ISBN 0-7216-4026-5

2 147 98765432

FOREWORD to the FIRST EDITION

This book is an outline of the means by which physics has provided us with a picture of the interior of the earth and of the development of its surface features. Geophysics has progressed so rapidly that it is presumptuous for one person to attempt to cover the entire field—even that part relating to the "solid" earth. Indeed, almost every chapter in what follows has been the subject of recent collections of papers. Nevertheless, anyone who has attempted to present the material to students will realize the need for a unified treatment. Even this does not lead to complete uniformity, for the individual subjects are often more conveniently treated with different notations and conventions. But at least a single author can draw attention to the similarities and differences between sections.

In a subject that is of interest to persons with varied backgrounds, it is necessary to draw compromises between the mathematical and the descriptive, and between the global view and the detailed study of local features. With the apparent success of the new global dynamics there is a great temptation to emphasize the description of surface features in terms of it. But if geophysics is to continue to progress, the geophysicist of the future must have a rather deeper grasp of the subject. This book is therefore frankly designed for the student who has completed two years or so of physics and mathematics, and who is now interested in turning to the study of the earth. A knowledge of vector analysis, for example, is presumed, whereas the geological terms employed are relatively few, and most of these are defined in the first chapter. Properties of global significance are emphasized, rather than the description of individual surface features.

The selection of references is always a difficulty, if the bibliographies are to be kept within reasonable limits. In a subject as international as geophysics, every student must have some appreciation of the contributions of such men as Gauss, Galitzin, Clairaut, Rayleigh, and Wadati. For the selection of recent references, it must be admitted that these have been chosen with the English-speaking reader in mind, and with an attempted balance between those dealing with extension of theory and new data, and those which are review articles.

Geophysics owes much of its recent progress to improvements in techniques and instrumentation, and no apology is given for including discussions on these aspects of the subject. At the same time, it owes a great deal to improved methods of data treatment, numerical filtering, time-series analysis and so on. It has not been possible to cover these fields, but the reader has been referred to an excellent series of papers on Theory and Computers in Geophysics.

I should like to thank all those who provided copy for illustrations; acknowledgement is given in the caption in each case. Most of the new illustrations were drawn by Mr. K. Khan.

Toronto G. D. GARLAND

FOREWORD to the SECOND EDITION

In preparing this Second Edition some eight years after the first, I have tried to do three things, in addition to the correction of obvious errors. First, there was the clear need for updating, since the majority of references in the First Edition date from 1969 or earlier. The intervening years have seen a continuation of great activity in the physics of the "solid" earth, with the elucidation of many details of global tectonics. This leads to the second change: in the First Edition, plate tectonics was introduced only in the final chapter. The elements of it are now discussed in the introductory chapter, and the concept is referred to throughout the remainder of the book.

The third modification results from the experience and comments of many persons who used the First Edition. The level of difficulty in the mathematical sections was admittedly variable, and it was difficult for a student, especially if his background were more geological than physical, to derive benefit from some sections, or even to know if he should be expected to. The book has therefore been redesigned, so that the more difficult mathematical aspects—for example, free oscillations in seismology and induction theory in geomagnetism—are in Appendices. In the body of the text, additional explanatory material and illustrations have been included, in the hope that all readers can find value in each section. This does not mean that the book has been stripped of mathematics; some new sections—for example, that on the flexure of the lithosphere—are themselves essentially mathematical, but at a level that is relatively straightforward.

I should like to thank all of those who wrote to point out errors or to make suggestions. In particular, it is a pleasure to acknowledge the help of the following reviewers, who made very specific suggestions for improvement: Dr. Rockne Anderson, NORDA, St. Louis; Dr. Allan Cox, Stanford University; Dr. Jon Galehouse, San Francisco State University; Dr. Glenn Julian, Miami University; Dr. Alan Leeds, University of Oregon; Dr. G. L. Maynard, Georgia Institute of Technology; Dr. James Peoples, University of Kansas; and Dr. John Thiruvathukal, Montclair State College.

Once again, I am most grateful to those who supplied illustrations and are acknowledged in the captions. The additional new drafting has been carried out, as in the First Edition, by Mr. K. H. Khan.

Finally, the preparation of this edition would not have been possible without the unfailing cooperation and friendly assistance of very many people in the Editorial and Production Departments of the W. B. Saunders Company. It is impossible to mention all, but I am particularly grateful to Joan Garbutt, Physics Editor, without whose constant encouragement the new edition would not have been possible.

G. D. GARLAND

CONTENTS

Gravimetry

Age and Thermal State

The Dynamic Earth

SOME INTRODUCTORY OBSERVATIONS 1

1.1 SURFACE FEATURES

The aim of geophysics, as applied to the body of the earth, is to determine the composition and state of the interior, and the manner in which internal processes produce the observed features of the earth's surface. In the following chapters of this book, we shall trace the main tools of geophysics, which involve the study of wave motions passing through the earth, or of some field—gravitational, magnetic, or thermal—of the earth. For most of these analyses there is some quantity, known from direct observations at the earth's surface, which provides a boundary condition. The purpose of this chapter is to establish these fundamental parameters.

The nearly spherical shape and mean radius of the earth are obvious starting points. As we shall show later (Chapter 11), the sea level surface departs from a sphere by only about one part in 300 of the equatorial radius. The radius itself is established through extensions of the method devised by Eratosthenes, that is, the measurement on the surface of a known number of degrees of latitude. Modern geodetic surveys have led to the adoption of the value of 6378.160 km for the equatorial radius. It is worth noting in passing that the deepest parts of the earth accessible to direct observation, in mines, are about four kilometers deep, and the deepest boreholes are about eight kilometers deep. The scope for the application of indirect methods of geophysics is thus very great.

On the outer surface, the most obvious feature is the difference between continents and oceans. The portion of the continents observable above sea level forms 29.2 per cent of the surface area, but if the shallow regions, known as continental shelves, which surround the continents are included, this figure is increased to almost 35 per cent. The topography of the continents has been known, at least in a general way, for some centuries. The land surface ranges from small depressions to a maximum elevation of about 8.8 km, with a mean height of just under 1.0 km. It is only in our own time that the full picture of the ocean floors is becoming known. In spite of the pioneer work of hydrographers such as Maury (1855), the details of the ocean floor topography were known only from scattered profiles, usually observed along the lines of trans-oceanic cables. The scientific oceanographic work of the past two or three decades has led to the more complete picture (Heezen and Menard, 1963), which on the one hand emphasizes such features as the continuity of the mid-ocean ridges (Fig. 1.1), and on the other gives an expression of the detailed topography (Fig. 1.2). The greatest oceanic depths are about 11 km, and the mean depth is just under 4.0 km.

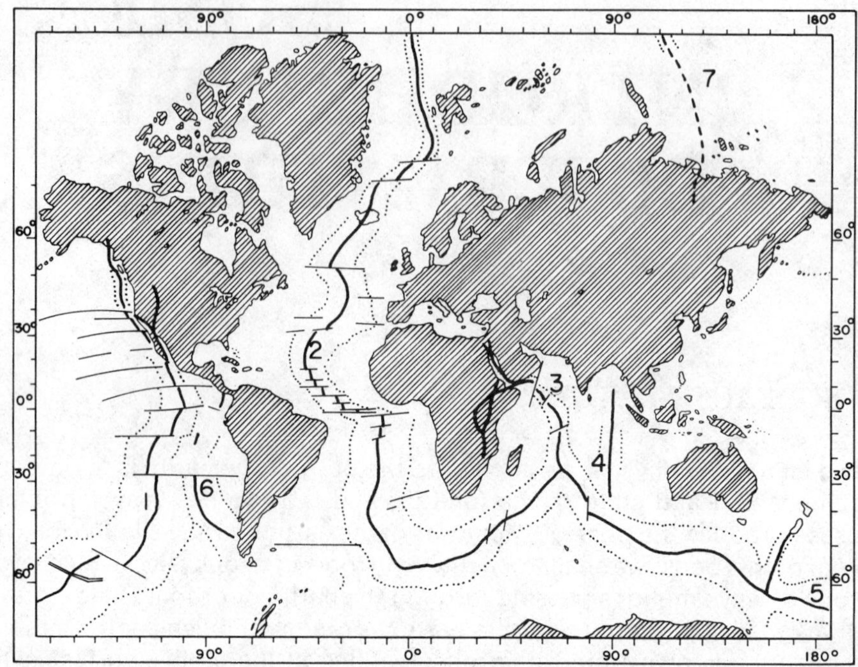

Figure 1.1 The world-wide system of ocean ridges and rises: (1) East Pacific Rise, (2) Mid-Atlantic Ridge, (3) Carlsberg Ridge, (4) 90°E. Ridge, (5) Pacific Antarctic Ridge, (6) Galapagos-Chile Rise, (7) Arctic Mid- Ocean Ridge. Heavy line shows the ridge axes, fine broken lines the edges. (After Matthews)

The continental land masses tend to be located at the corners of a tetrahedron, diametrically opposite ocean basins. This arrangement was at one time believed to have significance related to the origin of the surface features. However, if large horizontal motions are possible, the present distribution of the continents cannot be of great significance.

1.2 CHEMICAL COMPOSITION; MINERALS AND ROCKS

A remarkable fact concerning the distribution of elements in the outer part of the earth is the relatively great abundance of ten or eleven elements. Clarke

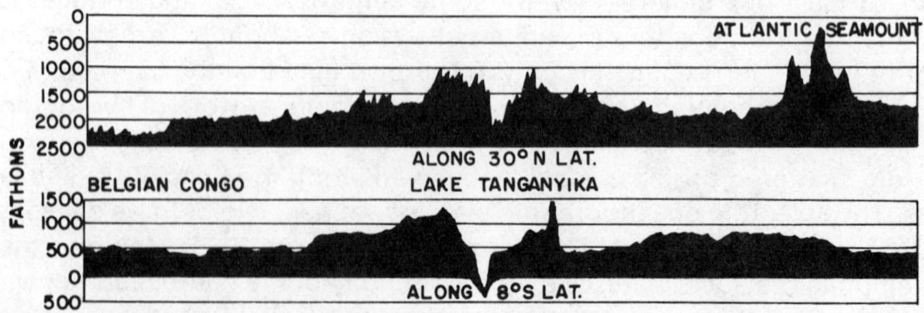

Figure 1.2 Cross-section of the Mid-Atlantic Ridge, showing the detail that is available, particularly in the vicinity of the median valley. For comparison, the cross-section of an African rift valley is shown. (After Heezen and Menard)

and Washington (1924)* determined the average composition of crustal rock; later work (Taylor, 1964) has not changed their abundances for the more common elements, which are:

	Per cent
Oxygen	46.60
Silicon	27.72
Aluminium	8.13
Iron	5.00
Calcium	3.63
Sodium	2.83
Potassium	2.59
Magnesium	2.09
Titanium	0.44
Hydrogen	0.14
Phosphorus	0.12
	99.29

Although many familiar elements (e.g., carbon, copper, zinc, tin) are missing from the list, over 99 per cent of the typical crustal rock has been accounted for.

The elements occur in minerals with definite chemical compositions and characteristic crystal structures; assemblages of minerals form rocks. It is sufficent for our purpose to outline the composition of only the most characteristic rocks of the continents and oceans. Rocks which have crystallized from a molten state are known as igneous, and are classified (Holmes, 1965) in a two-dimensional matrix, according to composition and texture. The coarse-grained types, which presumably cooled slowly and at great depths, range with decreasing content of SiO_2, from granite, through diorite and gabbro, to peridotite and dunite, a range also described as acidic (granite) to basic (gabbro) to ultrabasic (peridotite). Thus, granite consists of the minerals quartz (SiO_2), orthoclase ($KAlSi_3O_8$), and biotite or hornblende (both complex silicates), while peridotite consists of a pyroxene (e.g., $Ca(MgFe)Si_2O_6$) with olivine, which itself ranges from Mg_2SiO_4 (forsterite) to Fe_2SiO_4 (fayalite). A rock consisting entirely of olivine is known as dunite. With decreasing SiO_2 content, the rocks become characteristically darker in appearance, and of greater density. The fine-grained equivalents of these rocks are formed by rapid crystallization and are therefore typical of rocks extruded at the surface or on the sea floor. Basalt, the fine-grained equivalent of gabbro, is particularly important, as it is the characteristic rock of the oceanic crust. It is composed, typically, of a plagioclase ($CaAl_2Si_2O_8$) and a pyroxene.

Rocks deposited in water, known as sedimentary rocks, cover substantial portions of the continents, particularly in basins and over the continental shelves. Their chemical composition is usually fairly simple: sandstone consisting almost entirely of quartz, and limestone of calcite ($CaCO_3$). The shales are more complex, as they are composed of a variety of clay minerals. The fact that sedimentary rocks are normally deposited in successive horizontal layers, or strata, has been of great importance in the development of a geological time scale and in the study of later deformations.

*Clarke, F. W., and Washington, H. S. United States Geol. Surv. Prof. Paper 127 (1924).

The nuclei of all continents appear to consist of very old rocks, which have been so changed by pressure and heat, or metamorphosed, that their original nature is not always evident. A typical result is a coarsely foliated rock, gneiss (Fig. 1.3), in which alternate bands of dark and light minerals are seen. Gneisses may be formed from both igneous and sedimentary rocks, and are the predominant rock of the continental nuclei or shields.

A cross-section of a mountain-built continental area (Fig. 1.4) displays some of the complexities of geology, for which explanations must be sought in the interior of the earth. Of particular importance to geophysicists is the record of movements. That the crust has moved down is shown by the deposition of thick sequences of sedimentary rock; that it has at other times moved up is shown by deep erosion, and by the overthrusting of sections of it along fractures or faults. There is ample evidence for purely horizontal displacement also, and we shall return later to this question. The general impression conveyed by a continental profile is one of compressive deformation, which we shall show to be in contrast with the situation in the oceanic crust.

1.3 MASS, MOMENT OF INERTIA, AND ROTATION

The mass of the earth became known with the determination of G, the gravitational constant, and g, the acceleration due to gravity. It is 5.976×10^{27} gm; but a more meaningful figure is the mean density, 5.517 gm/cm^3. This is considerably greater than the mean density of crustal rocks, about 2.5 gm/cm^3, and is immediate evidence of an increase in density toward the center.

Study of the rotation of the earth about its axis provides important information on the moments of inertia of the earth and offers clues to possible mass redistributions within the earth. The earth rotates about an axis which is inclined at approximately 23.5° to the normal to the plane of the its orbit around the sun, or plane of the *ecliptic.* As a consequence of rotation, the earth's shape is that of a flattened sphere, and the attraction of the sun and moon on the resulting bulge around the equator produces a gyroscopic motion. The axis of rotation describes a cone in space, with a half-angle of 23.5° at its vertex, taking 26,000 years to complete one revolution. This motion is known as *precession;* it is detected (Fig. 1.5) through the measurement of the angles between the celestial poles (on the axis of rotation; the north celestial pole is currently near the star Polaris) and given stars. Analysis of the mechanics in detail (Appendix B) shows that the period of precession is related to the fractional difference between the moment of inertia of the earth about its rotation axis and that about an axis in the equatorial plane. Because of the spheroidal shape, the greater moment of inertia is that about the rotation axis. Denoting it by C, and the moment of inertia about an equatorial axis A, then observations on the precession yield the constant H, where

$$H = \frac{C - A}{C} = 0.00327293 \pm 0.00000075$$

A second relation between C and A is provided by the study of the earth's figure and its potential field (Chapter 11). Satellite observations have led to a remarkable improvement in our knowledge of these features, and it is now possible to determine the polar moment C with considerable precision. Recent determinations give

$$C = 0.3309\ MR_e^2$$

where R_e is the equatorial radius. For a uniform sphere, the moment of inertia about a diameter is 0.400 MR^2, so that the value of C provides further evidence of a concentration of mass near the center.

The motion of the axis of rotation in space is not precisely described by simply a smooth rotation about the surface of a cone. The axis moves in and out with small amplitude and with a period that is short compared to 26,000 years. The same effect can be seen in a spinning top, if it is given a slight knock while it is smoothly precessing. It is known as *nutation* and, in the case of the earth, it also follows from the attractions of the sun and moon.

Both precession and nutation refer to the orientation of the rotation axis in space. In addition, it is possible for the earth itself to "*wobble*" on its instantaneous axis of rotation. This is the motion seen most clearly when a flat disc is thrown into the air and set spinning about an axis making a small angle with the axis of figure. The rotation axis maintains a fixed direction, but the disc obvi-

Figure 1.3 Exposed rocks of the Canadian Precambrian shield, showing the foliated and heterogeneous character of the continental crust; a steeply dipping fault zone is visible near the middle of the photograph. (Courtesy of G. R. Guillet and the Ontario Dept. of Mines)

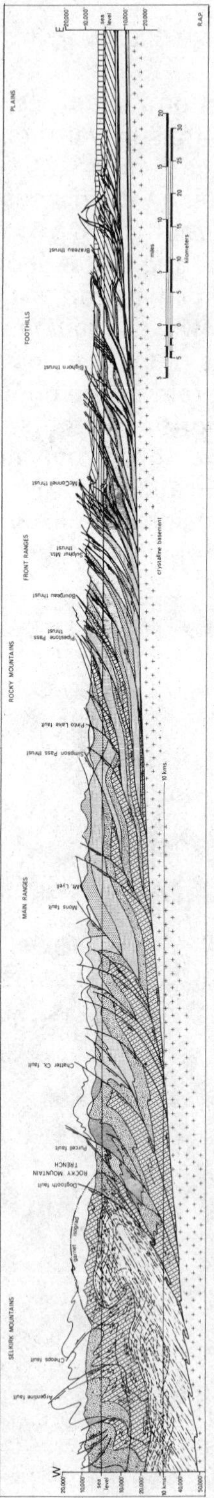

Figure 1.4 Geological cross-section of a portion of the Canadian Cordillera. Note the complex overthrusting of sedimentary formations toward the east of the profile, and the involvement of igneous rocks at the west. (Courtesy of R. A. Price and E. W. Mountjoy)

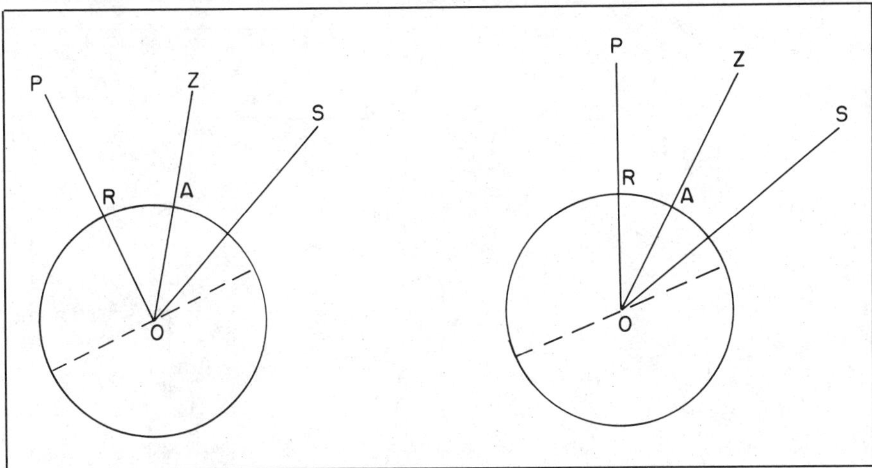

Figure 1.5 Precession and wobble. Precession alters the co-declination of a star, angle *POS*, while wobble (*Right*) changes the co-latitude of an observer *A*, which is the angle *POZ*. Point *P* is the north celestial pole; *R* is the pole of rotation. (After Munk and MacDonald)

ously wobbles about it, the axis of figure of the disc describing a cone about the rotation axis. In the case of the earth (Fig. 1.5) wobble alters the position of a point on earth relative to the pole of rotation; it is expressed quantitatively as a change in the co-latitude of the point.

For a rigid earth, theory predicts the existence of a free wobble of period 305 days. If the earth were completely fluid, its figure would immediately adjust to any axis of rotation and the period of wobble would be infinite. For the real earth, the effect is to increase the period to about 14 months, a fact pointed out by Chandler (1891), whose name has become associated with this wobble.

Redistribution of mass within the earth, including the oceans and atmosphere, can produce a forced wobble. For example, differences in air mass between summer and winter hemispheres produce a wobble with a period of 12 months, which must be separated from the Chandler wobble by spectral analysis.

Observations on the variation of latitude of fixed observatories have been internationally coordinated since 1905, forming one of the oldest examples of international cooperation in geophysics (Yumi, 1970). The results (Fig. 1.6) are displayed by a plot of the position of the pole of rotation (point *R* in Fig. 1.5) on a grid fastened to the earth's surface. While the motion contains both the Chandler and annual components, a period of rather more than one year is evident from the figure. Notice also that the amplitude of excursions varies over a time interval of 3 or 4 years. This is evidence that the free Chandler wobble is damped and must be re-excited. We shall return to the question of its cause in Chapter 27.

We have been looking at the direction of the rotation axis and its relation to the axis of figure. The magnitude of the angular velocity of rotation about the axis is also a quantity of great geophysical interest. It has been known for many years that there is a secular decrease in the magnitude, produced by the attraction of the moon and sun on tidal bulges (Chapter 14). There has been, and still is, some uncertainty as to the relative importance of tides in the oceans and the tide in the solid earth in decelerating the earth. The impinging of tides generated in the deep oceans upon shallower seas (Jeffreys, 1952) is believed to be one important factor. In any mechanism of deceleration involving the tides, the tide-producing body, sun or moon, is accelerated by attraction on it of the

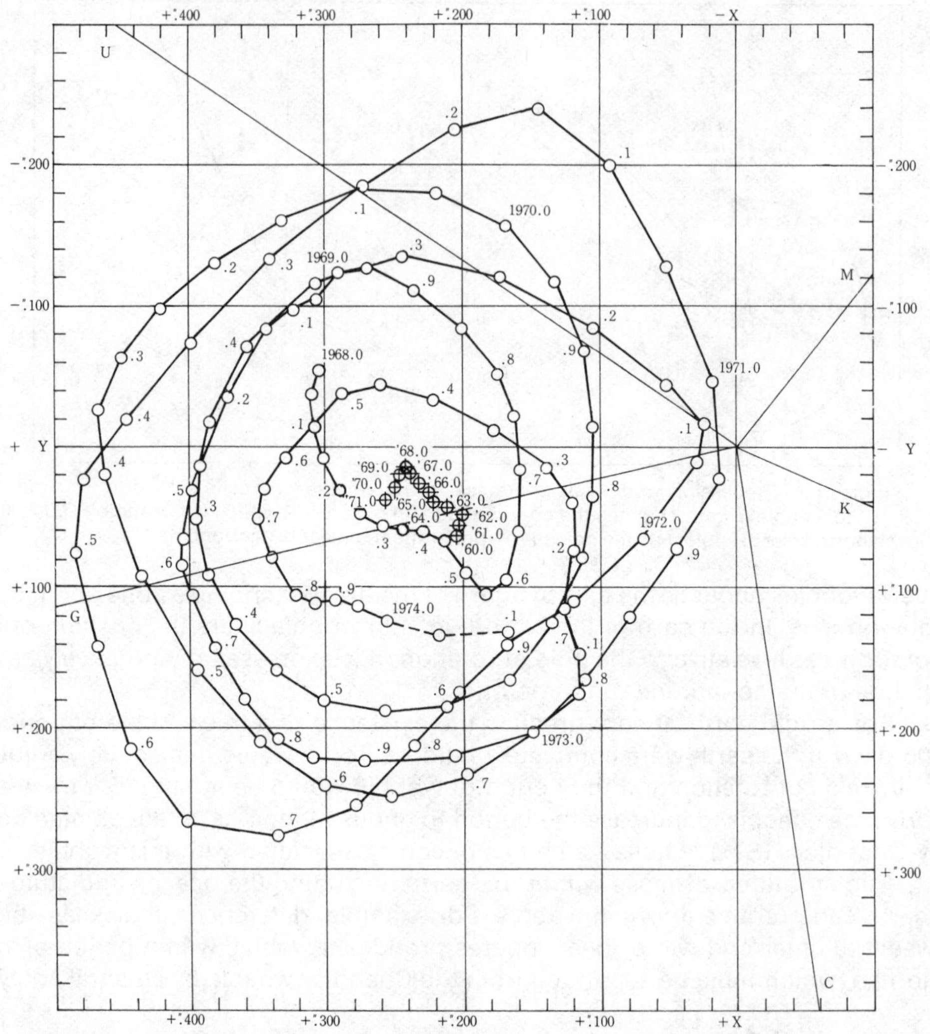

Figure 1.6 Wobble, as determined by the variation of latitude at five international stations. The curve shows the movement of the pole of rotation relative to a point fixed on earth; circles with crosses give yearly mean positions. Note the change in amplitude of wobble between 1971 and 1974. (Courtesy of S. Yumi, International Polar Motion Service)

bulges; in other words, angular momentum is transferred from the earth to the tide-raising body. The effect is greatest in the case of the moon; transfer of angular momentum to it results in its recession from the earth. Conversely, in earlier time, it must have been much closer to earth (Section 14.4).

The availability of time-keepers of high precision, in the form of atomic clocks, has confirmed that there are shorter-period variations in angular speed, of decade as well as annual duration (Fig. 1.7). Markowitz (1969) has given the amplitude of fractional changes in rotational speed as $\pm 70 \times 10^{-10}$ for the seasonal effects, and $\pm 500 \times 10^{-10}$ for those of longer duration. The latter cannot be explained on the basis of tidal effects, or of seasonal changes in the atmosphere. They appear to be caused by internal mass redistributions, of extremely short period geologically, and current thinking attributes them to motions in the fluid core. We shall refer to these observations later, when the question of the coupling of the core to the solid outer part of the earth is discussed.

1.4 SURFACE MOTIONS: PLATE TECTONICS

The concept that the outer part of the earth behaves as a relatively small number of rigid plates, with tectonic activity largely confined to their margins, is now so thoroughly diffused through the popular literature that it seems appropriate to include it as one of the basic earth parameters. The full implications can be brought out only as the evidence from seismology, gravity, the magnetic field, and other disciplines is developed in the succeeding chapters. But the kinematics of the plates and the general nature of the features which separate them can be discussed now.

The motions of the plates are intimately connected with large-scale displacements on the earth's surface, including the displacement of continents. Any historical summary must go back at least to the suggestions of Wegener as early as 1912 (Wegener, 1924) that the present continents have "drifted" through large distances. There are, however, great fundamental differences between Wegener's concept of continental rafts floating on a passive ocean floor and the new ideas. Furthermore, the assignment of credit for the establishment of plate tectonics as it now exists, even though it dates from as recently as 1960, is not easy. Certainly, the ideas owe more to the study of the ocean floors than to that of the continents themselves, and they were greatly stimulated by the suggestions of Hess (1962) and Dietz (1964) that sea floors spread from the mid-ocean ridges (Fig. 1.1). By this time, it was recognized that the occurrence of earthquakes is not random in location, but is highly concentrated along narrow lines (Figs. 6.2 and 6.3), suggesting that relative motions of the outer part of the earth are confined to these lines. Furthermore, the magnetic character of the ocean floor (Section 21.6) led Vine and Matthews (1963) to confirm sea-floor spreading by a physical test and to introduce the concept of

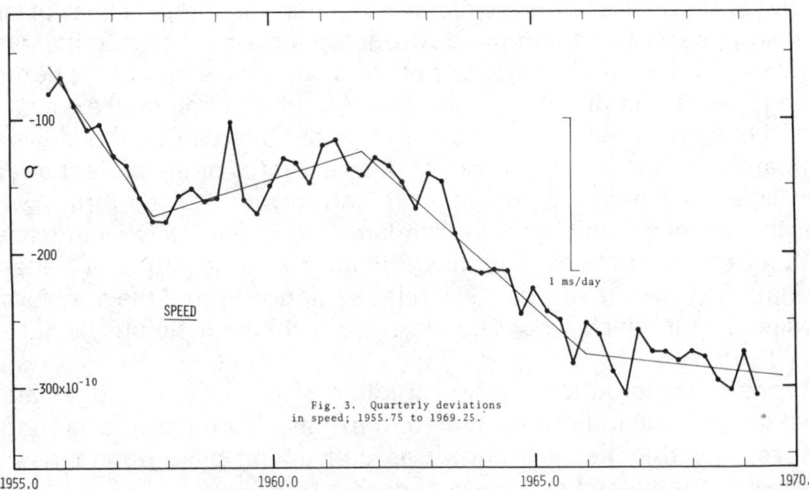

Figure 1.7 Quarterly deviations in speed, 1955.75 to 1969.25. The change in length of the day in terms of the proportional change in speed is also shown. (Courtesy of Wm. Markowitz, Nova University)

the spreading velocity, ranging from 1 to 8 cm/yr., of the sea floor from the ridges. All of these ideas on spreading involved the injection of new basaltic material into the oceanic crust along the ridge axes. A consequence is that older ocean floor must somewhere disappear, and it was presumed to go down into the earth in regions associated with oceanic trenches and deep earthquakes.

Away from the ridges and trenches, the outer part of the earth behaves in a rigid manner. This model of behavior had actually been proposed many years before, in the pioneer work of Barrell (1914), who introduced the term *"lithosphere"* for the rigid outer part of the earth, including continents, sea floor, and the material beneath, and *"aesthenosphere"* for a deeper, more yielding part of the earth. Barrell's research was based upon isostatic compensation (Chapter 12) and departures from it. He wrote: "The zone of compensation, being competent to sustain the stresses imposed by the topography and its isostatic compensation, must obey the laws pertaining to the elasticity of the solid state and is to be regarded as the nature of rock. Consequently, there may be extended to all of it the name of the lithosphere. . . ." Understanding of how sections of lithosphere could interact within the concept of ocean-floor spreading was advanced through the recognition by Wilson (1965) of three possible types of boundaries between these sections, on a plane earth. Morgan (1968) extended the study of the kinematics to a spherical earth. McKenzie and Parker (1967), in putting forth the major physical and geological consequences of lithospheric interactions, introduced the term *"plate"* for one section of lithosphere. LePichon (1968) synthesized the interaction of plates on a global scale, explaining thereby most of the observed major regions of tension, compression, and shear of the earth's surface. This was indeed remarkable, as his model encompassed only six large plates (Fig. 1.8). While the concept of relatively few rigid plates remains valid, the importance of minor plates is now recognized. A model by Morgan (1971) (Fig. 1.9) included six additional smaller plates, providing improved agreement with structure. In some areas, such as the eastern Mediterranean, still smaller plates play a role.

Let us return to the plane earth model, to introduce the types of plate boundaries noted by Wilson. Two of these, already noted, are the ridge, ideally a line of separation of two plates (Fig. 1.10), and the trench, toward which two plates converge. The trenches are often associated with arcuate lines of islands, and it is an observed fact that the down-going lithospheric plate impinges on the convex side of the arch (the shape of the Aleutians relative to the floor of the Pacific may be kept in mind). Plates in general are not bounded exclusively by ridges and trenches; there must be a third type of boundary, separating material moving parallel to the boundary, but at different rates or in different directions on each side, as Figure 1.9 suggests. Wilson introduced the term *transform fault* for this element, which plays a fundamental role in plate kinematics. The interaction of the two plates at a transform fault is in the nature of a shear; but note in Figure 1.10(a) that the sense of relative motion across the transform fault is opposite to that which would be observed if the fault simply displaced two sections of ridge.

In terms of the production or destruction of ocean floor, ridges are lines along which new ocean floor is created; they have been called accreting plate boundaries. The trenches are known as consuming plate boundaries, while along transform faults the surface is conserved.

Because of the rigid nature of plates, there are restrictions on the manner in which any of the boundary types may intersect at what is known as a *triple*

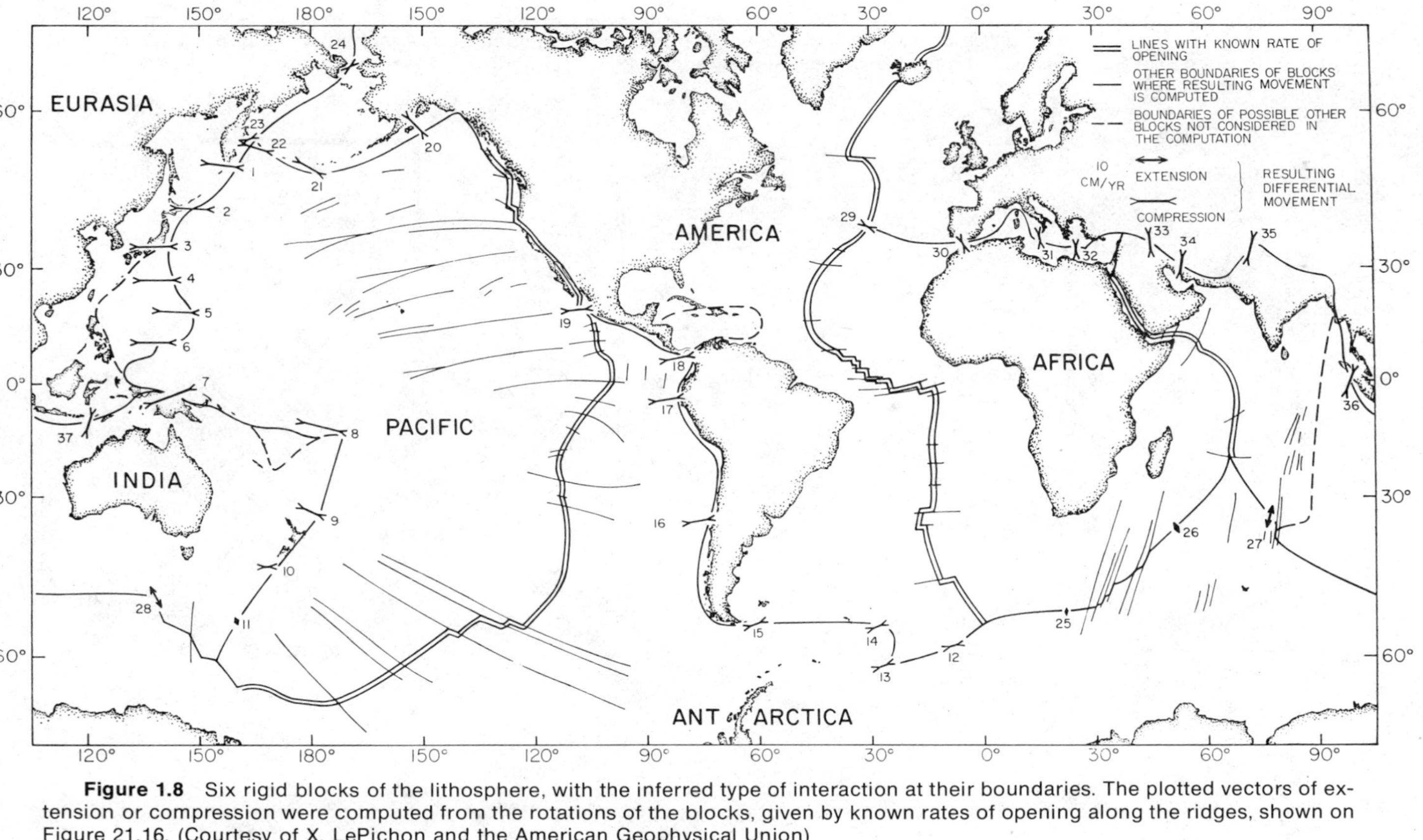

Figure 1.8 Six rigid blocks of the lithosphere, with the inferred type of interaction at their boundaries. The plotted vectors of extension or compression were computed from the rotations of the blocks, given by known rates of opening along the ridges, shown on Figure 21.16. (Courtesy of X. LePichon and the American Geophysical Union)

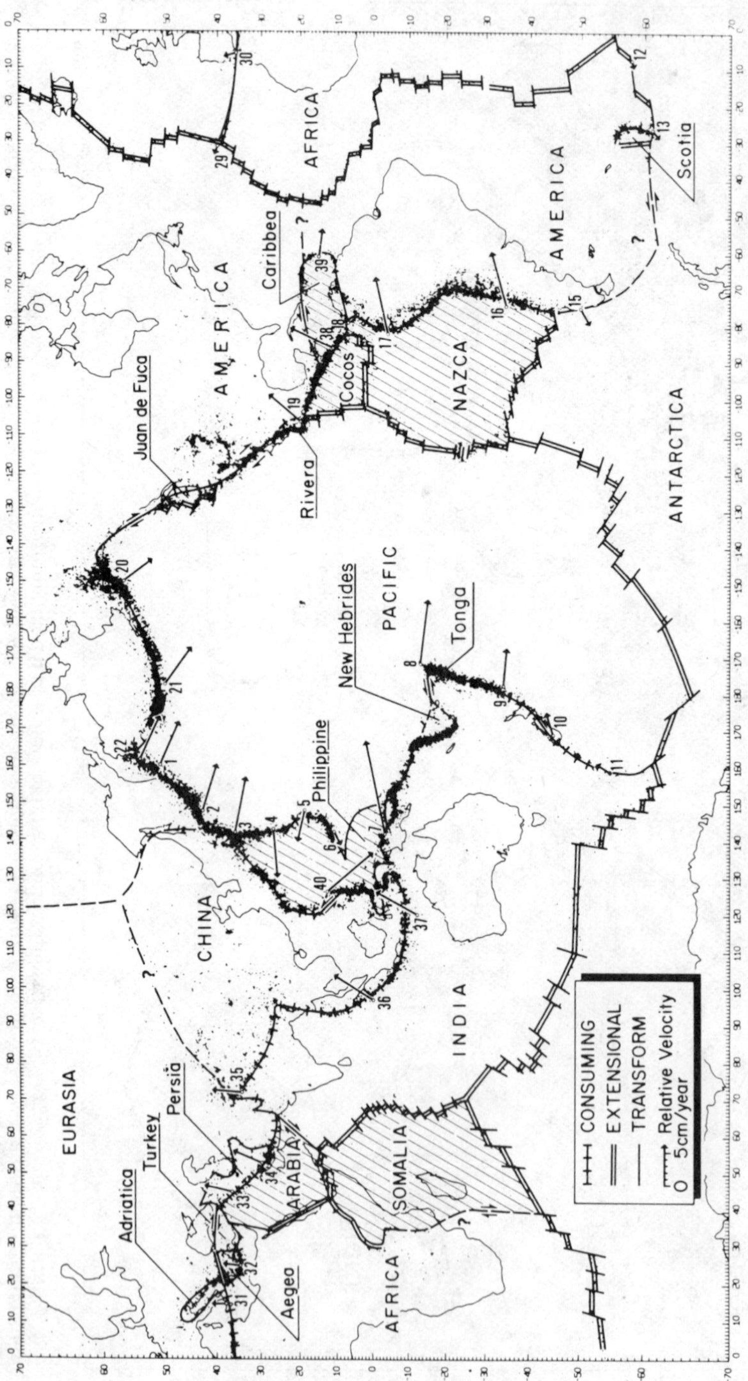

Figure 1.9 The twelve-plate model of Morgan (1971); hachured plates are those added to the six major plates of Le Pichon (1968). Small black dots are earthquake epicenters. Courtesy of X. LePichon and Elsevier Publishing Co.

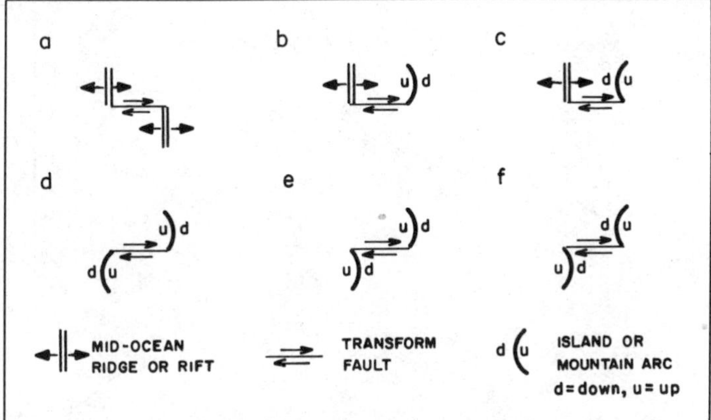

Figure 1.10 Possible types of transform faults, connecting combinations of ridges and trenches or island arcs. Arrows show direction of motion of material spreading from the ridges and descending beneath the convex side of arcs. (After Wilson)

junction, or point where three plates meet (Fig. 1.11). Across each boundary, there is a vector relative velocity for the motion between adjacent plates. In the ideal case, these vectors are normal to ridges and tangential to transform faults, but may be at any direction to the trenches. The vector sum of the three relative velocities at a triple junction must be zero, so that the vectors form a closed triangle (McKenzie and Morgan, 1969), and there are more stringent conditions if the triple junction is to remain stable. If the latter conditions, involving the orientation of the three features and the directions of motion, are not met, the character of the junction will change with time (York, 1973).

The chief modification required to extend the kinematics to a spherical earth is to invoke a theorem of Euler, that all motions of rigid plates on a sphere are rotations, and the relative motion of any two plates can be characterized by rotation about a fixed axis which intersects the sphere in two *poles of rotation.* The true relative velocity between the two plates is a (vector) angular velocity about the axis; a linear relative velocity may be measured on the earth's surface, but its magnitude will increase with distance from the axis of rotation. This recognition of the importance of rotations was fundamental in permitting reconstructions of plates before the most recent episode of spreading (Section 27.1).

The establishment of the pattern of motions in three dimensions has required a great deal of indirect evidence from regions of the earth not accessible to direct observations, and it is not yet complete. Figure 1.12 displays the chief features of plate tectonics in a vertical section. Notice that the lithosphere is absent at the ridges and thickens away from them until a thickness of the order of 100 or 200 km is reached. Because it is defined in terms of mechanical properties, the lithosphere can exist only where material has cooled until it is rigid. At the trenches, a thick section of lithosphere is pushed downward to depths of several hundred kilometers. The regions of down-going lithospheric slabs are known as *subduction zones,* or *Benioff zones,* in honor of the recognition of the arrangement of the foci of deep earthquakes along dipping conical surfaces (Benioff, 1949). Almost all lines of geophysical research have combined to determine the properties of the earth beneath ridges and subduction zones.

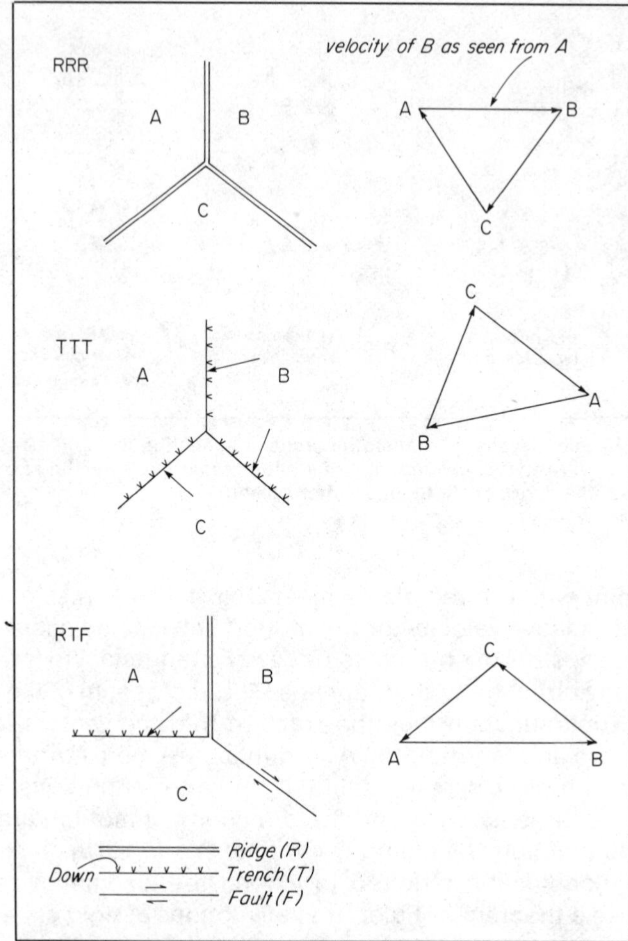

Figure 1.11 Three examples of stable triple junctions (left) and the corresponding velocity triangles (right). For none of these cases (whose real counterparts can be found on Fig. 1.9) is the junction stable for all orientations and velocities.

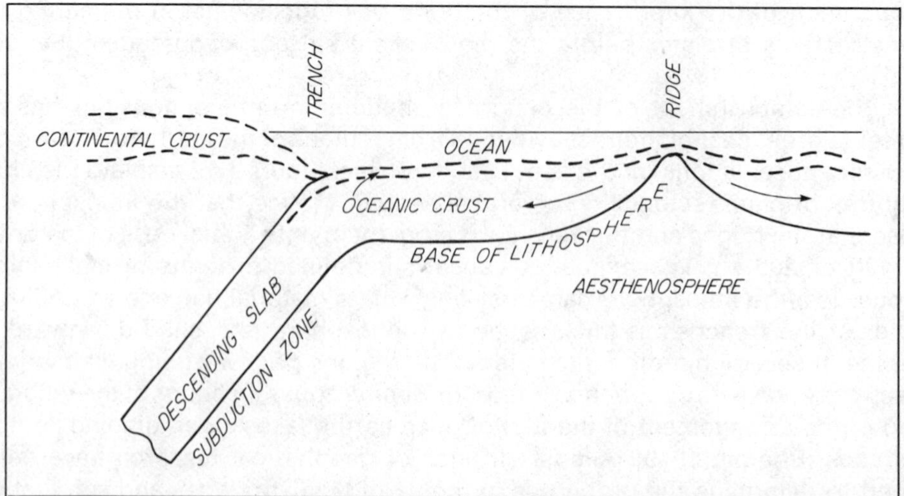

Figure 1.12 Idealized cross-section of the ridge–lithosphere–trench system.

1.5 OTHER PLANETS AND METEORITES

It is useful to compare the radius and mean density of the earth to the corresponding values for the other planets and the moon. In order of increasing distance of the bodies from the sun, these are:*

Body	Radius, km	Mean density, gm/cm^3
Mercury	2439	5.44
Venus	6052	5.24
Earth	6378.140	5.497
Moon	1738	3.3437
Mars	3397.2	3.9
Jupiter	71,398	1.3
Saturn	60,000	0.7
Uranus	25,400	1.2
Neptune	24,300	1.7
Pluto	2,500	?

It is obvious that there must be great differences in chemical composition between those planets with a density near 5 gm/cm^3 and those of lesser density. Obviously, too, size in itself does not produce a high density. We shall show later that the high mean density of the earth is related to the presence of a dense core; presumably, therefore, Mercury, Venus and Pluto also have cores. The general decrease in density with increasing distance from the sun is believed to result from outgassing of lighter elements in the case of the nearer planets.

The great wealth of physical information on the moon obtained through the Apollo missions (Levinson, 1970; Watkins, 1972) is outside of the scope of this book, but a few facts pertinent to the history of the earth-moon system should be mentioned. First is the great age of lunar soils and rocks, as determined by the methods (Chapter 23) established for dating terrestrial rocks. Dates as old as 4.6×10^9 years for the soils and in excess of 4.0×10^9 years for rocks have been obtained. These are of the same order as the age of the earth and rule out any possibility that the moon was torn from the earth during early history. We shall find, in Chapter 14, that there is evidence of lunar tides on earth at least as early as 2×10^9 years ago, so that the hypothesis of a later capture of the moon into orbit by the earth can also be ruled out. Geologically, rocks obtained from the lunar surface are broadly similar to rock types on earth: basalts in the lunar "seas" or maria and coarser-grained plagioclase-rich rocks known as anorthosites in the highlands. Bulk densities of lunar rocks range between 2.0 and 3.3 gm/cm^3, depending upon their mineralogy and, notably in the case of the basalts, their porosity. Notice how much closer to the mean density (above) is the surface density, as compared to the earth. Schreiber and Anderson (1970) showed that on the basis of seismic wave velocity alone, a number of cheeses would be consistent with lunar observations, but that they could be ruled out because of their low densities (about 1.1 gm/cm^3).

Turning to meteorites, we note that the average composition of the main types has frequently been used to suggest the composition of the earth's inte-

*Seidelmann (1977); lunar density from Bills and Ferrari (1977).

rior on the assumption that they are fragments of a similar planet. Two major classes, stony and iron (Mason, 1962) are recognized, and the stony meteorites are further classified into chondrites and achondrites. Chondrites are characterized by spheroidal structures, of the order of 1 mm diameter, known as chondrules. Chondrules normally consist of a single mineral, olivine or pyroxene, with the long axes of crystals oriented radially. Various theories for the origin of chondrules have been proposed, including crystallization from a melt, probably under high pressure of H_2O and CO_2 (Ringwood, 1959), or direct condensation from a cool dust cloud (Levin and Slonimsky, 1958). The origin of these strange features obviously holds a clue to the origin of the chondrites themselves. Chemically, the chondrites tend to be more homogeneous than the achondrites, which are stony meteorites lacking chondrules and having a lower content of alkali metals (e.g., sodium).

Urey and Craig (1953) give the following average composition, in terms of oxides and elements, based on 94 reliable analyses of chondrites*:

	Per cent		Per cent
SiO_2	38.04	Na_2O	0.98
MgO	23.84	Cr_2O_3	0.36
FeO	12.45	MnO	0.25
Fe	11.76	P_2O_5	0.21
FeS	5.73	K_2O	0.17
Al_2O_3	2.50	TiO_2	0.11
CaO	1.95	Co	0.08
Ni	1.34	Fe (total)	25.07

This is a trend similar to that found in composition of the average crustal rock, given in Section 1.2, and the similarity to basic rocks of the earth is amplified by the presence of olivine and pyroxene as characteristic minerals. However, there are important differences, notably in the abundances of iron, including free iron, and nickel. (Thirteen per cent of the average stone meteorite consists of an iron-nickel-cobalt metallic phase.)

The achondrites lack free iron and nickel. While they are variable in mineralogical composition, there is a class, composed of pyroxene and plagioclase, which is similar to terrestrial gabbro.

The iron meteorites are of great interest, because of the suggestion that the earth's core is of similar composition. Brown and Patterson (1947) found, as the average of 107 analyses, the following percentages of the three chief elements: Fe, 90.78 per cent; Ni, 8.59 per cent; Co, 0.63 per cent. There is some evidence that the absolute nickel content of iron-meteoritic material is higher (perhaps 11 per cent) since the proportion tends to increase with the mass of the meteorite.

The majority of iron meteorites contain coarse octahedral crystals of an iron mineral, kamacite. Other minerals, particularly the iron–nickel taenite, are frequently developed along lattice planes, so that, on a cut surface, a pattern of intersecting lines, known as Widmanstätten structure, is observed. The large

*Urey, H. O., and Craig, H. "The composition of the stone meteorites and the origin of the meteorites," *Geochim. et Cosmochim. Acta,* **4,** 36 (1953).

crystals are evidence of slow but appreciable cooling, and the Widmanstätten structure must be considered in any theory of the origin of iron meteorites.

The whole problem of the origin and history of meteorites is one of great interest, but also of very considerable complexity. From the point of view of geophysics, the key factors are their ages, the overall chemical composition, and the question of whether stony and iron meteorites could represent the remnants of the mantle and core of a planet similar to the earth. It was once thought that the structure of iron meteorites was evidence of formation under high pressure, and that they could have come from the core of a planet of terrestrial size. Recent work suggests that meteorites come from parent bodies, probably in the asteroid belt, whose dimensions were of the order of hundreds rather than thousands of kilometers. Perhaps the most striking advance in knowledge has come from the dating of five events (Anders, 1963, 1964) in the life history of meteorites: the synthesis of the elements, melting of the parent bodies after accretion, cooling of the parent bodies, breakup, and fall of the meteorites. Some of the dates are based upon radioactive decay, or exposure to cosmic rays, and we shall discuss them when we consider the dating of terrestrial rocks. A significant feature of the dates is that they bracket the period of cooling of the parent bodies. Further evidence of the rate of cooling has been provided by a study of the Widmanstätten structure of the iron meteorites. Wood (1967), on the basis of the diffusion in the iron-nickel system necessary to produce the kamacite-taenite pattern, has concluded that the rate of cooling was between 2 and 40°C per million years. When we consider the thermal state of the earth, we shall see that this cooling rate is far greater than can be obtained in the central region of a body of terrestrial size; this is probably the strongest argument in favor of relatively small parent bodies. A second argument is that any body as large as the earth, or even the moon, would require so much energy to break up that it would be heated to the point where any original structure, such as the Widmanstätten pattern, would be destroyed. Evidence in favor of a large parent body would be the presence of high-pressure forms of minerals. Diamonds have been found in meteorites, but these could have been formed during transient periods of high pressure, produced by impact shock (Anders and Lipschutz, 1966). Thus we cannot conclude that meteorites come from a planet of dimensions similar to those of the earth, but we can accept them as a sampling of the relative abundance of elements in the solar system.

The proportion of finds of iron to stony meteorites, for example, could be expected to indicate the relative abundance of iron. Foreseeing a result from Chapter 3, we may take the volume of the earth's core to be about 17 per cent of the volume of the earth; because of its greater density, the core has a mass very nearly one-third that of the whole earth. The total mass of iron meteorites might therefore be predicted to be one-half that of stony meteorites. In fact, of all "finds" (i.e., meteorites collected after being recognized on earth, but not seen to fall), iron meteorites account for over half in number but have a much greater total mass. On the other hand, in terms of observed "falls," iron meteorites account for only about 6 per cent of the total number, and less than 10 per cent of the total mass. The apparent contradiction could be explained in terms of a decreasing frequency of iron falls, but it is usually attributed to the fact that, once a meteorite has fallen, it is much more easily recognized as such if it is iron. If this is the case, and if the proportion of the mass of iron meteorites in the observed falls is characteristic of their true abundance, the parent bodies must have had an iron content significantly smaller than the earth's.

1.6 THE HUMAN IMPACT OF VOLCANISM AND SEISMICITY

Just as the deformation of geological strata provides evidence of past distortions of the earth's crust, so the occurrence of volcanoes and earthquakes provides proof that the earth is still in a very active state (Fig. 1.13). The geophysicist uses volcanoes as a means of studying the composition of the interior, and earthquakes as a source of elastic waves; before we turn to these rather coldly scientific applications, it is well to reflect upon the very great impact these terrestrial phenomena can have on human existence.

The toll of some of the great earthquakes of historic time is difficult to comprehend. One of the first destructive shocks of which there are detailed records was that which occurred in Lisbon on All Saints' Day, 1775. All of the large buildings in the city were destroyed, and between 30,000 and 70,000 persons were killed, many of them in churches, within 6 minutes. The earthquake was one of a series which occurred in Spain, Portugal, and northwestern Africa. There has long been a controversy concerning the area over which the Lisbon earthquake itself was felt (Davison, 1936), but it included at least the continent of Europe and northern Africa.

Japan has had its share of disastrous earthquakes, and has many times been swept by the sea-waves or tsunamis generated elsewhere. For purely seismological damage, few earthquakes can compare with that of 1923, which devastated the cities of Tokyo and Yokohama, although Tokyo was almost 60 miles from the source. In the two cities and the surrounding country 100,000 persons lost their lives. Recent years have seen major earthquakes in Morocco (Agadir) in 1960, in Chile in 1960, in Alaska in 1964, and in Peru in 1970. The death tolls were very different, a fact which emphasizes not only the importance of the location of the source relative to population density, but also the effect of the quality of construction. In Agadir, over one-third of the population of 33,000 are estimated to have perished, mostly in the rubble to which poorly constructed buildings were reduced in a few seconds.

The toll from active volcanoes is undoubtedly much lower; nevertheless, there have been some spectacularly tragic examples of the danger of building cities at the foot of apparently extinct volcanoes. The destruction of Pompeii and perhaps 2000 of its citizens by the eruption of Vesuvius, in A.D. 79, is well-known. Less familiar, perhaps, is the fact that an eruption of Vesuvius in 1631 (the tenth since that of 79) killed an estimated 18,000 people in its vicinity. In 1883 the spectacular eruption of Krakatoa occurred, resulting in the almost complete alteration of the topography, the formation of enormous clouds of dust, and the production of a tsunami similar to those generated by earthquakes. The casualties in the case of Krakatoa were almost entirely the result of the tsunami, over 35,000 persons being drowned in Sumatra and Java.

In our own century, the eruption of Mont Pelée, Martinique, in 1902 has been one of the most destructive. The volcano was quiet for over 50 years, until small eruptions began in the spring of that year. After a few weeks of moderate activity, an explosion of great violence swept gas-laden lava over the town of St. Pierre, killing 30,000 persons. A remarkable feature of the eruption (and possibly a contributing factor to the violence of the main outburst) was the formation of a solid plug of lava in the vent. The plug was slowly forced out and stood above the cone for several months.

These examples serve to emphasize the reality and power of the external effects of processes within the earth. Man cannot control the phenomena, but at the very least an understanding of the causes can lead to a minimizing of their destructiveness. Already, the most dangerous zones for earthquake activity

Figure 1.13 Volcanic processes in action: the birth of the volcanic island Surtsey, on the Mid-Atlantic Ridge. Photograph taken by S. Thorarinsson, University of Iceland, on December 5, 1963, when the island was three weeks old and about 1 km in diameter. Courtesy of S. Thorarinsson.

have been recognized so that man's activities can make allowance for them; through international cooperation warning can be given of the generation of tsunamis; volcanic eruptions have been foretold, and there is hope that the forecasting of earthquakes will become a reality.

BIBLIOGRAPHY *(Chapter 1)*

Anders, E. (1963) Meteorite ages. In *The Moon, meteorites and comets.* Vol. IV, ed. G. P. Kuiper. Univ. of Chicago Press, Chicago.

Anders, E. (1964) Origin, age and composition of meteorites. *Space Science Reviews,* **3**, 583.

Anders, E., and Lipschutz, M. E. (1966) Critique of paper by N. L. Carter and G. C. Kennedy, "Origin of diamonds in the Canyon Diablo and Novo Urei meteorites." *J. Geophys. Res.*, **71**, 643.

Barrell, Joseph (1914) The strength of the earth's crust. *J. Geol.*, **22**, 28–48; 145–165; 209–236; 289–314; 441–468, 537–555; 655–683; 729–741.

Benioff, H. (1949) Seismic evidence for the fault origin of oceanic deeps. *Bull. Geol. Soc. Amer.*, **60**, 1837–1856.

Brown, H., and Patterson, C. (1974) The composition of meteoritic matter. II. The composition of iron meteorites and of the metal phase of stony meteorites. *J. Geol.*, **55**, 508.

Bills, Bruce G. and Ferrari, Alfred J. (1977) A lunar density model consistent with topographic, gravitational, librational and seismic data. *Journ. Geophys. Res. 82,* 1306–1314.

Chandler, S. (1891) On the variation of latitude. *Astronomical Journal,* **11**, 83.

Clarke, F. W., and Washington, H. S. (1924) The composition of the earth's crust, *U.S. Geol. Surv. Prof. Paper 127.*

Davison, C. (1927) *The founders of seismology.* Cambridge Univ. Press, Cambridge.

Dietz, R. S. (1961) Continent and ocean evolution by spreading of the sea floor. *Nature,* **190**, 854–857.

Heezen, B. C., and Menard, H. W. (1963) Topography of the deep-sea floor. In *The sea,* Vol. III, 233, ed. M. N. Hill. Interscience, New York.

Hess, H. H. (1962) History of the ocean basins. In *Petrologic studies—Buddington Memorial Volume,* Geol. Soc. Amer., 599–620.

Holmes, A. (1955) *Principles of physical geology.* Nelson, London.

Jeffreys, H. (1962) *The earth* (4th ed.). Cambridge Univ. Press, Cambridge.

Keen, M. J. (1963) *An introduction to marine geology.* Pergamon Press, Oxford.

LePichon, X. (1968) Sea-floor spreading and continental drift. *J. Geophys. Res.,* **73**, 3661–3697.

Levin, B. Y., and Slonimsky, G. L. (1958) Question of the origin of meteoric chondrules. *Meteoritka,* **16**, 30.

Levinson, A. A., ed. (1970) *Proceedings of the Apollo 11 Lunar Science Conference* (3 volumes). Pergamon Press, New York.

Markowitz, W. (1970) Sudden changes in the rotational acceleration of the earth and secular motion of the pole. *Proc. of the NATO Advanced Study Institute on Earthquake Displacement Fields and the Rotation of the Earth,* ed. A. E. Beck, L. Mansinha and D. E. Smylie. D. Reidel, Dordrecht.

Mason, B. (1962) *Meteorites.* Wiley, New York.

Matthews, D. H. (1967) Mid-ocean ridges. In *Dictionary of geophysics,* ed. S. K. Runcorn. Pergamon Press, London.

Maury, M. F. (1855) *The physical geography of the sea.* S. Low, London.

McKenzie, D. P., and Morgan, W. J. (1969) Evolution of triple junctions. *Nature,* **224**, 125–133.

McKenzie, D. P., and Parker, R. L. (1967) The North Pacific: an example of tectonics on a sphere. *Nature,* **216**, 1276–1280.

Morgan, W. J. (1968) Rises, trenches, great faults and crustal blocks. *J. Geophys. Res.,* **73**, 1959–1982.

Morgan, W. J. (1971) Plate motions and deep mantle convection. In *Hess Volume,* ed. R. Shagam. Geol. Soc. Amer. Mem. 132.

Munk, W. H., and MacDonald, G. J. F. (1960) *The rotation of the earth.* Cambridge Univ. Press, Cambridge.

Price, R. A., and Mountjoy, E. W. (1970) Geologic structure of the Canadian Rocky Mountains between Bow and Athabasca Rivers—a progress report. *The Geologic Assoc. of Canada, Spec. Paper No. 6.*

Ringwood, A. E. (1959) On the chemical evolution and densities of the planets. *Geochim. et Cosmochim. Acta,* **15**, 257.

Schreiber, E., and Anderson, O. L. (1970) Properties and composition of lunar materials: Earth analogies. *Science,* **168**, 1579–1580.

Seidelmann, P. K. (1977) Summary of the IAU (1976) system of astronomical constants. *International Astronomical Union, Information Bulletin 37.*

Taylor, S. R. (1964) Abundance of chemical elements in the continental crust: a new table. *Geochim. et Cosmochim. Acta,* **28**, 1273.

Urey, H. C., and Craig, H. (1953) The composition of the stone meteorites and the origin of the meteorites. *Geochim. et Cosmochim. Acta,* **4**, 36.

Vine, F. J., and Matthews, D. H. (1963) Magnetic anomalies over oceanic ridges. *Nature,* **199**, 947–949.

Watkins, C., ed. (1972) *Lunar Science III.* Lunar Science Institute, Houston (824 pp.).

Wegener, A. (1924) *The origin of continents and oceans,* E. P. Dutton and Co., New York.

Wilson, J. T. (1965) A new class of faults and their bearing on continental drift. *Nature,* **207**, 343–347.

Wood, J. A. (1967) The early thermal history of planets: evidence from meteorites. In *Mantles of the earth and terrestrial planets,* ed. S. K. Runcorn. Interscience, London.

York, D. (1973) Evolution of triple junctions. *Nature,* **244**, 341–342.

Yumi, S. (1970) Polar motion in the recent years. *Proc. of the NATO Advanced Study Institute on Earthquake Displacement Fields and the Rotation of the Earth,* ed. A. E. Beck, L. Mansinha and D. E. Smylie. D. Reidel, Dordrecht.

THE FUNDAMENTAL PRINCIPLES OF ELASTIC WAVE THEORY 2

2.1 SEISMOLOGY AS A GEOPHYSICAL DISCIPLINE

It is appropriate to begin the consideration of the earth's interior with seismology, for in many ways that discipline provides the most certain information on those parts of the earth which cannot be directly examined. At the very least, it provides a framework within which other evidence may be discussed, by revealing the broad divisions of the earth into core, mantle and crust. Literally, the word *seismology* means the study of earthquakes, but it has come to include the study of elastic waves, originally from earthquakes, but also from artificial explosions, and of all the parameters that can be deduced from the propagation of these waves.

The range of information that is available through seismology is very great, and in this regard, the discipline is in contrast to the study of a particular force field of the earth, such as the gravitational field. For example, we shall show that in the very distribution of earthquakes, which represent sudden releases of stress within the earth, there is evidence of the nature of the interior. The distribution is by no means random, particularly as regards the depth below the surface at which shocks can occur. When an earthquake does occur, waves of elastic vibration of various types are radiated, both through the earth and around the surface. From the travel times of these waves, their velocities as a function of depth can be determined; since the waves are reflected and refracted at boundaries within the earth, these boundaries can be located. The nature of the arrivals at seismic stations can be used to indicate the mechanism of the earthquake source, and this in turn leads to a knowledge of the stress system which produced it. Additionally, if the earthquake is of sufficient magnitude, the earth as a whole may be set into oscillations of various modes, and the periods of these free oscillations can yield information.

Let us begin with an examination of a seismogram (Fig. 2.1) of a moderately strong earthquake in Turkey that was recorded in northwestern Canada, an arc distance of approximately 75°. The diagram represents 24 hours of recording, with time increasing from left to right on each line; small ticks on the lines are minute marks. Since the seismogram is from a vertical-component instrument,

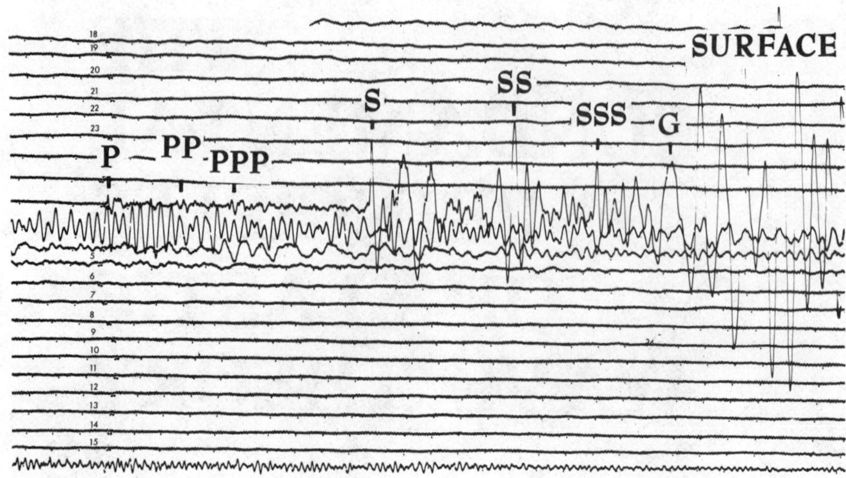

Figure 2.1 Record of a magnitude-6 earthquake (in Turkey, March 28, 1969) produced by the vertical-component long-period seismograph at Yellowknife ($\Delta = 74.93°$). (Courtesy of Earth Physics Branch, Dept. of Energy, Mines and Resources, Canada)

departures of the trace from a straight line indicate that there is vertical motion of the ground at the station site. Apart from very small oscillations, the ground is quiet until the first arrival of waves, at the instant marked P. Thereafter, the ground is in motion for about three hours, but with a character which changes with time. For almost 10 minutes after P, the disturbances are of small amplitude and high frequency; the instant marked S corresponds to a great increase in amplitude, with longer period, while later both amplitude and period increase again. Other instants marked by a temporary increase in amplitude are PP, PPP, SS, and SSS. All of these times denote the arrival of separate wave trains, known as *phases,* from the earthquake source. If we had for comparison the records of horizontal component ground motion for the same station and date, we would see differences in the relative amplitudes, on different components of the various phases, indicating that the different wave trains carry different types of particle displacement. For example, the arrivals P, PP, and PPP would be found to be much less distinct on the horizontal component records, while S, SS, and SSS would be strong. A closer examination of the record suggests that the general shape, or character, of P, PP, and PPP is similar, as is the character of S, SS, and SSS. We might surmise that each group represents similar waves which have traveled to the seismograph station by different paths.

The existence of different types of waves, traveling by more than one path from source to receiver, is of fundamental importance in seismology, and it was predicted by elasticity theory before any observations were available.

2.2 STRESS AND STRAIN

We shall develop the theory of elasticity to the extent necessary to show the existence of elastic waves of different types. In view of the fact that several examples of non-elastic behavior will be discussed in this and later parts, a word of explanation is required. At one time (not so many years ago) developments based on elastic behavior were used to interpret many characteristics of the "solid" earth, including seismic wave transmission. Experience has shown that this approach was extremely successful in predicting and interpreting the characteristics of seismic waves, even though it is now generally accepted that non-elastic phenomena play a major role in other earth processes. The difference appears to be in the time domain involved; seismic waves have characteristic periods of a few seconds, and travel times through the earth of a few minutes at most. For these short times, developments based upon elasticity theory appear to come very close to the truth. We shall find it necessary to discuss departures from perfect elasticity only as they apply to the attenuation of seismic waves with distance in the earth.

The development begins most conveniently with the concept of stress in an extended elastic body. In elementary treatments, stress as a force per unit area is introduced, but normal and shearing stresses are usually considered separately, and in geometries which leave no doubt as to the directions of the forces involved. To describe the "stress at a point" in an extended body is less direct. Let us take (Fig. 2.2) axes x_1, x_2, and x_3 (which we may refer to as 1, 2, and 3 axes), and isolate in the body a small tetrahedron with the point in question at one corner and with three faces parallel to the coordinate planes. The matter of the tetrahedron is in static equilibrium, under the forces and moments exerted by the surrounding matter, across the four faces. Across each face there will, in general, be a normal and a shear force; in the case of the three principal faces of the tetrahedron, each shear force can be resolved into components parallel to the axes. We define the nine components of stress at the point P to be the limiting values of the ratios of these nine forces to the areas of the corresponding faces, as the tetrahedron shrinks to zero volume. The components of stress are designated by subscripts, as in p_{12} (Fig. 2.2), with the first subscript indicating the plane concerned, and the second the direction of the force which results from that stress component. In general, the stress at P is denoted p_{ij}. It will be recognized that a component such as p_{11} is related to a normal force, while p_{12}, for example, gives rise to a shear.

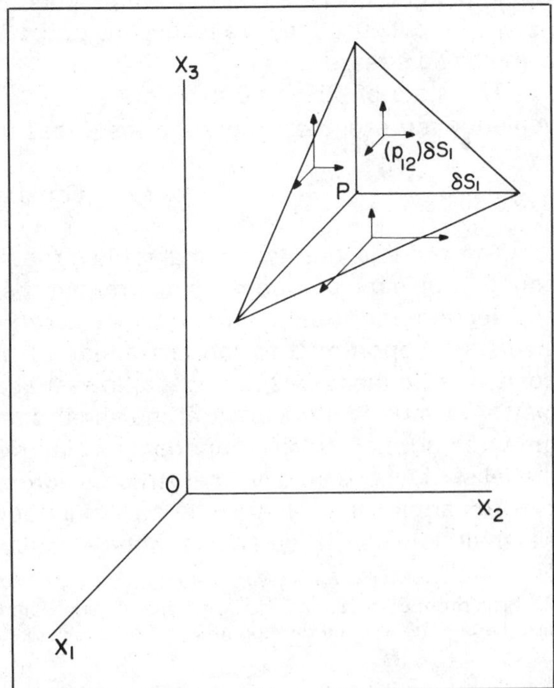

Figure 2.2 The specification of stress at a point involves 9 quantities: 3 normal and 6 shear components.

Because of the static equilibrium of the tetrahedron, the stress components associated with the sloping face are not independent. If we take a new set of axes 1′, 2′, and 3′, in which 1′ is perpendicular to the sloping face of the tetrahedron, it is not difficult to show, for example, that

$$p_{1'2'} = \sum_{i=1}^{3} \sum_{j=1}^{3} a_{i1'} a_{j2'} p_{ij} \qquad 2.2.1$$

where $a_{i1'}$ is the cosine of the angle between the i and 1′ axes; and so on. Equation 2.2.1 is nothing more than the transformation rule for second-order *tensors,* of which stress is an example.

We shall find, in what follows, a number of expressions of the form of Equation 2.2.1, and it is convenient to introduce immediately the usual tensor notation, in which summation signs are omitted, but are implied whenever a subscript is repeated in one term of any expression.

Returning to the equilibrium of the matter in the tetrahedron, we have not yet satisfied the second condition for static equilibrium. This requires the balancing of moments, which leads (the development is now more straightforward with a rectangular prism than with our tetrahedron) to

$$p_{ij} = p_{ji} \qquad 2.2.2$$

so that there are, in fact, only six independent components of stress to be defined at each point of a body.*

In this discussion of equilibrium, it will be noted that no mention has been made of forces, such as weight, which act throughout the body of the matter, rather than across surfaces. However, such forces diminish with the volume of the isolated piece of matter, and therefore vanish more rapidly than forces which depend on areas. In following sections, body forces may frequently be neglected because they are locally insignificant compared to forces associated with elastic stresses.

The state of stress at any point (x_1, x_2, x_3) in an elastic body may be conveniently represented by forming the stress quadric

$$p_{ij} x_i x_j = \text{constant}.$$

The quadric has two remarkable properties: first, the coefficients in its equation give the components of stress at the point for any orientation of axes, and second, the square of the radius vector from the origin to any point on it is inversely proportional to the magnitude of normal stress across a plane perpendicular to the radius vector. It follows that one set of axes (the axes of figure) may always be found such that the shear components of stress vanish, and, of the three normal stress components which remain, one is the greatest and one the least of all possible normal stresses across planes through the point.

The application of stress to an elastic body produces deformations, which we shall consider to be always infinitesimally small. It is necessary, in consid-

*The proof of Equation 2.2.2 forms one of the problems based on this chapter; in later sections also, certain parts of the development will be left as exercises.

ering the new position of points in a body after the application of forces, to separate any rigid-body rotation from the elastic deformation itself. This is usually done in the following way. We consider a point P, now at coordinates x_i, which as a result of rotation and deformation suffered a displacement a_i. Let a neighboring point be Q, at $(x_i + y_i)$.

The displacement of Q can always be written

$$u' = u_i - \zeta_{ij}y_j + e_{ij}y_j$$

where

$$\zeta_{ij} = \frac{1}{2}\left(\frac{\partial u_j}{\partial x_i} - \frac{\partial u_i}{\partial x_j}\right)$$

$$e_{ij} = \frac{1}{2}\left(\frac{\partial u_j}{\partial x_i} + \frac{\partial u_i}{\partial x_j}\right) \qquad 2.2.3$$

It is evident that

$$\zeta_{ij} = -\zeta_{ji};\ \therefore\ \zeta_{ij} = 0,\ i = j$$

$$e_{ij} = e_{ji}$$

By considering the contributions to the displacement of Q of the different terms of ζ_{ij} and e_{ij}, and their effects on the line PQ, the reader may show that ζ_{ij} represents a pure rotation (ζ_{23} being the rotation about the x_1-axis), that e_{11} represents a simple fractional elongation parallel to the x_1-axis, and that e_{23} represents the shear strain in planes perpendicular to the x_1-axis. In the latter case, it must be noted that the angle of shear, or the angle by which a right angle is deformed, is $2e_{23}$, since e_{23} and e_{32} contribute equally to this deformation.

The tensor e_{ij}, with six independent components, is the strain tensor, defining the strain at any point in a body under stress.

We note that the tensor e_{ij} gives also the familiar cubical *dilatation*, or volume strain resulting from a change in hydrostatic stress. Consider a cube, with faces parallel to the coordinate planes, now of unit size. Before straining, its edges were $1 - e_{11}$, and so forth. The fractional change in volume was

$$\theta = 1 - [(1 - e_{11})(1 - e_{22})(1 - e_{33})] \qquad 2.2.4$$

or

$$\theta = e_{11} + e_{22} + e_{33} = e_{ii}$$

because products and squares of the strain components are neglected in the infinitesimal strain theory.

Substituting from Equation 2.2.3, we find the dilatation in terms of displacements as

$$\theta = \frac{\partial u_1}{\partial x_1} + \frac{\partial u_2}{\partial x_2} + \frac{\partial u_3}{\partial x_3} \qquad 2.2.5$$

2.3 ELASTICITY AND THE MODULI

A perfectly elastic body obeys a generalized Hooke's Law, in which each component of stress is a linear function of the components of strain. In general, this would be expressed by an array of six equations, with 36 constants:

$$\begin{aligned} P_{11} &= Ae_{11} + Be_{22} + Ce_{33} + De_{12} + \cdots \\ P_{22} &= Ge_{11} + \cdots \\ &\cdot \\ &\cdot \\ &\cdot \end{aligned} \qquad 2.3.1$$

Imposing conditions of symmetry reduces the number of independent constants. For example, if the material concerned is crystalline, it will have symmetries determined by the crystal class to which it belongs. The higher the degree of symmetry, the smaller the number of independent constants. In the limiting case of perfect isotropy, it may be shown that only two constants survive; in fact, these are the constants designated A and B above. All other constants of the array, C, D, G, and so forth, either vanish or are expressible in terms of A and B.

The assumption of isotropy is, of course, an approximation, but not as poor a one as may be supposed. The rocks of at least the outer part of the earth are indeed crystalline, but if the crystals are randomly oriented, a condition of isotropy is approximated in volumes which contain many crystals. Obviously there could be departures from this condition in the case of foliated rocks.

Substitution of the constants in Equations 2.3.1 gives

$$\begin{aligned} P_{11} &= Ae_{11} + B(e_{22} + e_{33}) \\ &= B\theta + (A - B)e_{11} \\ P_{23} &= (A - B)e_{23} \\ &\cdot \\ &\cdot \\ &\cdot \end{aligned} \qquad 2.3.2$$

The quantity $(A - B)$ relates stress to strain for pure shear, but since e_{23} is one-half the angle of shear, $(A - B)$ is equal to 2μ, where μ is the usual modulus of shear; the constant B is known as Lamé's constant, and is usually designated λ (Lamé introduced both λ and μ, but λ has no other name.) A compact form of expressing all of the stress-strain relations is then

$$p_{ij} = \lambda\theta\delta_{ij} + 2\mu e_{ij} \qquad 2.3.3$$

where δ_{ij}, (known as the "Kronecker δ") is simply an operator such that

$$\begin{aligned} \delta &= 1, \qquad i = j \\ \delta &= 0, \qquad i \neq j \end{aligned}$$

The other familiar elastic moduli may be expressed in terms of λ and μ by considering the application of special types of stress, and solving Equation

2.3.3. If a hydrostatic pressure dp ($dp_{11} = dp_{22} = dp_{33}$) is applied, the bulk modulus, or modulus of incompressibility, K, is given by

$$K = -dp/d\theta \tag{2.3.4}$$

Adding the three pertinent equations of 2.3.3 leads immediately to

$$K = \lambda + \tfrac{2}{3}\mu \tag{2.3.5}$$

If one longitudinal stress only, say p_{11}, is operative (as in the stretching of a wire), Young's modulus E is given by

$$E = p_{11}/e_{11} \tag{2.3.6}$$

In this case Equation 2.3.3 leads to

$$E = \frac{\mu(3\lambda + 2\mu)}{\lambda + \mu} \tag{2.3.7}$$

For the same situation, Poisson's ratio σ is given by

$$\begin{aligned} \sigma &= -e_{22}/e_{11} \\ &= \frac{\lambda}{2(\lambda + \mu)} \end{aligned} \tag{2.3.8}$$

Algebraic manipulation permits us to express all moduli in terms of E and σ, which, in the case of static loading experiments, are the values most easily measured on samples. Thus

$$\begin{aligned} \lambda &= \frac{E\sigma}{(1+\sigma)(1-2\sigma)} \\ \mu &= \frac{E}{2(1+\sigma)} \\ K &= \frac{E}{3(1-2\sigma)} \end{aligned} \tag{2.3.9}$$

The third equation of 2.3.9 shows that as σ approaches the value $\frac{1}{2}$, the material becomes incompressible. Some values of the elastic moduli for important rock types are shown in Table 2.1, in which the quantities actually measured are indicated.

TABLE 2.1 EXAMPLES OF VALUES OF ELASTIC CONSTANTS FOR DIFFERENT ROCK TYPES.

Rock	K 10^{12} dynes/cm^2	E 10^{12} dynes/cm^2	μ 10^{12} dynes/cm^2	σ
Limestone (Knoxville, Tenn.)		0.621*	0.248	0.251*
" (Nazareth, Pa.)	0.342*			
Granite (Quincy, Mass.)	0.132*	0.416*	0.197*	0.055
Gabbro (Quebec)	0.659*	1.08*	0.438	0.219*
Dunite (New Zealand)		1.52*	0.60*	0.27
Dunite (Balsam Gap, N.C.)	0.892*			

Measured quantities are starred; unstarred values are computed using the relations between the constants, assuming isotropic conditions. (Extracted from Francis Birch: Compressibility; Elastic Constants, in *Handbook of Physical Constants,* ed. S. P. Clark Jr., Mem. 97, Geol. Soc. Amer., 1966.)

2.4 EQUATIONS OF MOTION

As the purpose of this chapter is essentially to establish the existence of elastic waves, we pass directly to a consideration of the equations of motion of a point within an elastic body under stress. We consider that, at a point $P(x_1x_2x_3)$ in such a body, the stress, measured from an initially unstressed state, is p_{ij}. Let P form the corner, closest to the origin, of a rectangular prism of matter, of sides δx_1, δx_2, δx_3. The matter outside the prism exerts forces on all faces, but if the stress field were uniform, there would be no unbalanced external force acting on the prism, other than body forces. The presence of a gradient in any of the stress components leads to such an unbalanced force. For example, the component parallel to x_3, resulting from shear forces transmitted across faces perpendicular to x_2, is

$$\left[\left(p_{23} + \frac{\partial p_{23}}{\partial x_2}\,\delta x_2\right) - p_{23}\right]\delta x_1 \delta x_3$$

Considering all faces, the component of force parallel to x_3 is the sum

$$(\partial p_{j3}/\partial x_j)\;\delta x_1 \delta x_2 \delta x_3$$

If we let f_i be the components of acceleration of P, X_i the components of "body force" at P (X_i is a force per unit mass, and dimensionally equivalent to an acceleration), and ρ the density, the equations of motion are immediately seen to be

$$\rho f_i = \frac{\partial p_{ij}}{\partial x_j} + \rho X_i \qquad 2.4.1$$

However, it is more useful to express these in terms of the strains, using the results of the previous section.

Then

$$\rho f_i = \frac{\partial}{\partial x_j}\,(\lambda\theta\delta_{ij} + 2\mu e_{ij}) + \rho X_i \qquad 2.4.2$$

Substituting for e_{ij}, and recalling the significance of δ_{ij}, we find

$$\rho f_i = \frac{\partial}{\partial x_i}(\lambda\theta) + \frac{\partial}{\partial x_j}\left\{\mu\left(\frac{\partial u_j}{\partial x_i} + \frac{\partial u_i}{\partial x_j}\right)\right\} + \rho X_i \qquad 2.4.3$$

We shall deal here only with the case of homogeneous material, for which λ and μ are constant. In this case, the terms may be regrouped in the form

$$\rho\frac{\partial^2 u_i}{\partial t^2} = (\lambda + \mu)\frac{\partial\theta}{\partial x_i} + \mu\nabla^2 u_i + \rho X_i \qquad 2.4.4$$

where $\nabla^2 = \partial^2/\partial x_i^2$, and the partial derivative on the left has been written for the acceleration.

At this point, we recall the general form of the wave equation, for any physical quantity y which is time- and space-dependent:

$$\partial^2 y/\partial t^2 = c^2\nabla^2 y \qquad 2.4.5$$

The familiar one-dimensional solution is

$$y(x, t) = f(x - ct) + F(x + ct) \qquad 2.4.6$$

in which f and F are determined, from a very broad range of permissible functions, by the initial and boundary conditions, and in which c is the wave velocity.

An examination of Equation 2.4.4 suggests that, if body forces (ρX_i) were neglected, it would resemble the wave equation, except that it would still contain two rather different terms on the right side. But this is hardly surprising, when we recall from elementary experience that there are, in solids, two kinds of wave motion: those which represent the propagation of a change in density (sound waves), and those which involve a distortion or change in shape. If we further recall that the dilatation θ is related to change in density, and that it is obtained from the displacements u_i by Equation 2.2.5, a method of reducing Equation 2.4.4 is suggested. We operate on the latter equation with $\partial/\partial x_i$, and add the three resulting equations. This leads to

$$\rho\frac{\partial^2\theta}{\partial t^2} = (\lambda + 2\mu)\nabla^2\theta \qquad 2.4.7$$

This is now identical to the wave equation, which shows that a dilatation, or change in density, can indeed be propagated, and with a velocity

$$\alpha = \left(\frac{\lambda + 2\mu}{\rho}\right)^{1/2} \qquad 2.4.8$$

However, an alternative operation on Equation 2.4.4 is possible. Waves which involve a distortion of the material may be expected to produce a relative rotation of neighboring particles. In the language of vectors, rotation is expressed by the curl of the vector displacement, and we might therefore investigate the behavior of curl **u** or ($\nabla \times \mathbf{u}$). Let us, in fact, take the curl of both sides of Equation 2.4.4 (each term of which is really a vector). After a permissible change in the order of operations, this yields:

$$\rho \frac{\partial^2}{\partial t^2} (\nabla \times \mathbf{u}) = (\lambda + \mu) \left(\nabla \times \frac{\partial \theta}{\partial \mathbf{x}_i}\right) + \mu \nabla^2 (\nabla \times \mathbf{u}) \qquad 2.4.9$$

It is left as an exercise to show that the first term on the right of Equation 2.4.9 vanishes, leaving

$$\rho \frac{\partial^2}{\partial t^2} (\nabla \times \mathbf{u}) = \mu \nabla^2 (\nabla \times \mathbf{u}) \qquad 2.4.10$$

which is once again the wave equation. In this case, the rotation, or curl of the particle displacement, is propagated with the velocity

$$\beta = (\mu / \rho)^{1/2} \qquad 2.4.11$$

Although we have not shown the uniqueness of our two solutions, these in fact are the only two waves that can exist in an unbounded, homogeneous elastic body. The essential properties of the two waves were recognized by Poisson. The first, traveling with velocity α, represents the propagation of changes in density of the material; it is the familiar sound wave, in which the particle displacement is in the direction of propagation. The second, traveling with velocity β, is a wave which represents a rotation of material without change in volume, in which particle displacement is in planes perpendicular to the propagation direction; it is known as a rotational or shear wave. A more detailed treatment of the elastic wave equations may be found in Båth (1968).

Early in the history of instrumental seismology it was recognized that the sound or longitudinal wave from an earthquake arrives first, because α is always greater than β. That wave was denoted *P (primus)* and the shear wave was called *S (secundus)*. This nomenclature is universal in geophysics, and we shall use it henceforth.

Finally, we note that β vanishes with μ, indicating that *S* waves cannot be propagated in a fluid. Although our development above was for an elastic solid, no inconsistency is found in the development for *P* waves if μ is allowed to vanish; *P* waves are propagated through fluids with the resulting value of α.

Problems

1. Show (with a diagram) that $p_{ij} = p_{ji}$.

2. Establish the two important characteristics of the stress quadric, referred to in Section 2.2.

3. To establish that ξ_{23}, for example, represents a rotation about the 1-axis, show on a diagram, viewed in the 2–3 plane, the projected positions of two neighboring points *P* and *Q* in an elastic body, before and after straining; then show that ξ_{23} is the rotation of the projection of the line *PQ*.

4. Establish the relations between λ and μ, and between α and β, when $\sigma = \frac{1}{3}$.

BIBLIOGRAPHY *(Chapter 2)*

Båth, Markus (1968) *Mathematical aspects of seismology.* Elsevier, Amsterdam.
Poisson, S. D. (1827) Note sur les vibrations des corps sonores. *Ann. de Chimie,* **36**, 86.

BODY WAVE SEISMOLOGY 3

3.1 TIME-DISTANCE CURVES AND THE LOCATION OF EPICENTERS

We refer to *P* and *S* waves as body waves, to distinguish them from a series of surface waves which will be met later. In this chapter we shall discuss the main features of the earth that can be determined from the passage of body waves, generated by an earthquake or large explosion, through the earth to an observing station. For the moment, we will consider an earthquake to be simply a generator of waves, postponing a more detailed study of earthquake mechanisms. Similarly, we take a seismograph station to be simply a receiver capable of timing the arrival of a *P* or *S* wave.

If the velocities of *P* and *S* are to be used to provide information about the earth's interior, the travel times of the waves must be accurately determined. When an earthquake occurs, waves from it are recorded at known locations, at known times (although the latter are not free of error). For velocities to be determined, the source must be located in time and space. The clue to the location of the source lies in the fact that α is everywhere greater than β; a graph of arrival time against arc distances for *P* and *S* waves consists of two diverging curves (Fig. 3.1). We note here, incidentally, that travel-time curves are always plotted against arc distance, although the body waves travel through, rather than around, the surface of the earth. Once time-distance curves are accurately established, the time interval $S - P$ measured at any station gives a unique indication of arc distance to the earthquake, and distances determined from three or more stations locate the source uniquely. Similarly, the *P* or *S* travel-time curves, together with the observed arrival times at any station, permit an estimate of the origin time of the earthquake. The actual location of the source could be achieved by drawing arcs around the stations on a globe or on a suitable projection of the earth's surface. For the present, we shall consider only earthquakes which occur close to the earth's surface, so that the source is effectively at the *epicenter,* which is the point on the earth's surface vertically above the true source or *focus*. The complications arising from a focus at greater depth will be discussed later in this chapter.

The weakness in the above method is obvious; a time-distance curve must be established before epicenters can be located, but this location is essential for construction of the curve. Indeed, the early history of seismology dealt largely with the successive improvement of time-distance curves and epicenter location. The situation was not so hopeless as might appear, for some earth-

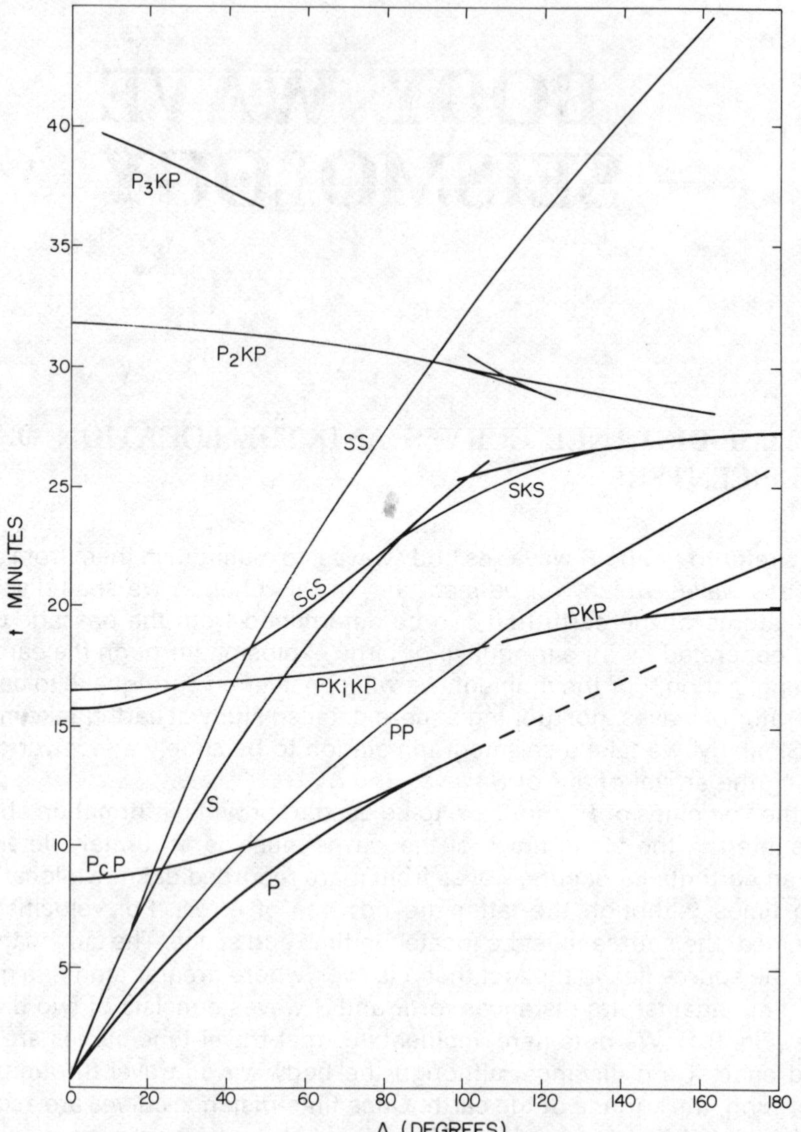

Figure 3.1 Travel times for some of the more important seismic wave paths. The behavior of *PKP* is shown in more detail in Figure 3.16. Times are basically those of Jeffreys and Bullen (1940), with the addition of PK_iKP from Engdahl et al (1970) and P_3KP from Buchbinder (1972). For P_mKP the arc distance is 360° less the angle plotted.

quakes do occur which can be rather precisely located through direct observation of effects at the earth's surface or by very close seismograph stations. The first time-distance curves were constructed from observations of waves from such earthquakes. By 1907, Wiechert and Zöppritz had constructed time-distance curves which permitted the determination of epicenters with reasonable accuracy.

Revision of time-distance curves was accomplished through a least-square technique developed by Jeffreys (1931). In this method, the residuals, or differences between observed arrival times of *P* or *S* and their computed arrival times, based on the preliminary epicenters and preliminary travel-time curves, are minimized by the simultaneous adjustment of epicenter coordinates, origin time, and travel-time curve. The number of equations involved becomes very great as more earthquakes and stations are utilized, but Jeffreys devised a method of treating the observations in groups. This analysis, in collaboration with Bullen, culminated in the Jeffreys–Bullen Tables (1935, 1940), which give the adjusted travel times, not only of *P* and *S*, but also of reflected and refracted waves, as functions of the arc distances Δ. Bullen (1965) has described the production of these tables, including the derivation of the correction for the elliptical shape of the earth.

It is difficult to overemphasize either the achievement of the J-B Tables, produced in an era before high-speed computers were available, or their importance to geophysics over a period of thirty years. They have provided the basic information for body wave seismology, which in turn yielded earth models that are still taken as standards for reference. As would be expected, improvements in observing technique, especially timing, and the availability of large artificial explosions, have permitted corrections to the J-B Tables to be made. In 1968, new tables for *P* waves were published by Herrin.

3.2 THE EFFECT OF BOUNDARIES

The theory developed in Chapter 2 applied to an unbounded elastic solid. Before proceeding further it is necessary to investigate the behavior of body waves in the vicinity of the boundary between two media of different properties. We shall take both the boundary and all wave fronts to be planar; this is not a serious restriction, because it is approximated if the radius of curvature of a non-planar boundary is large compared to a wavelength, and if the observer is not too close to the source.

Observation suggests that an incident wave of either type is in general reflected and refracted at a boundary, as in optics, but with the added complication that new waves of both types (*P* and *S*) are produced (Fig. 3.2). Our problem is to find the directions of these waves and to outline the method by which their amplitudes are determined.

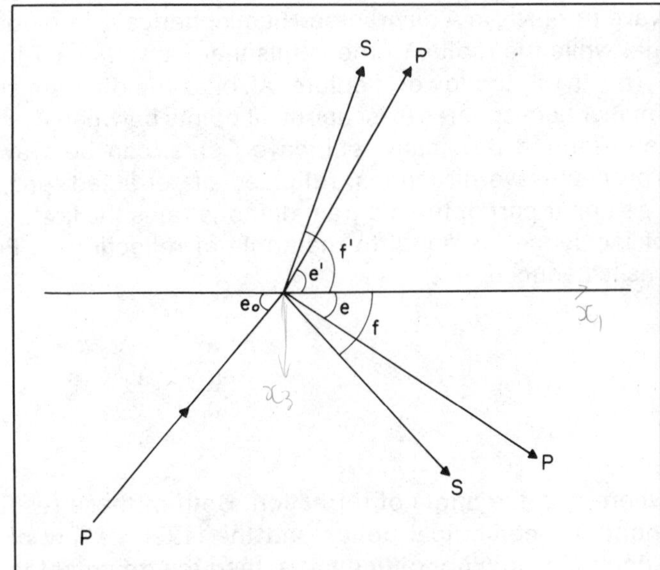

Figure 3.2 Four waves generated by a *P*-wave incident on the boundary between two solids. The angles of emergence are discussed in the text, and in Appendix C.

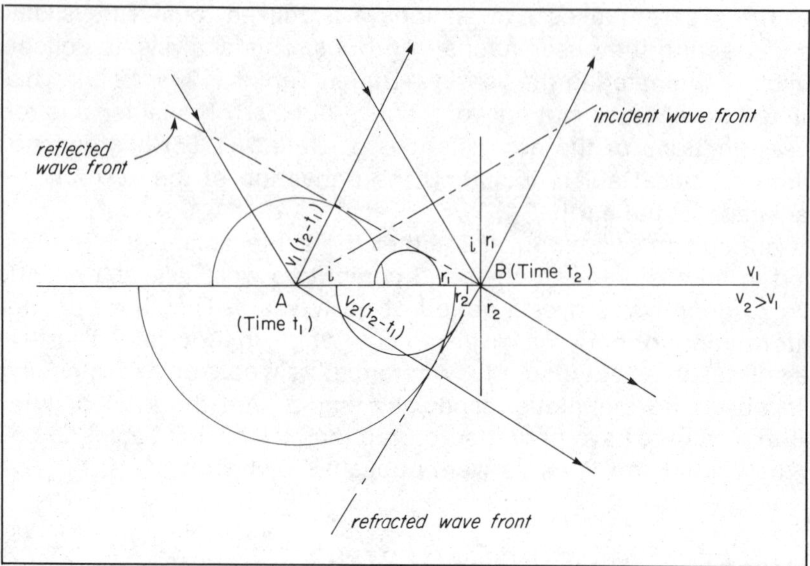

Figure 3.3 Construction of reflected and refracted wave fronts at a boundary by Huygens' principle.

A technique often used in geometrical optics, based on *Huygens' principle,* can in fact be used to infer the new directions. Huygens' principle states that any point, say on a boundary, reached by the incident ray may be treated as a new source, about which new hemispherical wave fronts expand on each side of the boundary. Since each of these elementary wave fronts corresponds to only an infinitesimal amount of energy, the determination of a physically realistic wave front consists of locating a surface to which an infinite number of them are tangent. Figure 3.3 indicates the application to reflection and refraction of waves, of the same type as the incident wave. An incoming plane wave strikes the boundary point A at time t_1, whereupon A becomes active as an infinitesimal source. By the time the incident wave front reaches B, at time t_2, wave fronts from A have spread hemispherically into both media. Note, however, that while the radius of the hemisphere is $v_1(t_2 - t_1)$ in the upper medium, it is $v_2(t_2 - t_1)$ in the lower medium. Also, as the diagram suggests, proportionately smaller hemispheres exist about all points between A and B, and sloping planes, representing physically real wave fronts, can be drawn tangentially to these. The new wave directions, reflected or refracted, are normal to these planes. The upper part of the diagram demonstrates the law of reflection, that the angle of incidence i is equal to the angle of reflection r_1. From the lower part, it is easily deduced that

$$\frac{\sin i}{\sin r_2} = \frac{v_1}{v_2} \qquad 3.2.1$$

where r_2 is the angle of refraction. Both of these results are identical to those found in geometrical optics and the latter, or law of refraction, is known as *Snell's law.* It is not difficult to extend the treatment to the case of reflection or refraction of a wave of different type (e.g., reflected S from an incident P), leading to a generalized form of Snell's law:

$$\frac{V}{\sin i} = \text{constant} \qquad 3.2.2$$

where V stands for either α or β on either side of the boundary, and i is the angle between the corresponding ray (incident, reflected, or refracted) and the normal on the same side.

The disadvantage of the purely geometrical approach using Huygens' principle is that no information on the relative amplitudes of the four new waves is obtained. In most cases, one or two of the four possible waves from the boundary predominate over the others. Qualitatively, this follows from the fact that energy of particle vibration will not be transferred into planes perpendicular to the original direction. For example, a P wave incident normally on a boundary will not produce reflected or transmitted S waves, because they would require just such a transfer of energy.

The quantitative determination of amplitude may be illustrated by consideration of the special case of a P wave incident normally on the boundary $x_3 = 0$, separating regions of velocity and density (α_1, ρ_1) and (α_2, ρ_2) respectively. The boundary conditions are that stress and displacement must be continuous for all points on the boundary, at all times. Assuming, as above, that no S waves are generated, all displacements are entirely in the x_3 direction. Let us write, for particle displacement,

$$\begin{aligned} u_3 &= A_1 \exp ik\,(\alpha_1 t - x_3) \text{ incident wave} \\ u_3 &= A_2 \exp ik\,(\alpha_1 t + x_3) \text{ reflected wave} \\ u_3 &= A' \exp ik\,(\alpha_2 t - x_3) \text{ transmitted wave} \end{aligned} \qquad 3.2.3$$

which represent incident and transmitted waves in the $+x_3$ direction, and the reflected wave in the opposite direction.

The stress is given (Equation 2.3.3) by

$$p_{33} = \lambda\theta + 2\mu \frac{\partial u_3}{\partial x_3}$$

which reduces in this simple case to

$$p_{33} = (\lambda + 2\mu)\frac{\partial u_3}{\partial x_3} = \rho\alpha^2 \frac{\partial u_3}{\partial x_3}$$

The continuity of stress at $x_3 = 0$ for all time therefore requires

$$-\rho_1\alpha_1^2 A_1 + \rho_1\alpha_1^2 A_2 = -\rho_2\alpha_2^2 A' \qquad 3.2.4$$

Similarly, the continuity of displacement at all times implies a second relation between A_1, A_2, and A'. It is most convenient to express this in terms of the particle velocity, $\partial u_3/\partial t$, since velocity must be continuous also. Then

$$\alpha_1 A_1 + \alpha_1 A_2 = \alpha_2 A' \qquad 3.2.5$$

Equations 3.2.4 and 3.2.5 are solved to give

$$\frac{A_2}{A_1} = \frac{\rho_2\alpha_2 - \rho_1\alpha_1}{\rho_2\alpha_2 + \rho_1\alpha_1}$$

$$\frac{A'}{A_1} = \frac{\alpha_1}{\alpha_2}\frac{2\rho_1\alpha_1}{\rho_2\alpha_2 + \rho_1\alpha_1} \qquad 3.2.6$$

We have thus obtained explicit expressions for the ratio of the amplitudes of the reflected and transmitted waves in terms of the incident amplitude. Notice that if $\rho_1\alpha_1 > \rho_2\alpha_2$, the expression for A_2/A_1 becomes negative. This implies a change in phase upon reflection, a positive pulse being reflected as a negative one, for example. Reflection at a free surface always involves a phase change, since in this case $\rho_2 = \alpha_2 = 0$.

Reflections at nearly normal incidence do not arise as frequently in physics of the earth as in seismic prospecting, where Equations 3.2.6 are extensively employed. Nevertheless, they are basic to marine seismic profiling of the sea floor; some reflections from the earth's core fall into this category; and in the study of the crust with explosions, normal incidence reflections are sometimes obtained.

The extension to oblique incidence introduces two complications: all four new waves (P and S, reflected and transmitted) must be considered, and the boundary itself is no longer a wave front. The development is outlined in Appendix C.

3.3 RAY GEOMETRY AND THE INVERSION OF TIME-DISTANCE CURVES

We shall make use first of the Snell's law relationship deduced above. Furthermore, we shall consider initially that part of the earth which lies beneath the surface layers. For most of this region, the smoothness of the time-distance curves suggests that velocities of P and S vary smoothly with depth. The curvature of the time-distance curves is due partly, of course, to the fact that time is plotted against arc distance Δ, whereas the waves do not travel around the surface; however, the observed curvature requires also that velocity, over most of the earth's radius, be an increasing function with depth. To investigate some principles of ray geometry, it is convenient to begin with an earth model consisting of thin shells of uniform velocity, each shell with a velocity higher than that above (Fig. 3.4). A ray is then successively refracted away from the normal (i.e., the radius to the point of refraction) such that

$$\frac{\sin i_1}{v_1} = \frac{\sin i'_1}{v_2} \qquad 3.3.1$$

or

$$\frac{r_1 \sin i_1}{v_1} = \frac{r_1 \sin i_1'}{v_2}$$

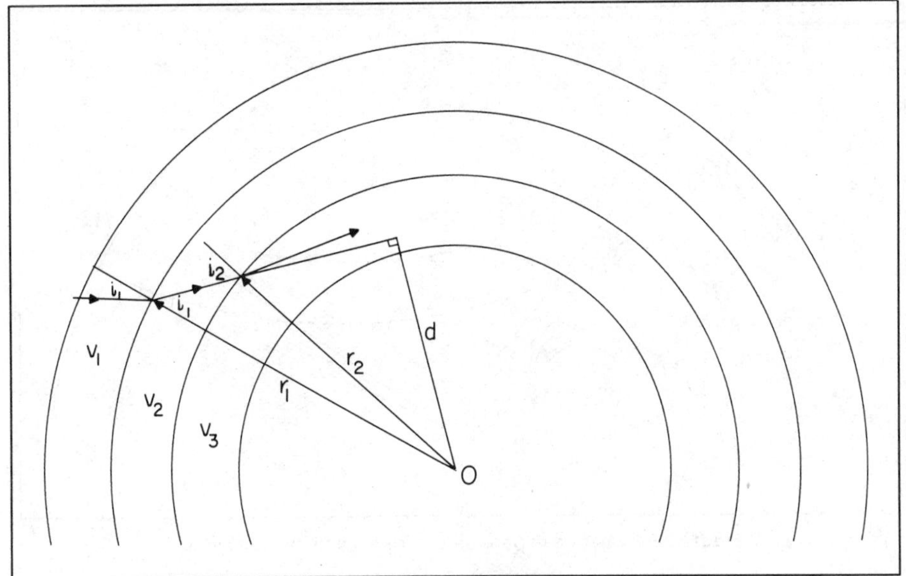

Figure 3.4 Successive refraction of a wave in a layered sphere.

But, from the figure, $r_1 \sin i'_1 = r_2 \sin i'_2 = d$

$$\therefore \qquad \frac{r_1 \sin i_1}{v_1} = \frac{r_2 \sin i_2}{v_2}, \text{ and so forth}$$

or

$$\frac{r \sin i}{v} = p \tag{3.3.2}$$

where r is the radius to any point on a ray, i is the angle between the ray and the radius at that point, and p is a constant for the given ray, known as the *ray parameter*.

The parameter p bears a remarkable relationship to the travel-time curve. Let T be the travel time over an arch distance Δ, for which the ray has an initial inclination i_0 (Fig. 3.5). Consideration of the incremental quantities for a neighboring ray shows that

$$\sin i_0 = \frac{v_0(\frac{1}{2}\,\delta T)}{r_0(\frac{1}{2}\,\delta\Delta)} \tag{3.3.3}$$

where r_0 is the outer radius.

$$\therefore \qquad \frac{r_0 \sin i_0}{v_0} = p = \frac{dT}{d\Delta} \tag{3.3.4}$$

In other words, the parameter p is precisely the slope of the time distance curve at a distance Δ from the source.

The problem of inversion is the determination of v as a function of r, from a knowledge of p as a function of Δ. We proceed by considering any point P on a given ray (Fig. 3.6). In terms of distance s along the ray, we may write

$$p = \frac{r \sin i}{v} = \frac{r^2}{v}\frac{d\theta}{ds} \tag{3.3.5}$$

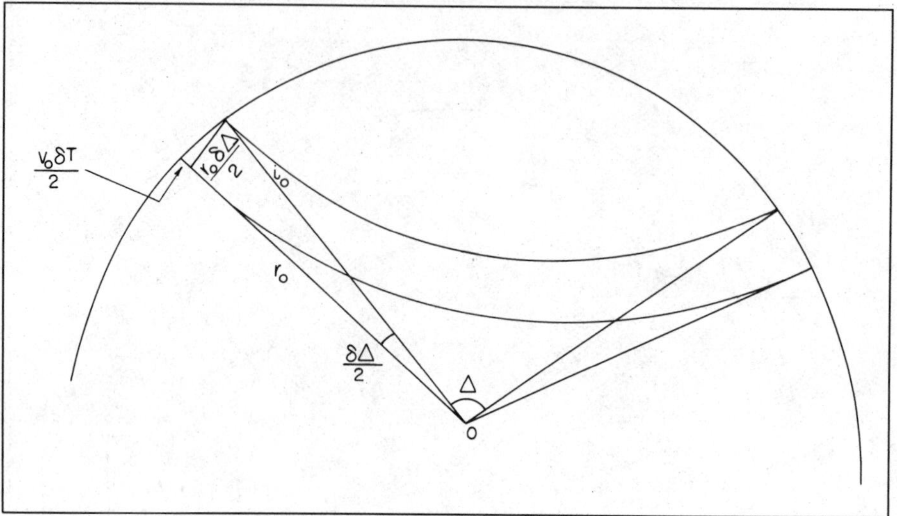

Figure 3.5 Change in parameters for a small increase in Δ.

But $(ds)^2 = (dr)^2 + r^2(d\theta)^2$

$$\therefore \qquad (d\theta)^2 = \frac{(dr)^2 p^2 v^2}{r^4 - r^2 p^2 v^2} \qquad 3.3.6$$

To simplify, let

$$\eta = r/v$$

Then

$$d\theta = \pm p r^{-1} (\eta^2 - p^2)^{-1/2}\, dr \qquad 3.3.7$$

We note that at the deepest point (i.e., midpoint) of the ray, sin $i = 1$, and η at

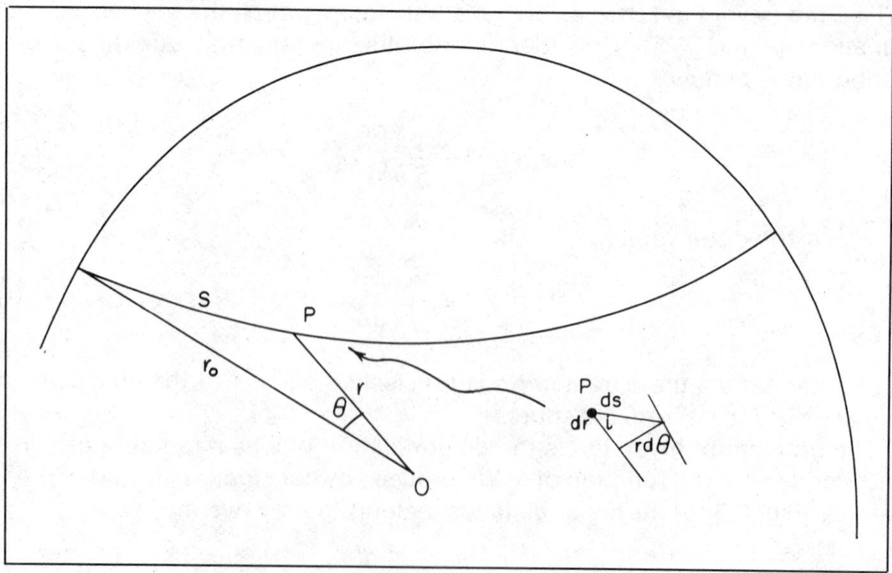

Figure 3.6 Quantities involved in ray geometry.

that point is equal to the value of p for the ray. Integrating along the ray between the deepest point of the ray and the outer radius gives

$$\Delta/2 = p \int_{r_{\mathrm{mid}}}^{r_0} r^{-1}(\eta^2 - p^2)^{-1/2}\, dr \qquad 3.3.8$$

Equation 3.3.8 is an integral equation for η, hence v, as a function of r, since p is a known function of Δ. It was so recognized by Herglotz (1907) and Wiechert, but the reduction to an explicit equation presented some complexities. Bullen (1965) quotes a method attributed to Rasch, which is more direct than those employed previously.

We recast Equation 3.3.8 as an integral over η:

$$\Delta = \int_{\eta_{\mathrm{mid}}}^{\eta_0} 2pr^{-1}(\eta^2 - p^2)^{-1/2} \frac{dr}{d\eta}\, d\eta \qquad 3.3.9$$

and apply the operation

$$\int_{\eta_1}^{\eta_0} dp\,(p^2 - \eta_1^2)^{-1/2}$$

to both sides, where the subscript 1 refers to values at a radius r_1, the deepest point of the ray emerging at Δ_1. Notice that this second integration, which appears for the moment to complicate rather than simplify matters, involves integration over a bundle of rays, down to the ray whose bottom is at radius r_1. Then

$$\int_{\eta_1}^{\eta_0} \Delta(p^2 - \eta_1^2)^{-1/2}\, dp = \int_{\eta_1}^{\eta_0} dp \int_{\eta_{\mathrm{mid}}}^{\eta_0} 2pr^{-1}\{(p^2 - \eta_1^2)(\eta^2 - p^2)\}^{-1/2} \frac{dr}{d\eta}\, d\eta \qquad 3.3.10$$

The order of integration on the right may be changed, as suggested graphically in Figure 3.7, where η is plotted against p. For any given ray (i.e., any given p),

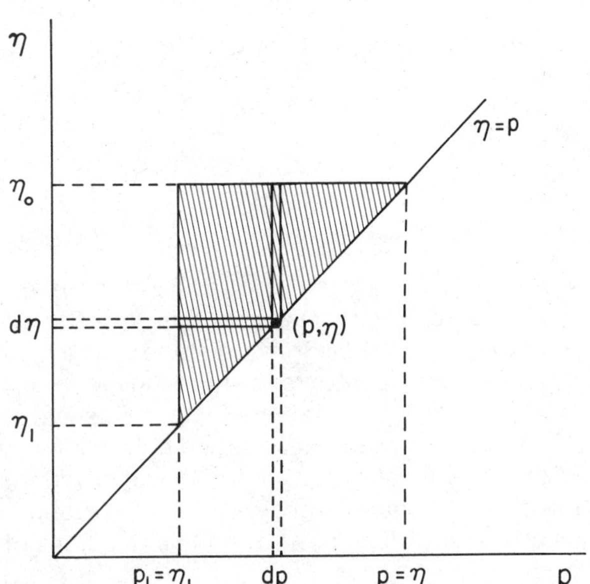

Figure 3.7 Domain of integration for the Herglotz-Wiechert inversion.

the minimum value of η is p, at the bottom of the ray path. The domain of double integration for the right side of Equation 3.3.10 is therefore the triangular shaded region. Change of order implies summing horizontal rather than vertical elements, and the diagram shows the appropriate limits. Thus,

$$\int_{\eta_1}^{\eta_0} \Delta(p_1 - \eta_1^2)^{-1/2}\, dp = \int_{\eta_1}^{\eta_0} d\eta \int_{p=\eta_1}^{p=\eta} 2pr^{-1}\{(p^2 - \eta_1^2)(\eta^2 - p^2)\}^{-1/2} \frac{dr}{d\eta}\, dp \qquad 3.3.11$$

We integrate by parts on the left, and prepare to carry out the integration with respect to p on the right:

$$\left[\Delta \cosh^{-1}\left(\frac{p}{\eta_1}\right)\right]_{\eta_1}^{\eta_0} - \int_{\eta_1}^{\eta_0} \frac{d\Delta}{dp} \cosh^{-1}\left(\frac{p}{\eta_1}\right) dp$$

$$= \int_{\eta_1}^{\eta_0} 2r^{-1} \frac{dr}{d\eta}\, d\eta \int_{p=\eta_1}^{p=\eta} p\{(p^2 - \eta_1^2)(\eta^2 - p^2)\}^{-1/2}\, dp \qquad 3.3.12$$

The integral over p on the right can be reduced to a standard form by the substitutions $x = p^2$, $a = \eta_1^2$, and $b = \eta^2$; then we have

$$\frac{1}{2}\int_a^b \frac{dx}{(x-a)^{1/2}\,(b-x)^{1/2}}$$

A further substitution, $x = a \cos^2 \theta + b \sin^2 \theta$, leads to

$$\frac{1}{2}\int_0^{\pi/2} \frac{2 \sin\theta \cos\theta\, (b-a)\, d\theta}{(b-a)^{1/2} \sin\theta\, (b-a)^{1/2} \cos\theta} = \frac{\pi}{2}\ !$$

The first term on the left of Equation 3.3.12 vanishes at both limits ($\Delta = 0$ when $p = \eta_0$; $\cosh^{-1} p/\eta_1 = 0$ when $p = \eta_1$); rewriting the second term as an integration over Δ, and using the above result for the integration on the right, we have

$$\int_0^{\Delta_1} \cosh^{-1}\left(\frac{p}{\eta_1}\right) d\Delta = \int_{\eta_1}^{\eta_0} \pi r^{-1} \frac{dr}{d\eta}\, d\eta$$

or

$$\int_0^{\Delta_1} \cosh^{-1}\left(\frac{p}{\eta_1}\right) d\Delta = \pi \log_e\left(\frac{r_0}{r_1}\right) \qquad 3.3.13$$

We have finally arrived at a result which may be evaluated numerically, given only a curve of p against Δ. In this equation, η_1 is the value of p for the ray that emerges at Δ_1; hence it is the slope of the travel-time curve at Δ_1. The left side can therefore be integrated numerically to any prescribed value of Δ_1, by obtaining values of p at intermediate points and evaluating $\cosh^{-1}(p/\eta_1)$. The unknown in the equation, r_1, is then determined. This is the radius to the midpoint of the ray emerging at Δ_1, and the velocity at the midpoint is given by $v_1 = r_1/\eta_1$. Inversion, therefore, consists of the evaluation of Equation 3.3.13 for successively greater values of Δ_1, until the velocities of P or S are known over a range of depths.

Inverse theory has recently become very popular in geophysics. The example of the Herglotz-Wiechert method shows that inversion was recognized at least as early as 1907. Note, however, that nothing in the above development permits the determination of the uncertainty in the velocity distribution resulting from incomplete or uncertain $p-\Delta$ relations. The estimate of uncertainty is a key feature of modern theories of inversion (Backus and Gilbert, 1970).

3.4 CASES OF SPECIAL VELOCITY DISTRIBUTIONS

It has been implied in the above development that not only are there no discontinuities in velocity, but also that the velocity increases monotonically with depth. In this case, p decreases with Δ, and p in the range of integration is greater than η_1. There are some important characteristics of the behavior of p, and of the nature of the time-distance curve, for certain special velocity distributions, which need not involve a discontinuity in velocity.

The case of an abrupt increase in the velocity gradient is shown in Figure 3.8(a). Rays which penetrate to the lower region are more sharply curved, and can emerge at a smaller value of Δ than rays which leave the source at a shallower inclination. The parameter p decreases monotonically as we consider increasingly steep rays, but, as Figure 3.8(b) indicates, the same value of Δ may be met for three values of p. Interpreting p as the slope of the time-distance curve permits us to construct Figure 3.8(c), which indicates the triplication of a pulse over a certain range of Δ, corresponding to three paths by which rays may reach a given point.

A second case of importance is that of a discontinuity at which velocity

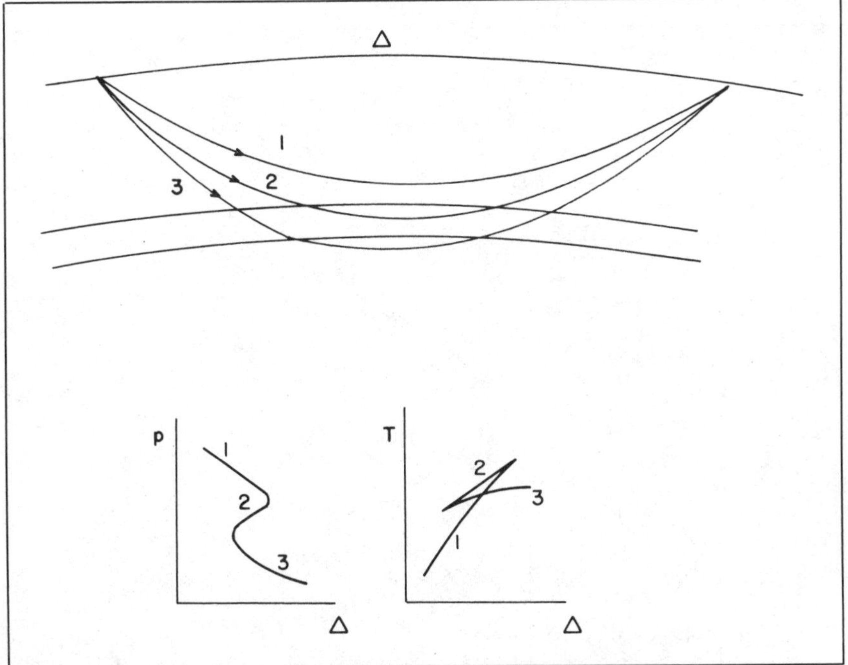

Figure 3.8 Triplication of a pulse when there is a region of high velocity gradient. The upper diagram (a) indicates ray paths, while the lower curves show the behavior of (b) p and (c) T. For the branch 2, Δ decreases as the steepness of the ray leaving the source increases; this portion of the time-distance curve is *concave upwards.*

decreases, but below which velocity again increases with depth. The expressions introduced above cannot, of course, be evaluated across the discontinuity, but we may consider separately rays which do and do not reach the discontinuity (Fig. 3.9). Ray geometry suggests that there will be a shadow zone, or range of values of Δ for which no ray arrives. For smaller values of Δ, the behavior of p is normal (Fig. 3.9); for larger Δ, rays may reach a given point by two paths, corresponding to two values of p. The resulting travel-time curve is also shown in Figure 3.9.

The complexity of the $p-\Delta$ curve in these cases can make interpretation very difficult. Observations are usually available only at a limited number of distances from the source, and there may be ambiguity in the way points on the $p-\Delta$ plot are connected, as Figure 3.13 will indicate. For this reason, a number of workers (Bessenova *et al.*, 1974, 1976; Kennett, 1976) prefer to introduce, as an additional parameter of the time-distance curve, the *intercept* τ of the tangent to the curve at any point (Fig. 3.10(a)). The quantity τ plotted against p yields a curve which is single-valued and monotonically decreasing, and which exhibits discontinuities only when there is a low-velocity region, as shown in Figure 3.10(b). There is much less uncertainty in producing the function $\tau(p)$ than in producing $p(\Delta)$ from separated observations, and, as the above authors show, the former function may be inverted in a manner analogous to that developed by Herglotz and Wiechert, to give velocity as a function of radius. The method appears to reduce the uncertainty in the location of low-velocity zones, although the actual velocity within such zones must remain poorly defined, since no refracted rays from a source above the zone ever "bottom" in the zone. The discontinuity in $\tau(p)$ corresponding to a low velocity zone is essentially a measure of the time spent by a ray in traversing the zone, and, in

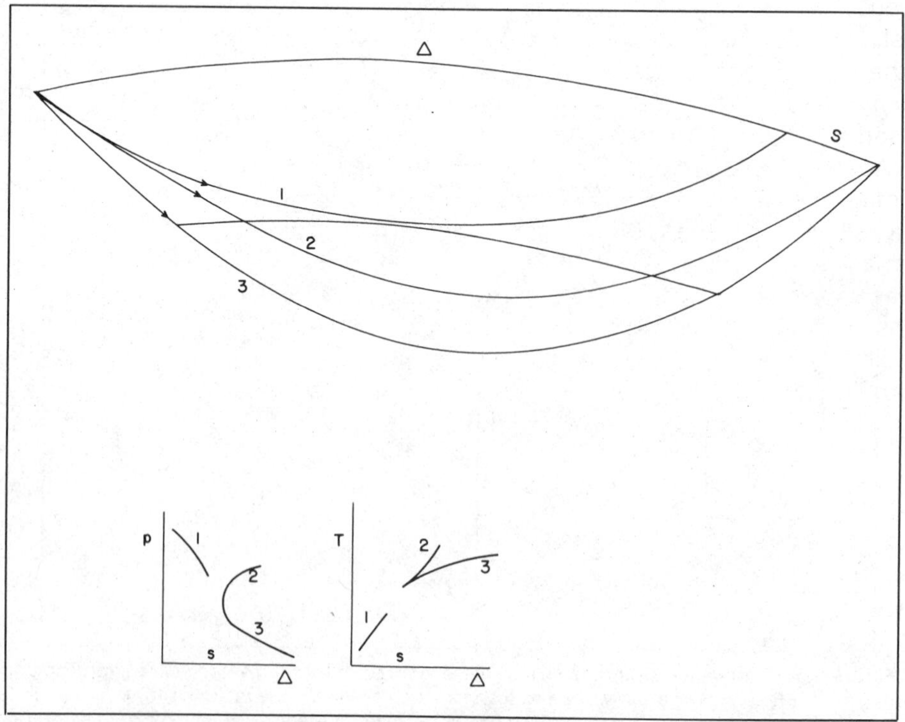

Figure 3.9 Creation of a shadow zone, *s*, by a region of low velocity.

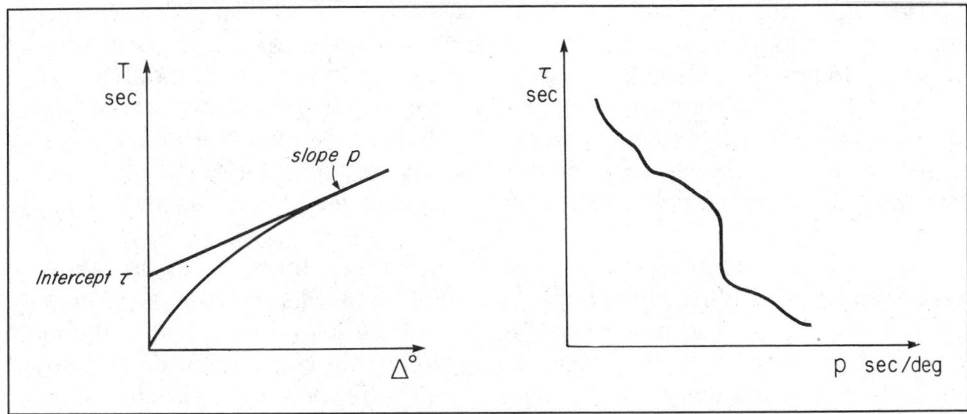

Figure 3.10 The intercept τ and, right, its variation with ray parameter p.

the absence of other information, the greatest detail one is really justified in presenting in a velocity-depth interpretation is the integrated ray *slowness* (reciprocal of velocity) over the depth range corresponding to the zone.

3.5 THE MAJOR DISCONTINUITIES AND RESULTING PHASES OF SEISMIC WAVES

The theory of Section 3.3, when applied to the best available travel-time curves, with recognition of the complications mentioned in Section 3.4, leads to curves of the velocities of P and S as functions of radius. We emphasize that, because of the discontinuities in velocity associated with the layers of the crust, the method is always applied to an earth which is "stripped" to the base of the crust, by the subtraction of appropriate times at both source and receiving stations. The methods of investigating the surface layers will be discussed in a later chapter.

Application of the method of inversion suggested, rather early in the twentieth century, the presence of important discontinuities within the earth. As a result of reflections from these discontinuities, seismic waves may travel from a source to a station by more than one path. The station then receives a series of arrivals of both P and S types; these arrivals, corresponding to different paths, are known as phases.

The most striking discontinuity deep within the earth is the boundary of the central core, the discovery of which provides a remarkable example of scientific prediction and confirmation. While suggestions of a central region of high density, to explain the mean density of the earth, were made in the nineteenth century, Oldham (1906) first observed that a P wave arriving at the antipodes of an earthquake was late, in comparison with the expected arrival time. He proposed the existence of a *core* of lower velocity than the outer region, and predicted the presence of a shadow zone, as explained in Section 3.4. Gutenberg (1912) verified that there was a shadow zone for P between $\Delta = 105°$ and $\Delta = 143°$ with strong arrivals, corresponding to a caustic in optics, just beyond 143°. Gutenberg estimated the depth to the core boundary as 2900 km, and also determined P velocities within the core. In turn, he predicted that both P and S would be reflected from the boundary, and should be observed. These reflected phases were not observed for many years (they arrive at a time which, for many earthquake records, is disturbed by surface waves), but by 1939 Jeffreys was able to use their travel times to fix the depth to the core boundary as (2898 $\pm$ 4) km. The agreement with Gutenberg's estimate is remarkable, as is also the fact that this figure has stood unchallenged until very recently. The region of the earth outside of the core, extending to the base of the outer crust, was given by Wiechert the German name *Mantel* and has become known in English as the *"mantle."*

The shadow zone of the core is not complete, there being arrivals of *P* waves, of small amplitude, through the entire zone. Miss Lehmann (1935) suggested that these arise from an inner core, of higher velocity, within the main core. Waves travelling in the outer core could be intercepted by the boundary of this inner core, at a depth of about 5120 km, and reflected, or refracted through the inner core, so as to arrive in the shadow zone. Miss Lehmann's proposal of an inner core became widely accepted with the demonstration of Jeffreys (1939b) that diffraction of energy into the shadow zone could not explain the amplitudes of arrivals observed throughout the zone.

We may now turn to the designation of phases which follow the various paths (Fig. 3.11). The letters *P* or *S* are used to designate each leg of a reflected or refracted wave outside of the core; the subscripts *c* and *i* denote reflections from the outer and inner core boundaries respectively, while the letters *K* and *I* indicate *P* waves through the outer and inner core. A wave of *S* type through the outer core has not been observed. We note also the presence of phases, such as *PP*, reflected from the earth's surface.

While there have been recent suggestions of seismological discontinuities within the region of steep velocity gradient in the upper mantle, sharp enough to produce reflections (Anderson, 1967 and Section 3.7 below), the generally accepted picture has been one in which the major discontinuities are the outer surface, base of the crust, and boundaries of the outer and inner cores. In view of the nonspherical shape of the earth, a word of explanation is necessary regarding the precise meaning of the depth to the major discontinuities. Because travel times are corrected for the ellipticity, all interpretations based on them give depths which would be found in a sphere of volume equal to that of the earth.

3.6 DERIVATION OF PROPERTIES FROM THE VELOCITIES

Recognition of the presence of major discontinuities within the earth, and application of the Herglotz-Wiechert method of inversion within each region, has led to velocity distributions, an example of which is shown in Figure 3.12. We shall discuss later certain recent observations which suggest revisions in the vicinity of the boundaries; the immediate aim is to determine whether the density and elastic constants can be extracted from the velocities α and β. At first this appears to be an indeterminate problem, since at each depth, two elastic constants and the density are the unknowns, and only two parameters are known. However, Adams and Williamson (1925) proposed a remarkably direct method of solution, applicable through a region, of uniform composition, which is in a state of hydrostatic stress. Below the surface layers, the stress is very probably nearly hydrostatic, and the Adams-Williamson method has been of fundamental importance in geophysics.

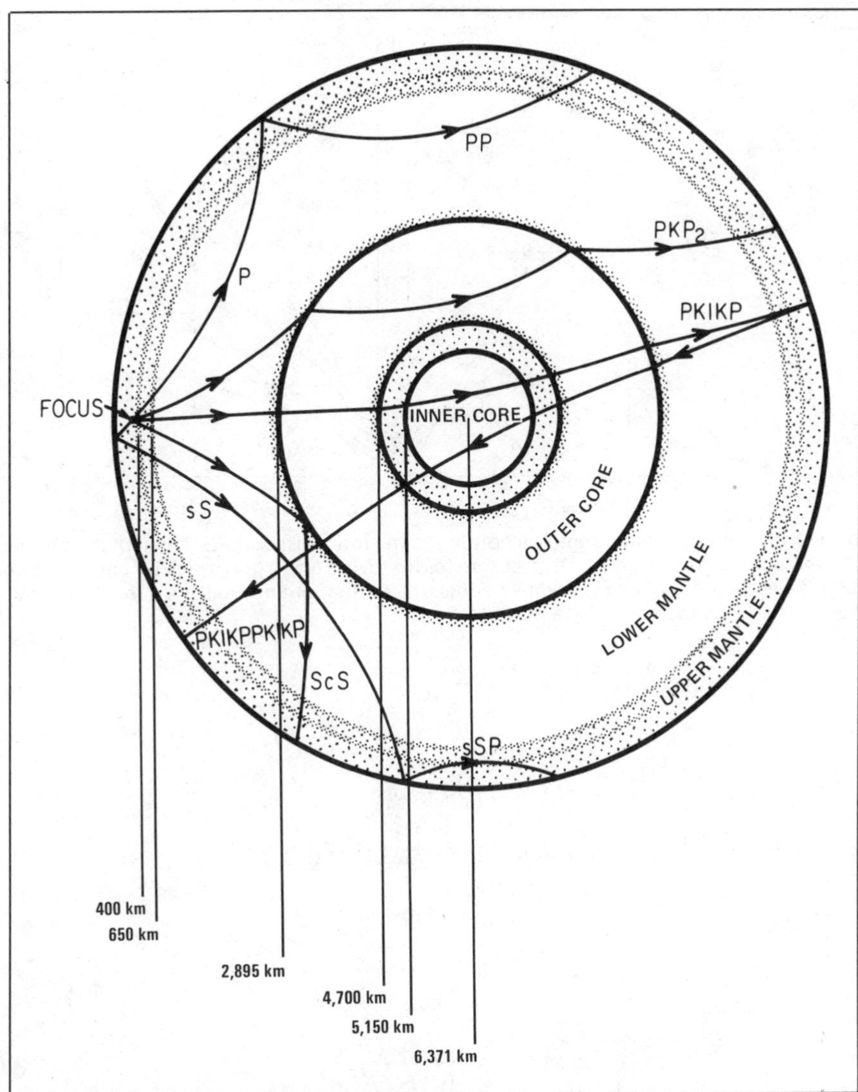

Figure 3.11 Labels of seismic waves which have penetrated to the mantle, core or inner core. The stippling indicates possible complexity in the upper mantle. (Courtesy of United States National Academy of Sciences)

At radius r in the region assumed, the gradient of hydrostatic pressure is simply

$$dp/dr = -g\rho \qquad 3.6.1$$

where ρ is the density at that depth, and g is the acceleration due to gravity at the same depth. Potential theory shows that g results only from the attraction of the mass m within the sphere of radius r:

$$g = Gm/r^2 \qquad 3.6.2$$

where G is the gravitational constant. We wish to determine ρ; therefore writing

$$d\rho/dr = \frac{d\rho}{dp} \cdot \frac{dp}{dr}$$

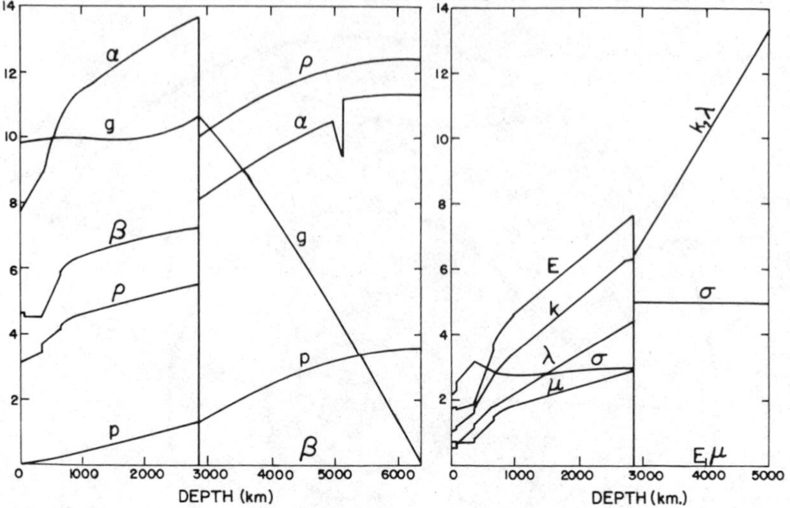

Figure 3.12 Variation with depth, according to the model Haddon-Bullen I, of the parameters α and β (km/sec), density (ρ, gm/cm^3), pressure (p, 10^{12} dynes/cm^2) gravity (g, 10^2 cm/sec). Young's modulus (E), incompressibility (k), Lamé's constants (λ and μ) and Poisson's ratio. The units for the moduli are 10^{12} dynes/cm^2; those of σ, 0.1.

we note that

$$d\rho/dp = \rho/K$$

where K is the bulk modulus. Also, from 3.6.1 and 3.6.2,

$$dp/dr = -Gm\rho/r^2$$

Therefore

$$d\rho/dr = \frac{(\rho)}{K}\frac{(-Gm\rho)}{r^2} \qquad 3.6.3$$

But the relations between the elastic moduli permit us to write

$$K/\rho = \alpha^2 - \tfrac{4}{3}\beta^2 \qquad 3.6.4$$

so that

$$d\rho/dr = \frac{-Gm\rho}{r^2(\alpha^2 - \frac{4}{3}\beta^2)} \qquad 3.6.5$$

The gradient of density is thus determined, in terms of velocities, and in terms of ρ itself and of m. Equation 3.6.5 can be applied successively to regions of uniform composition of the earth, provided that calculations begin at the top of the mantle, where ρ must be assumed, and m is taken as the mass of the earth less the mass of the crust. Values of ρ and m are then determined at the base of the first region, and the calculation is repeated.

Within the mantle, the separate determination of ρ permits the determination of λ and μ, and therefore of all other elastic constants, at each depth.

There is no direct method of determining the discontinuity in density across the core boundary, but the variation in density within the core is rather well limited by the known total mass, and moment of inertia, of the earth. The Adams-Williamson method has been applied most extensively by Bullen (1965), who has derived several earth models; examples of the variation of the velocities, density, and moduli are shown in Figure 3.12.

There are serious limitations to the application of the simple theory outlined above without further constraints on the density, but we shall leave the discussion of these, and of the composition of the mantle and core, to Chapter 26. We may observe, however, that the variation in elastic properties and density through the mantle is reasonably consistent with that to be expected in a silicate, such as olivine, under pressure, except in the depth range from 500 to 1000 km, where all properties change very rapidly with depth. In this range a change in composition, or of phase, may be implied. The density of the core is consistent with a composition dominantly of iron, as was suggested by the occurrence of iron meteorites, and the thermal conditions may permit the inner core to be solid, and of a composition similar to that of the liquid outer core (Jacobs, 1953).

3.7 RECENT DEVELOPMENTS

Seismic velocities in the mantle and core obtained by inversion of the T-Δ curve stood for a number of years without modification, but there have been recent suggestions, based on body-wave studies, that the velocity distribution near the top of the mantle, and near the boundaries of core and inner core, requires modification. These are in addition to the evidence from surface waves and free oscillations, which we discuss later, and they follow directly from characteristics of the p-Δ curve, or from the amplitudes of reflected phases.

The greatly increased number of seismograph stations, including arrays of seismometers, that have become available in recent years have yielded much more precise estimates of p, the slope of the time-distance curve. Johnson (1967) presented a detailed curve of p versus Δ from nearby earthquakes recorded at an array of stations in Arizona (Fig. 3.13(a)). As we have seen, a break in the p-Δ curve is evidence of a decrease in velocity; triplication of the values of p is evidence for a very large velocity gradient. The method of inversion cannot be applied to a region in which velocity is decreasing with depth, but of several models which are consistent with the p-Δ curve, Johnson found that the one which most closely yielded the observed travel times was that shown in Figure 3.13(b). We note the pronounced minimum of P-velocity in the upper mantle, at a depth of about 100 km. For eastern North America, Massé (1973) has inferred the presence of a low velocity zone from the travel times of waves from the Early Rise explosion in Lake Superior. His evidence is essentially twofold: the presence of reflections from the base of the supposed zone, and the rate of decrease of amplitude with distance from the wave refracted along the high velocity lid above it. The low velocity channel was found, however, to be much more restricted in depth range, and less severe in velocity variation, than that proposed by Johnson for western North America. The zone extends from 94 to 107 km in depth, and the velocity in it does not fall significantly below 8.0

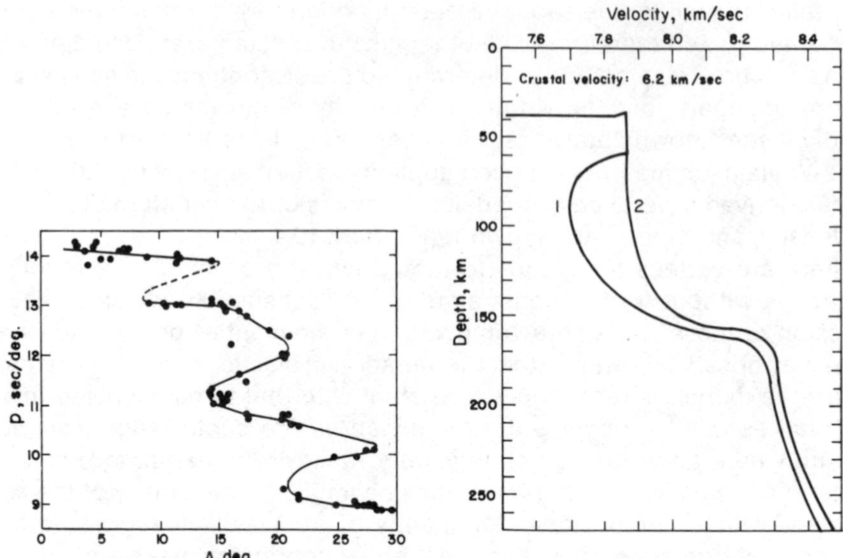

Figure 3.13 Complex variation of the parameter p, from detailed observations by Lane Johnson. The interpretation which gives the best fit to observed travel times is curve 1 on the right, although curve 2 would also be compatible with the $p-\Delta$ variation. The interpretation corresponding to curve 2, with no low-velocity zone, assumes that the p-Δ curve is connected by the broken line.

km/sec. Massé proposed that this narrow low-velocity zone is typical of the tectonically less active eastern half of the continent. This "low-velocity layer" will be discussed again, but it is noteworthy that evidence for it can be found from the p-Δ relation alone.

The most recent estimates of the radial variation of wave velocities, and the properties derived from them, are not based on the study of body waves independent of other constraints. Information from surface waves and from free oscillations is combined with body-wave studies, as we shall see in Chapter 8. The most significant discoveries relating to the mantle beneath the low-velocity zone, based on body waves, are the presence of two regions of either very steep velocity gradient or discontinuity in velocity, and the existence of important lateral variations in velocity. Niazi and Anderson (1965), in an analysis of p-Δ curves, found evidence for discontinuities in velocity gradient at depths of 320 and 640 km. Other studies have confirmed that the velocity variation at these depths is anomalous. For example, Green and Hales (1968), in an analysis of travel times from large explosions in Lake Superior (Project Early Rise), inferred the velocity-depth relation shown in Figure 3.14. A comparison with the curve given above for the variation of P-wave velocity from the earth's surface to the core (Fig. 3.12) shows immediately that any discontinuities near the depths of 360 and 640 km must be very minor compared to the first-order discontinuity at the core boundary. Nevertheless, these small discontinuities, or intervals of very steep gradient, are significant in the study of the composition and state of the mantle. We shall see (Chapter 26) that they are probably related to a change of "phase" or atomic packing, instead of to a change in chemical composition.

Lateral variations in velocity at a given depth in the upper mantle appear to be related to the gross features of plate tectonics. Descending slabs of relatively cool lithosphere can represent regions of significantly higher velocity. In an analysis of travel times from the nuclear explosion "Longshot" in the Aleutian arc (Fig. 3.15), Abe (1972) found that waves traversing the slab arrived 3 seconds early, at a given distance, compared with waves which did not cross the slab. Sleep (1973) showed that these observations were compatible with the velocity distribution shown in Figure 3.15. At first it may appear paradoxical that a slab of material from near the earth's surface, where the velocity is lower, represents a high-velocity region when it descends to depths of over 200 km. But the lithospheric slab responds immediately to the pressure characteristic of the depth, while it requires some millions of years to come to the temperature of the surroundings (Chapter 27); the net effect is that its velocity is higher than that of normal mantle at the same depth.

It is becoming very evident that lateral variation in velocity at any depth, to the extent of at least 1 per cent, is the rule rather than the exception in both the upper and lower mantle. The existence of *arrays* of seismometers (Section 5.5) permits the investigation of heterogeneities on a global scale. However, although the pattern of velocity variations in the upper mantle appears to be related to the gross features of plate tectonics, the significance of lower mantle lateral variations is not yet clear. Kanasewich et al. (1973), in a study of waves which reached their lowest point in the lower mantle beneath Hawaii, detected velocity anomalies which they associated with the surface expression of a *plume* (Section 27.7). More work will be required to establish the global "noise-level" of lower mantle velocities. A promising approach appears to be the study of residuals in the difference in arrival times of body-wave phases, such as P_cP-P and S_cS-S, against the standard tables (Jordan and Lynn, 1974). At epicentral distances of over 30°, the direct and core-reflected waves sample very similar paths in the upper mantle, so that time residuals in the difference are controlled largely by lower mantle conditions. Jordan and Lynn identified a

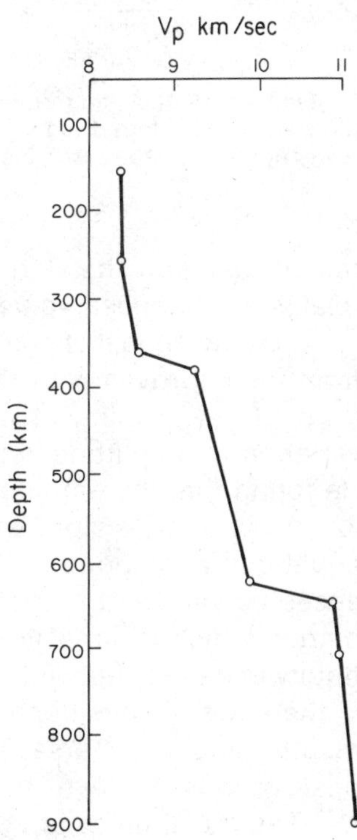

Figure 3.14 Detail of upper mantle *P*-wave velocity determined by Green and Hales (1968).

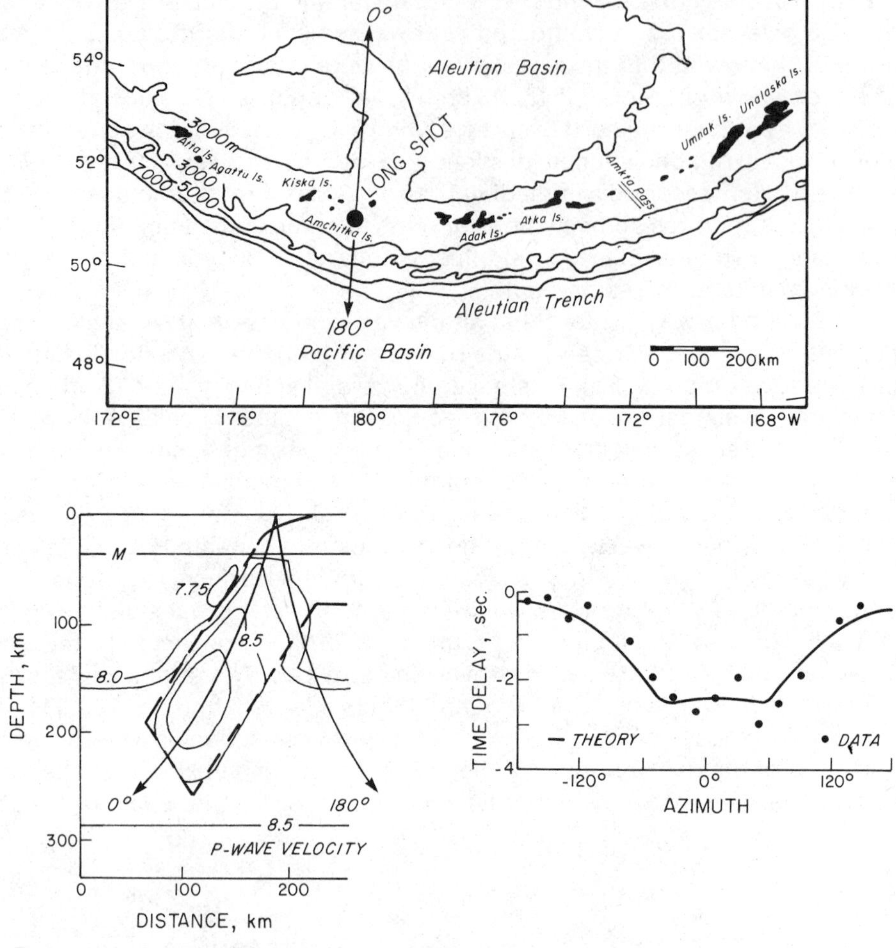

Figure 3.15 Location of explosion Longshot in the Aleutians (top); paths for azimuth 0° and 180° relative to inferred high-velocity slab (lower left) and early arrival as a function of azimuth (lower right). After Abe (1972) and Sleep (1973).

region, beneath the Caribbean, with a velocity contrast of at least 1 per cent relative to the mean value at any depth.

New evidence of the nature of the boundary of the outer core has come from the improvement of the world network of seismograph stations, and also from the availability of large explosions. Buchbinder (1968) has studied the variation in amplitude, with distance Δ, of the reflected phase P_cP (Fig. 3.11). He found that the amplitude-distance curve, which displays a minimum at $\Delta = 32°$, was not consistent with the computed reflection amplitudes (derived from equations analogous to 3.2.11) for a solid-liquid interface, if the previously accepted values of α and density were employed. A model proposed by Buchbinder, which is consistent with the observed amplitudes, provides no discontinuity in density between the lower mantle and core. Such a model might arise if there were considerable mixing of the core material with the lowermost mantle, and vice versa, but this proposal has not yet been tested by other observations.

The travel times of P_cP could be expected to throw light on the form of the outer surface of the core. Theoretically, this should have the form of a particular surface of constant density within the earth, which will be discussed in the study of the gravitational field. However, the suggestion has been made that there is "topography" on this theoretical surface. Buchbinder concluded that there is no evidence for topography greater than ±5 km in height, and that topography less than this could not be resolved with the present observations.

The nature of the velocity distribution at the boundary of the inner core was, until recent years, rather uncertain. Once again as a result of the improved facilities for recording, Bolt (1962, 1964) has shown that the arrival of *PKIKP*, in the main shadow zone, is frequently preceded by a small but definite phase. Bolt's analysis is outlined in Figure 3.16. The presence of the early phase requires a special branch *GH* on the arrival time curve, which is otherwise similar to the theoretical case shown in Figure 3.8. This branch could be produced by a region of constant velocity, bounded by two discontinuities in velocity. The wave which traverses the intermediate region, and whose arrival defines *GH*, was named *PKHKP* by Bolt. Following Bolt's original publication, it has been

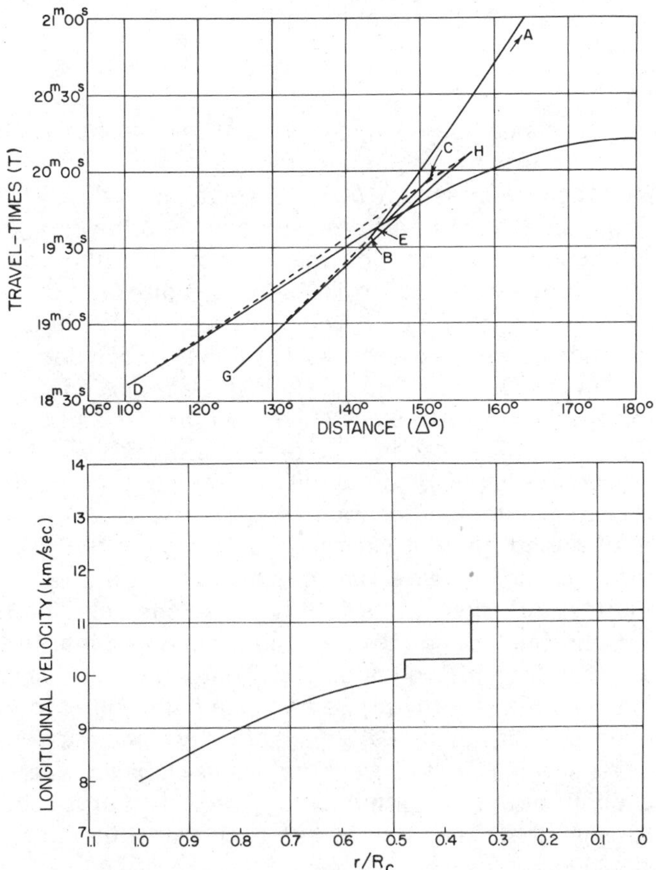

Figure 3.16 Bolt's analysis of travel times for waves refracted near the boundary of the inner core. The inferred velocity distribution is shown below, with depth plotted as a fraction of the radius of the outer core; discontinuities are proposed at depths of approximately 4700 km and 5100 km. The arrival-time graph involves two cusps of the type shown in Fig. 3.8. Waves penetrating to the region between the two discontinuities arrive at times given by the portion marked GH.

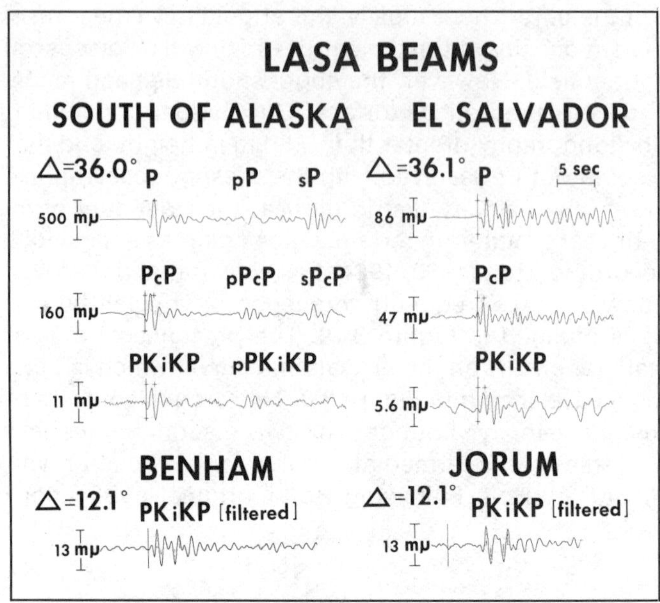

Figure 3.17 Seismograms showing the arrival of the phase PK_iKP reflected from the inner core. (Courtesy of E. R. Engdahl)

recognized that other seismologists (e.g., Hai, 1961) had observed the *PKHKP* phase, particularly on records of deep earthquakes.

The arrival of the new phase *PKHKP* is observed near the distant side of the main shadow zone, in a range of Δ from 125° to beyond the edge of the zone at 143°. Recently, considerable attention has been paid to the near side of the zone, with Δ in the range from 103° to 110°. In this region, the diffracted *P* wave is observed, followed by a series of arrivals, then by a phase whose arrival-time curve strongly suggested that it was PK_iKP, the wave reflected from the inner core (Bolt, O'Neill, and Qamar, 1968). Definitive evidence on the existence of PK_iKP was provided by Engdahl *et al.* (1970), on the basis of array-processed data (Fig. 3.17). The ability to detect and isolate an arrival which has twice crossed the major core-mantle discontinuity, as well as suffering reflection, is a tribute to modern seismological techniques, while the existence of this reflection, for waves of period about 1 second, is evidence of the sharpness of the boundary of the inner core (below the transition region). The train of arrivals between the diffracted *P* wave and PK_iKP has also been studied by Hai (1963) and by Husebye and Madariaga (1970). All authors agree that these waves do not result from structure in the outer core, but from multiple paths in the mantle. Hai proposed that energy may arrive by a path similar to *PP* (Fig. 3.11), but with reflection at the lower side of a discontinuity in the upper mantle, rather than at the earth's surface. Bolt, O'Neill and Qamar labelled such a phase P_dP, where *d* is the depth of the discontinuity involved. They proposed, on the evidence of the arrivals, that there may be two such discontinuities at depths between 360 and 420 km.

Some evidence for limited inhomogeneity in the outer core is provided by measurements and $dT/d\Delta$ for the diffracted *P* wave, obtained with large arrays or networks of stations (Husebye and Toksöz, 1968). There is a suggestion of a decrease in the velocity gradient at a depth of about 3900 km. A very tentative

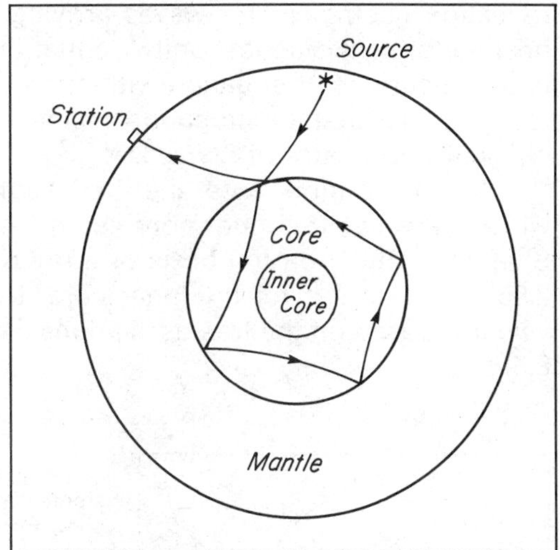

Figure 3.18 Ray path for the phase P_4KP.

explanation is that the core at this depth contains solid material in a fluid matrix, an arrangement known to possess a lower velocity than the fluid alone under similar conditions.

On the other hand, one of the most remarkable recent discoveries in seismology points to relative uniformity in the outer core. This is the recognition of waves multiply reflected at the core-mantle boundary from beneath (Fig. 3.18). These new phases have been recognized on the records from both nuclear explosions and earthquakes (Buchbinder, 1972; Qamar and Eisenberg, 1974), a typical arrival being shown in Figure 3.19. The designation adopted for them is P_mKP (or S_mKS), where m represents the number of path segments within the core. Waves up to P_7KP have been recognized. The very existence of these multiply-reflected phases, with arrival times close to those predicted, shows that there cannot be large-scale lateral heterogeneity in the outer core. Because of the large discontinuity in velocity at the core-mantle boundary, the

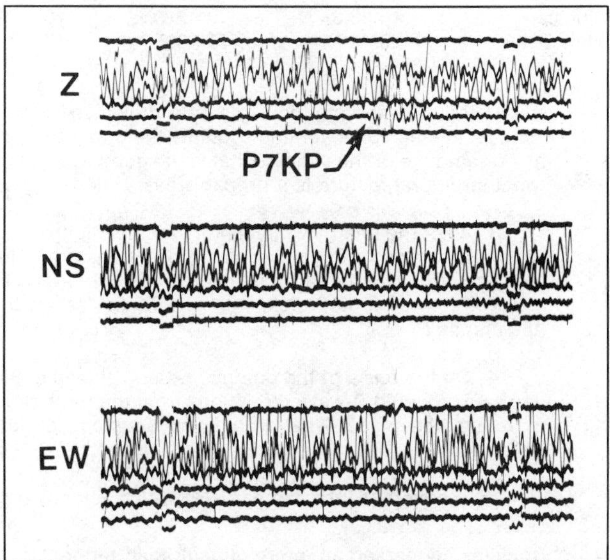

Figure 3.19 Example of a P_mKP arrival on an earthquake seismogram. (After Buchbinder, 1972)

reflection coefficient for waves arriving from beneath at angles of incidence near the critical is almost unity, so that the amplitude ratio, of P_7KP to P_4KP say, is a measure of attenuation within the core. The conclusions that have been reached are that attenuation is small, the outer core being an extremely efficient propagator of *P* waves.

As for the inner core, the first reported observation of the phase *PKJKP*, which traverses the inner core as an *S*-wave, was given by Julian, Davies and Sheppard (1972) on the basis of array observations. They obtained a value of 2.95 ± 0.1 km/sec for the inner core *S*-wave velocity. Other estimates of this velocity, based on the free oscillations of the earth, will be discussed in Section 8.5.

Problems

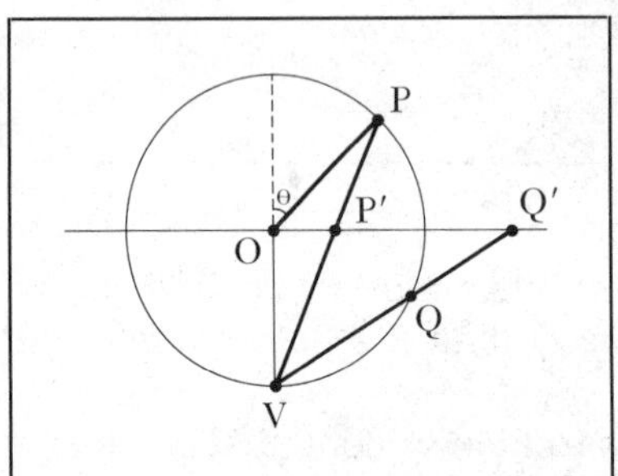

1. In the location of earthquake epicenters, the stereographic projection is very often used. As shown in the diagram, *V* (at the south pole) is the vertex of the projection onto the plane through *O*, perpendicular to *OV*. A point *P* in the northern hemisphere projects to *P′*; point *Q* projects to *Q′*, on the line *VQ* produced. How may *P′* be located? The method of epicenter determination requires the drawing of arcs of known radius Δ about a series of stations. Show that the projection of a circle, on the sphere, of radius Δ about *P*, is a circle of radius $(R \sin \Delta)/(\cos \theta + \cos \Delta)$ about a point distant $(R \sin \theta)/(\cos \theta + \cos \Delta)$ from *O* along the meridian of *P*. Note that on the projection, the circle is not centered about *P′*.

2. Readings of the arrival times (universal time) of *P* and *S* for an earthquake were as follows:

Station	Latitude	Longitude	*P* (or *PKP*) h	m	s	*S* (or *SKS*) h	m	s
Nemuro	43°19′42″N	145°35′12″E	10	43	15	10	49	23
Mizusawa	39°08′03.4″N	141°08′10″E	10	43	55	10	50	37
Scoresby Sund	70°29′N	21°57′W	10	44	16.0	10	51	14
Atlanta	33°26′N	84°20′15″W	10	44	38	10	51	55
Upsala	59°51′30″N	17°37′36″E	10	46	02.9	10	54	28
Dourbes	50°05′45.7″N	4°35′39.2″E	10	46	58	10	56	18
Trieste	45°38′34″N	13°45′14″E	10	47	28	10	57	40
New Delhi	28°41′N	77°13′E	10	48	10	10	58	31

Using a stereographic projection drawn to some convenient scale (say R = 10 cm), locate the epicenter and estimate the uncertainty of your position. Determine also the origin time of the shock. What is the geographical area of the earthquake, and with what structural feature is it probably associated.?

3. All of the energy which enters the mantle from a near-surface earthquake in the crust must pass through a cone, whose sides are defined by the critical angle of incidence. What fraction of a hemisphere is subtended by this cone for *P*-waves in the case of a crust 40 km thick, with $\alpha = 6.5$ km/sec in the crust, and $\alpha = 8.0$ km/sec in the mantle?

4. On the basis of the energy partition shown in Figure C2, at what distance from the source would *P*-wave reflections from the base of a crust 35 km thick be most likely to be observed? What would you conclude about the strength of near-vertical reflections?

*5. Show that for a flat earth, in which velocity increases linearly with depth, the

*Problems so marked are more difficult than others, and may require theory developed in the Appendices.

seismic rays are arcs of circles. Can you deduce the ray shape for a spherical earth, in which velocity varies linearly with depth?

6. An earthquake occurred near the Tonga Islands (epicenter 17°42′S, 178°48′W), and good readings of *pP* (see Section 6.1) as well as *P* were observed at the following stations:

Station	Latitude	Longitude		*P*			*pP*	
			h	m	s	h	m	s
Byrd Station	80°01′S	119°31′W	17	14	43.0	17	16	39
Tucson	32°18′35″N	110°46′56″W	17	16	03.8	17	18	03
Uinta Basin	40°19′18″N	109°34′07.3″W	17	16	25	17	18	28

Using diagrams, and the velocity distribution given in Figure 3.12, make a preliminary estimate of the depth of focus of the earthquake from the *pP–P* interval.

7. Show that for a uniform sphere of radius *a*, in which the *P*-wave velocity is α, the travel-time curve is given by

$$T = 2a/\alpha \sin(\Delta/2)$$

Could you invert this travel-time curve by the methods of Section 3.3 to obtain α?

8. Show that Snell's Law as derived in Section 3.2 is consistent with Fermat's Principle, in that the time required for energy to travel between two points on opposite sides of a boundary is a minimum, compared to all possible paths, when Snell's Law is obeyed.

BIBLIOGRAPHY *(Chapter 3)*

Abe, Katsuyuki (1972) Seismological evidence for a lithospheric tearing beneath the Aleutian arc. *Earth Plan. Sci. Letters,* **14**, 428–432.

Adams, L. H., and Williamson, E. D. (1923) Density distribution in the earth. *J. Wash. Acad. Sci.,* **13**, 413.

Backus, G., and Gilbert, F. (1970) Uniqueness in the inversion of inaccurate gross earth data. *Phil. Trans. Roy. Soc. Lond. A,* **266**, 123–192.

Bessenova, E. N., Fishman, V. M., Ryaboyi, V. Z., and Sitnikova, G. A. (1974) The tau method for inversion of travel times. I: Deep seismic sounding data. *Geophys. J.,* **36**, 377–398.

Bessenova, E. N., Fishman, V. M., Shnirman, G. M., Sitnikova, G. A., and Johnson, L. R. (1976) The tau method for inversion of travel times. II: Earthquake data. *Geophys. J.,* **46**, 87–108.

Birch, F. (1966) Compressibility; elastic constants. In *Handbook of physical constants,* ed. S. P. Clark, Jr. Mem. 97, Geol. Soc. Amer.

Bolt, B. A. (1952) Gutenberg's early *PKP* observations. *Nature,* **196**, 122.

Bolt, B. A. (1964) The velocity of seismic waves near the earth's center. *Bull. Seism. Soc. Amer.,* **54**, 191.

Bolt, B. A., O'Neill, M., and Qamar, A. (1968) Seismic waves near 110°: is structure in core or upper mantle responsible? *Geophys. J.,* **16**, 475.

Buchbinder, Geotz G. R. (1968) Properties of the core-mantle boundary and observations of P_cP. J. Geophys. Res., **73**, 5901.

Buchbinder, Geotz G. R. (1972) Travel times and velocities in the outer core from P_mKP. *Earth Plan. Sci. Letters,* **14**, 161–168.

Bullen, K. E. (1965) *An introduction to the theory of seismology* (3rd Ed.). Cambridge Univ. Press, London and New York.

Carder, Dean S. (1964) Travel times from central Pacific nuclear explosions and inferred mantle structure. *Bull. Seism. Soc. Amer.,* **54**, 2271.

Engdahl, E. R., Flinn, E. A., and Romney, C. F. (1970) Seismic waves reflected from the Earth's inner core. *Nature,* **228**, 852–853.

Ergin, K. (1952) Energy ratio of the seismic waves reflected and refracted at a rock-water boundary. *Bull. Seism. Soc. Amer.,* **42**, 349.

Green, R. W. E., and Hales, A. L. (1968) The travel times of *P* waves to 30° in the central United States and upper mantle structure. *Bull. Seism. Soc. Amer.*, **58**, 267–289.

Gutenberg, B. (1914) Uber Erdbebenwellen VII A. Boebachtungen an Registrierungen von Fernbeben in Göttingen und Folgerungen über die Konstitution des Erdkörpers. *Nachr. der Konig, Gesell. der Wiss. zu Göttingen, Math. Phys. Kl.* **1**.

Gutenberg, B. (1944) Energy ratio of reflected and refracted seismic waves. *Bull. Seism. Soc. Amer.*, **34**, 85.

Haddon, R. A. W. and Bullen, K. E. (1969) An earth model incorporating free earth oscillation data. *Phys. Earth Plan. Int.*, **2**, 35.

Hai, N. (1961) Propagation des ondes longitudinales dans le noyau terrestre d'après les seismes profonds des îles Fidji. *Ann. de Géoph.*, **17**, 60.

Hai, N. (1963) Propagation des ondes longitudinales dans le noyau terrestre. *Ann. de Géoph.*, **19**, 285.

Herglotz, G. (1907) Uber das Benndorfsche Problem der Fortplanzungsgeschwindigkeit der Erdbebenstrahlen. *Phys. Zeit.*, **8**, 145.

Herrin, E. (1968) Introduction to the 1968 Seismological Tables for *P* phases. *Bull. Seism. Soc. Amer.*, **58**, 1193.

Husebye, E. and Madariaga, R. (1970) The origin of precursors to core waves. *Bull. Seism. Soc. Amer.*, **60**, 939.

Husebye, E., and Toksöz, M. Nafi (1968) Structure of the earth's core (abstr.). *Trans. Amer. Geoph. Un.*, **49**, 284.

Jacobs, J. A. (1953) The earth's inner core. *Nature* **172**, 297.

Jeffreys, H. (1932) An alternative to the rejection of observations. *Proc. Roy. Soc. A*, **137**, 78.

Jeffreys, H. (1939a) The times of P_cP *and* S_cS. *Mon. Not. R. A. S., Geophys. Suppl.*, **4**, 537.

Jeffreys, H. (1939b) The times of core waves. *Mon. Not. R. A. S., Geophys. Suppl.*, **4**, 548.

Jeffreys, H., and Bullen, K. E. (1935, 1940) *Seismological tables.* Brit. Assoc., Gray-Milne Trust.

Johnson, Lane R. (1967) Array measurements of *P* velocities in the upper mantle. *J. Geophys. Res.*, **72**, 6309.

Jordan, T. H., and Lynn, W. S. (1974) A velocity anomaly in the lower mantle. *J. Geophys. Res.*, **79**, 2679–2685.

Julian, B. R., Davies, D., and Sheppard, R. M. (1972) *PKJKP. Nature,* **235**, 317–318.

Kanasewich, Ernest R., Ellis, Robert M., Chapman, Chris H. and Gutowski, Paul R. (1973) Seismic array evidence of a core boundary source for the Hawaiian linear volcanic chain. *Journ. Geophys. Res.* **78**, 1361–1371.

Kennett, B. L. N. (1976) A comparison of travel time inversions. *Geophys. J.*, **44**, 517–536.

Knott, C. G. (1899) Reflection and refraction of elastic waves with seismological applications. *Phil. Mag.* (5), **48**, 64.

Lehmann, I. (1935) P^1. *Pub. Bur. Cent. Seism. Internat. A*, **14**, 3.

Massé, R. P. (1973) Compressional velocity distribution beneath central and eastern North America. *Bull Seism. Soc. Amer.*, **63**, 611–935.

Niazi, Mansour, and Anderson, Don L. (1965) Upper mantle structure of western North America for apparent velocities of *P* waves. *J. Geophys. Res.*, **70**, 4633.

Oldham, R. D. (1906) Constitution of the interior of the earth as revealed by earthquakes. *Quart. J. Geol. Soc.*, **62**, 456.

Qamar, A., and Eisenberg, A. (1974) The damping of core waves. *J. Geophys. Res.*, **79**, 758–765.

Sleep, N. H. (1973) Teleseismic *P*-wave transmission through slabs. *Bull. Seism. Soc. Amer.*, **63**, 1349–1373.

Wiechert, E., and Zöppritz, K. (1907) Uber Erdbebenwellen II. *Nach. der König. Gesellsch. der Wissensch. zu Göttingen, Math-Phys. Kl.*, **529**.

SURFACE WAVES 4

4.1 INTRODUCTION

We have, in the previous chapters, considered the effect of boundaries between media of different elastic properties only insofar as the reflection and refraction of body waves are concerned. In this chapter we consider a new class of elastic waves, which can exist only in the vicinity of such boundary surfaces.

Seismic surface waves play an important role in geophysics. Not only are they frequently the most prominent disturbance on a seismogram, but they provide a tool for investigating the velocity structure in the outer part of the earth which complements the information available from body waves. Because the treatment of surface waves in the most general case is fairly complex, we shall restrict our consideration to the basic types. More extensive developments will be found in Ewing, Jardetzky, and Press (1957), Haskell (1953), and Harkrider (1970).

4.2 WAVES ALONG A FREE SURFACE

Rayleigh (1885), whose development is outlined in Appendix C, proposed that a special kind of wave can be associated with a "free" surface, such as the earth-air interface. His result was that this new wave traveled with a velocity of approximately 0.9β and should be expected, at short distances from the source, not long after the arrival of the *S*-wave. The particle motion contains both vertical and horizontal components, but at any point is constrained to the vertical plane containing the direction of propagation of the wave. The theory predicts (Appendix C) that the motion is elliptical, with the major axis of the ellipse vertical, and at small depths it is retrograde, in that the motion at the top of the ellipse is opposite to the direction of propagation. Both components of motion decrease with depth below the surface. Figure 4.1 shows the variation of particle trajectory with depth, as observed in one of the very few experiments where this was possible (Dobrin, Simon, and Lawrence, 1951). Notice how the sense of particle motion reverses at a depth near 40 feet, which was about one-eighth of a wavelength.

A wave with the above characteristics is known as a free Rayleigh wave. The fact that it can play a role in the succession of waves received from an earthquake was established by Lamb (1904), in a paper which provides the basis of much of modern seismology. Lamb determined the displacements at a point on a free surface, due to a concentrated force of given time variation, at a distant point. His solutions (Fig. 4.2) clearly show the arrival of *P*, *S*, and a single prominent pulse which is the Rayleigh wave. On the real earth, the free Rayleigh wave is very often obscured by waves associated with surfaces of discontinuity within the earth, as we investigate in the following section.

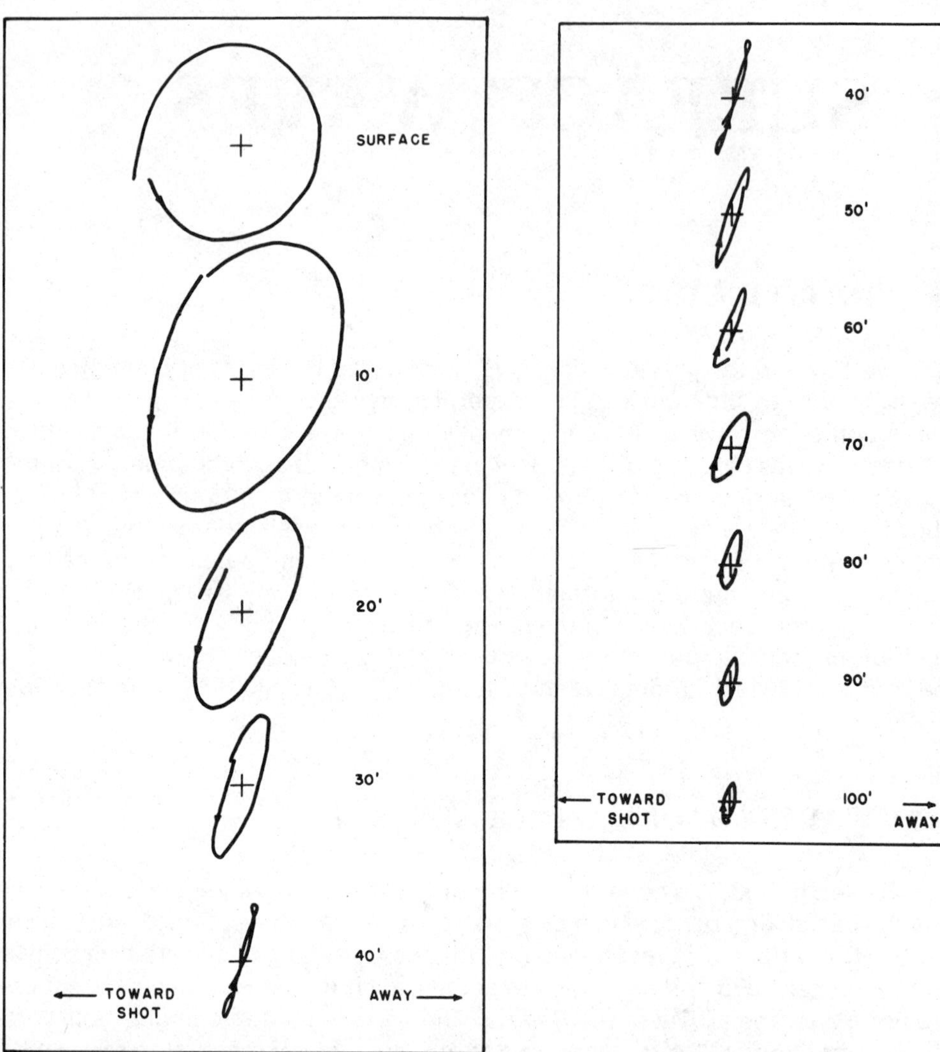

Figure 4.1 Rayleigh-wave particle motion from a small explosion as recorded by buried detectors at various depths. (After Dobrin *et al.*, 1951; courtesy of the American Geophysical Union)

4.3 THE EFFECT OF LAYERING

It was noted, shortly after the work of Rayleigh and Lamb, that observed surface waves frequently display two characteristics not provided for in the above case: the waves contain a horizontal component of particle motion, transverse to the propagation direction, and they arrive in a train rather than as a single pulse. The conclusion of an analysis by Love (1911) was that horizontally polarized shear waves could be propagated through a surface layer. The result (Appendix C) was that the wave velocity c must lie in the range

$$\beta' < c < \beta$$

where β' and β are the S-wave velocities in the layer and substratum, respectively. This indicates that for the assumed type of wave to be propagated, the

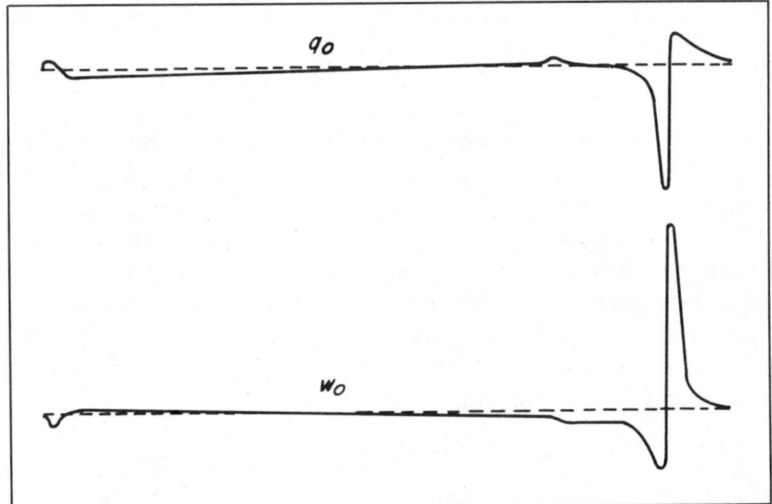

Figure 4.2 Lamb's solution for the vertical (w_0) and radial horizontal (q_0) displacements at the surface of a half-space, produced by a blow at a distant point. Time increases to the right, starting with the onset of P; S is the very small event before the large Rayleigh oscillation at the right.

crustal layer must have a velocity for S-waves lower than that of the substratum. In addition, the development showed that c depends not only upon β, β', and layer thickness, but also upon the wavelength. We have therefore met our first example, among elastic waves, of *dispersion,* which is precisely the condition that the velocity is dependent upon wavelength (or frequency).

Waves associated with a layer, such as Love waves, may be viewed as the result of constructive interference of plane waves, reflected successively from the top and bottom of the layer. To illustrate this, we consider the situation shown in Figure 4.3, in which the wave portrayed is a horizontally polarized S-wave. If the angle of incidence θ is greater than the critical angle, no energy will be transmitted to the medium beneath, and as none is lost to the space above, energy is trapped in the layer. In the simple case of a waveguide with perfectly reflecting boundaries (no phase change upon reflection), any disturbance y may be written as the sum of disturbances produced by upward and downward traveling waves of equal amplitude:

$$y = A \sin [\omega t - (ax_1 + bx_3)] + A \sin [\omega t - (ax_1 - bx_3)] \qquad 4.3.1$$

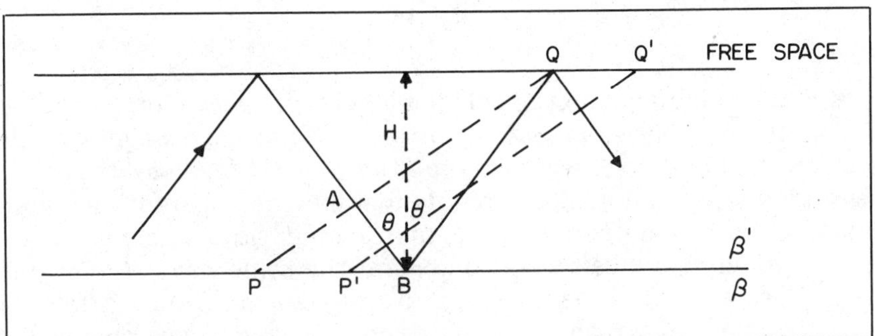

Figure 4.3 Representation of a wave confined to a layer by successive reflections. PQ and $P'Q'$ are two positions of the wave front.

where a and b are determined by the angle of incidence. The sum

$$y = 2A \cos bx_3 \sin(\omega t - ax_1) \qquad 4.3.2$$

represents a standing wave in the x_3 direction, but a traveling wave in the x_1 direction. The elastic wave case is the same in principle, but complicated by phase changes upon reflection at each boundary. In the figure, PQ and $P'Q'$ represent successive positions of the trace of the wave front, which advances with the wave velocity c introduced above. If PQ is a plane of constant phase, energy which has traveled the additional distance ABQ, and then been reflected at Q, must have suffered a phase change of $2n\pi$, where $n = 1, 2, 3 \ldots$. The phase change associated with the increased distance is $(2\pi/\lambda_0)ABQ$, where λ_0 is the true wavelength of the interfering waves. Thus,

$$(2\pi/\lambda_0)ABQ - \phi_1 + \phi_2 = 2n\pi \qquad 4.3.3$$

where ϕ_1 and ϕ_2 are respectively the changes in phase upon reflection at the base and top of the layer. Now $ABQ = 2H \cos\theta$; also, λ_0 is related to the wave number k, since $k = 2\pi/\lambda$, where λ, the wavelength measured along the surface, is given by $\lambda = \lambda_0/\sin\theta$. Equation 4.3.3 therefore becomes

$$\frac{2kH\cos\theta}{\sin\theta} - \phi_1 + \phi_2 = 2n\pi \qquad 4.3.4$$

or, since $\sin\theta = \beta'/c$,

$$2kH\sqrt{c^2/\beta'^2 - 1} - \phi_1 + \phi_2 = 2n\pi \qquad 4.3.5$$

The phase changes ϕ_1 and ϕ_2 depend upon the velocities concerned and the angle of incidence θ. They may be obtained through an analysis of the type discussed in Appendix C for amplitudes, but we shall here simply take the result that, for S-waves, ϕ_2 is zero, and ϕ_1 is given by

$$\tan(\phi_1/2) = \frac{\mu\sqrt{1 - c^2/\beta^2}}{\mu'\sqrt{c^2/\beta'^2 - 1}} \qquad 4.3.6$$

The condition for constructive interference finally becomes

$$kH\sqrt{c^2/\beta'^2 - 1} - n\pi = \tan^{-1}\frac{\mu\sqrt{1 - c^2/\beta^2}}{\mu'\sqrt{c^2/\beta'^2 - 1}} \qquad 4.3.7$$

which is the defining equation of velocity in terms of wave number.

This method of development is not restricted to Love waves, but can be applied, for example, to P-waves trapped in a liquid layer overlying a solid.

Returning to Equation 4.3.5, we note that there is apparently a multiplicity of solutions, depending upon the value of n. In other words, a series of waves of different wave number k can have, in general, the same wave velocity c. These different waves are known as modes, and are characterized by different numbers of horizontal nodal planes (planes of no particle displacement) within the layer. Unless they are specifically designated as higher modes, the Love waves normally considered are those of the first mode, with no nodal plane. Figure 4.4

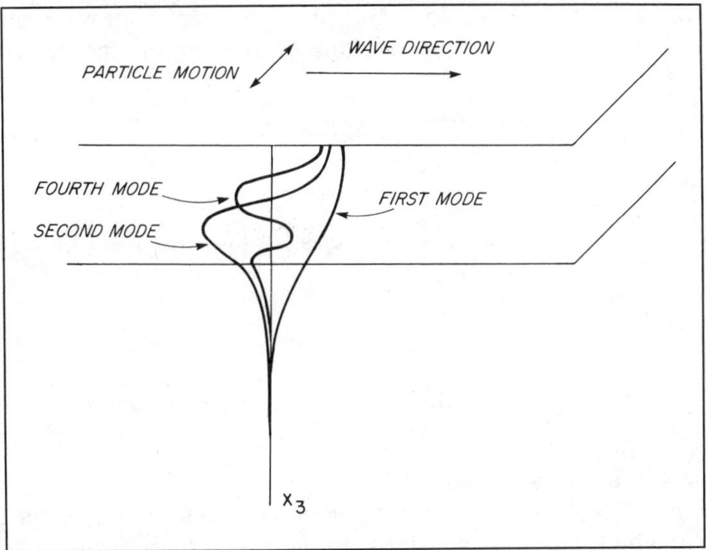

Figure 4.4 Sense of particle motion, and variation of amplitude with depth, for Love waves of different modes.

summarizes the particle motion and variation of amplitude with depth for Love waves.

While Rayleigh waves and Love waves are the best known types of surface waves in seismology, there are other important types. We note first that waves of the Rayleigh type can exist also in layered media, or in a medium containing a velocity gradient, although it is the fact that they alone can exist at the surface of a uniform half-space which distinguishes them from all other types. Stoneley (1924), in a pioneer analysis, showed that under suitable conditions waves (now known as Stoneley waves) may be guided along the interface between solid layers. There is also a series of imperfectly trapped waves in layers, produced by the successive reflection of waves at less than the critical angle of incidence. These waves, known as leaky modes (Phinney, 1961), lose energy into the material above or below, and are therefore attenuated more rapidly than perfectly trapped waves.

The characteristic of true surface waves which is of overwhelming importance in geophysics is that the energy is spread in two dimensions rather than in three, as in the case of body waves. This accounts for their relative prominence on records, particularly those of distant earthquakes.

4.4 SURFACE WAVES AND DISPERSION

All surface waves except the Rayleigh wave propagated along the free surface of a half-space exhibit dispersion. Where dispersion is present, the quantity c introduced above is the phase velocity, or velocity with which a pure harmonic note is propagated. When a spectrum of frequencies is produced by a source, their interference produces a pattern which is also propagated through the medium, and it is this pattern or modulation which is recognized as a signal. It propagates with a velocity known as the signal or group velocity.

As a simple case, consider the interaction of two waves traveling in the x

direction, of the same amplitude A, but of slightly different wavelengths (λ and λ') and velocities (c and c'). The resultant displacement can be written as

$$u = A \sin \frac{2\pi}{\lambda}(x - ct) + A \sin \frac{2\pi}{\lambda'}(x - c't)$$

$$= 2A \sin \pi \left\{\left(\frac{1}{\lambda} + \frac{1}{\lambda'}\right) x - \left(\frac{c}{\lambda} + \frac{c}{\lambda'}\right) t\right\} \cdot \cos \pi \left\{\left(\frac{1}{\lambda} - \frac{1}{\lambda'}\right) x - \left(\frac{c}{\lambda} - \frac{c}{\lambda'}\right) t\right\} \quad 4.4.1$$

or, very nearly,

$$u = 2A \cos \pi \left\{\delta\left(\frac{1}{\lambda}\right) x - \delta\left(\frac{c}{\lambda}\right) t\right\} \sin \frac{2\pi}{\lambda}(x - ct) \quad 4.4.2$$

where λ and c are mean values.

Equation 4.4.2 shows that the displacement pattern consists of a modulated wave in which the maxima of the amplitude are advancing with a signal or group velocity U, where

$$U = \frac{\delta(c/\lambda)}{\delta\left(\frac{1}{\lambda}\right)} = \left[c - \lambda \frac{dc}{d\lambda}\right]_{\lambda} \quad 4.4.3$$

The above simple derivation can be extended to any number of component frequencies, and Equation 4.4.3 will be found to be generally valid.

A typical computed dispersion curve for Love waves is shown in Figure 4.5.

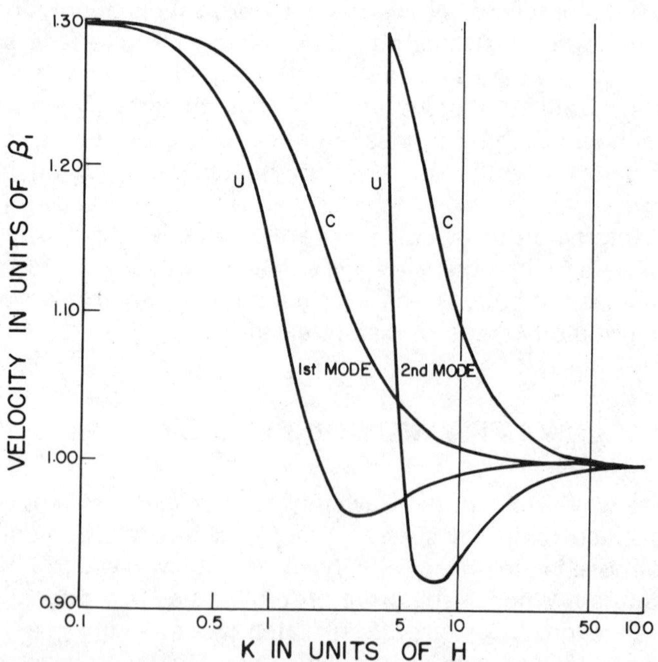

Figure 4.5 Computed dispersion curves for Love waves in a layer over a substratum whose velocity is 1.297 times as great. U is group velocity and c is phase velocity. (After Ewing, Jardetzky, and Press)

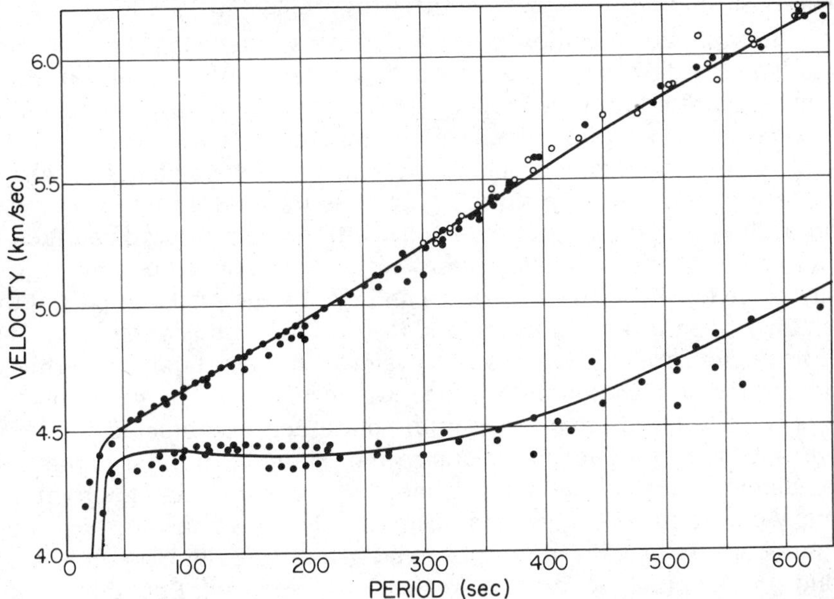

Figure 4.6 Compilation of observed phase velocity (*upper curve*) and group velocity variations for Loves waves. Points represented by open circles are from free oscillations. The solid lines represent computed velocities for a model similar to the Pacific case of Figure 4.9. (After Anderson and Toksöz)

We note that, as expected, c varies from β' to β, becoming asymptotic to each value. The general form of the variation of group velocity curve can be predicted from Equation 4.4.3. Where c is independent of λ, U equals c; a minimum of U occurs when $dc/d\lambda$ is greatest (i.e., where dc/dk has its greatest negative value). The illustration also shows the dispersion curve for the second mode, as defined above. We note that the same range of values of c is covered by the second mode as by the first, but the curve is displaced to larger values of k (shorter wavelengths).

As far as the arrival of signals at a distant station is concerned, it is the group velocity curve that is significant. Figure 4.5 indicates that arrivals would begin with the longest periods, followed by a mixture of long-period and short-period components, which would gradually merge into the last arrival, a relatively pure harmonic signal, corresponding to the group velocity minimum. The latter arrival characteristically has prominent amplitude, and is known as the Airy phase. In the real earth, because of complexities in structure near the surface and the presence of a velocity gradient in the mantle, dispersion curves rarely have this simple shape. Observations on Love waves, for example, are summarized in Figure 4.6.

4.5 STUDY OF THE EARTH BY SURFACE WAVES

The principle employed in the investigation of the outer part of the earth by surface waves is the comparison of observed dispersion curves with those computed for various models of velocity distribution. We have seen above an example of a curve computed for a simple model; the calculation can be extended to a model consisting of many layers, but it remains to discuss the establishment of the observed curve.

The derivation of dispersion curves from seismograms has, in fact, proceeded through an evolution over the past two decades. For the simplest method, and that first employed, only a single station and a source at some distance are required. From the dispersed train of surface waves observed on the one seismogram, a group velocity curve may be established by dividing the arrival times of oscillations of different periods into the distance. Each oscillation is here considered as a signal, which has traveled from the source to the station at the appropriate group velocity. If the station records surface waves from a number of earthquakes located in a small region, a good average group velocity curve for the path concerned can be constructed. Examples of this approach have been given by Kovach and Press (1961), who constructed Rayleigh wave dispersion curves, for waves traveling from earthquakes in the South Pacific and Indian Oceans to Pasadena, California.

It is remarkable that so much information can be obtained from a single station, but there are two limitations to the method. The first is that it is the group velocity curve which is determined, whereas the parameters of the system really control the variation of phase velocity, as we saw for Love waves in Equation 4.3.7. To pass from an observed group velocity curve to the corresponding phase velocity curve requires an integration (cf. Equation 4.4.3), with an attendant ambiguity. Secondly, the measurement of group velocity from source to receiver assumes that all frequencies are generated simultaneously. The earthquake source is now known to be rather complex (Chapter 6) and quite able to introduce phase shifts between the different frequencies that are radiated.

Recent work on surface waves has therefore emphasized the direct determination of phase velocity, and the avoidance of measuring times from the source. Press (1956) showed that from a group of three or more stations, direct measurement of the arrival times of the peaks or troughs of a dispersed train yields phase velocities. We consider (Fig. 4.7) three stations in an arrangement known as a tripartite array, and assume that the distances between stations are sufficiently small that the character of the arrivals does not change appreciably as the wave train passes the array. On the other hand, the distances must be large enough to permit the time differences between arrivals at the three stations to be determined with small error, and it is found that the optimum dis-

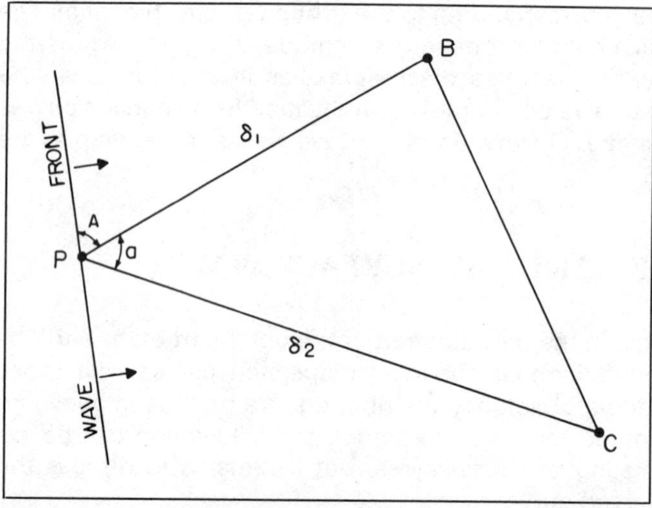

Figure 4.7 Determination of phase velocity by means of a tripartite array.

tance is about one wavelength (a few hundred kilometers, for surface waves). A particular peak or trough travels across the array with the phase velocity c appropriate to its mean period. It is important to note the distinction between this approach, which utilizes a short portion of the path and traces a given portion of the wave train, and the group velocity method. With the notation of Figure 4.7, the phase velocity c and angle of approach A are determined by

$$c = \frac{\delta_1 \sin A}{\Delta T_1} = \frac{\delta_2 \sin (A + a)}{\Delta T_2} \quad 4.5.1$$

where ΔT_1 and ΔT_2 are the measured time intervals between stations P and B, and P and C, respectively. By working with features corresponding to oscillations of different period, a complete phase velocity curve is constructed.

The above procedure utilizes visual examination of the records to trace a particular peak or trough from station to station. A more objective approach is to compute the Fourier spectrum of the arrivals at each station, to yield the relative phases of different frequency components. Shifts in phase are introduced by the differences in phase velocity, and from them the phase velocity curve can be computed (Ben-Menahem and Toksöz, 1963; Anderson, 1965; Knopoff, Mueller, and Pilant, 1966).

Finally, it has been possible to apply the same procedure to a single station, in those cases where a train of surface waves passes around the earth on a great-circle path and is reobserved at the same station. This remarkable technique, which combines the convenience of a single station approach with the desirable features of the phase-velocity method, has been utilized by Toksöz and Anderson (1966), Dziewonski and Landisman (1970), and others. Figure 4.8, from Dziewonski and Landisman, shows beautiful trains of Rayleigh and Love waves as observed in Australia from an earthquake in the Kurile Islands. Because of the velocity gradient in the mantle, the Rayleigh as well as Love waves show dispersion. The dispersed trains alternate in length, depending upon whether they arrive by the long or short section of the great circle path.

The general instrumental requirements for surface wave studies are the availability of instruments capable of recording over the wide range of frequencies which must be considered, and provision for converting the seismograms to digital form for computation.

The interpretation of dispersion curves in terms of structure has usually been carried out by comparison with curves computed for models of the earth, in which layers of different elastic properties are assumed. This process of determining a model from the observations, known as inversion, has been shown by Knopoff (1961) to be non-unique. In other words, different models can yield identical dispersion curves, unless other information is available to assist in determining some parameters. Furthermore, Love waves depend only upon the shear-wave velocity; and while Rayleigh waves in principle are influenced by P-wave velocities, it turns out that their dispersion curves are relatively insensitive to them. Most of the results of surface wave studies are therefore presented as models of S-wave variation with depth.

Figure 4.9 displays three types of velocity distribution (Dorman, 1969). The Canadian shield type is characterized by a thick crust, below which the shear velocity reaches a value of 4.7 km/sec, but passes through a broad and shallow minimum. The Alpine type, also continental, displays a much more severe minimum, while the curve for the Pacific shows a thin crust, with fairly pro-

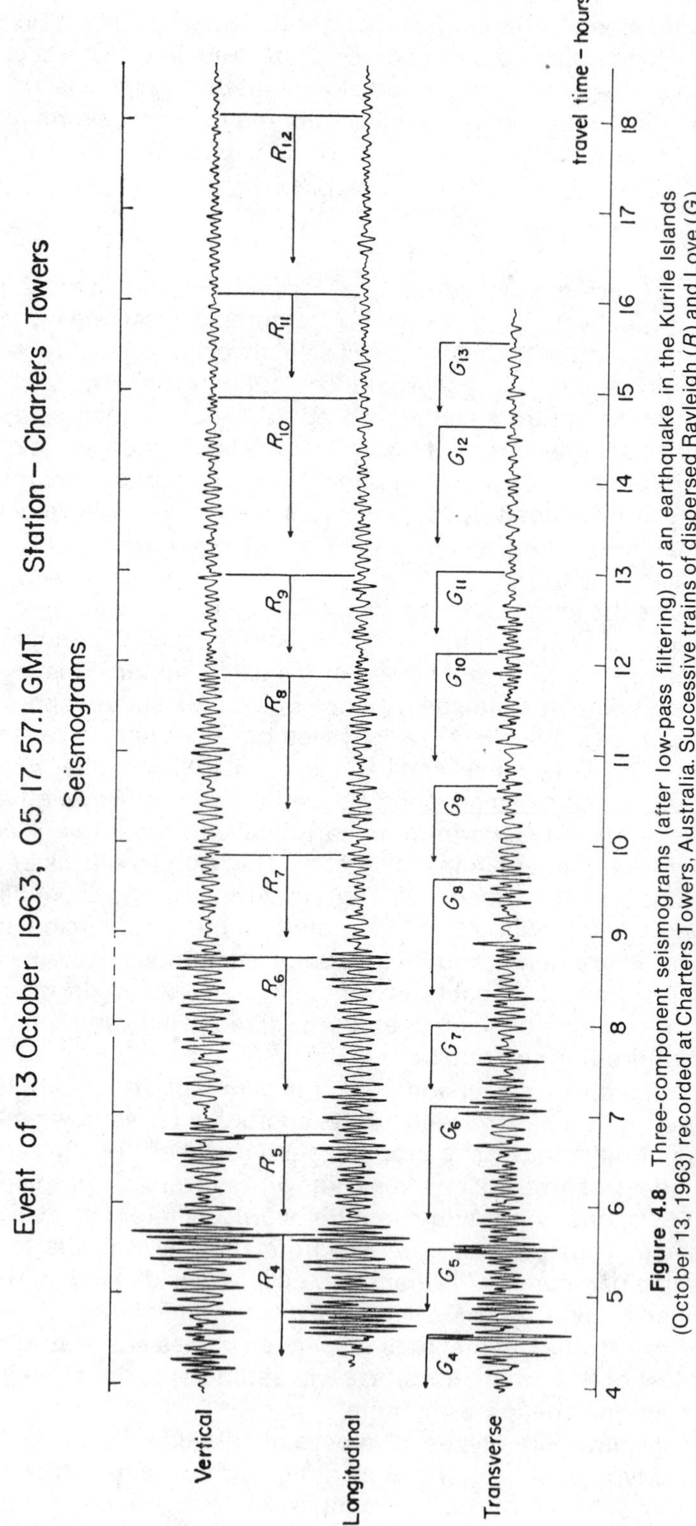

Figure 4.8 Three-component seismograms (after low-pass filtering) of an earthquake in the Kurile Islands (October 13, 1963) recorded at Charters Towers, Australia. Successive trains of dispersed Rayleigh (*R*) and Love (*G*) waves are indicated, the even orders having traveled by the longer great-circle path. (From A. Dziewonski and M. Landisman, 1970; courtesy of the authors and the *Geophysical Journal*.)

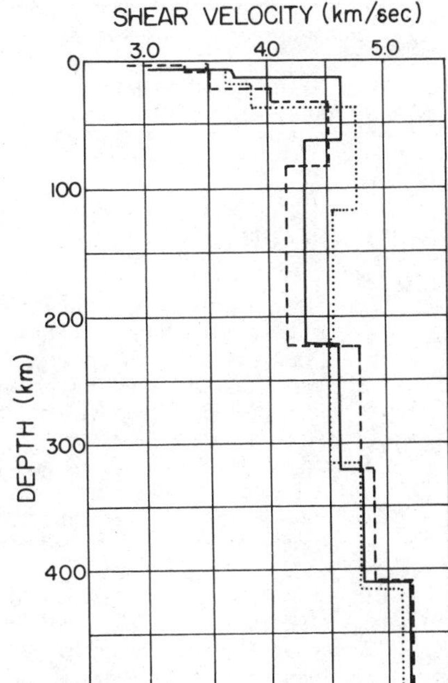

Figure 4.9 Models of the shear wave velocity distribution in the upper mantle. Dotted curve is Canadian shield type, broken curve Alpine, and solid curve Pacific (after Dorman).

nounced minimum. The minima associated with the Alpine and Pacific studies are centered at about 150 km in depth, while that for the Canadian shield is somewhat deeper. These minima are an expression of the low-velocity layer, which has been discussed earlier in connection with body waves. The important point is the variability in its depth and intensity, not only between the mantle beneath continents and oceans, but also between the mantle beneath different tectonic areas. Such regional differences have a bearing upon the great circle path analysis mentioned above. It becomes important to choose paths which are purely oceanic or (more difficult) purely continental. We shall show later that the mechanical properties of the mantle above, in, and below the low-velocity region may be quite different; it is convenient to adopt the term "lithosphere" for that portion of the earth, including crust and uppermost mantle, which lies above the low-velocity zone.

The great advantage of surface waves is their ability to sample the outer part of the earth over long paths, and to provide an average structure for parts of the earth which are not readily accessible to body-wave studies. Higher modes of waves trapped in layers are much more sensitive to the fine structure of the velocity model (Kovach, 1965). Future investigations will therefore probably see an increasing use of these modes to study regional departures from the average global picture.

Problems

1. Show that the elliptical particle motion of the Rayleigh wave (Equation C.2.8, Appendix C) is in fact retrograde.

*2. It was noted in Section 4.3, in connection with Love waves, that there is a series of solutions of Equation 4.3.7, because of the periodic nature of the tangent function. That is, a given phase velocity c can be characteristic of different wave numbers k, each of which is related to a particular Love wave mode. Investigate the behavior in depth of the particle displacement (Equation C.2.10, Appendix C) and show that increasing modes have successively larger number of nodal planes within the layer.

3. Use Equation 4.4.3 to confirm, for selected points, that the group velocity curve of Figure 4.5 is actually that corresponding to the phase velocity curve shown. If you were given only the group velocity curve, could you deduce the phase velocities?

BIBLIOGRAPHY

Anderson, Don L. (1965) Recent evidence concerning the structure and composition of the earth's mantle. In *Physics and chemistry of the earth.* Vol. 6, ed. L. H. Ahrens, Frank Press, S. K. Runcorn, H. C. Urey. Pergamon Press, Oxford.

Ben-Menahem, Ari, and Toksöz, M. Nafi (1963) Source mechanism from spectrums of long-period surface waves 2, the Kamchatka earthquake of November 4, 1952. *J. Geophys. Res.,* **68**, 5207.

Dobrin, M. B., Simon, R. F., and Lawrence, P. L. (1951) Rayleigh waves from small explosions. *Trans. Amer. Geophys. Un.,* **32**, 822–832.

Dorman, James (1969) Seismic surface-wave data on the upper mantle. In *The earth's crust and upper mantle,* ed. Pembroke J. Hart. *Amer. Geophys. U., Monograph* **13**, 257.

Dziewonski, A., and Landisman, M. (1970) Great circle Rayleigh and Love wave dispersion from 100 to 900 seconds. *Geophys. J.* **19**, 37–91.

Ewing, W. Maurice, Jardetzky, W. S., and Press, Frank (1957) *Elastic waves in layered media.* McGraw-Hill, New York.

Harkrider, David G. (1970) Surface waves in multilayered elastic media II. Higher mode spectra and spectral ratios from point sources in plane layered earth models. *Bull. Seism. Soc. Amer.,* **60**, 1937.

Haskell, N. A. (1953) Dispersion of surface wave in multilayered media. *Bull. Seism. Soc. Amer.,* **43**, 17.

Knopoff, Leon (1961) Green's function for eigenvalue problems and the inversion of Love wave dispersion data. *Geophys. J.,* **4**, 161.

Knopoff, L., Mueller, S., and Pilant, W. L. (1966) Structure of the crust and upper mantle in the Alps from the phase velocity of Rayleigh waves. *Bull. Seism. Soc. Amer.,* **56**, 1009.

Kovach, R. L. (1965) Seismic surface waves: some observations and recent developments. In *Physics and chemistry of the earth,* ed. L. H. Ahrens, Frank Press, S. K. Runcorn, H. C. Urey. Pergamon Press, Oxford.

Kovach, R. L., and Press, F. (1961) Rayleigh wave dispersion and crustal structure in the eastern Pacific and Indian Oceans. *Geophys. J.,* **4**, 202.

Lamb, H. (1904) On the propagation of tremors over the surface of an elastic solid. *Phil. Trans. Roy. Soc. A,* **203**, 1.

Love, A. E. H. (1911) *Some problems of geodynamics.* Cambridge Univ. Press, London.

Phinney, Robert A. (1961) Leaking modes in the crustal waveguide. The Oceanic PL wave. *J. Geophys. Res.,* **66**, 1445.

Press, Frank (1956) Determination of crustal structure from phase velocity of Rayleigh waves. Part I: Southern California. *Bull. Geol. Soc. Amer.,* **67**, 1647.

Rayleigh, Lord (1885) On waves propagated along the plane surface of an elastic solid. *Proc. Lond. Math. Soc.,* **17**, 4.

Stoneley, R. (1924) The elastic waves at the surface of separation of two solids. *Proc. Roy. Soc. A,* **106**, 416.

Toksöz, M. Nafi, and Anderson, Don L. (1966) Phase velocities of long-period surface waves and structure of the upper mantle, I: Great circle Love and Rayleigh wave data. *J. Geophys. Res.,* **71**, 1649.

SEISMOMETRY 5

5.1 THE INSTRUMENTAL REQUIREMENTS FOR SEISMOLOGY

We have postponed a discussion of instruments until both body and surface waves were considered, in order to appreciate the requirements. While body-wave studies emphasized the determination of the arrival times of phases, the analysis of surface waves demanded the registration of dispersed trains. Thus any instrument, or combination of instruments, for the recording of surface waves must possess a frequency response window of adequate width.

A seismograph is any instrument which provides a visual record of some characteristic of ground motion during the arrival of seismic waves. The heart of the instrument is the seismometer or detector, which responds to one of the following: ground displacement, velocity or acceleration; it must produce a signal, usually electric, which may be recorded. Normally, a seismometer responds to a particular component of ground disturbance, and for a complete description of the arrival three instruments, corresponding to three coordinate directions, are required at each station.

It is evident that a general requirement for seismological instruments is adequate coupling to the earth. This is usually accomplished by mounting the seismometers on piers, which are preferably isolated from the observatory building, and if possible, constructed on rock outcrop. A second requirement is the placing of accurately controlled time marks on the record. While this appears straightforward, it is in fact only very recently that absolute timing accuracy to about 0.1 second has been achieved at a majority of seismograph stations of the world.

The early history of seismometry has been described by Dewey and Byerly (1969).

5.2 THE SEISMOGRAPH EQUATION

Regardless of the type of transducer employed in a seismometer, it must respond to ground displacement or its time derivatives, and this appears to require that there be a fixed reference point. No point of the seismometer remains truly fixed during the arrival of a seismic wave, but a mass of large inertia, loosely coupled to the frame of the instrument, remains nearly so. We begin with a simple type of horizontal seismometer, consisting of a pendulum free to rotate about an axis which is slightly inclined to the vertical (Fig. 5.1).

The plane containing the pendulum, when it is at equilibrium position, is known as the neutral plane. We take θ as the rotation of the boom from the

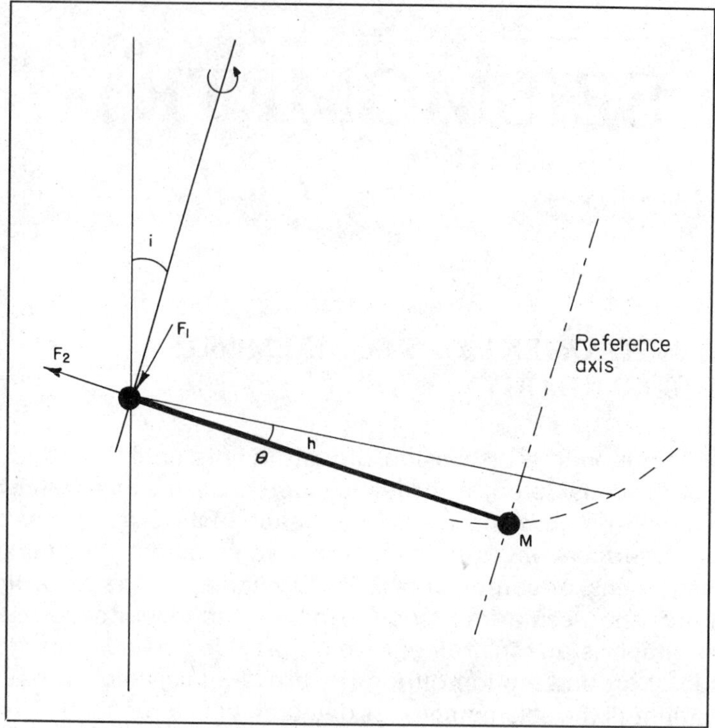

Figure 5.1 The horizontal seismograph. A pendulum of total mass M is free to rotate about an inclined axis.

neutral plane, h as the distance from the hinge axis to the center of mass of the pendulum, and u as the displacement of the frame perpendicular to the neutral plane. The forces exerted by the frame on the pendulum can be represented by a force F_1, perpendicular to the neutral plane, and F_2, in the neutral plane and perpendicular to the axis; a force parallel to the axis does not contribute to the rotation. All of these forces are exerted on the pendulum through the supporting pivot or hinge; the only other force acting on the suspension is the weight Mg, which acts through the center of mass of the pendulum.

Then, in the standard notation for time differentiation,

$$\left.\begin{aligned} F_1 &= M(\ddot{u} + h\theta) \\ F_2 &= Mg \sin i, \text{ very nearly} \\ &= Mgi, \text{ if } i \text{ is small} \end{aligned}\right\} \qquad 5.2.1$$

If we consider the rotation of the frame about an axis through the mass center parallel to the axis of rotation, we have

$$Mk^2\theta = -F_1h - F_2h\theta \qquad 5.2.2$$

where k is the radius of gyration of the suspension about this axis.

Substitution for F_1 and F_2 from 5.2.1 leads to

$$(k^2 + h^2)\,\ddot{\theta} + ghi\dot{\theta} = -hu$$

or

$$\ddot{\theta}+\omega^2\theta=-\ddot{u}/l \quad 5.2.3$$

where

$$\omega^2=gi/l$$

$$l=(k^2+h^2)/h$$

Equation 5.2.3 is evidently the equation of (undamped) forced oscillations with ω, the natural angular frequency of the pendulum, being determined by the inclination i of the axis. The quantity l, which becomes equal to h in the case of a point mass on a massless boom, is known as the reduced pendular length.

In practice there will always be some damping, so that a term proportional to $\dot{\theta}$ must be added to Equation 5.2.3. A standard form is then

$$\ddot{\theta}+2\lambda\omega\dot{\theta}+\omega^2\theta=-\ddot{u}/l \quad 5.2.4$$

where λ is the damping constant.

A vertical component seismometer consists of a mass at the end of a light boom, suspended by an inclined spring, and free to move in a vertical plane. An equation identical to 5.2.4 is obtained for it, with θ representing the rotation of the boom in its vertical plane, and ω determined by the force constant of the spring and the geometry.

In practice the angle θ itself is seldom recorded, but it is instructive to consider the solution of Equation 5.2.4 for a given ground motion. We suppose first that a linear displacement x is recorded, where x is a simple multiple of θ, given by

$$x=l\theta V_s \quad 5.2.5$$

in which V_s is known as the static magnification. This relationship would arise, for example, if small rotations θ were magnified by means of an optical lever.

Then, for a harmonic ground motion of the form

$$u=a\cos pt \quad 5.2.6$$

we have

$$\ddot{x}+2\lambda\omega\dot{x}+\omega^2x=V_sp^2a\cos pt \quad 5.2.7$$

of which the solution is

$$x=aV_sp^2\{(\omega^2-p^2)^2+4\lambda^2\omega^2p^2\}^{-1/2}\cos(pt-\delta) \quad 5.2.8$$

where $\tan\delta=2\lambda\omega p(\omega^2-p^2)^{-1}$

Equation 5.2.8 is the familiar solution for forced oscillations, displaying a resonant peak at $p=\omega$. In the absence of damping, the peak becomes infinite, but with increasing λ, the response is smoothed out.

Two important special cases follow from Equation 5.2.7. If the natural period of the seismometer is great compared to the period of the seismic waves, $\omega \ll p$, and the equation reduces to

$$\ddot{x} = -V_s\ddot{u} \qquad 5.2.9$$

so that x behaves as u, and the instrument is a displacement meter. If, however, the natural period is small compared to the period of the seismic waves, $\omega \gg p$, and we have

$$\omega^2 x = -V_2\ddot{u} \qquad 5.2.10$$

so that x records essentially the ground acceleration, and the instrument is an accelerometer.

Regardless of the type of response, the ratio of the amplitude of trace displacement to that of ground displacement at any frequency is known as the dynamic magnification.

Virtually all seismographs in use today record the seismometer motion electromagnetically with a technique invented by Galitzin (1914). The seismometer pendulum carries a permanent magnet, free to move relative to coils mounted on the frame, or vice versa; induced emf causes current to flow through a galvanometer connected to the coils, and the deflections of the galvanometer are recorded photographically. A modern seismograph is shown in Figure 5.2.

The induced emf may be written

$$E = -C_1\dot{\theta}$$

where C_1 is a constant determined by the electromagnetic and mechanical properties of the seismometer (i.e., magnetic field strength at the coils, number of turns, and so forth).

If we let ϕ be the angular deflection of the galvanometer, we may write the counter emf generated by the galvanometer as $-C_2\dot{\phi}$, where C_2 is determined by the electromagnetic properties of the galvanometer (number of turns and area of coil, and magnetic field strength at the coil).

The current in the seismograph-galvanometer circuit, for circuit resistance R, is then

$$\frac{E - C_2\dot{\phi}}{R}$$

and the torque on the galvanometer suspension is

$$C_2\,\frac{(E - C_2\dot{\phi})}{R}$$

The equation of motion of the galvanometer may be written

$$K\ddot{\phi} + \phi D + b\dot{\phi} = \frac{C^2E}{R} - \frac{C_2^2\dot{\phi}}{R} \qquad 5.2.11$$

where K is the moment of inertia of the suspension about its rotation axis, D is the elastic torque constant of the suspension, and b is a constant representing

Figure 5.2 Interior of a modern seismograph station. The pier on the left holds 3 short-period seismometers (rear corner) driving galvanometers which reflect onto the drum in the foreground. Long-period seismometers and galvanometers are on the pier to the right. Both piers are isolated from the floor of the vault. (Photograph courtesy of Earth Physics Branch, Dept. of Energy, Mines and Resources, Canada)

damping torques, other than electromagnetic, on the suspension. Substituting for E and rearranging,

$$K\ddot{\phi} + [b + C_2^2/R]\,\dot{\phi} + D\phi = -\frac{C_1C_2}{R}\,\dot{\theta} \qquad 5.2.12$$

In more standard form

$$\ddot{\phi} + 2hn\dot{\phi} + n^2\phi = -\frac{C_1C_2}{KR}\,\dot{\theta} \qquad 5.2.13$$

where $n^2 = D/K$ and

$$2hn = b/K + \frac{C_2^2}{KR}$$

The quantity actually recorded is now a simple multiple of ϕ, determined by the length of an optical lever. Equation 5.2.13 is similar in form to Equation 5.2.4, except that the coupling is to seismometer velocity, whereas the seismometer was coupled to ground acceleration. It is to be expected that the galvanometer will exhibit maximum deflections, for a given torque, when the frequency of the signal is close to its natural angular frequency n. Considering the overall system of seismometer and galvanometer, we would predict a double-peaked response curve, with maxima at the natural frequencies of both the seismometer and galvanometer. This is indeed the case, and the principle was used by Press, Ewing, and Lehner (1958) in the design of a long-period instrument which has now become standard. The construction of seismometers of the suspension type becomes difficult for natural periods longer than about 30 seconds, but galvanometers with natural periods of 90 seconds or greater are obtainable. The response of a Benioff seismograph with two galvanometers is shown in Figure 5.3.

However, to return to the derivation of Equation 5.2.13, it will have been noticed that the galvanometer motion influences the current in the circuit, and therefore the forces acting on the seismometer suspension. In principle it is not valid to consider the seismometer motion independent of the galvanometer; in practice, however, the reaction of the galvanometer through the circuit to the seismometer is sufficiently small to be neglected.

It is also necessary to consider in more detail the damping of both seismometer and galvanometer. Damping to some degree is desirable, to remove the extreme peaks of resonances and to attenuate the free vibration of the suspension that would normally follow an impulse arrival. On the other hand, overdamping reduces the sensitivity to an undesirable degree. Most seismographs are adjusted to a damping which is approximately critical, that is, the systems have just ceased to be oscillatory. As is well known, this condition is given by $\lambda = 1$ in Equation 5.2.7, or $h = 1$ in Equation 5.2.13. It is seen that, for the galvanometer, h is determined by fixed constants and the value of the circuit resistance R. The same is true of the quantity λ for the seismometer, and in this regard the external circuit must be considered when the seismometer motion is derived. It is usual to connect the seismometer and galvanometer in a circuit consisting of two series resistors and a shunt, so that each element "sees" the correct external resistance for the desired degree of damping.

The overall response of a seismometer-galvanometer system is always determined experimentally, rather than by calculation. While the seismometer

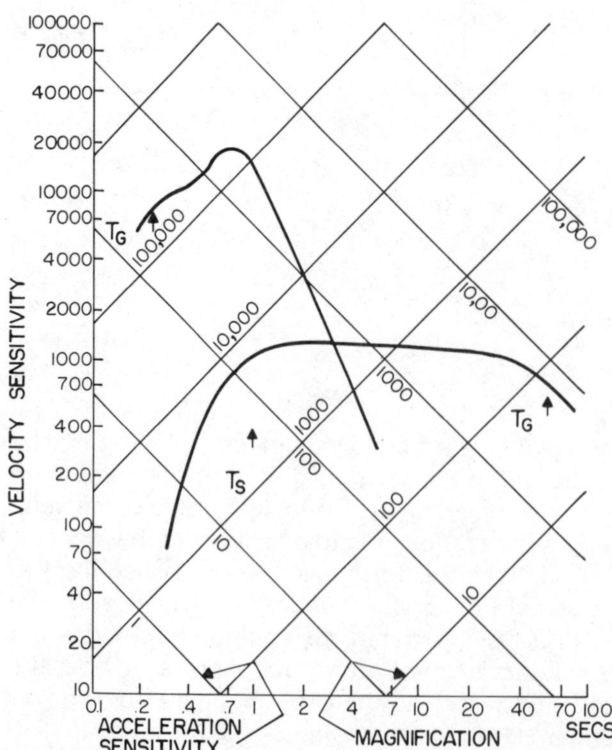

Figure 5.3 Response of the same seismometer (period 1 sec) when connected to galvanometers of period 0.25 sec and 60 sec. Note that magnification is shown by the graticule inclined upward to the right. Velocity and acceleration sensitivities give trace deflection for unit ground velocity and acceleration respectively. (After Willmore)

may be placed on a shaking table and subjected to known disturbances, there is a more convenient method described by Willmore (1959) for the calibration of electromagnetic seismographs. In this method, the seismometer coil is made to be one arm of an impedance bridge, which is driven, over a range of frequencies, by a signal generator. The change in galvanometer deflection between the clamped and unclamped conditions of the seismometer gives the output of the seismometer in response to a known driving force as a function of frequency.

5.3 STANDARD SEISMOGRAPHS AND SEISMOGRAMS

We may now return to the consideration of the requirements of a seismograph station, for the recording of both body and surface waves. What has become known as a "standard station" contains three components of short-period instruments (0.5 to 1.0 second natural period), with a magnification, in aseismic areas, as great as 500,000. The rapid response of short-period instruments provides records on which the onset of body-wave phases may be accurately read. A standard station also contains three components of long-period instruments of the Press-Ewing type described above, whose magnification is of the order of 10,000. Time marks are placed on all records by means of clocks whose error is not greater than 0.05 second. The growth of a world-wide network of stations of this type has come about chiefly as a result of the desire to detect nuclear explosions, and to distinguish them from earthquakes.

A portion of a seismogram from a long period instrument, of a moderately distant earthquake, has already been seen in Figure 2.1. For completeness, it is important that any record indicate the component recorded, the clock correction, and the sense of ground motion corresponding to an upward deflection on the trace. We noted, on the sample shown, the time marks at minute intervals, the arrival of body-wave phases, and the prominent train of surface waves. A point which we have not yet stressed is the presence of background noise, even before the arrival of *P* (although this would be relatively greater on a record from a short period instrument) and the fact that, after the arrival of *P*, the trace does not become completely quiet between the arrivals of the identified phases. The record is much more complicated than the solution obtained by Lamb (Fig. 4.1), presumably because the inhomogeneity of the real earth allows scattered energy to reach the station by a number of paths in addition to those corresponding to the main phases.

Seismology owes much to international cooperation in the exchange of data. Original seismograms are normally kept at the observatories where they were produced, but microfilm copies are filed at World Data Centers in Washington and Moscow. Routine determination of epicenters, and the publication of bulletins, is carried out on the basis of arrival times of body-wave phases, read at cooperating stations throughout the world and reported to a central agency. At the present time this work is conducted by the United States Coast and Geodetic Survey, and by centers set up under the International Association of Seismology and Physics of the Earth's Interior at Strasbourg and Edinburgh.

5.4 THE STRAIN SEISMOGRAPH

It was recognized by Benioff (1935) that a seismograph other than the pendulum type possessed advantages in the recording of very long period dis-

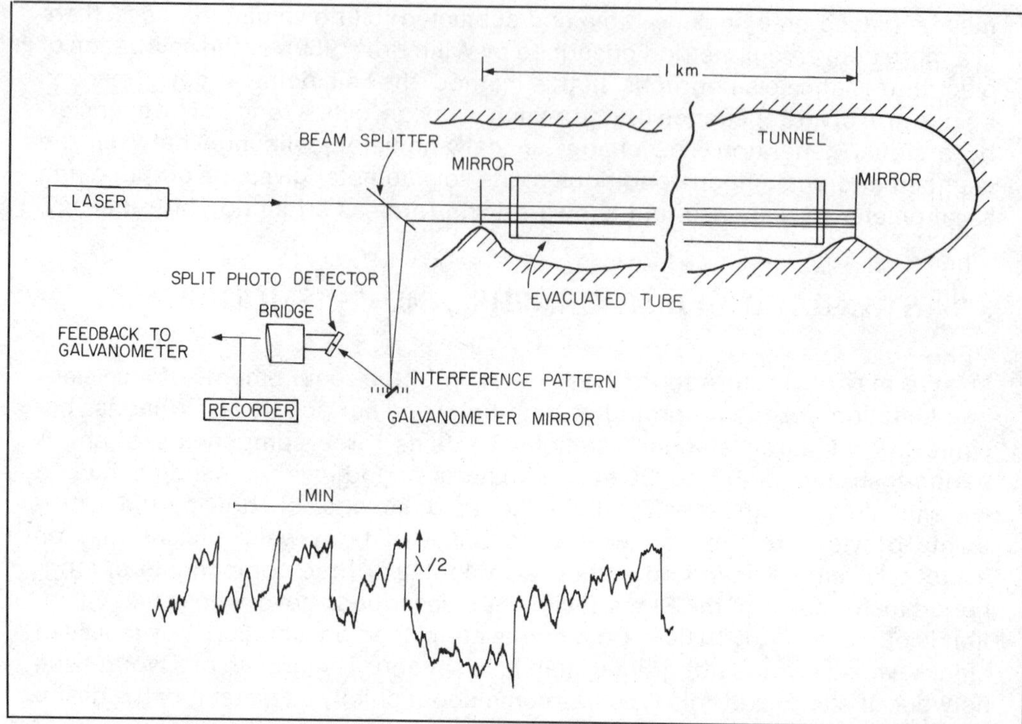

Figure 5.4 Schematic diagram of a laser strainmeter (top) and tracing of a record from Steven's Pass, Washington, (bottom). The quantity recorded is the feedback current required to lock the galvanometer to a particular fringe; vertical displacements (equivalent to a displacement of one-half wavelength of the laser light or 30×10^{-10} m.) occur when the galvanometer jumps to an adjacent fringe. (After Vali 1970)

turbances. Benioff constructed the first high-sensitivity instrument to measure linear strain in the earth, but it is fair to say that the full significance of instruments of this type has only recently become appreciated. Interest in very long period instruments has grown because of the recognition of the importance of free oscillations of the earth (Chapter 8), which have periods of minutes rather than seconds.

The essential measurement with the strain seismograph is the relative displacement between piers on the earth's surface, some tens of meters distant from each other. Originally this was accomplished with a rigid rod fastened to one pier, supported on free bearings and brought sufficiently close to the second pier for some type of transducer to measure the changes in distance between the free end of the rod and the second pier. One such installation consisted of a quartz tube, 61 meters long, mounted in an abandoned mine tunnel to reduce temperature effects, and carrying at its free end one plate of a condenser (Alsop, Sutton, and Ewing, 1961). The second condenser plate was mounted on the pier adjacent to this end of the rod, so that relative displacements produced a change in capacitance which was recorded. Increased sensitivity could be achieved by the use of interferometry to measure these relative displacements, but the availability of lasers permits the direct measurement of strain between mirrors attached to piers, eliminating the need for the quartz tube (Vali, Krogstad, Moss, and Engel, 1968). In one type of laser strain seismograph, light from the laser passes through a beam splitter (Fig. 5.4) to an interferometer consisting of a nearby, partially reflecting mirror and a distant mirror;

after multiple reflection, it passes back to the beam splitter and then to the detector. A pattern of interference fringes is set up at the detector, and this pattern moves with any change in the length of the interferometer. In the detector, a photoelectric system locks onto a particular fringe of the pattern, converting its motion to a record of length variation. Vali (1970) has described such an instrument, installed in a tunnel at Stevens Pass, Washington, in which the length of the interferometer is 1 kilometer. For an optical path of this length, an evacuated tube with end windows must be placed between the mirrors, as otherwise variations in atmospheric conditions would alter the velocity of light sufficiently to produce spurious indications of length change. The strain sensitivity claimed for systems of this type is 1 part in 10^{12}, a factor of about 10^3 higher than that of the quartz rod strain seismometer.

The response of the strain seismograph to surface waves (true or apparent) was derived by Benioff (1935). Consider the rod to extend a distance L in the x-direction, from an origin located at the free pier. At coordinate x, let the ground displacement be ζ, at an angle β to the x-axis. Then the total earth strain over the length L is

$$y = \int_0^L \cos\beta\ \partial\zeta/\partial x\ dx \qquad 5.4.1$$

For plane waves, whose wavelength is long compared to L, we may write

$$y = L\cos\beta\ \partial\zeta/\partial x \qquad 5.4.2$$

If the ground displacement arises from a longitudinal wave propagating at angle β to the rod, the displacement is of form

$$\zeta = f\left(t - \frac{x\cos\beta}{c}\right)$$

where c is the velocity along the surface, and

$$y = -L/c\ \cos^2\beta\ \partial\zeta/\partial t \qquad 5.4.3$$

On the other hand, if the displacement arises from a transverse wave, the equation for ζ is

$$\zeta = f\left(t - \frac{x\sin\beta}{c}\right)$$

and

$$y = -\frac{L}{c}\sin\beta\cos\beta\ \partial\zeta/\partial t \qquad 5.4.4$$

or

$$y = \frac{L}{c}\sin\alpha\cos\alpha\ \partial\zeta/\partial t$$

where α is the angle between the propagation direction and the x-axis.

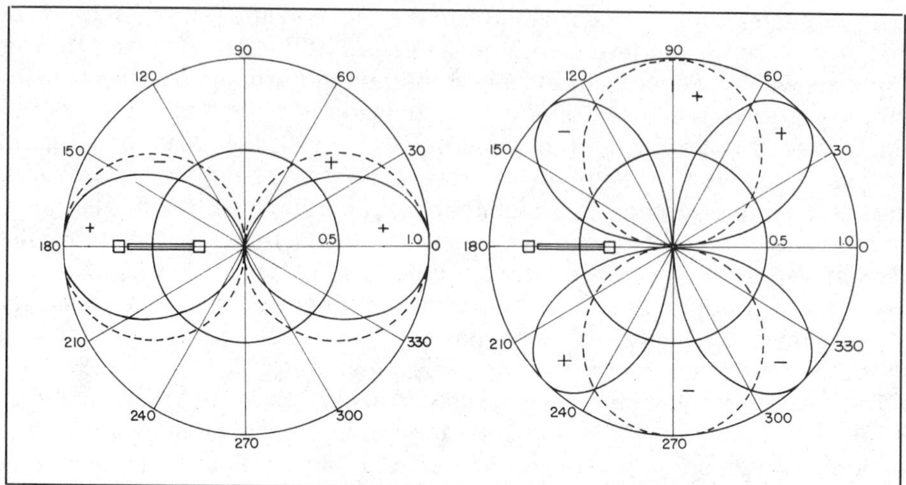

Figure 5.5 Response of the strain seismograph, oriented as shown, to longitudinal (*left*) and transverse waves traveling in the azimuths indicated on the outer circle. The dotted curves give the response of a pendulum seismograph. (After Benioff)

Equations 5.4.3. and 5.4.4 therefore give the response of the strain seismograph in terms of the angles α or β, wave velocity, and particle velocity. We note, in particular, the directivity pattern, $\cos^2 \beta$ or $\sin \alpha \cos \alpha$, which is characteristic of this instrument (Fig. 5.5). The fact that particle velocity appears in the equations indicates that the response does decrease with increasing wave period, but less severely than in the case of the pendulum instrument, which, as we have seen, responds to ground acceleration at long periods. It should be emphasized that even this decrease occurs only when the strain arises from a traveling elastic wave. There is nothing instrumental to prevent a strain seismometer from responding to a long-term change in local earth strain, which might be associated, for example, with a change in tectonic stress in the region.

5.5 THE ARRAY

We noted in Section 5.3 the presence of noise on seismograms. Where body waves from small events at great distances are to be recorded, the signal on a single seismograph may be lost in this noise. In order to improve signal-to-noise ratio, arrays of seismometers have been designed, in such a way that their outputs can be combined with chosen time delays. The result bears some similarity to a grating as used in spectroscopy, or to the receivers used in radio astronomy.

The noise on seismograms may be of various types. It may be produced near the station by man or by wind, or over the sea by pressure variations associated with storms. Seismic surface waves from these sources are known as microseisms. On the other hand, as the sensitivity of a seismograph is increased, it is found that there is a background of noise which probably represents true body waves from many small earthquakes. The principle of the array is to remove the first type of noise, by virtue of its surface wave characteristics. A body wave, arriving at near vertical incidence from a distant source, has an apparent velocity along the earth's surface which is large compared to the velocity of surface waves, and it is this difference which permits the discrimination.

Suppose a plane wave with apparent surface velocity V and wavelength λ arrives at an array at azimuth α (Fig. 5.6). In the notation of the figure, the time delay at the Kth seismometer, relative to the reference, is

$$\tau_K = -r_K \cos(\theta_K - \alpha)/V \qquad 5.5.1$$

For an array to be tuned to an arrival with particular V and α the outputs must be delayed in processing so as to compensate for these phase delays. The processing, of course, will involve recording on magnetic tape, digitizing, and high-speed computation. Let us suppose that a trial velocity U and trial azimuth β are employed in processing. Then the delay introduced at the Kth seismometer is

$$\tau'_K = -r_K \cos(\theta_K - \beta)/U \qquad 5.5.2$$

and the phase shift in the signal from it is

$$\begin{aligned} \mu_K &= \omega(\tau_K - \tau'_K) \\ &= -2\pi r_K \left[\cos(\theta_K - \alpha) - \frac{V}{U}\cos(\theta_K - \beta)\right] \Big/ \lambda \end{aligned} \qquad 5.5.3$$

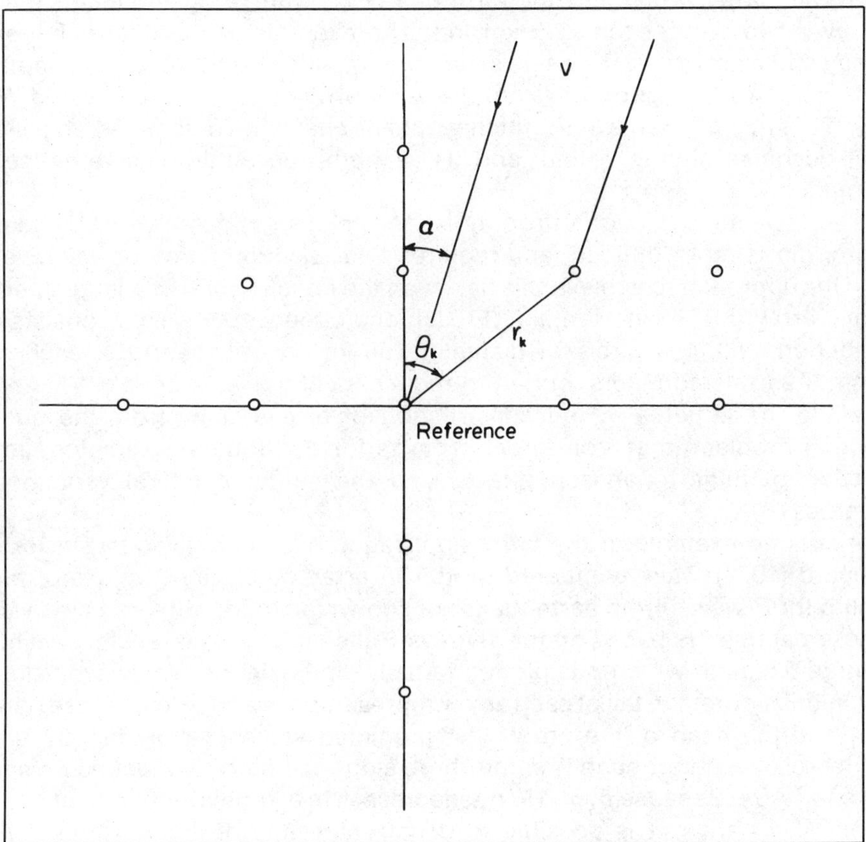

Figure 5.6 A wave incident upon an array of seismometers.

If the signal at the reference point is cos ωt, the processed signal at the Kth seismometer is

$$\sigma_K = \cos(\omega t + \mu_K) \tag{5.5.4}$$

The processed outputs may be combined in different ways. One procedure is to smooth the square of the sum of outputs; this involves computing, for each point in time, the quantity

$$Z = \lim_{T\to\infty} \frac{1}{2T} \int_{-T}^{T} \left(\sum_K \sigma_K \right)^2 dt \tag{5.5.5}$$

which reduces to

$$Z = \tfrac{1}{2}\{[\Sigma \cos \mu_K]^2 + [\Sigma \sin \mu_K]^2\} \tag{5.5.6}$$

The squared sum exhibits a maximum when $\mu_K = 0$, i.e., when U and β are chosen equal to V and α. The array is then sensitive to an arrival with given apparent surface velocity, arriving from a given azimuth. We note also that if the true local body-wave velocity is known, the cosine of the angle of emergence of the arrival at the earth's surface can be determined from the ratio of true to apparent velocity. Somers and Manchee (1966) have described the *tuning* of the Yellowknife array, which has the form of a cross with 19 seismometers distributed over two arms, each 22.5 km long. An example of the determination of velocity and azimuth on the same array, with resultant signal-to-noise improvement, is shown in Figure 5.7 (from the work of Weichert, Manchee, and Whitham, 1967) in which an event, which would not have been detected on a single trace, becomes clearly visible, and its azimuth and angle of emergence are determined.

The separation of signal from noise in the case of noise which is itself of deep origin is more difficult, and requires finer discrimination between velocities. One approach to achieve this has been the construction of a large aperture seismic array (LASA) in Montana (Frosch and Green, 1966). This consists of a distribution, within a circle of diameter 200 km, of 21 subarrays, each consisting of 25 seismometers, buried to reduce local noise.

While the impetus in the construction of arrays came from the nuclear detection problem, they hold great promise for earthquake seismology in the detection of hitherto unseen phases and the study of lateral variations in properties.

A striking example of the latter application has been given by Davies and Sheppard (1972). They compared the LASA array-determined values of p and azimuth for P-waves from earthquakes of known location, with predicted values of these parameters based on the Jeffreys-Bullen tables. The results are shown in Figure 5.8(a), in which p is plotted radially, and azimuth clockwise from the top. On this figure, the tail of each arrow represents the array-determined parameters, and the head of the arrow, the predicted values. Arrow length, therefore, represents heterogeneity somewhere along the path connecting an earthquake to LASA. Because p, or $dT/d\Delta$, decreases in a known way with increasing epicentral distance, it is possible to construct a map of the world as "seen" seismically from LASA; this is shown in Figure 5.8(b). Comparison of the two

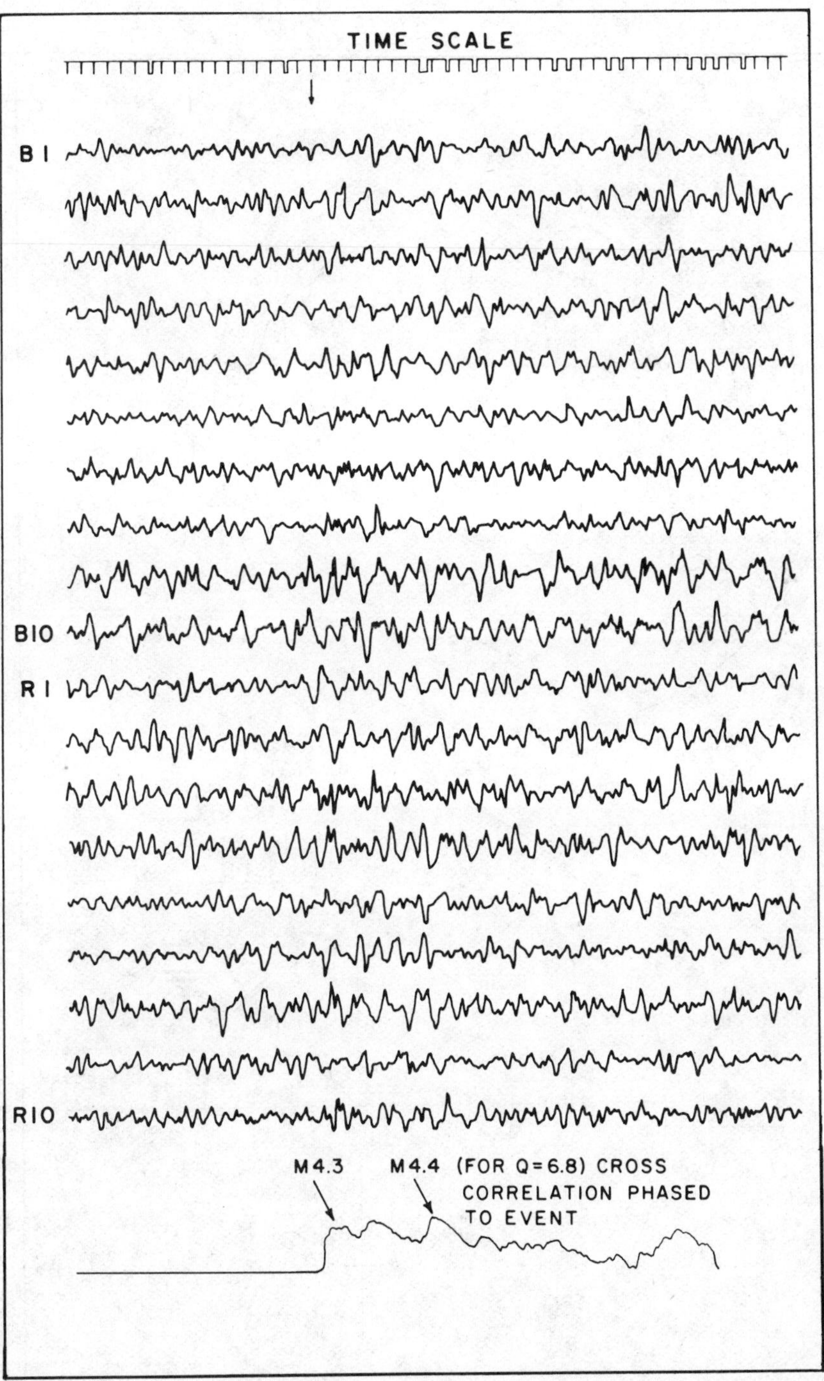

Figure 5.7 Improvement in signal of a magnitude-4 event recorded at the Yellowknife array. Traces *B* 1-10 and *R* 1-10 are from the individual seismometers of the two crossed lines; the bottom trace is the processed output obtained with the array tuned to an azimuth of approximately 290° and an apparent velocity of 21 km/sec. (Courtesy of K. Whitham and the Earth Physics Branch, Dept. of Energy, Mines and Resources, Canada)

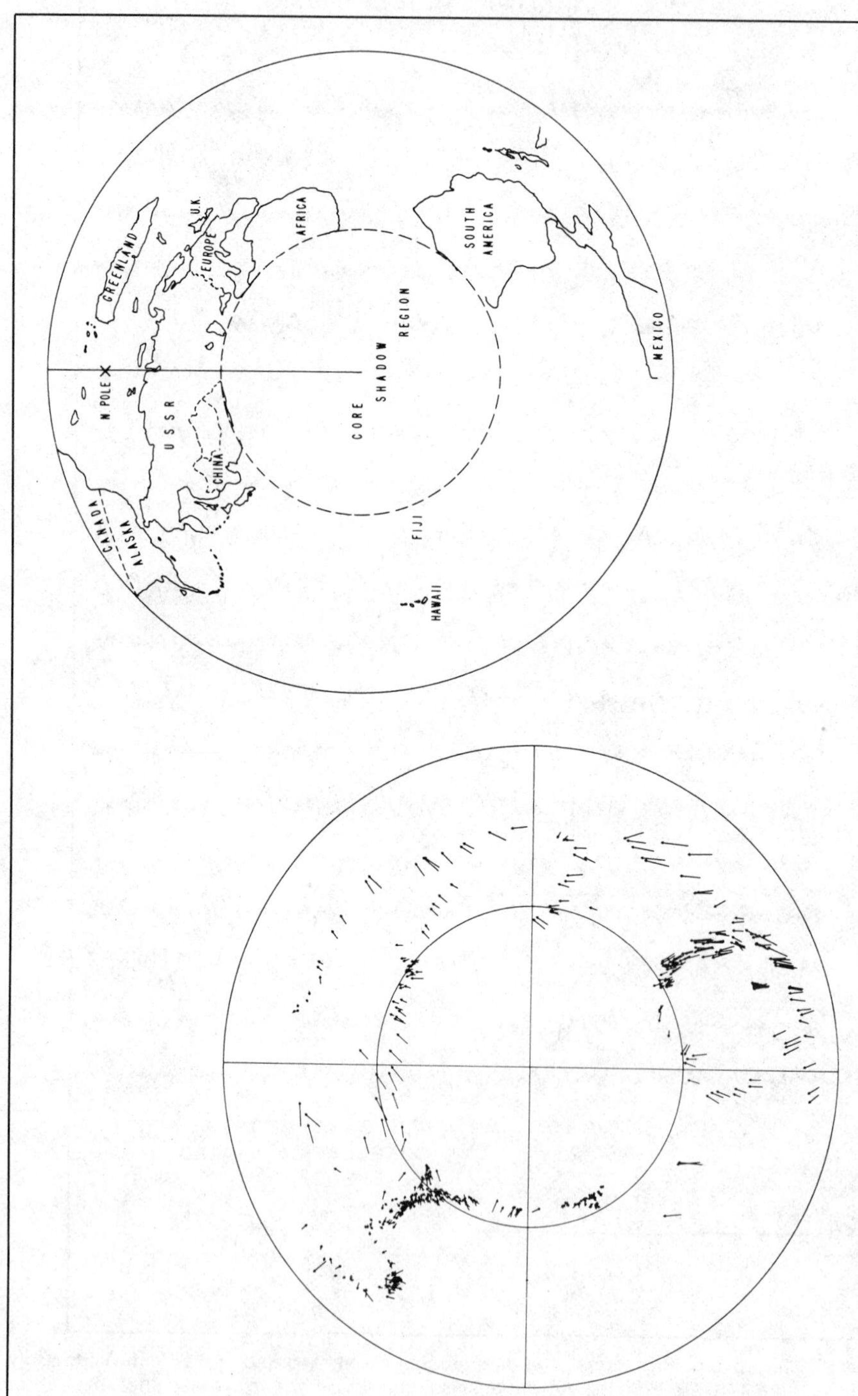

Figure 5.8 Heterogeneity in the earth as seen through the LASA array. (a) Arrivals are plotted as arrows on the azimuth of their arrival, at a radius proportional to slowness, out to a slowness of 10 seconds/degree. The head of each arrow represents the azimuth and slowness predicted from known epicenters and standard tables; the arrow tails represents the array-observed values. (b) Earthquake source regions of the earth as seen from LASA on the plot used in (a). (From D. Davies and R. M. Sheppard 1972; courtesy of the authors and *Nature.*)

diagrams then permits the identification of the seismic source region which is responsible for a given arrow pattern. Arrows which lie tangentially to circles in Fig. 5.8(a) represent anomalies predominantly in direction of arrival, presumably caused by lateral refraction at velocity discontinuities along the path, while arrows lying radially represent residuals predominantly in p, perhaps caused by velocity anomalies near the turning point of rays. Davies and Sheppard identify some over-all bias due to conditions near LASA itself, but they argue that the greater part of all residuals is produced by conditions near the source or along the path as a whole. An interesting feature of Figure 5.8 is the suggestion of features of different scale length. Notice that, in the case of earthquakes in Eurasia (azimuth 0° to 90°), the residuals are slowly varying, while for earthquakes in the southwestern Pacific (azimuth 315° approximately) rapid changes in residual are found. The latter may be more characteristic of conditions near the source, but we must emphasize that the precise location of anomalous regions along the paths remains uncertain.

Problems

1. In the seismograph equation for free oscillations (Equation 5.2.7 with right side put equal to zero) the nature of the response is critically dependent upon the magnitude of the damping constant λ. Show that for $\lambda < 1$, the displacement varies with time as an oscillatory motion with exponentially decreasing amplitude, and that for $\lambda > 1$, the motion is not oscillatory.* Find the complete expression for x as a function of time when $\lambda = 1$ (critical damping).

2. Figure 5.3 apparently gives three responses of a seismometer–galvanometer system as a function of frequency: magnification, velocity sensitivity and acceleration sensitivity. Confirm that the graticules of the diagram are correctly placed to give these quantities.

3. Confirm the distance of the earthquake recorded in Figure 2.1 (by means of the time-distance curve, Figure 3.1), and deduce the origin time of the earthquake.

4. The Yellowknife array (latitude 62°28′42″N, longitude 114°28′42″W) observes that the signal of a P-wave arrival is a maximum when the array is tuned to an azimuth of 10° east of north. The wave has an apparent surface velocity of 9.5 km/sec. The local true velocity of P-waves is 6.0 km/sec. Estimate the approximate geographic location of the source.

BIBLIOGRAPHY

Alsop, Leonard E., Sutton, George H., and Ewing, Maurice (1961) Free oscillations of the earth observed on strain and pendulum seismographs. *J. Geophys. Res.*, **66**, 631.

Benioff, H. (1935) A linear strain seismograph. *Bull. Seism. Soc. Amer.*, **25**, 283.

Davies, D., and Sheppard, R. M. (1972) Lateral heterogeneity in the earth's mantle. *Nature*, **239**, 318–323.

Dewey, J., and Byerly, P. (1969) The early history of seismometry (to 1900). *Bull. Seism. Soc. Amer.*, **59**, 183.

Frosch, R. A., and Green, P. E. Jr. (1966) The concept of a large aperture seismic array. *Proc. Roy. Soc. A*, **290**, 368.

Galitzin, B. (1914) *Vorlesungen über Seismometrie*. Teubner, Leipzig.

Press, Frank, Ewing, Maurice, and Lehner, Francis (1958) A long-period seismograph system. *Trans. Amer. Geophys. Un.*, **39**, 106.

Somers, H., and Manchee, E. B. (1966) Selectivity of the Yellowknife seismic array. *Geophys. J.*, **10**, 401.

Vali, Victor (1970) Some earth strain measurements with laser interferometer. In *Earthquake displacement fields and the rotation of the earth*, ed. L. Mansinha, D. E. Smylie, and A. E. Beck, Springer-Verlag, New York.

Vali, V., Krogstad, R. S., Moss, R. W., and Engel, René. (1968) Some observations of strains across the Kern River fault using laser interferometer. *J. Geophys. Res.*, **73**, 6143.

Weichert, D. H., Manchee, E. B., and Whitham, K. (1967) Digital experiments at twice real-time speed on the capabilities of the Yellowknife seismic array. *Geophys. J.*, **13**, 277.

Willmore, P. L. (1959) The application of the Maxwell impedance bridge to the calibration of electromagnetic seismographs. *Bull. Seism. Soc. Amer.*, **49**, 99.

THE EARTHQUAKE SOURCE 6

6.1 DEPTH OF FOCUS

All of our previous discussion on the travel times of seismic waves, and on the location of epicenters, were based on the assumption that earthquakes occurred close to the earth's outer surface. Indeed, this view was held for many years, but it is obviously of great importance, if we are to understand the nature of the deep interior, to know if earthquakes can occur at depth.

The history of the realization that deep-focus earthquakes do exist is another example of scientific insight, prediction and verification. Turner (1922), during the determination of epicenters, found cases in which the arcs of radius Δ determined from distant stations, and based on the standard time-distance curves of that era, failed to intersect. For the same earthquakes he noted that the core phase *PKP* was early, and he suggested that these earthquakes occurred at depth within the mantle. Wadati (1928) complemented these observations by noting that, for some earthquakes, *P* arriving at stations near the epicenter was late. Additional, and most important, evidence for deep–focus earthquakes is based on a fact pointed out by Stoneley (1931), that deep earthquakes simply cannot produce large amplitude surface waves. The displacement in surface waves decreases with depth, and it is a fundamental principle in the theory of vibrations that a mode cannot be stimulated by the application of a force at nodal points for that mode.

The quantitative determination of focal depth is based on an additional reflected phase (Fig. 6.1) whose importance was noted by Stechschulte (1931). The conditions for reflection at the outer surface can be satisfied at two points, one of which is near the epicenter; the corresponding phase is labelled *pP* (or

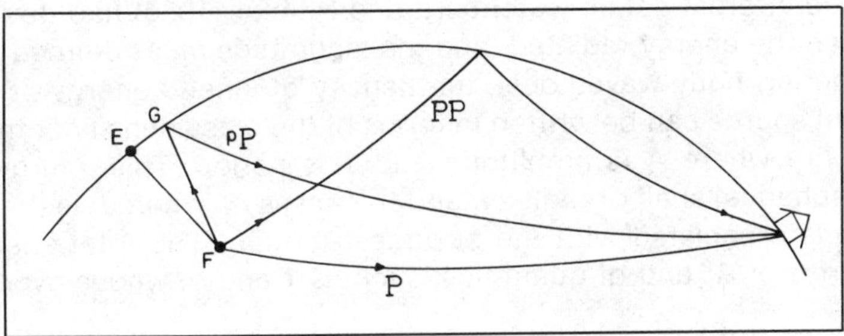

Figure 6.1 The reflected phase *pP* from a deep earthquake. Observations of the time interval $(pP - P)/2$ give the distance *GF*. Point *F* is the focus, *E* the epicenter, of the earthquake.

sS). If this phase can be identified on a record, the time interval $(pP - p)/2$ can be used to indicate the (sloping) distance of the focus below the surface, and this can be corrected to give the true depth of focus. Stechschulte's method was used to determine the focal depths of a sufficient number of earthquakes for the construction of time-distance curves for various depths. In modern practice, depth of focus is usually determined by fitting to the best of these curves during epicenter location, so that a determination can be based on P and S arrivals alone. The accuracy of the resulting focal depth is less than that of horizontal position, but is usually considered to be about ± 25 km.

The general distribution of earthquakes with depth can be illustrated by the following table of the earthquakes (essentially those of the period 1918–1946) considered by Gutenberg and Richter (1954) in their study "Seismicity of the Earth."*

Depth (km)	100	150	200	250	300	350	400	450	500	550	600	650	700
Number	412	187	137	78	26	41	45	20	35	39	57	25	9

The depth of about 750 km appears to be an absolute maximum for the occurrence of sudden failure within the earth. Furthermore, deep-focus earthquakes are even less random in their distribution than earthquakes of normal depth, as will be illustrated by a comparison of Figure 6.3 (which shows the global distribution of epicenters of deep earthquakes) with Figure 6.2, which is based on all earthquakes for the period shown. Very deep shocks are restricted to certain definite belts, notably the margin of the Pacific Ocean. Within these belts, the depth of focus increases toward the continental side. We shall have occasion to return to this type of figure when we consider the tectonic significance of earthquakes.

6.2 MAGNITUDE AND FREQUENCY

Richter (1935) set up a scale of *magnitudes* for earthquakes that has been of great practical application. The magnitude m is defined as the logarithm, to the base 10, of the amplitude in microns of the largest trace deflection that would be observed on a defined short-period seismograph, of static magnification 2800, at a distance of 100 km from the epicenter. Richter also constructed, empirically, tables to convert to observations at other distances.

We shall discuss, in a subsequent section, recent ideas on the manner in which energy is released in an earthquake. Before the nature of the source had been studied in great detail. Gutenberg and Richter (1956) had derived a relation between the energy radiated, and the magnitude m, as defined above.

Considering body waves only, the density of kinetic energy at distance h from a point source can be written in terms of the mass density of the material, and of $(A/T)^2$, where A is amplitude and T is period. This energy is spread through a spherical shell of radius h and thickness $n\lambda$ where n is the number of wave periods associated with the strongest motion. The total energy is then given in terms of A^2 and of quantities, such as T and n, whose average values

*Gutenberg, B., and Richter, C. F. *Seismicity of the Earth and Associated Phenomena.* Princeton University Press, Princeton, N.J., 1954.

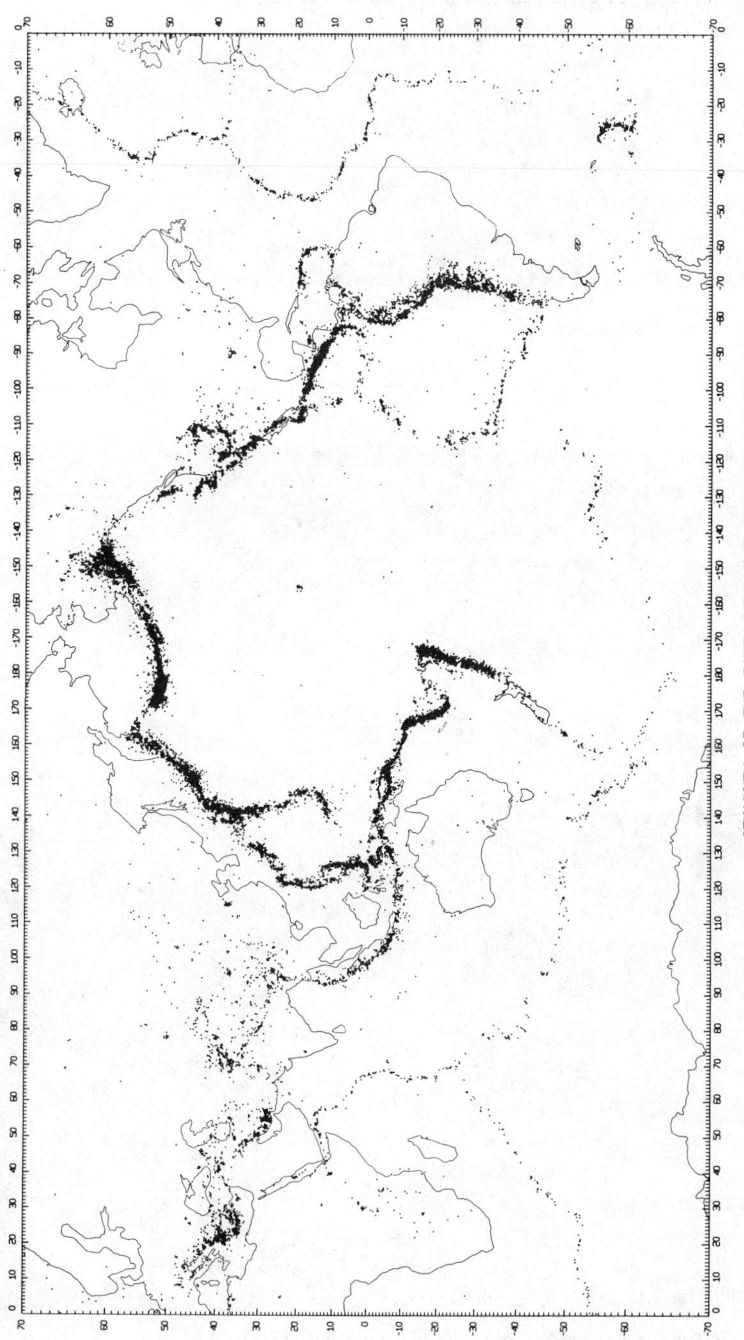

Figure 6.2 Distribution of earthquakes in the depth range 0–100 km. (Courtesy of Muawia Barazangi and James Dorman, and the Seismological Society of America)

Figure 6.3 Distribution of earthquakes in the depth range 100–700 km. (Courtesy of Muawia Barazangi and James Dorman, and the Seismological Society of America)

are obtained from observation. Since m is proportional to log A it follows that the radiated energy E (in ergs) can be written in the form

$$\log_{10} E = a + bm \qquad 6.2.1$$

where a and b are constants to be determined.

The years following the original investigation of Gutenberg and Richter have seen a great deal of research upon the determinations of the constants in this equation. One problem is that the above analysis deals only with body waves, so that the estimate of total earthquake energy involves assumptions on the relative energy carried by body waves and surface waves. The same type of measurement, of maximum amplitude, mean period and duration, can be applied to surface waves, and an alternate estimate of magnitude, denoted M, can be obtained. Again, one is dealing with a fraction of the radiated energy. It is important to note that the magnitude is, ideally, a unique property of an earthquake. Unfortunately, the estimates of magnitude are not, and it has become usual to speak of "body wave magnitude" and "surface wave magnitude." The comparison of the estimates obtained is of great importance in the explosion-earthquake discrimination problem, and recent work (Basham, 1969; Evernden et al, 1971) has taken advantage of the availability of large explosions in attempts to make the scales consistent. It is agreed that there are significant differences between the estimates m and M, with the further complication that the difference depends on the size and nature of the source. There is still not agreement on which estimate is preferable. The use of body waves has the advantage that the method can be applied to earthquakes at all depths; public announcements of magnitudes on the Richter scale normally refer to m. Deep focus earthquakes, which generate very small surface waves, can hardly be evaluated by estimates based on surface-wave amplitude. However, there has been a tendency to use the surface-wave magnitude M in the study of explosions.

Båth (1966), in summarizing the available comparisons of energy, M and m, suggested that energy in ergs be computed from:

$$\log E = 12.24 + 1.44M \qquad 6.2.2$$

and that, on the average for earthquakes of normal depth,

$$m = 0.56M + 2.9 \qquad 6.2.3$$

Again, we emphasize that m and M are estimates, whose interrelationship is complicated and not yet fully developed.

As might be expected, small earthquakes are much more abundant than great ones, as the following tabulation, after Gutenberg and Richter (1954), shows*:

	Magnitude	No./yr.
Great earthquakes	8	1.1
Major earthquakes	7–7.9	18
Destructive earthquakes	6–6.9	120
Damaging earthquakes	5–5.9	800
Minor earthquakes	4–4.9	6,200
Smallest generally felt	3–3.9	49,000
Sometimes felt	2–2.9	300,000

*Gutenberg and Richter, *op. cit.*

However, in spite of the relative abundance of small earthquakes, much more energy *in toto* is released by the great earthquakes. Examination of Equation 6.2.1 indicates that the energy release of a magnitude 8 shock, for example, exceeds that of one of magnitude 2 by a factor of more than 10^{10}, while the smaller shocks are only 3×10^5 more frequent. The average total release of energy per year is about 12×10^{26} ergs.

In accounts of earthquake damage, reference is often made to a scale of *intensities,* graduated from I to X or XII. It is important to differentiate the idea of intensity, introduced by M. S. Rossi of Italy and F. A. Forel of Switzerland, from that of magnitude, which, as discussed above, refers to energy release at the source. The intensity scale (Wood and Neumann, 1931) is a measure of observed effects at a site on the earth's surface. It depends not only upon epicentral distance and depth of focus, but also upon soil conditions at the site, and while useful in gauging the destructive effect of a shock, it is not closely connected to conditions at the source.

6.3 DIRECTION OF MOTION IN FAULTING

Virtually all earthquakes of significance appear to be the result of the sudden release of tectonic stress within the earth, that is, the release of those same stresses which produce the changes in the outer part of the earth that result in the observed structures of continents and oceans. Some earthquakes are known to have been associated with movement or faults which can be observed at the earth's surface, and this raises the question of whether the nature of the stress release can be determined from any characteristics of the seismic waves.

Suppose that a fault consists of motion on a vertical plane striking north-south, that the westerly block moves northward, and that the first motions originate from a small region around the origin *O*. We would intuitively expect that, within an area small enough for the curvature of the earth to be neglected, stations in quadrants northwest and southeast of *O* would experience a compression as the sense of the first *P* arrival, while stations northeast and southwest of *O* would experience initial dilatations. The four quadrants would be divided by the traces of two vertical planes: the fault plane, and a plane through *O* perpendicular to the fault plane. In this simplified situation, therefore, a plot of the sense of the first motion of *P* would allow two planes to be determined, and an ambiguity would remain as to which was the fault plane.

Byerly (1938) and Byerly and Evernden (1950) showed that the principle can be extended to arrivals on a sphere, with a completely analogous result. There is a fundamental complication in the case of the earth, in that the seismic ray paths are curved, and a wave starting on one side of the fault plane at the source can be observed, at large distance, on the other side of the extension of the plane. This difficulty can be avoided if, for any given earthquake, arrivals are plotted, not at the true position of each station, but at the station's "extended position," which is the intersection with the sphere of a fictitious straight ray departing from the source in the same direction as the true ray to that station. Tables have been compiled to permit this construction, and a stereographic projection of the earth was frequently used to display the results (Fig. 6.4(A)).

In this diagram, we note that two circles, which are the projections of circles on the sphere, divide the surface of the sphere into four regions, corresponding

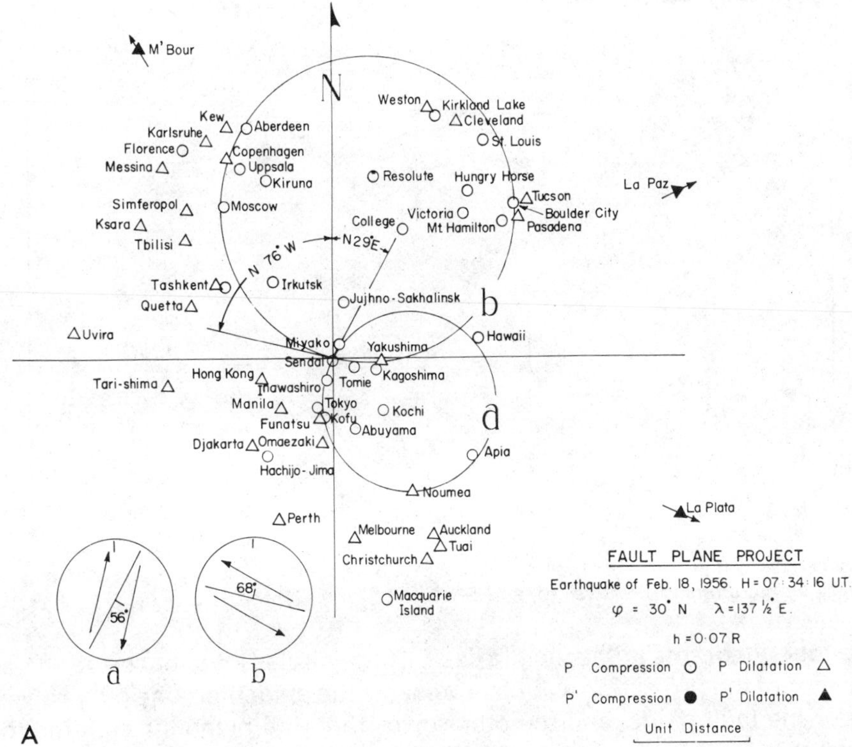

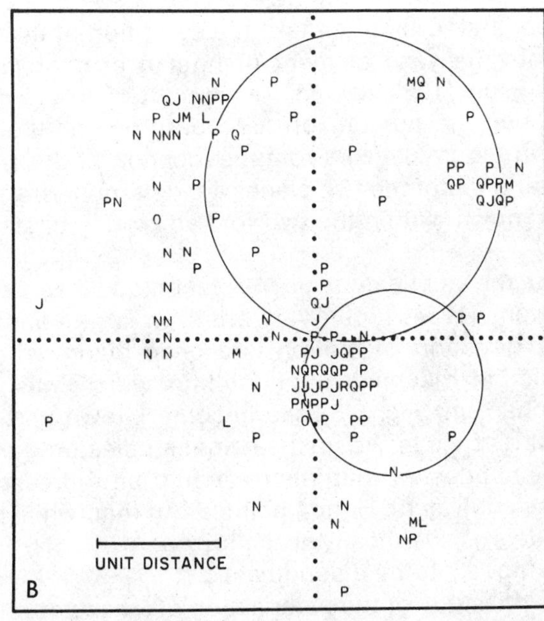

Figure 6.4 (A) Fault plane solution in which the two circles were drawn by eye to separate compressions and dilatations. The two possible fault planes, with the sense of motion, are shown at lower left. (B) Fault plane solution by computer for the same case as that shown in Fig. 6.4(A), with additional observations. Because of the close spacing of stations it has been necessary to group the observations: *N, M, L* represent one, two, and three dilations; *P, Q, R* represent one, two, and three compressions; *O* represents conflicting observations, *J* more than three observations. (Courtesy of J. H. Hodgson and the Earth Physics Branch, Dept. of Energy, Mines and Resources, Canada)

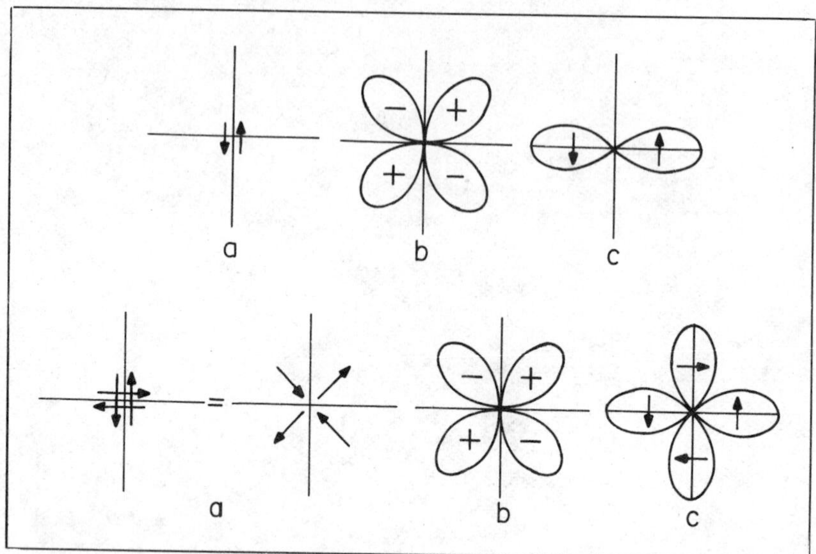

Figure 6.5 Single couple (*Top*) and double couple fault mechanisms, with their radiation pattern of first motions for *P*-waves (b) and *S*-waves (c).

to the four quadrants in the simple case. These circles intersect orthogonally at the center of the projection, which represents the epicenter; one circle gives the trace of the fault plane, and the other gives that of a plane through the focus, normal to the fault plane and to the motion. In an actual determination, one plotted the sign of first arrivals (assuming these are accurately reported by the stations); one then fitted the two circles in such a way that compressions and dilatations were separated by them and the condition at the center was satisfied. There was obviously an element of human error in fitting the circles. Wickens and Hodgson (1967) devised a computer program to perform this operation, minimizing the number of discordant points; the solution by their method is shown in Figure 6.4(B). From the location of the circles on the projection, the strike and dip of the two planes are determined, and the direction of faulting becomes known, within the two-fold ambiguity of stating which is the fault plane.

The history of the application of this technique has been described by Hodgson and Stevens (1964, 1969). Very briefly, a large number of determinations of apparent direction of motion had been made, when Honda (1961) questioned the simple mechanism of faulting which was assumed in the method. This mechanism (Fig. 6.5) has become known as the single-couple model. As we shall discuss in the next section, studies on the theory of failure had independently suggested that failure should be represented by a double couple, or its stress equivalent. Honda pointed out that, while the radiation of *P* from the two models is identical, the polarization of *S* is different, and this should permit the models to be distinguished.

Unfortunately, the study of the polarization of *S* has not yet advanced to the stage where this has been accomplished. Most seismologists now favor the second model, but there appear to be cases in which the single–couple model is favored by the observations. It is important to note that, while the point of view corresponding to the two models is quite different, the actual planes determined by the observations are identical. The seismologist who has adopted the first model thinks in terms of the motion which follows failure; one of his "nodal planes'' is the plane along which this motion has taken place. One who has adopted the second model thinks in terms of the release of a system of stresses; his nodal planes bisect the angles between two principal stresses. The elementary theory of failure suggests that if shear failure takes place, one of these planes will be the plane of failure, so that the model will in fact lead to the same set of ''fault planes,'' with the same ambiguity, as the first model.

The first extensive application of the method to tectonic studies was that of Isacks, Oliver, and Sykes (1969). The earthquakes used were those associated with the margins or ridges of oceans, and the results (Fig. 6.6) are of great significance in the study of the tectonics of the ocean floor. We shall return to a detailed discussion of the implied motions later, but it is noteworthy that the directions of maximum compressive or tensile stress, rather than the fault planes, are aligned with the seismic zone.

There is one point relating to the results in general which merits comment. In the earlier fault plane studies, a predominance of strike-slip displacement was found for the largest faults, as opposed to motion which is up the fault plane (thrust faulting) for the upper block, or down (normal or gravity faulting). Strike-slip displacement is indicated, for either model, if both planes are shown to be nearly vertical. Stevens and Hodgson (1968), in an examination of 437 solutions, found that this was the case for over 70 per cent of them. However, they also found that only 10 of all 437 solutions were well defined by the observations, so that any statement on the statistics of the type of motion must be treated with extreme caution. The study of Isacks, Oliver, and Sykes took advantage of the discovery by Sykes (1967) that first motions are more reliably recorded by long-period instruments than on the short-period records used by earlier investigators.

Sykes also introduced an alternative way of displaying the distribution pattern of first arrivals: the equal-area projection of the lower hemisphere of the

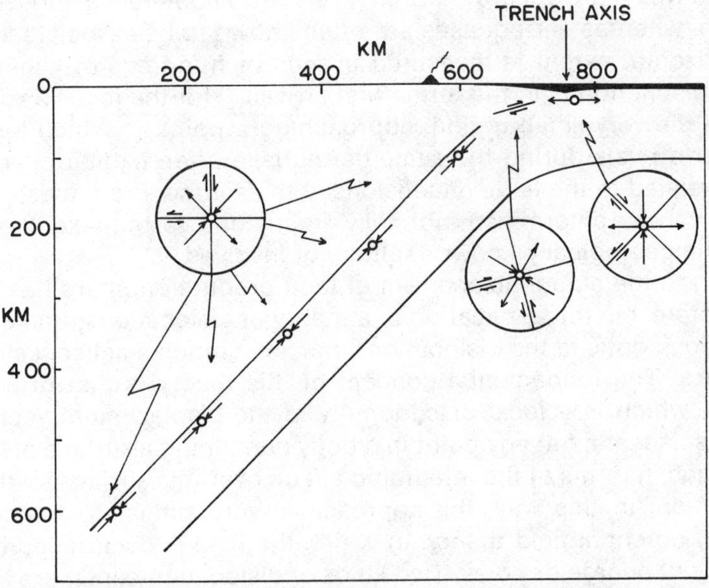

Figure 6.6 Representation of direction-of-motion results obtained from earthquakes associated with the seismic zone beneath South America. Note that the fault plane is parallel to the zone in only one case; in all others, the axes of greatest compression or tension are aligned with the zone. (After Isaacks, Oliver, and Sykes)

focal sphere, which is a sphere drawn completely around the focus. In this projection, the traces of the fault plane and auxiliary plane appear as curved lines across the projection, passing near the center if the planes are steeply dipping, and near the perimeter if they are almost horizontal. When only the result of a mechanism determination is desired, as opposed to the complete statistics, it is usual to shade the entire sectors corresponding to initial compressions. Thus, Figure 6.7, from the study by McKenzie (1972) of the Mediterranean area, shows at a glance the mechanisms for 21 earthquakes. With a little practice, one can recognize the principal types of faulting. For example, the shock of 1941.11.25, between the Azores and Gibraltar, displays two nearly vertical planes. Whichever is chosen as the fault plane, the motion must be horizontal and right lateral. In the earthquake of 1966.7.4, the planes are also quite steep, but downdip or normal faulting is implied. The outstanding feature of the diagram is the preponderance of overthrusting from the south, on east-west planes, between Gibraltar and Sicily, in complete agreement with the general features of global tectonics.

6.4 THE NATURE OF FAULTING

The main conclusion of the previous section is that while the sense of seismic wave arrivals at large distances yields information of tectonic significance, it does not by itself completely describe the nature of faulting. We must ask whether other parameters of the seismic waves could be used to determine, for example, the area of fault plane involved, and the velocity with which failure is propagated in this plane (the latter quantity being known as the rupture velocity). Before turning to theories of faulting in detail, we should clarify the distinction between the focus of an earthquake and the source region. Determination of the position of the focus from the first arrivals of body waves leads to positions which are accurate within a very few kilometers at most in horizontal location, whereas earthquakes are often known to be associated with faults whose horizontal extent is measured in tens or hundreds of kilometers. The key to the distinction is in the term "first arrivals," for the focus as determined by these is the very small region, approaching a point, in which failure originates. Later arrivals during the same earthquake often exhibit characteristics which are related to the finite dimensions of the source. From these, the extent of faulting can be determined; normally, in any one earthquake, the complete length of the geologically known fault is not involved.

Turning to the actual mechanism of fault production, there has been considerable interest in the application of a theory of dislocations (Steketee, 1958), which are analogous to the dislocations met, on a much smaller scale, in solid-state physics. The fundamental concept of this theory is that of a center of *dislocation,* which is a local discontinuity in the displacement vector u_i. The stress or displacement at any point in a body containing a surface of failure can be determined in terms of the integrated effect, over this surface, of the centers of disolocation. In this way, the approach is very similar to the solution of problems in potential field theory, in which the field is determined by surface distributions of charge or poles. Two kinds of dislocation center are found; one involves a discontinuity in displacement normal to the surface, and is associated with dilatational stress; the second involves discontinuities in displacements parallel to the surface, and is associated with two shearing couples. Two

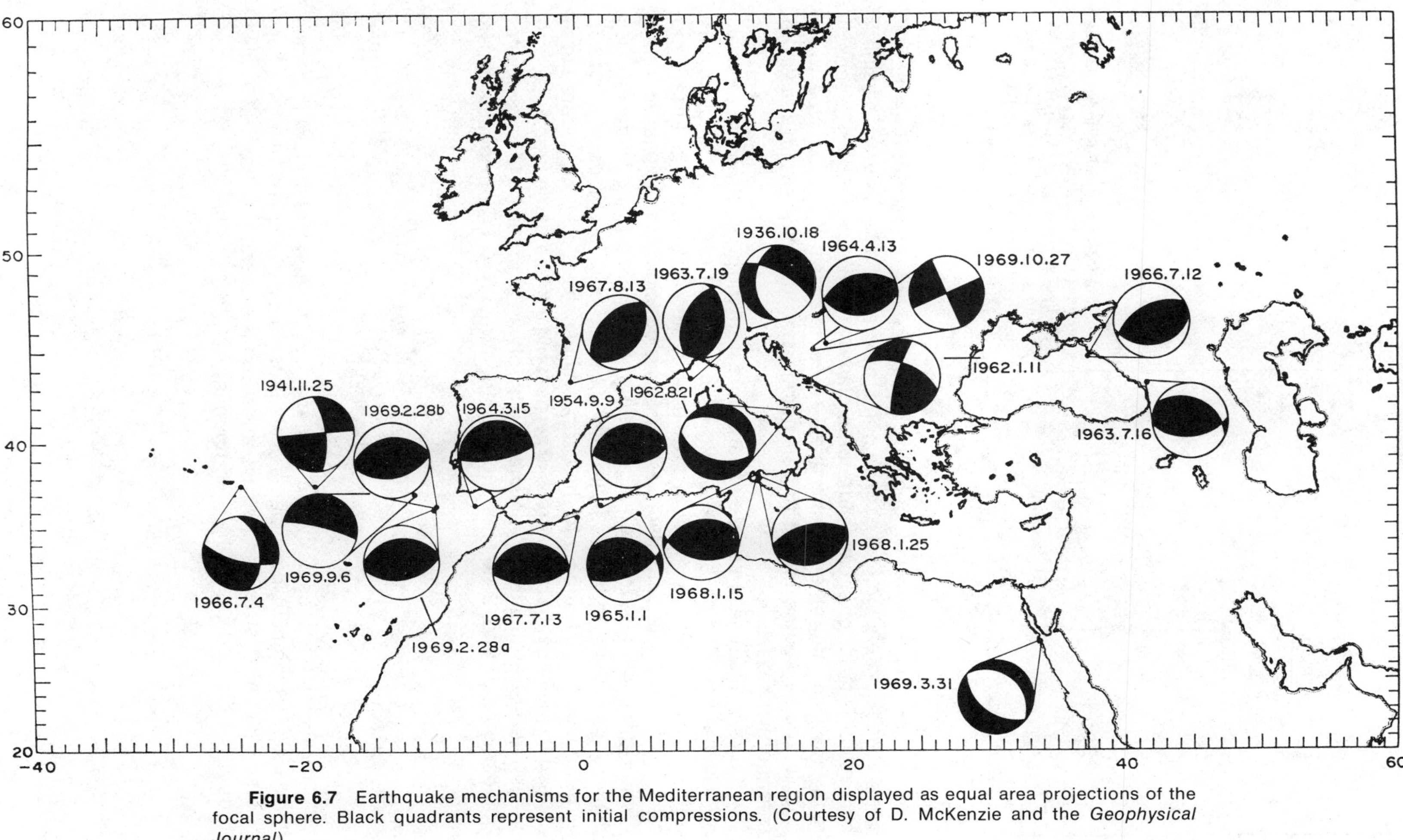

Figure 6.7 Earthquake mechanisms for the Mediterranean region displayed as equal area projections of the focal sphere. Black quadrants represent initial compressions. (Courtesy of D. McKenzie and the *Geophysical Journal*)

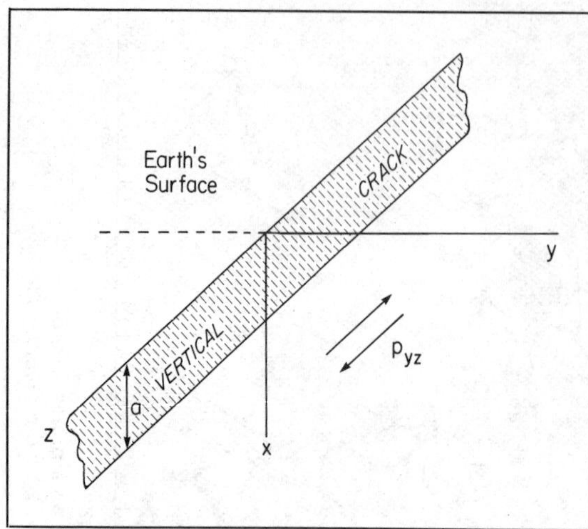

Figure 6.8 Notation for a fault model (after Knopoff). A crack of width 2*a* lies in the x − z plane. The y − z plane may represent the earth's surface, in which case the crack models a vertical fault of depth extent *a*.

couples are always involved, because a requirement of the theory is that the body be in static equilibrium at each point, before and after failure. The significant result is that, for shear failure on a plane, the dislocation theory leads directly from an integration of the effect of centers of the second type to the double couple source model of seismology, and this is a strong reason for favoring that model.

Knopoff (1958) has treated a shear failure from the point of view of a surface on which an applied shear stress field is held equal to zero. His analysis illustrates a further useful analogy to problems in electrostatics. The surface of a crack (Fig. 6.8) in an extended elastic body is considered to extend to infinity in the *z*-direction, and to be of width 2*a* in the *x*-direction. The plane $x = 0$ is a plane of symmetry; in many applications it will represent the earth's surface, with the axis of *x* pointing into the earth. A shear stress p_{yz} is considered to be applied to the body, but this stress is zero on the surface of the crack. For pure shear stress, the dilatation vanishes; it is left as a problem to show, using Equations 2.2.5 and 2.4.4, that, if a rotation vector **A** is introduced, where

$$\mathbf{A} = \text{curl } u_i \tag{6.4.1}$$

then

$$\text{curl } \mathbf{A} = 0 \tag{6.4.2}$$

The vector **A** is thus derivable from a scalar potential, and plays the same role as the electric field in problems of electrostatics. In the particular case considered, the shear stress and strain vanish on the surface of the crack; also, throughout the body, only the *z*-component of u_i is nonvanishing. Therefore,

$$e_{yz} = \frac{1}{2}\left(\frac{\partial u_y}{\partial z} + \frac{\partial u_z}{\partial y}\right) \tag{6.4.3}$$

$$= \tfrac{1}{2}\partial u_z / \partial y$$

and, on the surface,

$$\partial u_z / \partial y = 0$$

Since the components of **A** are $(\partial u_z/\partial y, -\partial u_z/\partial x, 0)$, this result shows that **A** is normal to the surface at the crack. This is the condition on the electric field at the boundary of a perfect conductor, so that Knopoff was able to solve for **A** and the displacements in terms of known solutions from electrostatics for the field around a conducting strip.

The displacement turns out to be

$$u_z = \pm A_0(y^2 + a^2)^{1/2} \tag{6.4.4}$$

At great distances, the displacement is $\pm A_0 y$, so that A_0 is simply p_{yz}/μ. In the absence of a crack, the displacement is $\pm A_0 y$ everywhere, so that faulting, or the release of stress, produces the relative displacement

$$A_0[(y^2 + a^2)^{1/2} - y] \tag{6.4.5}$$

This relative displacement corresponds to what has been called "elastic rebound" in some descriptions of earthquakes. Equation 6.4.5 suggests that if the ground displacement is determined following an earthquake, a plot of displacement against distance from the fault should permit the determination of both A_0 and a. The latter is the depth to which faulting extends, while the former gives the stress release produced by the crack.

Knopoff extended his analysis to determine the energy release, by integration of the energy density (curl $u_i)^2$. The result is

$$E = \tfrac{1}{2}\pi\mu A_0^2 a^2 \tag{6.4.6}$$

There have been a number of other investigations of the theory of faulting, as has been discussed by Chinnery (1969). In all derivations of stress release and energy release, there are two constraints. The stress release, or "stress drop," cannot be greater, for shallow faults, than the shear strength of rocks near the earth's surface; if it turns out to be much less, the question is why failure occurred. Secondly, the energy released cannot be greater than the energy appropriate to the magnitude of the earthquake associated with the faulting. Chinnery points out that in connection with stress drop considerable scatter is shown in the results obtained from studies of various faults. Quoted values range from 0.7 to 239 $\times$ 10^6 dynes/cm^2, which is puzzling in view of the usually quoted value of 1.5 $\times$ 10^9 dynes/cm^2 for the strength, as measured in the laboratory, of crustal rocks. Obviously the strength in critical zones of the outer part of the earth is much less than that measured in the laboratory.

We return next to the problem of the finite size of the source. Analyses such as that above can determine the depth extent of a near-surface fault. This is found (Chinnery, 1969) to range up to 26 km. The observed length of major faults associated with earthquakes may be 450 km. How much of the area of the fault plane is active during a particular earthquake?

Studies of the effect of finite sources (Hirasawa and Stauder, 1965; Savage, 1965) have shown that, while the pattern of first arrivals is not altered from that corresponding to an infinitesimal source region, there is an important alteration in the frequency domain for both body waves and surface waves. In particular, there are minima of amplitude at a series of frequencies, determined by the fracture length and rupture velocity. This can be seen directly in the simple case of an observer, close to a plane of faulting and in line with the fault, who receives arrivals from a fracture which begins at some distance from the observer and propagates toward him with velocity v over a length b. The stopping of the crack produces a signal of opposite sense to that of the starting; minima of amplitude will occur if radiation from the "stopping phase" interferes destructively with that from the "starting phase," and for this, the condition for a P wave is

$$b\left\{\frac{1}{v}-\frac{1}{\alpha}\right\}=\frac{n\cdot 2\pi}{\omega} \qquad 6.4.7$$

where ω is the frequency of the wave. There will therefore be a series of frequencies of minimum amplitude. For other positions of the observer, the geometry is somewhat more complicated, but it turns out that spectral analysis of waves at two or more stations permits both the distance b and the rupture velocity v to be determined.

That such minima exist, in the case of surface waves, is shown in Figure 6.9, which is from the work of Ben-Menahem and Toksöz (1963). The figure illustrates the spectra of Rayleigh waves from the Kamchatka earthquake of 1952, and from these the authors deduced a fault length of 700 km, with a rupture velocity of 3 km/sec in the direction N.146°W. Direct proof is provided, therefore, of faulting over a very long horizontal distance in a single earthquake. Because the fault associated with the Kamchatka earthquake is located under the Pacific Ocean (the epicenter as determined by first motions was at 52.6°N, 160.3°E), direct observations were not possible. Confirmation of the length of the active fault was provided by the distribution of aftershocks which occurred in the month following the main earthquake; the zone of these aftershocks extended for 675 km from the Kamchatka shock along the indicated direction of the fault. In an earlier study of the Mongolian earthquake of 1957, the same authors (1962) determined a fault length of 560 km, with a rupture velocity of 3.5 km/sec, in an azimuth of 100°. The effects of the Mongolian earthquake (epicenter 42.25°N, 99.4°E) were observable on the ground, where displacements of several meters were traced along a fracture exposed at the surface for 270 km. Both the Kamchatka and Mongolian earthquakes were of magnitude near 8, but the length of active fault associated with them is still remarkable.

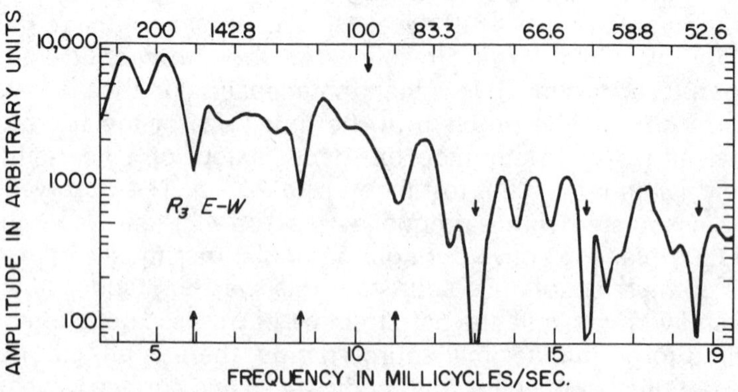

Figure 6.9 Spectrum of Rayleigh waves from the Kamchatka earthquake, showing the amplitude minima produced by the finite size of the source and rupture velocity. The scale at top shows period in seconds. (After Ben-Menahem and Toksöz)

6.5 DETECTION, DILATANCY, AND PREDICTION

Two problems which are currently of great interest, and in which the study of source mechanisms is basic, are the detection of large explosions and the prediction or forecasting of earthquakes. Much of the recent advance in the theory and instrumentation of seismology is due to the first problem, while the solution of the second would go far toward reducing earthquake losses.

An explosion equivalent to one megaton of TNT releases 10^{15} calories of energy. If all of this appeared as the energy of seismic waves, it would be equivalent to an earthquake of magnitude 6.3 approximately. In fact, a considerable fraction of the energy appears as heat and is absorbed near the source. Moreover, the mechanisms may be expected to be quite different: explosions normally occur at or near the earth's surface, and represent a pure dilatation in a small volume, rather than shear failure over an extended plane. Techniques of differentiating explosions from earthquakes have included the study of first motions, the complexity of the body-wave signals, the efficiency of generation of surface waves, and the ratio of long-period to short-period energy. An explosion should produce an initial compression everywhere, in contrast to the quadrant pattern of earthquakes. Evernden (1969) concludes that this is probably the case, but at the lower magnitudes of interest, detection of first motion is difficult. The argument for a less complex body-wave signal on the part of an explosion rests partly on the fact that a source below the earth's surface can produce phases reflected from the surface above it, while a surface source obviously cannot. It was at one time believed that the processed signal from an array, as illustrated in Figure 5.7, would be quite different for an explosion as compared to an earthquake. This is no longer considered to be an absolute criterion, and interest has turned to the production of long-period surface and body waves. The relatively large source dimensions associated with earthquakes cause these sources to be much more efficient generators of long period energy, particularly in the *S* waves. The absence of such energy appears to be the most useful criterion for the detection of explosions.

By the forecasting of earthquakes, we refer to the establishment of the probability that a shock will occur in a given region within a given time interval. This problem has not yet been solved. It is known, of course, that the absolute probability of the occurrence of earthquakes of given magnitude can be fairly accurately delineated over the earth's surface, but probability within a stated time interval is a different matter. Present research is directed toward the determination of some parameters of the source region that change in a measurable way before an earthquake (Hagiwara, 1969). Elastic stress and strain are obvious ones, and strain is now continuously measured at a number of stations in active areas (Pakiser, Eaton, Healy and Raleigh, 1969). Changes in the properties of material along a fault which is about to become active may lead to changes in *P* or *S* velocities, which could be monitored by the use of small explosions. While small earthquakes are often thought of as releasing accumulated strain, and thereby reducing the probability of a major earthquake, it has been suggested that very small earthquakes, or microshocks, are associated with the spread of dislocations along the fault zone, and these may be true forerunners of a major shock. There appear, in fact, to be two very different types of behavior: steady slippage along a fault, without major earthquakes, and a state of being locked, with occasional major shocks. Different sections of the San

Andreas fault in California exhibit the two types (Allen, 1968), possibly because of differences between the rocks in contact at the fault.

The forecasting of earthquakes has been greatly influenced by a theory of failure known as *dilatancy,* or local increase in pore volume due to micro-cracks in the source region shortly before faulting takes place. The observation which led to the hypothesis was that the ratio of *P*-wave to *S*-wave velocity (V_p/V_s) changed noticeably in the Garm region of the USSR (Semenov, 1969) before an earthquake. Garm was a region which had been provided with instruments specifically to detect premonitory signals. The relevant observation was that the ratio V_p/V_s showed a decrease for up to three months, then recovery to a normal value, before local earthquakes. This decrease was interpreted to be due to the formation of small cracks in the source region, but the recovery has been variously interpreted. Nur (1972) has proposed that pore fluids play a very important role in determining the velocity ratio. In the case of porous dry rock, laboratory methods show that while both V_p and V_s increase with confining pressure, owing to the closing of cracks and pores, V_p increases more rapidly, and the ratio V_p/V_s therefore increases. However, for saturated rocks, the behavior is different. Filling of the pores with water increases V_p, almost to the value the rock would have without porosity, but has little effect on V_s. Nur proposed, therefore, that the recovery of V_p/V_s just before faulting was the result of diffusion of ground water into the dilatant region (Fig. 6.10). The ground water itself may play a triggering role in faulting; fluid pore pressure may weaken the rock and provide lubrication on the fault surface. Seismologists in the USSR, while accepting the dilatancy model, do not attribute the recovery of V_p/V_s to ground water. Rather, they propose that, as the time of failure approaches, micro-cracks tend to become aligned parallel to the main fault plane so that a redistribution of stress permits many of the earlier-formed randomly-oriented cracks to close.

One promising technique measures a completely different parameter, the geomagnetic field. Both the susceptibility and remanent magnetization of rocks are stress-dependent, and changes in these, and therefore in the local mag-

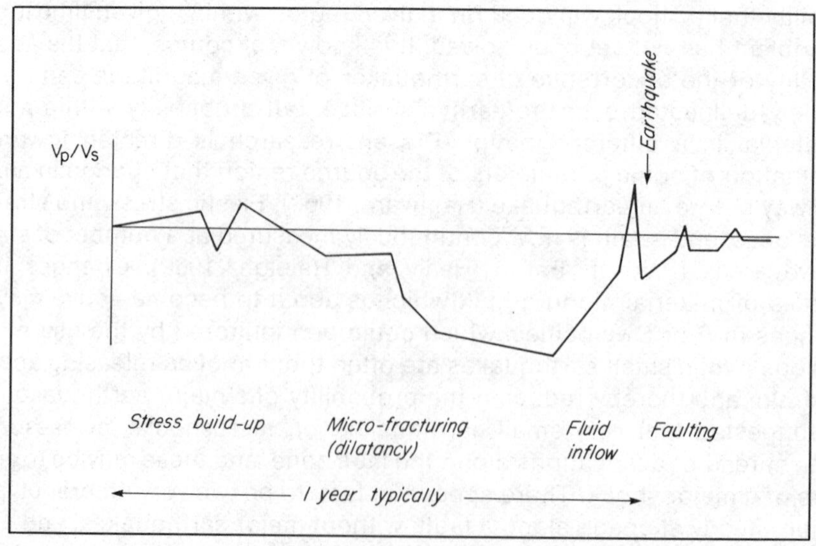

Figure 6.10 Typical variation in the ratio V_P/V_S observed before an earthquake, and its explanation in terms of dilatancy.

netic field, may precede earthquakes. The changes to be expected are very small, and their detection would require the best instrumentation (Chapter 16) as well as careful isolation from background noise in the magnetic field. However, one or two cases (Sect. 21.1) of otherwise unexplained magnetic events have been reported in advance of major earthquakes.

Undoubtedly, the most extensive program of earthquake prediction is in China (Bolt, 1974). The Chinese approach is not conditioned to any particular model of failure, but is based on the measurement of a great number of parameters, including the frequency of minor earthquakes, concentration of radon gas in well water, and change in level of the earth's surface. The world's first successful prediction of a major earthquake was that issued for the Haicheng earthquake of February 4, 1975 (Fung-Ming, Chu, 1976). There is no doubt that the earthquake, of magnitude 7.3, would have caused enormous loss of life in the heavily populated area, had a warning not been issued some nine hours before the shock. As a result of the warning, all people of the city were away from buildings at the time of the earthquake. This short-range prediction was based upon a great increase in the frequency of minor earthquakes, which, after the main event, were recognized to be foreshocks.

Finally, Mansinha and Smylie (1967) have apparently detected a relation between the Chandler wobble (Chapter 1) and large earthquakes, with the latter influencing the path of polar motion. There is even a suggestion in their analysis that a change in path precedes major earthquakes, which would be the case if the wobble were influenced by the accumulation of strain. If this is true, it could lead to the prediction of major earthquakes (without, however, the specification of the region of occurrence).

Problem

1. Direction-of-motion studies using the Byerly projection depended upon the interpretation of a diagram in the form of a stereographic projection (Fig. 6.4) on which there are two circles, separating regions of different sign of first arrival. What can you conclude, qualitatively, about the fault or auxiliary plane, (a) if one circle is so large that an almost straight-line section appears on the diagram, and, (b) if one circle is extremely small? What is the significance of the line, on the diagram, joining the epicenter to the other point of intersection of the circles? (This line is the projection of the "null axis.")

BIBLIOGRAPHY

Allen, C. R. (1968) In *Proceedings, Conference on the Geologic Problems of the San Andreas Fault System.* Stanford Univ. Pub., *Geol. Sci. No. 11*, 70.

Basham, P. W. (1969) Canadian magnitudes of earthquakes and nuclear explosions in Southwestern North America. *Geophys. J.* **17**, 1–13.

Båth, Marcus (1966) Earthquake energy and magnitude. In *Physics and Chemistry of the Earth,* **7** (ed. L. H. Ahrens, Frank Press, S. K. Runcorn, and H. C. Urey); Pergamon Press, Oxford.

Ben-Menahem, Ari, and Toksöz, M. Nafi (1962) Source mechanism from spectra of long-period seismic surface waves 1, the Mongolian earthquake of December 4, 1957. *J. Geophys. Res.,* **67**, 1943.

Ben-Menahem, Ari, and Toksöz, M. Nafi (1963) Source mechanism from spectra of long-period seismic surface waves 2, the Kamchatka earthquake of November 4, 1952. *J. Geophys. Res.,* **68**, 5207.

Bolt, B. A. (1974) Earthquake studies in the People's Republic of China. *Trans. Amer. Geophys. Un.,* **55**, 108–117.

Breiner, S. (1964) Piezomagnetic effect at the time of a local earthquake. *Nature,* **202**, 790.

Byerly, P. (1938) The earthquake of July 6, 1934: Amplitudes and first motion. *Bull. Seism. Soc. Amer.,* **28**, 1–13.

Byerly, P., and Evernden, J. F. (1950) First motion in earthquakes recorded at Berkeley. *Bull. Seism. Soc. Amer.,* **40**, 291.

Chinnery, M. A. (1969) Theoretical fault models. In *Symposium on processes in the focal region,* ed. Keichi Kasahara and Anne E. Stevens. *Pub. Dom. Obs. Ottawa,* **37**, 211.

Evernden, J. F. (1969) Identification of earthquakes and explosions by use of teleseismic data. *J. Geophys. Res.,* **74**, 3828.

Evernden, J. F., Best, W. J., Pomeroy, P. W., McEvilly, T. V., Savino, J. M. and Sykes, L. R. (1971) Discrimination between small-magnitude earthquakes and explosions. *J. Geophys. Res.,* **76**, 8042–8055.

Fung-Ming Chu (1976) An outline of prediction and forecast of Haicheng earthquake of M.7.3. In *Proceedings of the lectures by the seismological delegation of the People's Republic of China,* Jet Propulsion Laboratory, California Institute of Technology, Pasadena, California.

Gutenberg, B., and Richter, C. F. (1954) *Seismicity of the earth and associated phenomena.* Princeton Univ. Press, Princeton.

Hagiwara, Takahiro (1969) Prediction of earthquakes. In *The earth's crust and upper mantle,* ed. Pembroke J. Hart, *Amer. Geophys. Un., Monograph,* **13**, 174.

Hirasawa, Tomowo, and Stauder, William, S. J. (1965) On the elastic body waves from a finite moving source. *Bull. Seism. Soc. Amer.,* **55**, 237.

Hodgson, John J., and Stevens, A. E. (1964) Seismicity and earthquake mechanism. In *Research in geophysics,* ed. Hugh Odishaw, Vol. 2. M.I.T. Press, Cambridge, Mass.

Honda, H. (1961) The generation of seismic waves. In *A symposium on earthquake mechanism,* ed. J. H. Hodgson. *Pub. Dom. Obs. Ottawa,* **24**, 329.

Isaacks, B., Oliver, J., and Sykes, L. R. (1968) Seismology and the new global tectonics. *J. Geophys. Res.,* **73**, 5855.

Knopoff, Leon (1958) Energy release in earthquakes. *Geophys. J.,* **1**, 44.

Mansinha, L., and Smylie, D. E. (1967) Effect of earthquakes on the Chandler wobble and the secular polar shift. *J. Geophys. Res.,* **72**, 4731.

McKenzie, Dan (1972) Active tectonics of the Mediterranean region. *Geophys. J.,* **30**, 109–185.

Nur, Amos (1972) Dilatancy, pore fluids, and premonitory variations of ts/tp travel times. *Bull. Seism. Soc. Amer.,* **62**, 1217–1222.

Pakiser, L. C., Eaton, J. P., Healy, J. H., and Raleigh, C. B. (1969) Earthquake prediction and control. *Science,* **166**, 1467.

Richter, Charles F. (1935) An instrumental earthquake magnitude scale. *Bull. Seism. Soc. Amer.,* **25**, 1.

Savage, J. C. (1965) The effect of rupture velocity upon seismic first motions. *Bull. Seism. Soc. Amer.,* **55**, 263.

Semenov, A. N. (1969) Variations in the travel time of transverse and longitudinal waves before violent earthquakes. *Bull. Acad. Sci. U.S.S.R. Phys. Solid Earth,* **3**, 245–248.

Steketee, J. A. (1958) Some geophysical applications of the elasticity theory of dislocations. *Can. J. Phys.,* **36**, 1168.

Stechschulte, Victor C. (1931) Deep focus earthquakes. *Nature,* **128**, 673.

Stevens, Anne E., and Hodgson, John H. (1968) A study of *P* nodal solutions (1922–1962), in the Wickens-Hodgson catalogue. *Bull. Seism. Soc. Amer.,* **58**, 1071.

Stevens, Anne E. (1969) Worldwide earthquake mechanism. In *The earth's crust and upper mantle,* ed. Pembroke J. Hart. *Monographs, Amer. Geoph. Un.* Washington.

Stoneley, R. (1931) On deep-focus earthquakes. *Beitr. Geophys.,* **29**, 417.

Sykes, L. R. (1967) Mechanism of earthquakes and nature of faulting on the mid-oceanic ridges. *J. Geophys. Res.,* **72**, 2131.

Turner, H. H. (1922) On the arrival of earthquake waves at the antipodes and on the measurement of the focal depth of an earthquake. *Mon. Not. R. A. S., Geophys. Suppl.,* **1**, 1.

Wadati, K. (1928) On shallow and deep earthquakes. *Geophys. Mag.* (Tokyo), **1**, 162.

Wickens, A. J., and Hodgson, J. H. (1967) Computer re-evaluation of earthquake mechanism solutions 1922–1962. *Pub. Dom. Obs. Ottawa,* **33**, 1.

Wood, Harry O., and Neumann, Frank (1931) Modified Mercalli intensity scale of 1931. *Bull. Seism. Soc. Amer.,* **21**, 277.

SEISMOLOGY AND THE OUTER SHELL 7

7.1 THE IDENTIFICATION OF THE CRUST

By the early part of the twentieth century, preliminary time-distance curves had been established which showed that, for relatively short distances, the velocities of *P* and *S* were about 8.0 and 4.4 km/sec respectively. However, the distribution of stations was such that there was not a good determination of the curves within 100 or 200 km of the source. In 1909, Mohorovičić studied arrivals at short distances from an earthquake in the Kulpa valley (in present-day Yugoslavia) and showed that the time-distance curves must begin with segments of quite different slope. He suggested the presence of a surface layer, some 30 km thick, in which the velocities of *P* and *S* were 5.6 and 3.2 km/sec respectively. Today we refer to this outer shell as the *crust,* and the name of *Mohorovičić* has become attached to the discontinuity at its base.*

In the next section we shall consider the seismological methods appropriate for studies at short ranges. To continue from the historical point of view, we note that Conrad (1925) suggested that the crust itself is two-layered, and following this suggestion, Jeffreys (1937) attempted to identify the layers on the basis of their velocities. In contrast to the situation in the mantle, none of chemical homogeneity, hydrostatic stress, nor adiabatic temperature distribution can be assumed in the crust. The Adams-Williamson method is not applicable, and from the *P* and *S* velocities, only a combination of properties can be obtained. Jeffreys' approach was to tabulate the ratio K/ρ (which is given by $\alpha^2 - 4/3\beta^2$) and to compare its values with laboratory measurements. His values were the following:

	α km/sec	ρ km/sec	K/ρ c.g.s.
Upper crust	5.57	3.36	15.9×10^{10}
Lower crust	6.50	3.74	23.6×10^{10}
Mantle	7.76	4.36	34.9×10^{10}

Laboratory measurements for common rock types were

	K/ρ, c.g.s.
Obsidian	15.1×10^{10}
Granite	18.3
Diorite	22.8
Gabbro	27.8
Dunite	35.2

*The idea of a seismological crust was previously put forward by Milne; see Section 25.2.

On the basis of these values, Jeffreys gave the name "granitic layer" to the upper crust and "intermediate layer" (i.e., a layer with the properties of a petrologically intermediate rock) to the lower crust, and suggested that the uppermost mantle was similar to dunite.

Modern work has changed, in detail, the velocities used by Jeffreys, but has probably not contradicted his hypothesis that the crust varies, from top to bottom, between an acidic and an intermediate or basic composition. The important conclusion has been that layering, or the lack of it, is a local phenomenon, and it is meaningless to speak of crustal layers in any worldwide sense. On the other hand, the Mohorovičić discontinutiy or base of the crust has been observed so widely that areas in which it cannot be defined are considered to be anomalous.

7.2 TRAVEL TIMES AND LAYERS

We wish to establish the form of the arrival-time curve for a small portion of the earth consisting of one or more horizontal layers, in each of which the velocities are constant and the ray paths straight (Fig. 7.1). Energy from a source, which may be a local earthquake or an explosion, can reach detectors by a direct path, but it can also be critically refracted at the base of the layer, if the layer is underlain by higher-velocity material. It is found (as shown later in this section) that the critically refracted wave, or "*head wave,*" sets up a wave front in the layer, which corresponds to energy traveling back to the surface in a direction symmetrical to that of the wave incident on the boundary. In the case shown, consisting of a single dipping layer, observers to the right of the source observe two arrivals (say of P), one given by arrival time T_1, where

$$T_1 = x/V_1 \qquad 7.2.1$$

and one by time T_2 where

$$T_2 = \frac{2d}{V_1} \cos i_c + \frac{x \sin (i_c + \theta)}{V_2 \sin i_c} \qquad 7.2.2$$

where i_c is the critical angle of incidence, given by

$$\sin i_c = V_1 / V_2 \qquad 7.2.3$$

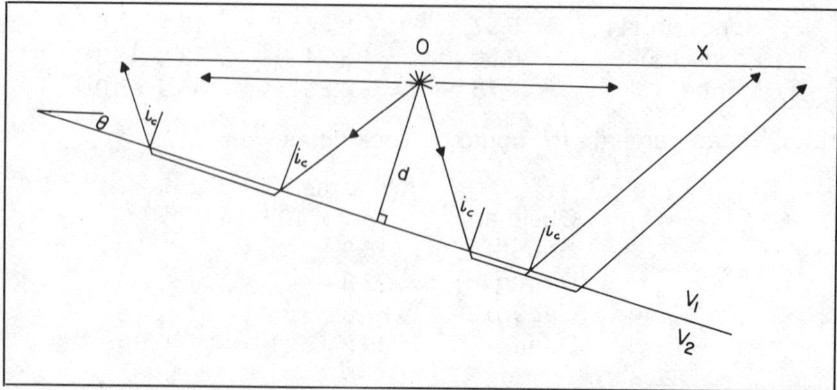

Figure 7.1 Refraction of a wave at a dipping interface.

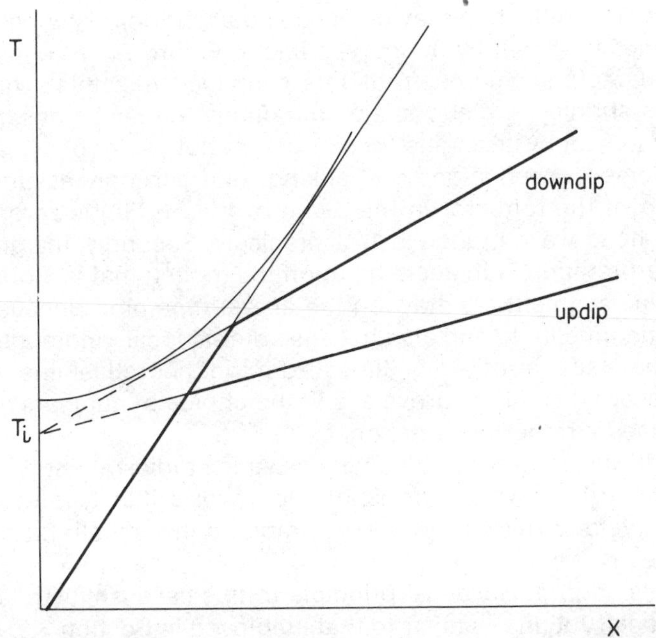

Figure 7.2 Idealized arrival-time curve for the case shown in Figure 7.1.

Equation 7.2.2 is obtained by writing down the times for segments of the path, and simplifying with elementary trigonometry. For observers up-dip from the source, the arrival time of the refracted wave is found to be

$$T_2 = \frac{2d \cos i_c}{V_1} + \frac{x \sin (i_c - \theta)}{V_2 \sin i_c} \qquad 7.2.4$$

The form of the arrival-time curves is shown in Figure 7.2, and a method of interpretation suggests itself. If observations in two directions from a source are available, and if arrival times plot as in Figure 7.2, a single dipping layer is suggested. Measurement of the three slopes yields values of V_1, i_c and θ, and thus of V_2. The intercept of the refracted branch is given by

$$T_i = \frac{2d \cos i_c}{V_1} \qquad 7.2.5$$

in which only d, the perpendicular to the layer, is unknown. The structure is thus completely determined.

In Figure 7.2, arrivals corresponding to the first arrival of energy at each seismometer are emphasized, as these are usually determined with the greatest certainty. When there are additional layers above the substratum, there will be additional straight-line segments to the arrival-time plot, as head waves may be generated at each interface. However, if the layers are thin, these segments may never appear as first arrivals. The method outlined above may be extended to any number of layers, using successive intercepts to determine all of the depths involved.

There is apparently a minimum value of x within which the head waves cannot be observed. At this point, which corresponds to the vanishing of the critically refracted portion of the ray path, the refraction merges with the reflected wave. The reflection may be observed at shorter distances; the locus of its arrival times is shown by the curved line in Figure 7.2.

Two key factors emerge from this simplified example. The first is that observations should be made in two directions, either by observing on both sides of a source, or by arranging for sources on both sides of a line of seismometers. If reversed observations are lacking, one can only assume that θ (the apparent dip of the refractor in the plane of the profile) is zero, and that the slope of the head wave branch is $1/V_2$ precisely. Secondly, the greatest uncertainty in interpretation is in deciding upon the model; that is, upon the number of straight-line segments to draw on the arrival-time plot. Because of errors in timing and identification, and also because of real local inhomogeneities, there will always be a scatter of arrival times from the theoretical lines, especially in the case of second and later arrivals, and the choice of numbers of layers in the model becomes a rather personal one.

As the discussion above indicates, seismic studies at short range depend greatly on the arrivals which correspond to waves critically refracted along the top of a high velocity layer, and it is desirable to investigate further the nature of this wave.

The application of Huygens' principle to this case is illustrated in Fig. 7.3, where the construction is similar to that employed in Section 3.2. A plane wave incident at the critical angle of incidence strikes the boundary point A at time t_1. By the time t_2, when the critically refracted wave, traveling with velocity v_2, reaches B, energy has spread from A in the upper medium to the hemisphere

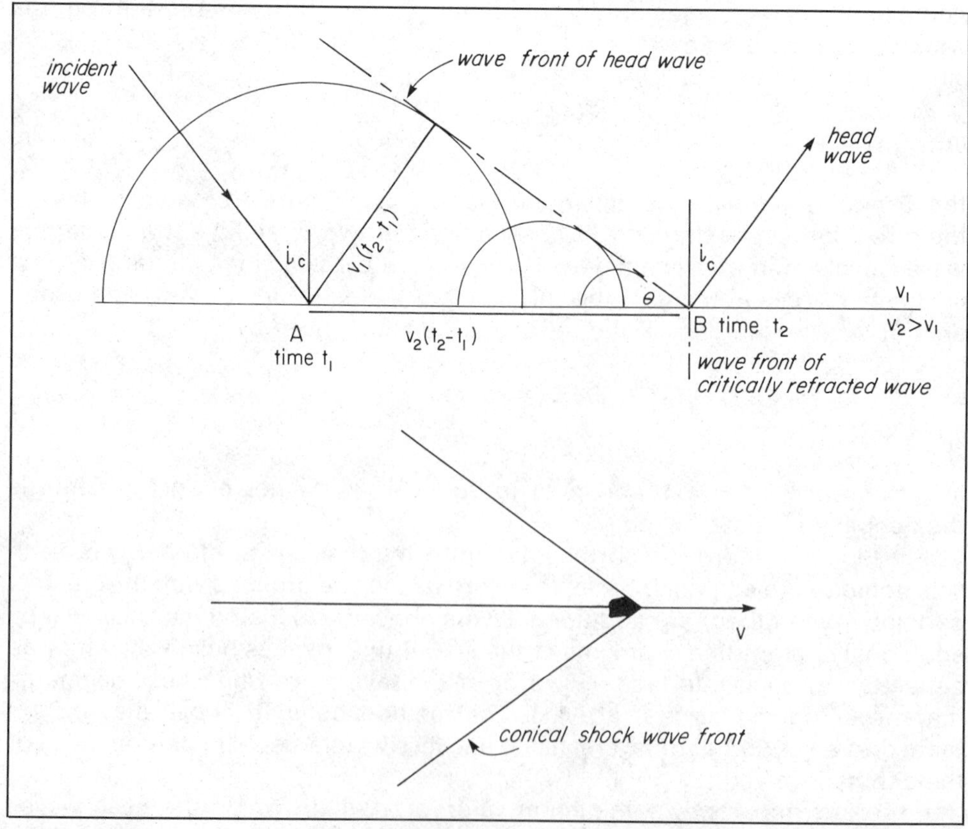

Figure 7.3 Construction of the headwave wavefront by Huygens' principle (top), and comparison with the shock front developed by a supersonic missile (bottom).

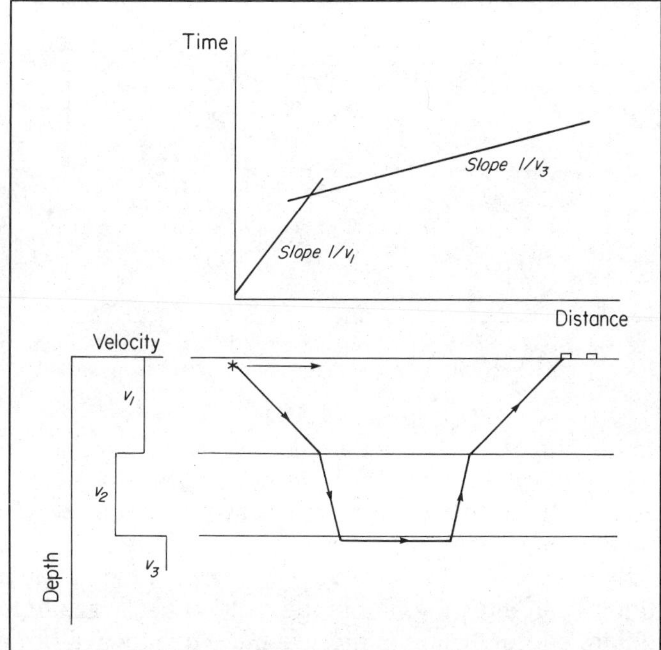

Figure 7.4 Arrival time plot (top) and ray paths (bottom) in the case of a low-velocity layer. The plot is schematic, since the minimum observation distance for the refracted arrival would be well to the right for the rays shown.

shown. Since an infinite number of hemispheres, of decreasing radius, exist between *A* and *B*, a sloping plane wavefront may be located as shown. This wavefront is inclined at an angle θ to the boundary; by the construction, $\theta = \sin^{-1} v_1/v_2$. But this is precisely the definition of the critical angle of incidence, i_c, showing that energy is propagated back into the upper medium along rays making the angle i_c with the normal.

As the lower part of Fig. 7.3 suggests, the new wavefront is analogous to the conical shock front which can be seen (and photographed) inclined from the head of a supersonic missile in air. For this reason, the critically refracted wave became known as the "head wave" quite early in the history of geophysics.

The difficulty with the geometrical optics approach is that if one assumes an incoming plane wave, upon critical refraction the cross section of the beam becomes vanishingly small, because every ray, after refraction, must be taken to travel in the boundary plane. It is impossible to reconcile this with the propagation of energy along the path. In fact, the critically refracted wave, or head wave, was observed in the field before its existence was shown theoretically. The theoretical treatment (Appendix C) confirms the existence of the wave, in the case of spherical, rather than plane, wavefronts diverging from a localized source.

Critical refraction is possible only when the lower layer has the higher velocity. If any layer in a sequence has a velocity less than the layer above it, refraction is toward the normal and no head wave can be produced at the upper boundary of the low-velocity layer. Waves may become critically refracted at some lower interface (Fig. 7.4), so that a time-distance graph of normal appearance is obtained. Notice that the graph itself contains no evidence of the low-velocity layer, and if the depths to boundaries beneath it are computed by the method discussed at the beginning of this section, these depths will be in error. A low-velocity layer within the crust is therefore elusive, just as a low-velocity zone in the range of continuous velocity variation in the upper mantle was shown

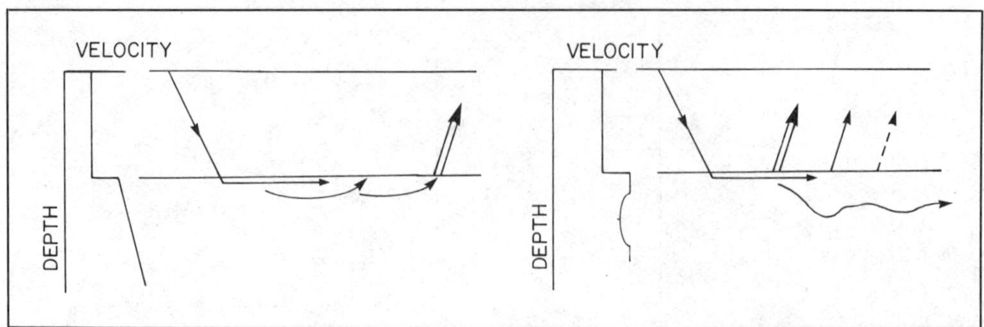

Figure 7.5 Schematic explanation of the variation in headwave amplitude in the cases of a velocity gradient (left) or low-velocity zone (right) below a critically-refracting boundary.

to be. The only recourse is to use some information in addition to arrival times; the amplitudes of arrivals can be used to infer the presence of a low-velocity zone, and for this reason the treatment of head waves to include amplitude estimates has received considerable attention. Figure 7.5 indicates qualitatively how the presence of a velocity gradient immediately beneath a velocity discontinuity can alter the amplitude of the head wave. If the velocity increases with depth, energy which might leave the head wave (for example, through scattering by irregularities in the boundary) is refracted back to the boundary. However, if the velocity begins to decrease toward a low-velocity zone, energy leaving the head wave becomes trapped in the low-velocity channel, and the amplitude of the head wave decreases rapidly with distance. For the longest crustal refraction profiles, amplitude and time calculations must take into account the spherical geometry of the earth's surface layers.

7.3 TECHNIQUES OF CRUSTAL SEISMOLOGY

Seismological investigations of the outer part of the earth began with the observing of earthquake waves at short distances, using conventional station instruments. An observation of arrival time, such as represented on Figure 7.2, thus corresponded to an event on a single seismogram trace. Modern work has tended to emphasize large explosions as sources of energy, for the obvious reason that they are controllable in time and position, and it has also adopted techniques of detection and recording that are closer to applied geophysics than to earthquake seismology. The great limitation of having only a single seismometer at each observing station is the difficulty in correlating a given event, particularly if it is not a first arrival, between stations. If a spread of closely spaced detectors is available at each station, the character of an arrival, particularly its local apparent velocity, can be determined, and correlation becomes easier. The ideal situation would be to have spreads of detectors cover the entire profile line, but this is rarely accomplished, because of the sheer cost of providing detectors at intervals of about 100 m over a line that may be 300 km long (Steinhart and Meyer, 1961).

The instrumentation itself usually incorporates some changes from that considered in Chapter 6. Portable seismometers having a natural frequency between 1 and 10 cps are commonly used, but these are coupled, through input transformers, to high-gain amplifiers. The amplifiers, which incorporate band-pass filters, are in turn coupled to small taut-wire galvanometers. As the galvanometers usually have natural frequencies much above those of the signals, the frequency response of the system is controlled by the seismometers and electronic filters, rather than by the galvanometers. Recording, if on visual records, is multiple-trace, with timing lines marked on the record, usually at intervals of 0.01 second. However, recording on magnetic tape is often employed.

Field procedures vary greatly depending upon local conditions, but the aim should always be to obtain a reversed profile, of sufficient length that the wave from the top of the mantle is obtained as a first arrival. For an average thickness of continental crust, this requires a profile at least 300 km in length. Explosions equivalent to several hundreds or thousands of pounds of TNT are necessary, and it is usually found to be most convenient to detonate these in water. In many cases it turns out to be desirable to leave the recording arrays fixed, and to arrange for a series of explosions at different points. Because travel times are unchanged if the ray direction is reversed, this procedure is equivalent to observing a single explosion at different distances.

The fact that seismic refraction work may be carried out at sea has been of the greatest importance in geophysics. In principle, the ocean acts simply as an additional layer of rather low velocity, and both source and receivers may be placed in the water. Certain precautions are necessary to avoid continued radiation from an oscillating bubble of gas following the detonation of the explosive; this is usually accomplished by arranging to detonate the explosive at such a depth that energy reflected from the ocean surface interferes destructively with these "bubble pulses." At the receiving end, detectors, known as hydrophones in this case, are trailed from a recording ship at a sufficient depth beneath the sea surface to avoid wave and other noise. If only a single ship is available, it is normally used to fire the charges, and the hydrophones are connected to sonobuoys, which incorporate radio transmitters for the relaying of signals to the ship. The operations have been described in detail by Shor (1963) and Hill (1963).

Seismic reflection work at sea has also added greatly to our knowledge of the oceanic crust, especially the layers of unconsolidated sedimentary rock. Techniques for continuous seismic profiling described by Hersey (1963) are still employed, but these have benefited from research on sources, receivers, and signal enhancement carried out for exploration geophysics (Dobrin, 1976). The observations are based essentially on normal incidence reflection (Section 3.2), which can produce a measurable signal at any boundary where the product of velocity and density changes. Two-way time for the reflected waves is a measure of depth, but (in contrast to refraction observations, in which velocities are determined) the velocity for conversion of time to depth must be obtained independently. The energy source is no longer limited to conventional explosives; electric sparkers and gas exploders towed behind the observing ship are now used much more frequently. Normal incidence reflections are observed on hydrophones also near the ship, but there is an advantage to observing as well the reflections at a broad angle of incidence, through the use of hydrophones connected to sonobuoys. This is chiefly because the average velocity to different depths may be computed. The travel time, by the reflected path, from a horizontal layer at depth z to a hydrophone at horizontal distance x from the source is easily shown to be

$$T = \left(\frac{x^2}{4} + z^2\right)^{1/2} \Big/ V$$

where V is the average velocity from the sea surface to the layer in question. A plot of T^2 against x^2 for a given reflection is linear and yields the value of V. Values of average velocity to different horizons permit the conversion of the time profiles to depth sections.

Perhaps the greatest obstacle to interpretation of the sections is the presence of spurious, curved horizons, arising from reflections at non-vertical incidence from strong reflectors of limited area. For example, a boulder buried in the sediment presents a curved surface, from which normal incidence reflections may be observed at several horizontal distances. With the boulder at depth z, the reflection time (and therefore inferred depth) varies with horizontal distance from the boulder according to

$$VT = 2\,(x^2 + z^2)^{1/2}$$

or

$$V^2T^2 - 4x^2 = 4z^2$$

The shape of the reflection on the depth section is therefore hyperbolic, and it may be misinterpreted as a curved, dipping interface.

In general, however, continuous reflection profiling at sea has yielded information with a good degree of certainty. Some of the implications for plate tectonics will be discussed in Chapter 27, where a striking section across an oceanic trench is shown (Fig. 27.13).

The situation may arise, in work either on land or sea, where it is not possible to observe arrivals along straight-line profiles. Partly for this reason, and partly to provide a more objective method of interpretation, Scheidegger and Willmore (1957) introduced a method of analysis known as the "time-term method." In this approach, it is presumed that there is a distribution of N stations, each of which may be occupied as a source or a receiver. It follows, from the development in Section 7.2, that if the dip is not too great, the travel time for a refracted arrival from the ith to the jth station may be written

$$t_{ij} = \frac{d_{ij}}{V} + a_i + a_j \qquad 7.3.1$$

where d_{ij} is the distance, V is the refractor velocity, and a_i and a_j are constants for the two stations. The approximation, of course, is in the neglect of the fact that a term a_i in fact depends on the azimuth of the path. For N stations, the total possible number of observation equations is $N(N-1)$ while the number of unknowns is $(N+1)$: the a_i's and V. It is therefore possible to obtain least-square estimates for the best values of the time terms (a_i) and V. The time terms are related to the depth to the refractor below each station, but the conversion to depth requires a detailed knowledge of the velocity above the refractor, which the method itself does not yield. The time-term method has been applied effectively in cases (Berry and West, 1966) where information from an array rather than a linear distribution of stations was available.

The same approach has been used at sea (Raitt et al., 1969), with some modifications. Because of the drift of shooting and observing ships, it is not usually possible to secure travel-time observations between identical points. In order to make the solution determinable, the time terms a_i have been assumed to vary smoothly with position, according to some polynomial of the coordinates. This is permissible, provided that the crustal conditions in the vicinity of each station, to which the a_i terms are related, do not change greatly within the area of the survey. The method has recently been used at sea to test for velocity anisotropy in the uppermost mantle (Raitt et al., 1969; Keen and Tramontini, 1970). As the study of anisotropy will undoubtedly become increasingly important, we shall outline the method of analysis. First, the behavior of Equation 7.3.1 under the assumption of a small perturbation in refractor velocity, δV, must be examined. We may write

$$t_{ij} = \frac{d_{ij}}{V_0} + [a_i + a_j]_{V_0} + \delta V\left[\frac{\partial a_i}{\partial V} + \frac{\partial a_j}{\partial V} - \frac{d_{ij}}{V^2}\right]_{V_0} \qquad 7.3.2$$

It is not difficult to show that

$$\left[\frac{\partial a_i}{\partial V}\right]_{V_0} = \left[\frac{\partial a_j}{\partial V}\right]_{V_0} = \frac{x}{V_0^2} \qquad 7.3.3$$

where x is the horizontal offset between the source or receiver and the point where the ray becomes critically refracted. The time equation is therefore

$$t_{ij} = \frac{d_{ij}}{V_0} + [a_i + a_j]_{V_0} + \frac{(2x - d_{ij})\,\delta V}{V_0^2} \qquad 7.3.4$$

In this equation, δV arises from anisotropy in the refracting mantle beneath the crust. Backus (1965), by solving the equation of motion for a slightly anisotropic medium, showed that the variation in velocity with azimuth ϕ is given by

$$\delta V = A \sin 2\phi + B \cos 2\phi + C \sin 4\phi + D \cos 4\phi \qquad 7.3.5$$

where the coefficients on the right are assumed to be constant for the area of the survey.

Only a curving refracting interface would contribute the identical terms to δV; if other terms are present, they arise from lateral inhomogeneities, rather than from anisotropy. Substituting δV from 7.3.5 into 7.3.4 then gives

$$t_{ij} = \frac{d_{ij}}{V_0} + [a_i + a_j]_{V_0} + \frac{2x - d_{ij}}{V_0^2}[A \sin 2\phi + B \cos 2\phi + C \sin 4\phi + D \cos 4\phi] \qquad 7.3.6$$

Sufficient observations of travel times in different azimuths must then be made to permit a least-squares determination of the velocity and the anisotropy coefficients in Equation 7.3.6. In order to reduce the number of unknowns, the offset distance x is normally assumed to be equal for all stations, and the a_i terms are assumed to vary smoothly, with position, as already described. With the determination of the constants *A, B, C,* and *D,* it is then possible to plot mantle velocity as a function of azimuth ϕ.

Laboratory measurements of velocities on rock samples are useful for comparison with velocities measured in the crust, provided that extrapolation to the appropriate pressure and temperature is possible. Table 7.1 gives an example of the range of measured velocity in common rock types, selected from the much

TABLE 7.1 EXAMPLES OF ROCK VELOCITIES MEASURED AT LOW PRESSURE AND TEMPERATURE, SELECTED FROM VALUES TABULATED BY PRESS IN CLARK (1966).*

	α	β
	(km/sec)	
Sandstone	1.4–4.5	
Limestone	1.7–7.1	2.8–3.5
Granite	5.1–6.0	2.9–3.2
Diorite	5.8	3.1
Basalt	5.1–6.4	2.7–3.2
Dunite	7.4–8.6	3.8–4.4

*Clark, S. P. *Handbook of Physical Constants,* Memoir 97. Geological Society of America, 1966.

more extensive list published by Clark (1966). (In accepting tabulated values of physical properties, great caution is required, because of the possible variation in character between individual samples of rocks labeled simply as "limestone" or "granite.")

It is evident, however, that the values of α and β increase with the density of the rocks, which are listed in the table in order of increasing density. At first this appears to be a paradox, because equations 2.4.8 and 2.4.11 show the velocities to depend inversely on the square root of density. But in almost all common rocks the proportionate increase in the elastic moduli toward the basic end of the scale is greater than the proportionate increase in density; the velocities therefore increase.

Laboratory measurements of velocities at pressures up to 10 kilobars (Birch, 1957, 1960) indicate that the proportional increase of velocity with pressure is greatest in the range from 0 to 1 kilobar, as the pore spaces are reduced. The effect is variable between samples, depending upon the initial porosity. For most igneous rocks, an increase of velocity of 20 to 30 per cent of the initial value is observed in this range. Above 1 kilobar, the increase is about 1 to 2 per cent per kilobar increase in pressure. The effect of increasing temperature is to decrease velocity, by (40 to 60) $\times 10^{-6}$ parts per °C increase.

7.4 THE RESULTS OF CRUSTAL SEISMOLOGY

To indicate the contribution of seismic investigations of the crust, we should either have to display a great number of individual results, or show only profiles based on statistical averages. This is because the pattern of velocity structure above the Mohorovičić discontinuity, and the depth to that discontinuity, are found to vary greatly between different regions. We shall show two examples and then pass on to compilations of results.

A typical array of refraction records is shown is Figure 7.6. It is from a crustal study in Colorado (Jackson, Stewart, and Pakiser, 1963) which forms part of a transcontinental study of the crust under North America (Pakiser and Zietz, 1965). We note that first arrivals at shorter distances define a velocity of about 6.1 km/sec in the crust, and that at greater distances they define a velocity of 8.0 km/sec for the uppermost mantle. It appears from the records, however, that a number of later arrivals could be defined, and that most of these would nowhere emerge as first arrivals. The selection of the number of layers within the crust therefore depends somewhat on the observer's choice. In the North American work referred to, an upper crust with a *P* velocity between 5.9 and 6.2 km/sec is shown for most areas, with the lower crust exhibiting a greater variation in properties.

Some idea of the possible complexity in the velocity structure of the continental crust is shown by the work of Berry and Fuchs (1973). These authors present profiles (Fig. 7.7) for three lines across the Canadian Precambrian shield, sampling two age provinces and the boundary between them. A low-velocity channel within the crust appears to be present in all cases. As discussed above, this could not be recognized by travel times alone, but was inferred by comparison of the observed seismograms with computed seismograms for layered structures. Two explanations of the low-velocity channel are possible: it could result from variations in rock pore pressure, as described in Chapter 6, or it could represent real differences in chemical composition, with acid rocks being trapped between more basic layers above and below.

Extensive coverage of continental arrays by long refraction profiles has also been accomplished in the U.S.S.R. A summary of the crustal sections which

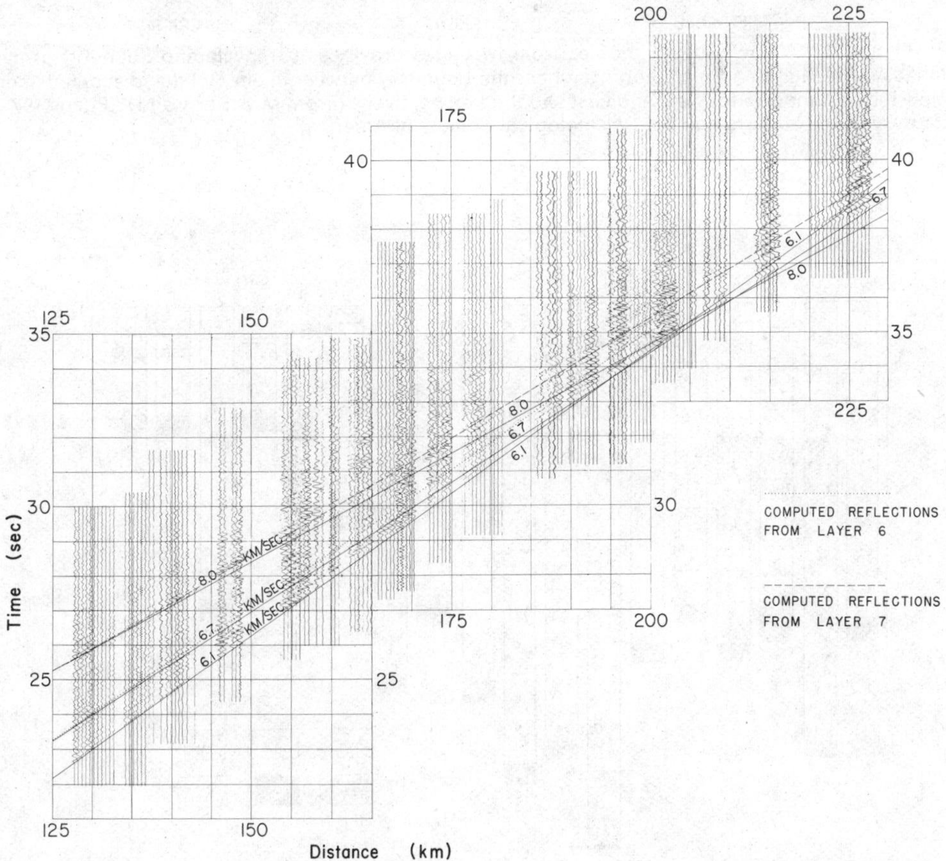

Figure 7.6 Seismograms from a profile across the Great Plains of eastern Colorado. Note how the wave through the upper crust, with velocity 6.1 km/sec, is defined by first arrivals out to a distance of 210 km, beyond which P_n emerges as a first arrival. Arrivals corresponding to the dashed lines may be reflections from discontinuities within the crust, or from the base of the crust. (Courtesy of L. C. Pakiser and the U.S. Geological Survey)

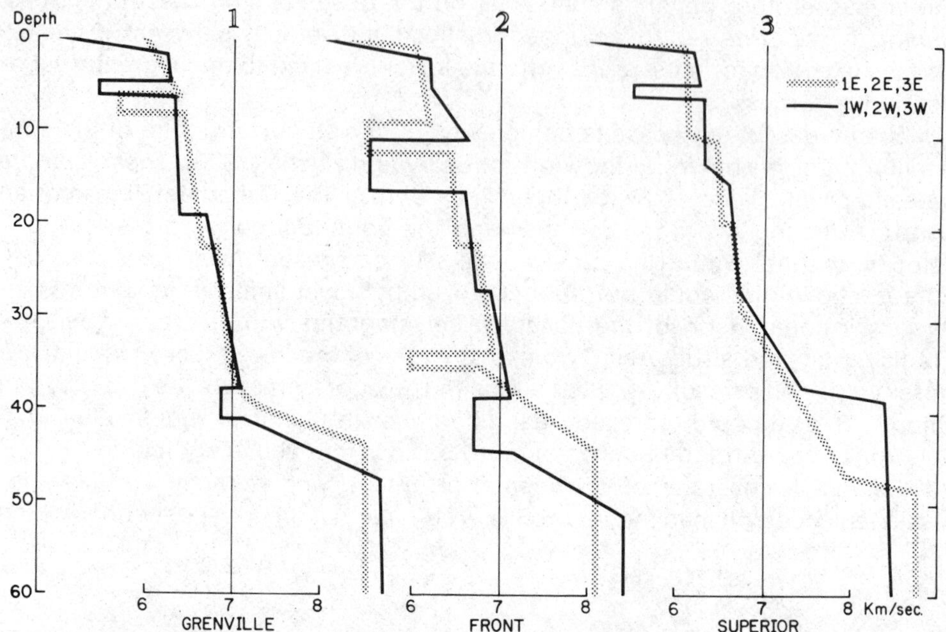

Figure 7.7 Velocity-depth profiles from two ages provinces (Grenville and Superior) of the Canadian Precambrian Shield and from near the boundary between them. Solid and stippled lines show the best models for west and east shooting respectively. (From M. J. Berry and K. Fuchs 1973, courtesy of the authors and the Seismological Society of America)

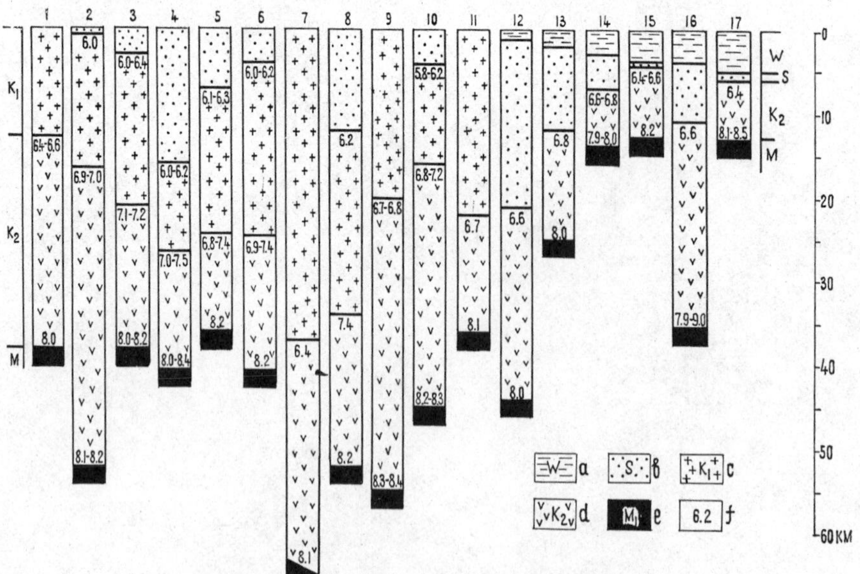

Figure 7.8 Models of the earth's crust in the U.S.S.R. from seismic observations; (a) water, (b) sedimentary rocks, (c) upper crust, (d) lower crust, (e) mantle, (f) velocity in km/sec. (Courtesy of I. P. Kosminskaya and the American Geophysical Union)

have been obtained in that country has been given by Kosminskaya et al. (1969) and is shown in Figure 7.8. We shall leave the question of the variation of crustal thickness until the gravitational field is considered, because this variation may be closely related to isostatic compensation of the topography. But the various distributions of velocity in the crust confirm that "layering," at least beneath the continents, is a regional rather than global phenomenon. It is certainly misleading to think of the "Conrad discontinuity," representing the top of the basic part of the crust, as occupying a role as major as that of the Mohorovičic discontinuity at the base of the crust.

Statistical considerations are made possible by a compilation of all available refraction results, prepared by McConnell and McTaggart-Cowan (1963). Average profiles of velocity, as produced by a computer for all continental and all oceanic profiles, are shown in Figure 7.9. We note immediately the thinner crust below the oceans; an examination of individual profiles would indicate

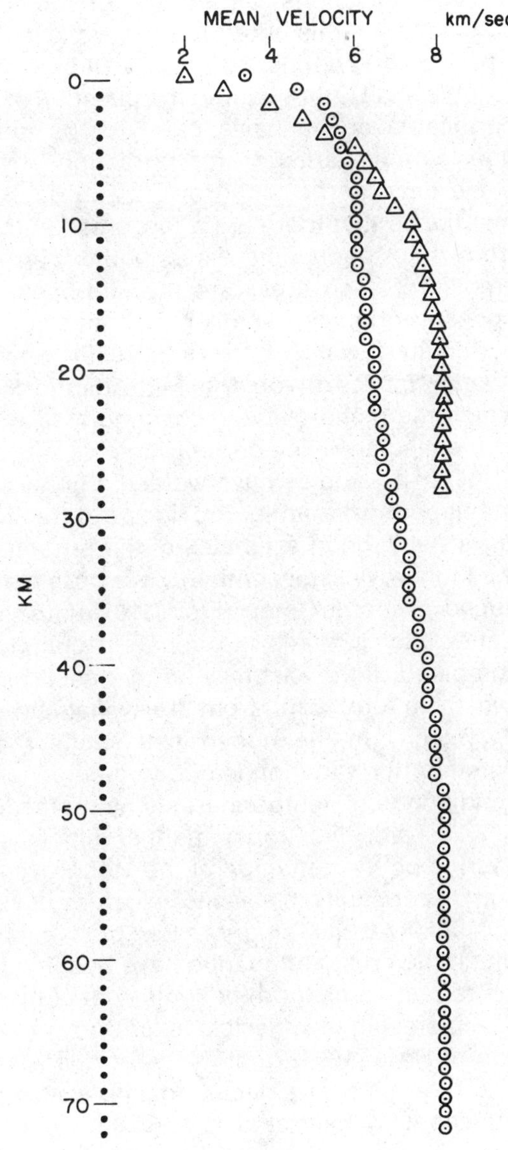

Figure 7.9 Mean value of *P*-wave velocity as a function of depth, for a compilation of continental (circles) and oceanic profiles. (After McConnell and McTaggart-Cowan)

rather less variation between different areas in the oceanic as compared to the continental case. Beneath the continents, the velocity increases, through an average crustal thickness of 40 km, from values appropriate to acid igneous rocks to those appropriate to basic rocks. The oceanic crust is characterized by a thin veneer of very low velocity material underlain by a thick layer ("Layer 2") of material with a velocity of 4.5 to 5.5 km/sec, and, for the remainder of its mean thickness of 9.0 km, by material ("Layer 3") with a velocity of approximately 6.7 km/sec. There has been uncertainty as to the exact nature of the material of Layers 2 and 3 (Raitt, 1963), but recent dredging of rocks from the Mid-Atlantic Ridge (Barrett and Aumento, 1970) has provided samples of basalt, metamorphosed basalt, and gabbro. Barrett and Aumento show that the depths from which these samples were dredged, and also their seismic velocities as measured in the laboratory, strongly suggest that Layer 2 is composed of basalt, and Layer 3 of metamorphosed basalt and gabbro. However, still more recent drilling by the *Glomar Challenger* (Aumento, Ade-Hall, and Keen, 1975) and direct observations with the use of submersibles (Bellaiche et al., 1974) in turn suggest that this interpretation is over-simplified. The composition of the oceanic and continental crusts will be considered again in Chapter 26.

We shall return later to the question of the boundaries between oceanic and continental crustal types, but it may be noted that these are areas which present considerable difficulty for refraction observations, because of the relatively steep dips on the Mohorovičić and other discontinuities. It has already been mentioned that there are a few areas of the earth in which no refraction arrival from the Mohorovičić discontinuity can be observed. The most striking examples of such areas are the mid-ocean ridges, beneath which the velocity appears to indicate a mixing of oceanic crust and mantle, with no sharp discontinuity (Ewing and Ewing, 1959). However, the observations of Keen and Tramontini (1970) on the Mid-Atlantic Ridge would limit the zone beneath which the Mohorovičić discontinuity is missing to about 20 km on either side of the median valley of the Ridge.

Crustal studies have yielded a great deal of information on the velocity of the uppermost mantle. Provided a refraction profile is of sufficient length, and is reversed, good estimates of the refraction velocity (known as P_n) just under the Mohorovičić discontinuity are possible. Herrin and Taggart (1962) have produced a contour map (Fig. 7.10) of the variation in P_n velocity beneath the United States. While the apparent complexity of the variation in the west as compared to the east may be the result of differences in the detail of coverage (which, in turn, result from the availability of sources), the results show that the P_n velocity in the uppermost mantle ranges between 7.8 and 8.3 km/sec. Whether this variation is due to differences in composition, or to differences in condition (e.g., temperature) is by no means yet established.

The fact that upper mantle velocity does show this variation has some bearing on the definition of the Mohorovičić discontinuity. It may be defined as that discontinuity in seismic velocities at which the *P* wave velocity first reaches a value of 7.8 km/sec or greater, but we must realize that, if there are regions in which the crust and mantle have become intermixed, the definition (as well as the refraction method) becomes difficult to apply.

Interest in the possible anisotropy of the P_n velocity arose from an observation of Hess (1964), that in the northeastern Pacific the available seismic profiles gave higher velocities for the east-west than north-south directions. Observations by Raitt et al. (1969), made expressly for the determination of aniso-

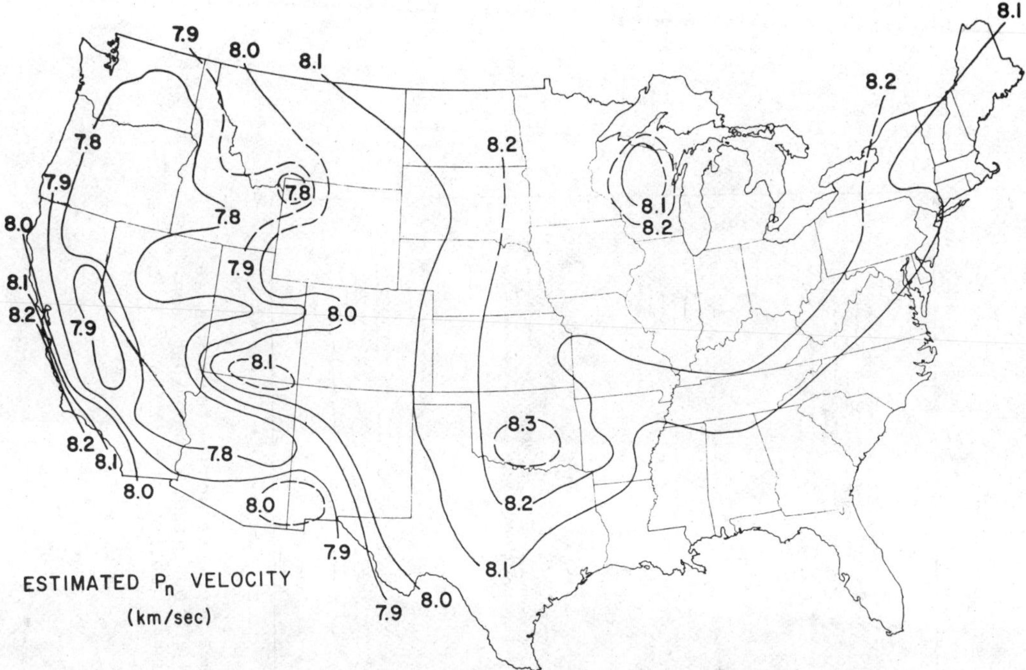

Figure 7.10 Variation in P-wave velocity in the upper mantle (P_n) beneath the United States. (Courtesy of Eugene Herrin and the American Geophysical Union)

tropy by the time-term method, confirmed that the east-west velocity is 8.6 km/sec, contrasted with a velocity of 8.0 km/sec in the north-south direction. Similar, but smaller, anisotropy was found in the vicinity of the Mid-Atlantic Ridge at 45°N by Keen and Tramontini (1970). There, the difference between maximum and minimum directions was 0.25 km/sec. In both cases, the direction of maximum velocity was approximately normal to the nearest oceanic ridge. Hypotheses that relate this anisotropy of the uppermost mantle to tectonics will be discussed in Chapter 27.

7.5 REFLECTIONS FROM CRUSTAL DISCONTINUITIES

The success of the use of near-vertical reflections in the seismic exploration of sedimentary layers by the geophysical prospector causes one to inquire into their use in the study of the crust. Both the hypothetical arrival time curve (Fig. 7.2) and the example (Fig. 7.6) show the locus of arrival times of predicted or observed reflected energy. The fact is that reflections are frequently observed at distances corresponding to, or just less than, the first appearance of the head wave. However, observations of reflections at very short distances from the source are much less frequent. Clowes, Kanasewich, and Cumming (1968) have detected deep crustal reflections on records which have been numerically filtered (Fig. 7.11) to enhance the signal-to-noise ratio. While the strongest reflections came from a discontinuity within the crust, weak events corresponding to reflections from the Mohorovičić discontinuity were observed. Near-vertical-incidence reflections have not yet been observed at enough points to

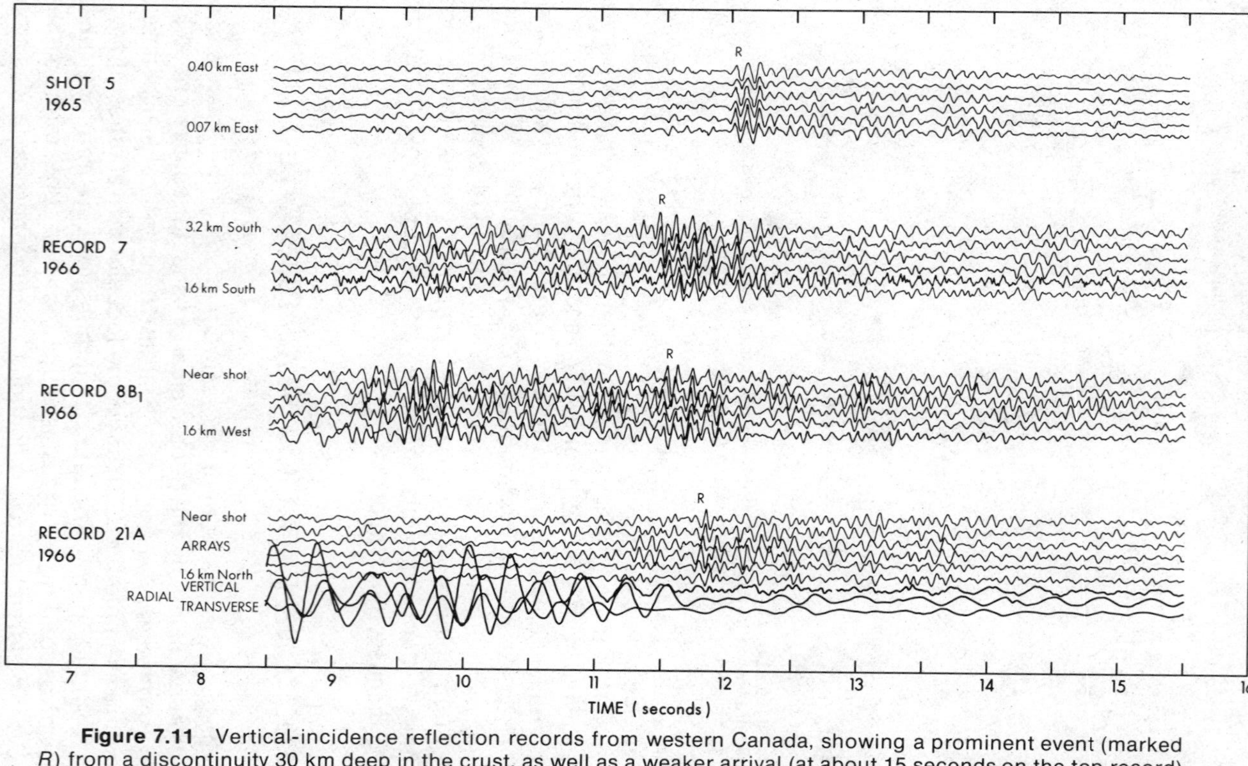

Figure 7.11 Vertical-incidence reflection records from western Canada, showing a prominent event (marked *R*) from a discontinuity 30 km deep in the crust, as well as a weaker arrival (at about 15 seconds on the top record) which has been interpreted as a reflection from the Mohorovičić discontinuity. (Courtesy of R. M. Clowes, E. R. Kanasewich, and G. L. Cumming)

add greatly to the mapping of the Mohorovičić discontinuity, but the fact that they exist implies that the discontinuity is sharp. In fact, Clowes, Kanasewich, and Cumming conclude that, if there is a gradational change in velocity at the base of the crust, the change must be accomplished within 0.5 km; otherwise, a reflection, even of the magnitude observed, would not have been produced.

7.6 THE EARTH AS A FILTER

The history of body-wave seismology has been one of successively extracting more information from the seismogram. Thus, in the earlier years, the time of arrival of phases was of prime importance. Later, detailed studies of the slope, $dT/d\Delta$, of the arrival-time curve became possible. To extract all of the information from the seismogram requires that consideration be given to amplitude also. This in turn requires that the source be considered, and that waves be not treated simply as an infinite periodic train.

The exploration geophysicist has examined the same problem in detail (Kanasewich, 1973; Claerbout, 1976), and has brought to an advanced stage the calculation of synthetic seismograms, which can be compared to the field records. In these calculations it is extremely convenient to consider the earth as a filter placed between the seismic source and the observing station. A linear filter is normally characterized by its time-domain response function, $g(t)$, to a unit "spike" impulse $\delta(t)$, where

$$\delta(t) = \infty, \qquad t = 0$$
$$= 0, \qquad t \neq 0$$

and

$$\int_{-\infty}^{\infty} \delta(t)\, dt = 1$$

The output of the filter to any other input $f(t)$ is then given by

$$h(t) = \int_0^t f(\tau) g(t-\tau)\, d\tau \qquad 7.6.1$$

The integral operation of Equation 7.6.1 is known as *convolution,* which is often represented formally as

$$h(t) = f(t) * g(t) \qquad 7.6.2$$

In geophysics, the output function $h(t)$ is given by the seismogram. The source function $f(t)$ may be known or assumed, but it is generally unknown. Properties of the earth are contained in the impulse response $g(t)$; these could include geometrical spreading of the energy, nonelastic attenuation, and the introduction of reflected or critically refracted phases at given times, with amplitudes given by the methods of Section 3.2 and Appendix C. If $g(t)$ can be extracted from the seismogram, it provides a clue to earth structure; reflecting horizons, for example, simply introduce spikes, of height proportional to the reflection coefficient, at discrete times. Conversely, one can compute synthetic seismograms by computing $g(t)$ for a given model, and performing the convolution for an assumed source shape.

The inverse operation, to obtain $g(t)$, is known as *deconvolution.* Deconvolution is most conveniently represented, and carried out, in the complex frequency domain, by taking the Laplace transform, defined by

$$F(i\omega) = \int_{-\infty}^{\infty} f(t)\, e^{-i\omega t}\, dt \qquad 7.6.3$$

Then convolution is achieved by the multiplication of transforms:

$$H(i\omega) = F(i\omega)G(i\omega) \qquad 7.6.4$$

and deconvolution by division:

$$G(i\omega) = \frac{H(i\omega)}{F(i\omega)} \qquad 7.6.5$$

The operations in practice are complicated by the presence of noise on the seismogram, and by the fact that functions such as $h(t)$ are normally approximated by digital values at discrete times. Applications of the approach to problems of exploration geophysics, where the nature of the source function or "wavelet" is well known, have been described by Sengbush, Lawrence, and McDonal (1961), Rice, (1962), and Robinson and Treitel (1964). Applications to large-scale problems have been few, partly because of the complexity of the earthquake source. However, Julian and Anderson (1968) have included amplitude calculations in studies of model earths, and Helmberger and Wiggins (1971) have calculated synthetic seismograms to assist in the interpretation of upper mantle structure beneath North America. The latter authors, working with seismograms from large explosions in Nevada, assumed that the waves recorded at the largest epicentral distances had passed beneath the complex region of the upper mantle, and they used these arrivals to determine the source function.

Problems

1. Show that the refraction calculations of Section 7.2 can be extended to any number of horizontal layers, with the intercept of the straight line representing arrivals from a wave critically refracted in the nth layer given by

$$(T_i)_n = 2 \sum_{1}^{n-1} \frac{h_j(V_n^2 - V_j^2)^{1/2}}{V_j V_n}$$

where h_j and v_j are the thickness and velocity of the jth layer.

2. The ray diagram of Figure C3 shows various converted head waves, such as *PSS* and *PPS*. Why are phases *SSP* or *PSP* not shown?

3. In a particular continental area, the crust is believed to consist of an upper layer, of thickness 25 km and *P*-wave velocity 6.0 km/sec, overlying a lower layer of thickness 15 km and *P*-wave velocity 7.0 km/sec. The upper-mantle *P*-wave velocity is about 8.2 km/sec. By calculating arrival-time curves for refracted arrivals from a surface explosion, show whether or not the head-wave from the lower layer can be ex-

pected as a first arrival. Deduce the best distances from the source to place detectors if critical-angle reflections from the interfaces are to be observed. Finally, estimate the total length of the line of detectors that would be required to define the upper-mantle velocity to within 0.05 km/sec on the basis of first arrivals, if times can be read to 0.01 seconds and distances are known within negligible error.

4. Another region of the crust consists of an upper layer, 20 km thick, with *P*-wave velocity 6.0 km/sec, underlain by a layer 8 km thick, with velocity 5.6 km/sec. Beneath the latter layer is mantle with velocity 8.0 km/sec. (a) Plot to scale the ray paths for all critically refracted waves. (b) What error would be made in estimating the depth to the mantle from a time-distance plot? (c) Compare the predicted minimum distances, for a mantle head wave arrival, in the real situation and in the erroneous interpretation.

5. A long, unreversed refraction profile in a continental area gave arrival-time observations as follows:

Distance, km	Time, sec	Distance, km	Time, sec
32	6.3	245	39.4
	8.3		39.7
70	12.5	290	44.4
130	22.2		45.4
	22.4		46.3
	23.5	350	51.5
175	29.1		54.1
	29.4		57.8
	30.0	430	61.7
	30.4		71.0
	30.6		
215	35.1		
	35.9		
	36.0		
	36.7		

Times were read to 0.05 sec, and are quoted to the nearest 0.1 sec. Using all of the information available, give an interpretation in terms of crustal structure beneath the profile. Would you modify your interpretation if you plotted only the first arrivals at each station?

BIBLIOGRAPHY

Aumento, F., Ade-Hall, J. M., and Keen, M. J. (1975) 1974—The year of the Mid-Atlantic Ridge. *Rev. Geophys. Space Physics,* **13**, 53–65.

Backus, D. L. (1965) Possible forms of seismic anisotropy of the uppermost mantle under oceans. *J. Geophys. Res.,* **70**, 3429.

Barrett, D. L., and Aumento, F. (1970) The Mid-Atlantic Ridge near 45N. XI. Seismic velocity, density, and layering of the crust. *Can. J. Earth Sci.,* **7**, 1117.

Bellaiche, G., Cheminee, J. C., Francheteau, J., Hekinian, R., Le Pichon, X., Needham, H. D., and Ballard, R. D. (1974) Rift Valley's inner floor: first submersible study. *Nature,* **250**, 558–560.

Berry, M. J., and Fuchs, K. (1973) Crustal structure of the Superior and Grenville provinces of the northeastern Canadian shield. *Bull. Seism. Soc. Amer.,* **63**, 1393–1432.

Berry, M. J., and West, G. F. (1966) An interpretation of the first arrival data of the Lake Superior experiment by the time-term method. *Bull. Seism. Soc. Amer.,* **56**, 141.

Cagniard, L. (1939) *Réflexion et réfraction des ondes séismiques progressives.* Gauthier-Villars & Cie., Paris.

Cerveny, V., and Ravindra, R. (1971) *Theory of seismic head waves.* University of Toronto Press, Toronto.

Claerbout, J. F. (1976) *Fundamentals of geophysical data processing with applications to petroleum prospecting.* McGraw-Hill, New York.

Clowes, R. M., Kanasewich, E. R., and Cumming, G. L. (1968) Deep crustal reflections at near-vertical incidence. *Geophysics,* **33**, 441.

Conrad, V. (1925) Laufzeitkurven des Tauernbebens vom 28 November 1923. *Mit der Erdbeben-Kommission der Wiener Akad. der Wissensch.,* **59**.

Dobrin, M. B. (1976) *Introduction to geophysical prospecting.* McGraw-Hill, New York.

Ewing, J., and Ewing, M. (1959) Seismic refraction measurements in the Atlantic

Ocean basins, in the Mediterranean Sea, on the Mid-Atlantic ridge and in the Norwegian Sea. *Bull. Geol. Soc. Amer.*, **70**, 291.

Ewing, M., Jardetzky, W., and Press, F. (1957) *Elastic waves in layered media.* McGraw-Hill, New York.

Heelan, P. A. (1953) On the theory of head waves. *Geophysics,* **18**, 871.

Helmberger, Donald, and Wiggins, Ralph A. (1971) Upper mantle structure of midwestern United States. *J. Geophys. Res.*, **76**.

Herrin, E., and Taggart, James (1962) Regional variations in P_n velocity and their effect on the location of epicenters. *Bull. Seism. Soc. Amer.*, **52**, 1037.

Hersey, J. B. (1963) Continuous reflection profiling. In *The sea* (Vol. 3) ed. M. N. Hill. Interscience, New York and London.

Hess, H. (1964) Seismic anisotropy of the uppermost mantle under oceans. *Nature,* **203**, 629.

Hill, M. N. (1963) Single-ship seismic refraction shooting. In *The sea* (Vol. 3), ed. M. N. Hill. Interscience, New York and London.

Jackson, W. H., Stewart, S. W., and Pakiser, L. C. (1963) Crustal structure in eastern Colorado from seismic-refraction measurements. *J. Geophys. Res.*, **68**, 5767.

Jeffreys, H. (1937) On the materials and density of the earth's crust. *Mon. Not. R.A.S., Geophys. Suppl.*, **4**, 50.

Julian, Bruce R., and Anderson, Don L. (1968) Travel times, apparent velocities and amplitudes of body waves. *Bull. Seism. Soc. Amer.*, **58**, 339.

Kanasewich, E. R. (1973) *Time sequence analysis in geophysics.* University of Alberta Press, Edmonton.

Keen, C., and Tramontini. C. (1970) A seismic refraction survey on the Mid-Atlantic Ridge. *Geophys. J.*, **20**, 473.

Kosminskaya, I. P., Belyaevsky, N. A., and Volvovsky, I. S. (1969) Explosion seismology in the USSR. In *The earth's crust and upper mantle,* ed. Pembroke J. Hart. *Amer. Geophys. Un., Monograph 13,* 195.

McConnell, Robert K. Jr., and McTaggart-Cowan, Gillian H. (1963) Crustal seismic refraction profiles, a compilation. *Scientific Report 8.* University of Toronto, Institute of Earth Sciences.

Mohorovičić, A. (1910) Das Beben vom 8, X, 1909. *Jahrb. des Meteorologischen Observ.* (Zagreb), **9**.

Pakiser, L. C., and Zietz, Isidore (1965) Transcontinental crustal and upper-mantle structure. *Rev. Geophys.*, **3**, 505.

Raitt, R. W. (1963) The crustal rocks. In *The sea* (Vol. 3), ed. M. N. Hill. Interscience, New York and London.

Raitt, R. W., Shor, G. G., Francis, T. J. G., and Morris, G. B. (1969) Anisotropy of the Pacific upper mantle. *J. Geophys. Res.*, **74**, 3095.

Rice, R. B. (1962) Inverse convolution filters. *Geophysics,* **27**, 4.

Robinson, E. A., and Treitel, S. (1964) Principles of digital filtering. *Geophysics,* **29**, 395.

Scheidegger, A. E., and Willmore, P. L. (1957) The use of a least-squares method for the interpretation of data from seismic surveys. *Geophysics,* **22**, 9.

Shor, G. G., Jr. (1963) Refraction and reflection techniques and procedure. In *The sea* (Vol. 3), ed. M. N. Hill. Interscience, New York and London.

Sengbush, R. L., Lawrence, P. L., and McDonal, F. J. (1961) Interpretation of synthetic seismograms. *Geophysics,* **26**, 138.

Steinhart, J. S., and Meyer, R. P. (1961) Explosion studies of continental structure. Carnegie Inst. of Washington, Publication 622.

Stoneley, R. (1924) The elastic waves at the surface of separation of two solids. *Proc. Roy. Soc. A,* **106**, 416.

FREE OSCILLATIONS OF THE EARTH 8

8.1 A NEW BRANCH OF SEISMOLOGY

One of the major advances in solid-earth geophysics came with the suggestion by Benioff (1954) that, on the record of the Kamchatka earthquake of 1952 as written by a strain seismometer (Section 5.4), there was an oscillation with a period of 57 minutes. Such a period would be much too great for it to be associated with a traveling elastic wave, and Benioff suggested that it represented a standing wave, or *free oscillation* of the earth. The free oscillations of a uniform elastic sphere had been studied many years before, and it was known that periods near that observed by Benioff were associated with certain modes.

The interpretation of the Kamchatka records was not unambiguous, but in the years since 1954 oscillations corresponding to a number of modes of vibration have been identified with certainty. It is usually agreed that the first records which clearly showed free oscillations were obtained from the large Chilean earthquake of 1960. Methods have now been developed for the calculation of the theoretical periods of oscillation for earth models, so that the observations may be used to check models of the earth based on body-wave seismology. In addition, the damping of the free oscillations can be observed, and this permits the non-elasticity of the earth to be determined. The use which has been made of a type of observation that was unavailable before 1960 is remarkable.

Let us begin by recalling the relationship between traveling waves and standing waves in a very simple one-dimensional case, say that of waves on a string. The deflection of the string due to a traveling wave is represented by

$$y = A \sin (\omega t - kx)$$

where A is amplitude, ω is the angular frequency, and k is 2π/wavelength. If reflection at the end of the string produces a wave of equal amplitude traveling in the opposite sense, the net deflection is

$$y = A \sin (\omega t - kx) \pm A \sin (\omega t + kx)$$

where the choice of sign ($\pm$) depends on whether or not there is a change of phase upon reflection. The expression for y can be rewritten as

$$y = 2A \begin{matrix}\sin\\ \cos\end{matrix} \omega t \begin{matrix}\cos\\ \sin\end{matrix} kx$$

in which the time and space variables are separated. The deflection is now represented by a standing wave, in which the points of zero deflection (nodes) and maximum deflection (antinodes) remain fixed in position. The wavelength of the interfering waves which produced the pattern is twice the distance between adjacent nodes. A string fixed at both ends vibrates, in its lowest mode or fundamental, with a node at each end, so that the wavelength is twice the length of the string; an infinite series of higher modes, or overtones, is possible, the only condition being that there remain a node at each end.

With this review, we turn to the earth and suppose that there exists some type of standing distortion in shear, with an antinode at each end of the diameter. What would be a rough estimate of the period of this oscillation? If we take the wavelength λ to be twice the earth's diameter and the mean value of β in the earth to be 6 km/sec, then from

$$f\lambda = \text{velocity}$$

we have

$$f = \frac{6}{2 \times 2 \times 6360} = 2.5 \times 10^{-4} \text{ sec}^{-1}$$

corresponding to a period of 4000 seconds. The analysis is over-simplified, because in spherical geometry we cannot take the wavelength as simply twice the diameter. However, it does indicate that if standing waves near the fundamental are to be found, the periods involved must be much greater than those dealt with in conventional seismology.

8.2 PATTERNS OF FREE OSCILLATIONS

The mathematical treatment of free oscillations, even for a stationary, uniform sphere, is involved; the procedure is outlined in Appendix C. Briefly, the results show that two kinds of standing waves are possible (Fig. 8.1): those with a radial component of particle displacement, known as coupled or spheroidal oscillations (denoted by S) and those with no radial component, known as toroidal oscillations (T). The geometry of the displacement field over the surface of the earth, expressible in spherical harmonics (Appendix A), is described by the *degree* l in latitude, and *order* m in longitude, which represent it (Fig. A2). In addition, the theory indicates that an oscillation with given (l, m) can, in general, be present as a *fundamental*, or as *overtones* with increasingly complex behavior in radius. The full designation of a free oscillation is therefore ${}_nS_l{}^m$ or ${}_nT_l{}^m$, where the subscript n gives the number of surfaces (along a radius) of null displacement. Theory permits the frequency of oscillation of any mode to be computed, not only for a uniform sphere but, most important for interpretation, for earth models of variable density and seismic wave velocity.

The numerical values obtained in Appendix C for the ratios of period to travel time of P or S waves in a uniform sphere may appear to be surprisingly non-integral. If free oscillations represent the interference of long wavelength traveling waves, why are not simple ratios obtained for the first few modes? One reason is that the wavelength of the interfering waves in the sphere is not what it may appear to be. Matomoto and Sato (1954) have pointed out that the

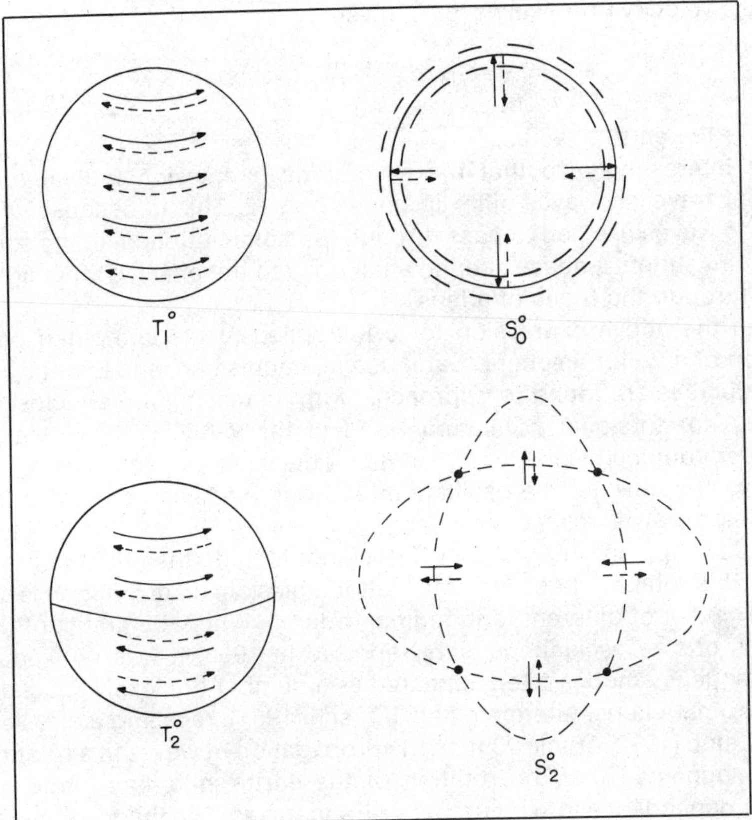

Figure 8.1 Simplest modes of oscillation of a sphere.

standing wave–traveling wave relationship of free oscillations may be readily seen by introducing an asymptotic form for the spherical harmonic $P_l\,(\cos\theta)$, valid at large values of l. The standing wave displacement may be written as

$$y = \Sigma\, F(p,r)\; P_l\,(\cos\theta)\;\exp ipt \qquad 8.2.1$$

where $F(p,r)$ is some function of radius r and angular frequency p. The asymptotic expression for $P_l\,(\cos\theta)$, due to Laplace (Appendix A), is

$$P_l\,(\cos\theta) = \left[\frac{2}{(l+{}^1\!/\!_2)\;\pi\sin\theta}\right]^{1/2} \sin\left[(l+{}^1\!/\!_2)\;\theta + \frac{\pi}{4}\right] \qquad 8.2.2$$

and when this is inserted in the expression for y, we can write

$$y = \Sigma\, F(p,r)\left[\frac{2}{(l+{}^1\!/\!_2)\;\pi\sin\theta}\right]^{1/2}\left\{\exp i\left[pt - (l+{}^1\!/\!_2)\;\theta + \frac{\pi}{4}\right] + \exp i\left[pt + (l+{}^1\!/\!_2)\;\theta - \frac{\pi}{4}\right]\right\} \qquad 8.2.3$$

This shows that, when l is sufficiently large for Laplace's formula to be valid, the pattern is produced by waves, traveling in the $\pm\theta$ directions, of wavelength

$$\lambda = 2\pi a/(l + {}^1\!/\!_2)$$

The phase velocity of these waves is given by

$$c = p\lambda/2\pi = ap/(l + {}^1\!/_2)$$

where a is the earth's radius.

It is interesting also that the asymptotic expression indicates that the equivalent traveling waves differ in phase by $\pi/2$. This is because one of the interfering waves has had to pass through the point antipodal to the source and at this point suffers a phase change analogous to the phase shift in light waves passing through the focus of a lens.

When the above expression for c is applied to the computed periods of oscillation of a uniform sphere (Sato, Usami, Landisman, and Ewing, 1963), the phase velocities are found to approach—with increasing l—very closely to the value of β for torsional oscillations, and to the value of the Rayleigh wave velocity for coupled oscillations (Fig. 8.2). This is an elegant demonstration of the equivalence of the free oscillations to an interference pattern of S-waves and Rayleigh waves, respectively.

The development in Appendix C does not include the effect of gravity, nor of the earth's rotation; because of the latter omission, no distinction is apparent between modes of different longitudinal order m. When the first confirmed observations of free oscillations were reported in 1960, it was noticed that the lowest frequency modes often appeared as a number of closely spaced lines in the spectrum. Pekeris, Alterman, and Jarosch (1961), recalling a classical treatment by Lamb (1932, Article 209) on the propagation of waves in a rotating basin of fluid, suggested that the rotation of the earth splits any mode having a longitude dependence into a number of frequencies. The theory of this splitting has been developed by several other authors (Backus and Gilbert, 1961), but is too complex to treat in detail here.

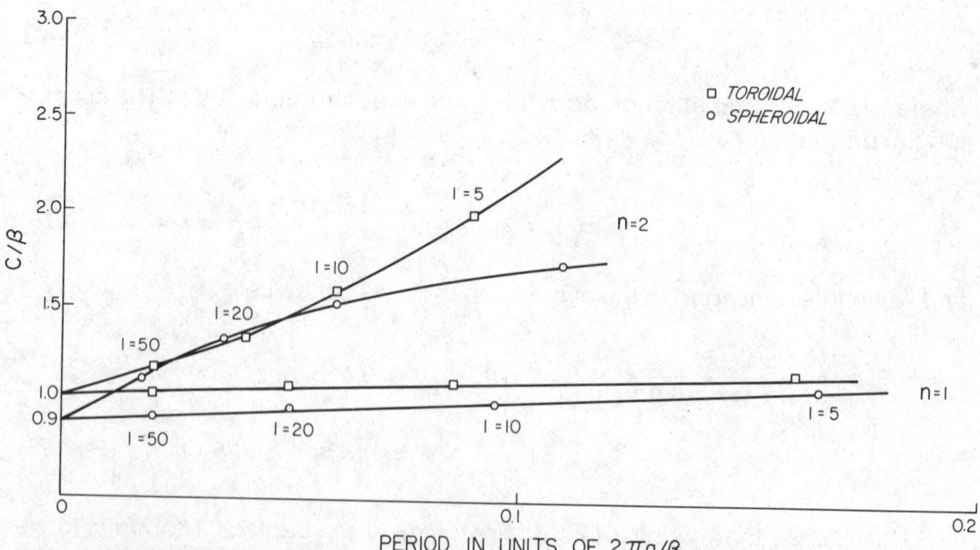

Figure 8.2 Convergence of the phase velocities of toroidal and spheroidal free oscillations in a uniform sphere, with increasing degree l, to the S-wave velocity and the Rayleigh wave velocity respectively. Parameter n indicates fundamental and first overtone in radial variation. (After Matumoto and Sato 1954).

Briefly, the effect of rotation can be taken into account by adding to the equation of particle motion terms corresponding to the centrifugal and Coriolis accelerations. The result of the additional terms is to produce coupling between the components of displacement, so that particles no longer oscillate in straight lines. Furthermore, the terms contribute in such a way that the frequency of oscillation in a traveling wave is less for waves traveling in the sense of, rather than opposed to, the rotation. This is the result obtained by Lamb for waves in the rotating basin.

Free oscillations of the earth which have a longitude dependence given by m can be considered to be the standing wave interference pattern produced by two waves, of equal wavelength, one traveling eastward (negative m) and one westward (positive m). Since the frequencies are different, the phase velocities of these two waves are different, and the "standing" wave pattern moves slowly westward. An observer at any point on the earth records two different frequencies for each value of m, but since a mode of degree l contains contributions from l values of m, he could in theory observe $2l + 1$ closely spaced frequencies. For toroidal oscillations, the frequency interval between the components is (Gilbert and Backus, 1965)

$$\left[\frac{m}{l(l+1)}\right]\omega$$

where ω is the earth's angular velocity of rotation.

A mode such as ${}_0T_2$, which for $m = 0$ has a period of 42.3 minutes, is split into five lines with periods 41.9, 42.1, 42.3, 42.5, and 42.7 minutes. As l increases, the shift in frequency rapidly becomes very small, and individual lines cannot be resolved. Splitting of the frequencies of oscillation by the earth's rotation has been called the seismological analogue of the Zeeman effect of particle physics.

The close relation between free oscillations and traveling waves is an important concept. Just as the former can be considered as interference patterns of the latter, so can traveling waves be represented as the superposition of a number of free oscillations (Landisman, Usami, Sato, and Massé, 1970).

We have indicated that, from the computed periods of free oscillation for a model, the phase velocity of the interfering traveling waves can be computed. This in turn allows the calculation of the group velocity of a signal to be obtained, and if the time variation of a source function is adopted, the displacement as a function of time at a distant point (i.e., the theoretical seismogram) may be constructed. The procedure is similar to that outlined in Section 7.6 for the calculation of theoretical seismograms for the crust and upper mantle, where the seismogram was considered to be the convolution of the source and a filter (the earth). Periods of free oscillations of an earth model are the characteristic periods of the whole earth as a filter, and a very satisfactory method of computing theoretical seismograms for waves involving the whole earth begins with the periods of free oscillations (Landisman, Usami, Sato, and Massé, 1970).

We show, in Figure 8.3, one very beautiful example of theoretical seismograms computed by Sato, Usami, Landisman, and Ewing (1963). The model adopted is a sphere with uniform solid mantle over a fluid core, with a time-varying torsional stress applied around the pole. The seismograms are computed from the periods of torsional oscillations and show the expected arrivals, at increasing distance, of transverse waves. Notice, first, how the resemblance

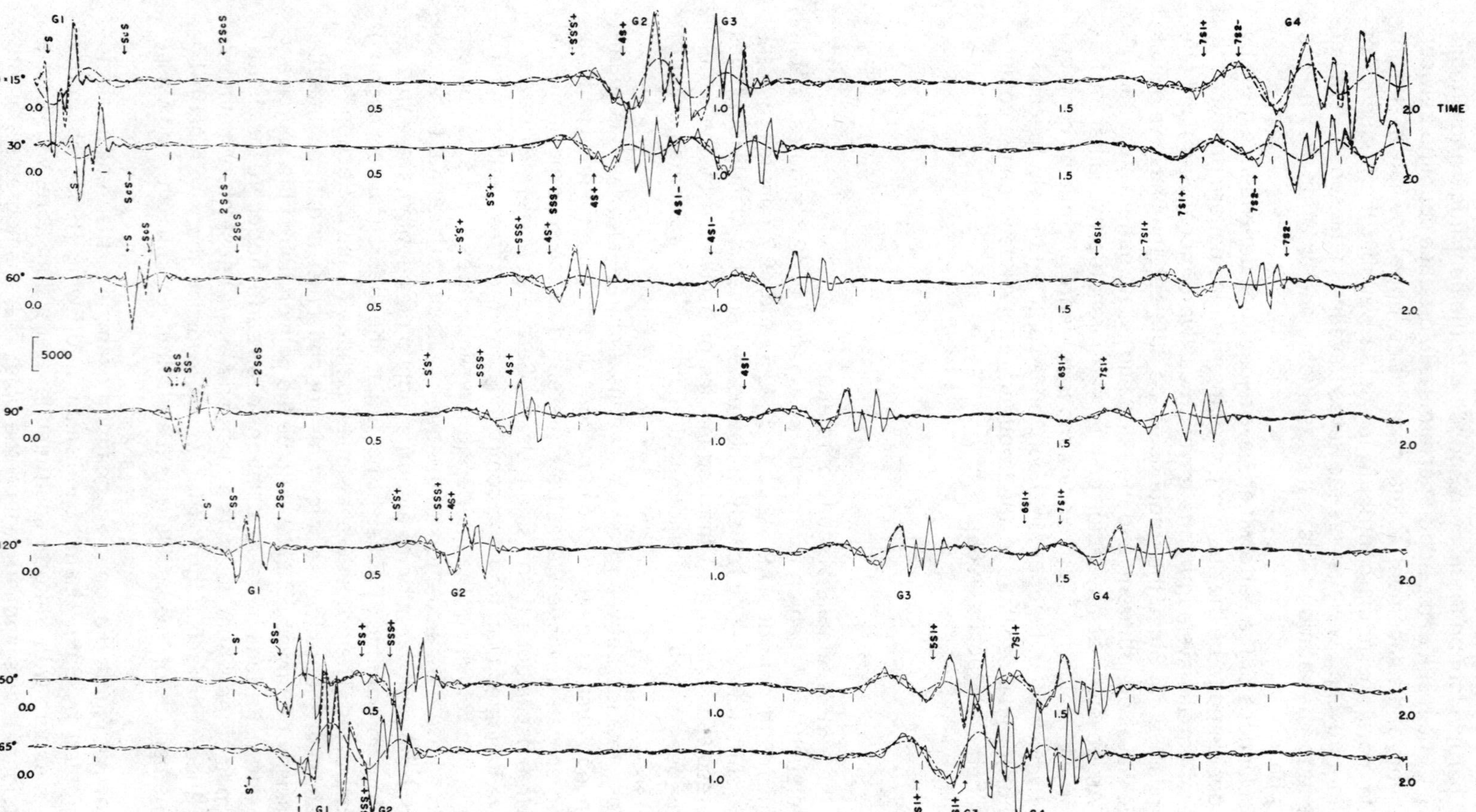

Figure 8.3 Theoretical seismograms at increasing arc distance θ, of displacement produced by a torsional stress around the pole, in a homogeneous sphere containing a liquid core. Displacements are synthesized from computed free oscillations, to ($n = 6$, $l = 60$) in solid lines, ($n = 1$, $l = 60$) in broken lines, ($n = 1$, $l = 10$) in chain lines. Arrows show arrival times of phases computed by ray theory. (From Sato et al 1963; courtesy of the authors and the *Geophysical Journal*)

to an actual seismogram increases as higher modes (up to $l = 60$) are included. Very obvious on the seismograms are the trains of long-period Love waves known as *G*-waves (after Gutenberg), with trains of alternate order circling the sphere in opposite senses. The presence of the liquid core is shown in two obvious ways: by the arrival of the reflected wave S_cS, and by the disappearance of *S* at the beginning of the shadow zone, between arc distances of 90° and 120°. A weak diffracted wave, marked *S'*, continues into the shadow zone. Altogether the figure provides a convincing demonstration of the equivalent nature of free oscillations and body waves.

8.3 THE OBSERVATIONS AND THEIR INTERPRETATION

Toroidal oscillations are, of course, observable only on horizontal component instruments. Since the periods range up to several minutes, strain seismometers are most suitable for their detection, but the torsional oscillations excited by great earthquakes have been observed on the records of pendulum seismometers (Alsop, Sutton, and Ewing, 1961). Spheroidal or coupled oscillations involve a radial displacement, and are most conveniently observed on long-period vertical seismometers. It has developed that among the most suitable instruments for their detection are the recording gravimeters set up to observe earth tides (Chapter 14), from which a number of results have been given by Slichter (1967).

As in the case of so many time-variable geophysical phenomena, the treatment of the records to obtain true power spectra is of critical importance. Slichter (1967) has presented spectra (Fig. 8.4) of the spheroidal oscillations from two major earthquakes, which show remarkable agreement in the location of the major peaks.

The splitting produced by rotation of the earth, if not resolved, produces a broadening of the spectral lines, which makes the determination of the central frequency much more difficult. In observations to date, there appears to have been more success in resolving the split frequencies of spheroidal oscillations, on gravimeter records, than those of toroidal observations.

The application of the results of free oscillation observations to the study of the earth has been through the comparison of the observed periods with those

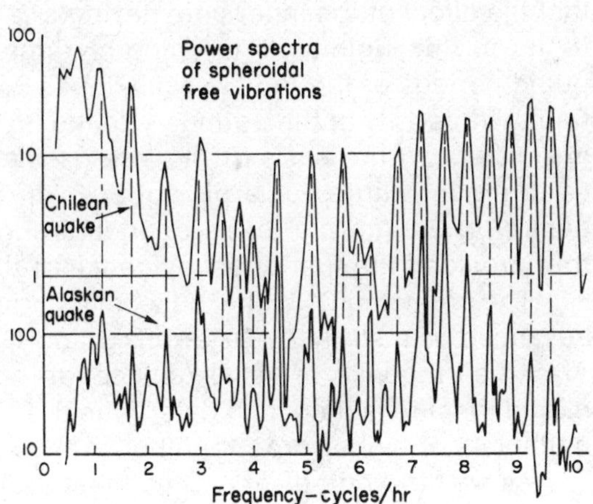

Figure 8.4 Spectra of gravimeter recordings of coupled oscillations. (Courtesy of L. B. Slichter and Pergamon Press)

computed for the different modes from model earths. The simplified treatment of Appendix C for a uniform sphere may be extended to an earth composed of spherical shells of differing properties, and methods of calculation of the periods have been given by several authors (Pekeris, Alterman, and Zarosch, 1961; MacDonald and Ness, 1961; Landisman, Sato, and Nafe, 1965). The effect of body forces must be included in these calculations, which are required to yield precise periods. For realistic models containing a fluid outer core, oscillations of torsional type are confined to the crust and mantle.

A very complete examination of the effect of varying earth parameters on the free periods was carried out by Haddon and Bullen (1969). These authors produced an earth model (shown in Fig. 3.12), defined by the radial variation of density and the seismic velocities, which yielded very close agreement with the observed free periods for all modes which had then been observed with reasonable certainty, and was comparable with most of the criteria used in previous models. The most striking departure of the model from those based on body-wave seismology was in the radius of the core, for which an increase of 20 km was proposed. As we have mentioned, the times of P_cP appeared to determine the core radius to within a very few kilometers, and the increase suggested by the free oscillation results was not at first expected.

However, the periods of free oscillations have been included with P and S travel times and the mass and moment of inertia, in a computer-based "Monte Carlo" test of randomly selected earth models, designed by Press (1968). The availability of high-speed computing reduces the dependence upon the direct inversion of geophysical data, in favor of the assumption of models and the direct test of their properties. Thus, out of five million models chosen at random and defined by radial variation of density, α and β, only those which passed the observational material listed above were considered acceptable. The limits on radial variation of density and velocities which were obtained suggested that the accepted curves, which have been shown in previous chapters, cannot be greatly in error. However, all successful models required an increase in the radius of the core of 5 to 20 km from the previously accepted value, and this was in agreement with the proposal of Haddon and Bullen. Free oscillations, being global phenomena, cannot provide the detail of body-wave studies, but they may yield better average values for certain parameters of the earth, simply because of the non-random distribution of the sources of body waves.

Other properties of the core and inner core may also be more readily determined from free oscillations than from body waves. Derr (1969) has shown that the effect of the inner core having a value of β of 3 km/sec, as contrasted to its having no rigidity, is to smoothly increase, by several seconds, the computed periods of the radial oscillations ${}_0S_0$, ${}_1S_0$, and ${}_2S_0$. The effect of a discontinuity in density at the boundary of the inner core is in the opposite sense, but is smaller. On the basis of the free oscillation periods, Derr has proposed an earth model with a value of β of 2.18 km/sec in the inner core, and a jump in density of 2.0 gm/cm^3 at its boundary (Fig. 8.5). Free oscillations have therefore provided the first direct evidence of the solidity of the inner core.

The refinement in the observed periods of free oscillations was greatly advanced in a study by Dziewonski and Gilbert (1972) of 84 recordings of the Alaska earthquake of 1964. Application of advanced filtering techniques permitted them to isolate, and determine the period of, 103 modes, including many multiplets of higher order number.

As we have noted earlier, the most satisfactory approach to an earth model

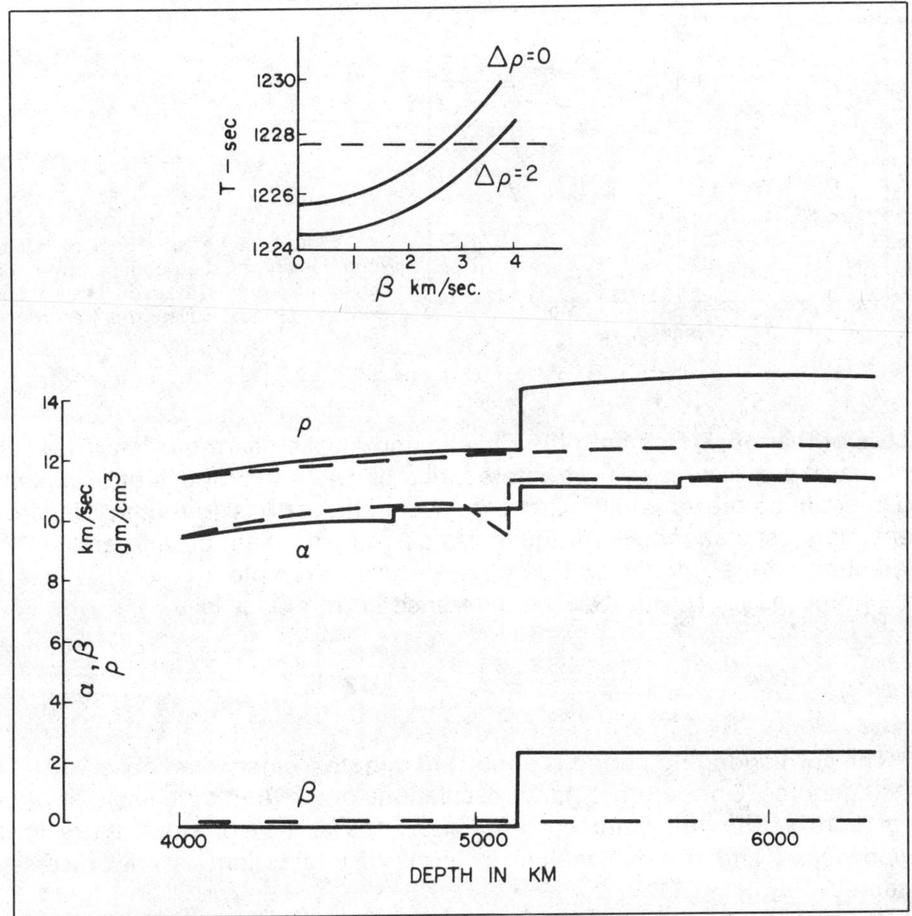

Figure 8.5 (*Top*) Variation of the period of $_0S_0$ with the value of β in the inner core and the jump in density at its boundary. The horizontal line shows the observed value (after Derr). (*Bottom*) Comparison of the inner core properties in the models of Haddon and Bullen (broken lines) and Derr (solid lines).

is now through the combined application of body-wave travel times, surface-wave dispersion, and free oscillation periods. This approach has been employed by Jordan and Anderson (1974) in a most thorough study of all available information. Starting with an earth model similar to that of Haddon and Bullen, but with the solid inner core of Derr, they adjusted the velocities and density to provide the optimum agreement with the three types of seismological information, including the free oscillation periods of Dziewonski and Gilbert mentioned above. Their final model is shown in Figure 8.6. A comparison between observed free oscillation periods and those computed for the model, for only a few of the modes used, is shown in Table 8.1. Noteworthy in the model is the minimum in V_s (β) in the upper mantle, although this was not included in the starting model. The radius of the core in the model is 3485 km. For the main features of a spherically symmetrical earth, the model of Jordan and Anderson is probably the best now available.

Oscillation frequencies have been reported (Bozzi Zadro and Caputo, 1968) which do not correspond to the theoretical frequency of any mode of the earth. In particular, Bozzi Zadro and Caputo list oscillations with periods in

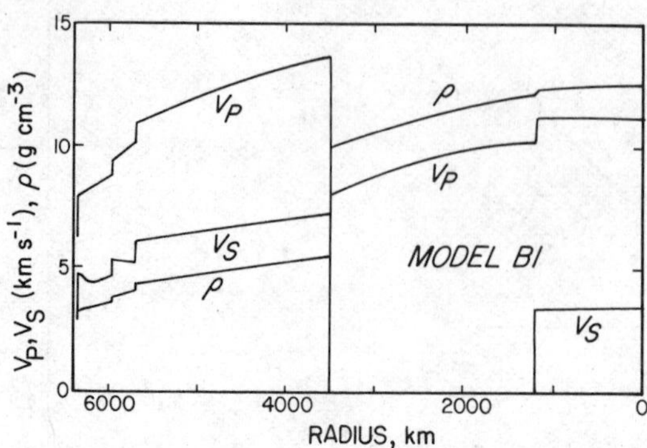

Figure 8.6 Final model of velocity and density variation obtained by Jordan and Anderson (1974) on the basis of body waves, surface waves and free oscillations.

excess of that of ${}_0S_2$, for both the Chilean and Alaskan earthquakes. They suggest that these may arise from interactions between the modes of oscillation, which could be present only if the response of the earth is non-linear. If there is interaction, oscillations with frequencies given by the sum or difference of the interacting natural modes will be observed. For example, one of the interactions proposed is ${}_0T_4$ with ${}_0S_4$. The difference in frequency is

$$\frac{1}{1305} - \frac{1}{1547} = 0.00012 \text{ Hz}$$

The corresponding period is about 140 minutes, close to an observed peak at 143 minutes. On the other hand, oscillations of the fluid core itself, or of the inner core within the outer core (Slichter, 1961), may produce these long-period peaks, and the explanation in terms of interactions is not universally accepted.

The free oscillations excited by a major earthquake persist for several days (at least 5 days for the Chilean earthquake), but their amplitude diminishes because of imperfections in elasticity within the earth. If spectra are computed for successive time intervals following the excitation, the damping of each mode can be determined. In many ways the procedure is more straightforward than is the determination of damping for traveling waves, which involves comparisons of amplitude between different stations. We shall refer to the observations of the damping of the free oscillations in Chapter 27.

TABLE 8.1 FREE OSCILLATION PERIODS FOR MODEL OF JORDAN AND ANDERSON (1974)

Mode	Observed (sec.)	Computed (sec.)
${}_0S_0$	1227.65	1227.61
${}_0S_2$	3233.30	3232.45
${}_0S_4$	1547.30	1545.82
${}_0S_7$	811.45	812.24
${}_0S_{63}$	147.09	146.92
${}_0T_2$	2640.63	2630.81
${}_0T_4$	1305.45	1303.53
${}_0T_7$	819.31	817.76
${}_0T_{46}$	176.85	176.93
${}_2T_8$	343.46	343.48

Problems

*1. It is desired to demonstrate free oscillations of a sphere in the laboratory. A steel sphere, 20 cm in diameter, is available, and Young's modulus for the steel is known to be 20×10^{11} dynes/cm^2. Assuming that Poisson's ratio is 1/3, compute the periods of the first few spheroidal ($l = 0$) coupled oscillations. What experimental technique would you suggest to excite and observe the oscillations?

2. An important measure of the earth's elasticity is the quantity Q, to be defined in Section 27.5. One measure of Q is the ratio $\omega/\Delta\omega$, where ω is a resonant frequency of oscillation, and $\Delta\omega$ is the width of the spectral peak (i.e., the width at the half-maximum points). Values of Q given by the longest period free oscillations are at least 1400. Suppose, however, that the five peaks of ${}_0T_2$ (Sect. 8.2) were not resolved in a frequency analysis. What value of Q would be obtained from this inferior analysis?

BIBLIOGRAPHY

Backus, G. E., and Gilbert, F. (1961) The rotational splitting of the free oscillations of the earth. *Proc. Nat. Acad. Sci. Wash.*, **47**, 362.

Benioff, H. (1958) Long waves observed in the Kamchatka earthquake of November 4, 1952. *J. Geophys. Res.*, **63**.

Bozzi Zadro, M., and Caputo, M. (1968) Spectral, bispectral analysis and Q of the free oscillations of the earth. *Supplemento al Nuovo Cimento*, **6**, 67.

Derr, J. S. (1969) Internal structure of the earth inferred from free oscillations. *J. Geophys. Res.*, **74**, 5202.

Dziewonski, A., and Gilbert, F. (1972) Observations of normal modes from 84 recordings of the Alaska earthquake of 1964 March 28. *Geophys. J.*, **27**, 393–446.

Gilbert, F., and Backus, G. (1965) The rotational splitting of the free oscillation of the earth, 2. *Rev. Geophys.*, **3**, 1.

Haddon, R. A. W., and Bullen, K. E. (1969) An earth model incorporating free earth oscillation data. *Phys. Earth and Plan. Int.*, **2**, 35.

Jordan, Thomas H., and Anderson, Don L. (1974) Earth structure from free oscillations and travel times. *Geophys. J.*, **36**, 411–459.

Lamb, H. (1882) On the vibrations of an elastic sphere. *Lond. Math. Soc. Proc.*, **13**.

Lamb, H. (1932) *Hydrodynamics (6th ed.)* Cambridge University Press, Cambridge.

Landisman, M., Sato, Y., and Nafe, I. (1965) Free vibrations of the earth and properties of its deep interior regions, Part I: Density. *Geophys. J.*, **9**, 439.

Landisman, M., Usami, T., Sato, Y., and Massé, R. (1970) Contributions of theoretical seismograms to the study of modes, rays and the earth. *Rev. Geophys.*, **8**, 533.

MacDonald, Gordon, J. F., and Ness, Norman F. (1961) A study of the free oscillations of the earth. *J. Geophy. Res.*, **66**, 1865.

Matomoto, T., and Sato, Y. (1954) On the vibration of an elastic globe with one layer. Vibration of the first class. *Bull. Earthq. Res. Inst.*, **32**, 247–258.

Pekeris, C. L., Alterman, Z., and Zarosch, H. (1961) Comparison of theoretical with observed values of the periods of free oscillations of the earth. *Proc. Nat. Acad. Sci.*, **47**, 91.

Press, Frank (1968) Earth models obtained by Monte Carlo inversion. *J. Geophys. Res.*, **16**, 5223.

Sato, Y., Usami, T., Landisman, M., and Ewing, M. (1963) Basic study on the oscillation of a sphere. Part V: Propagation of torsional disturbances on a radially heterogeneous sphere—Case of a homogeneous mantle with liquid core. *Geophys. J.*, **8**, 44–63.

Slichter, L. B. (1961) The fundamental free mode of the Earth's inner core. *Proc. Nat. Acad. Sci. Wash.*, **47**, 186.

Slichter, L. B. (1967) Earth, free oscillations of, in *The international dictionary of geophysics,* ed. S. K. Runcorn. Pergamon, Oxford, 331.

9 THE NATURE OF GRAVITY

9.1 THE APPROACH

The attraction exerted by the earth on a mass outside of itself, which is a special case of universal gravitation, causes the mass to accelerate vertically downward. We denote the resulting acceleration by g, which is a quantity approximately equal to 980 cm/sec^2. However, small variations in this quantity result from special characteristics of the earth's shape and internal density distribution, and it is the aim in this branch of geophysics to measure, isolate, and interpret these variations in g.

Newton's law of gravitation, applied to a small mass m located outside a spherical earth of radius R, gives

$$g = \frac{F}{m} = \frac{GM}{R^2} \qquad 9.1.1$$

where F is the force of attraction, M the mass of the earth, and G is the constant of gravitation. Equation 9.1.1 illustrates that the measurement of g and G determine the mass, or mean density, of the earth. In fact, an approximate value of the mean density became available with the determination of G by Cavendish in 1798. With a modern value of G, 6.67×10^{-8} c.g.s. units (Heyl, 1930), the mean density is found to be 5.52 gm/cm^3. This value of density is much greater than that corresponding to rocks at the earth's surface, and was the first clue to the variation in density which we have seen.

In succeeding chapters, we shall discuss the methods available for the measurement of gravity, to a precision of one part in 10^6 or 10^7 of its actual value. In geophysics the unit of acceleration is known as the gal (after Galileo), and we shall normally express quantities in milligals (1×10^{-3} cm/sec^2), which are nearly equivalent to parts per 10^6 of g itself.

The gravitational field of the earth is distorted by a number of factors, whose effects must be separated. To visualize the effects, it is convenient to contrast the actual earth with a non-rotating uniform sphere, whose gravitational field would obviously be a constant. On the real earth, g at sea level varies in a regular way from equator to poles, because of the acceleration related to rotation, and in a less regular way, because of departures of the shape of the sea-level surface from a sphere. However, g is rarely measured at sea level, but rather on the land surface at some known height above sea level. All of these effects, which are functions of the position of each station, must be removed in a series of reductions, before the more subtle variations due to internal density differences are revealed. We note that a radial variation in density would not produce differences in g; the fact that reduced measurements of g reveal differences between points on the earth is proof of lateral variations in density, at different depths within the earth.

The close relationship between the gravitational field and the shape of the sea level surface of the earth arises from the fact that the latter surface (free of the effects of winds and tides) is an equipotential surface of the field.

9.2 THE POTENTIAL

Certain properties of the potential play an essential role in the analysis of the gravitational field, and incidentally will be used later in connection with magnetic and thermal fields. We recall first that, in a conservative field of force, *a scalar function of position* U may be introduced, such that the components of force acting on a unit mass at any point are

$$X = \frac{\partial U}{\partial x}, \qquad Y = \frac{\partial U}{\partial y}, \qquad Z = \frac{\partial U}{\partial z} \qquad 9.2.1$$

Defined in this way, the potential U, at distance r from a mass m is

$$U = \frac{Gm}{r} \qquad 9.2.2$$

The quantity U is thus taken to be positive, and the difference in U between two points in the field is the negative of the difference in potential energy per unit mass.

In vector form, Equation 9.2.1 is equivalent to

$$\underline{F} = \nabla U \qquad 9.2.3$$

and an important property of the force field immediately follows:

$$\nabla \times \underline{F} = 0 \quad \text{everywhere} \qquad 9.2.4$$

If any closed surface S is taken about a distribution of attracting mass, the total flux, or number of field lines, crossing S may be written as

$$\int_s F_n \, ds = \int_V \nabla \cdot \underline{F} \, dv \qquad 9.2.5$$

where F_n is the component of F in the direction of the outward normal, and V is the volume contained with S. If S encloses no attracting mass, the integrals in Equation 9.2.5 vanish, and must remain zero if S shrinks about any point of V; thus

$$\nabla \cdot \underline{F} = 0 \quad \text{everywhere}$$

and we have Laplace's equation

$$\nabla^2 U = \frac{\partial^2 U}{\partial x^2} + \frac{\partial^2 U}{\partial y^2} + \frac{\partial^2 U}{\partial z^2} = 0 \tag{9.2.6}$$

at all points. On the other hand, if S encloses a total mass M, we easily establish that

$$\int_s F_n \, ds = \int_v \nabla \cdot \underline{F} \, dv = -4\pi GM \tag{9.2.7}$$

and at any point of V

$$\nabla \cdot \underline{F} = \nabla^2 U = -4\pi G\rho \tag{9.2.8}$$

where ρ is the density at the point, equation 9.2.7 expresses a theorem due to Gauss, and equation 9.2.8 is known as Poisson's equation.

Laplace's equation, or Poisson's equation, will be found to play the dominant role in studies of force fields, as the wave equation did in the analysis of seismic waves. We shall be involved with the space variation of quantities whose time variation can usually be neglected.

In geophysics we very frequently must find a solution for the potential when its values, or those of its derivatives, are given on some surface. This situation arises, for example, in the attempt to determine the internal mass distribution from measurements of gravity on the earth's surface. The potential has a remarkable property related to boundary conditions that may be shown as follows. Suppose that, within a volume V, surrounded by a surface S, there are two solutions of Laplace's equation, U_1 and U_2, but that on S, U_1 is identical to U_2. We let $U = U_1 - U_2$, where U also satisfies $\nabla^2 U = 0$, and U vanishes on S. An identity valid for any continuous function of position is:

$$\left(\frac{\partial U}{\partial x}\right)^2 + \left(\frac{\partial U}{\partial y}\right)^2 + \left(\frac{\partial U}{\partial z}\right)^2 = \nabla \cdot (U\underline{\nabla U}) - U\nabla^2 U \tag{9.2.9}$$

Let us apply this identity for U as defined above, write it for every point of V, and integrate. Then

$$\int_v \left[\left(\frac{\partial U}{\partial x}\right)^2 + \left(\frac{\partial U}{\partial y}\right)^2 + \left(\frac{\partial U}{\partial z}\right)^2\right] dv = \int_v \nabla \cdot (U\underline{\nabla U}) \, dv - \int_v U\nabla^2 U \, dv \tag{9.2.10}$$

The second integral on the right vanishes; the first is equivalent to

$$\int_s (U\underline{\nabla U}) \cdot \underline{ds},$$

which also vanishes. Evidently, therefore, at every point of V,

$$\left(\frac{\partial U}{\partial x}\right) + \left(\frac{\partial U}{\partial y}\right)^2 + \left(\frac{\partial U}{\partial z}\right)^2 = 0 \tag{9.2.11}$$

In other words, U is everywhere zero, and U_1 and U_2 are in fact identical.

The fact that solutions of Laplace's equation are completely determined by their values on a bounding surface can be both a help and a hindrance. It is a help in that, once a solution which satisfies a given boundary condition is found, there need be no concern about lack of generality. But the theorem leads also the conclusion that quite different mass distributions can produce identical potential fields through regions of space. This in turn results in an ambiguity of interpretation, which is characteristic of all measurements of force fields on the earth's surface, and which sets them apart, to some extent, from seismological measurements.

Other properties of the potential, in particular its expansion in series and its definition on a sphere will be developed as they are required.

Problems

1. Derive the identity quoted in Equation 9.2.9.

2. Many problems relating to the earth involve properties of the gravitational potential of spheres first pointed out by Newton. Confirm by direct integration that the potential of a spherical shell, at points external to itself, is the same as that of an equal point mass at its center. Show that the potential at an internal point is constant. Extend these results to give the potential and attractive force of a solid sphere, at internal and external points. Plot these quantities as a function of radial distance from the center.

3. Results of early investigations on the gravitational constant G were often stated in terms of the mean density Δ of the earth. Derive the relationship between Δ and G.

4. A favorite problem in mechanics is to confirm that the motion of a mass moving without friction in a tunnel drilled along any chord of a uniform sphere is simple harmonic. Do this; then find the period for a uniform sphere of the same radius and density as the earth. By referring to Figure 3.12, state how you would expect the behavior in the real earth to differ from the ideal.

BIBLIOGRAPHY

Heyl, Paul R. (1930) A redetermination of the constant of gravitation. *J. Res. Nat. Bur. Stand.*, **5**, 1243.

10 THE MEASUREMENT OF GRAVITY

10.1 ABSOLUTE MEASUREMENTS

In the measurement of a physical quantity over such a large area as the surface of the earth, there is often an important distinction between the precise determination of the quantity at a few points, and the measurement of differences in it from place to place. The latter type of measurement is characteristically simpler, and, at least until very recently, has been more adapted to the use of portable instruments.

As far as gravity measurements are concerned, a great deal of useful information about the form and internal mass distribution of the earth could be obtained from relative measurements only, if these were connected into a world-wide net. Indeed, until a few years ago the results of absolute measurements were considered to be of more interest to metrology (e.g., the determination of precise standards of pressure, temperature, and electric current) than to geophysics. This situation has changed with the availability of satellite orbits for the determination of certain characteristics of the field. It is the absolute acceleration field to which the satellite responds, and if satellite measurements are to be compared to ground measurements, the latter must be on an absolute datum.

Absolute measurements of g are based upon some physical equation in which g appears, together with times or lengths which can be measured to one part in 10^7. The most direct approach is the measurement of the acceleration of a freely falling body, but all measurements before 1950 were based on the periods of physical pendulums. The theory of the physical pendulum shows that for any point of support, A, there is a complementary point B, on the production of a line from A through the mass center, such that the periods of swing for small oscillations about axes through A and B are equal. Furthermore, the distance AB is the length l of an equivalent simple pendulum, such that

$$T_A = T_B = 2\pi \sqrt{\frac{l}{g}} \qquad 10.1.1$$

The reversible pendulum, as it is known, was employed by Kater (1818), Kühnen and Furtwängler (1906), Heyl and Cook (1936), and Clark (1940) for absolute determinations at London, Potsdam, Washington, and Teddington, respectively. Until second order effects are considered, the measurements are quite straightforward. We picture a pendulum, perhaps one-half meter long, with a uniform cross-section along its length, and constructed of metal or fused silica. Two pivots (knife edges, or planes which bear on fixed knife edges) are fastened to the pendulum. The position of the mass center is then altered, either by grinding one end of the pendulum or by adjusting the position of sliding bobs, until the equality of periods in the two positions is obtained. The only length measurement required is then the distance between the knife edges or planes.

Difficulties arise from small mechanical perturbations of the system, such as the sway of the support, or even the bending of the pendulum itself, during the swinging. While these can be evaluated, the method of determination based on falling bodies is now considered to be so superior that the measurements mentioned above are chiefly of historical interest. The exception is the measurement of Kühnen and Furtwängler at Potsdam; their value of 981.2740 gals, measured in the Geodetic Institute, was adopted as the base of the world gravity network, and gravity values were quoted on this standard until 1968.

In 1952, Volet described a free-fall apparatus in which a graduated rule was dropped, in an evacuated chamber, in such a way that it passed the optical axis of a camera. Photographs taken of the rule at known time intervals permitted the acceleration to be calculated. This type of apparatus was employed by Preston-Thomas et al. (1960) at Ottawa, and by Tate (1968) at Washington. Tate modified the apparatus to allow the graduated rule to fall in an enclosure, which itself was in almost free fall. Any effect of the residual air in the chamber on the rule was thereby completely eliminated.

If a small object is used, rather than a graduated rule, two optical systems are required, but this is compensated by the fact that a more symmetrical experiment is possible, in which the object is projected upward and observed to pass twice between the axes of these systems (Cook, 1967).

All of the above absolute determinations referred to one site only, since the apparatus was not portable. The development of a small portable absolute gravity instrument became possible with the availability of lasers (Faller, 1965), allowing direct interferometric tracking of an object in free fall. Figure 10.1 outlines the principles of the apparatus of Hammond and Faller (1967). The dropped object is a corner cube reflector, which has the property that light reflected from it emerges precisely parallel to the incident direction. Light from a laser is directed vertically upward by a beam splitter, so that it strikes the base of a falling cube. A portion of the laser beam passes through an identical stationary reference cube; after reflection, the two beams are mixed at the beam splitter. Use of the reference cube compensates for the phase shift within the falling cube, and the mixed beam exhibits an interference pattern which changes with time as the falling cube drops. The photomultiplier and counter permit the counting of fringes (each fringe corresponding to a fall of one-half wavelength) within any desired time interval. Hammond and Faller arranged to count through two intervals (Δt_1 and Δt_2), each beginning at a common time at which the cube had fallen about 10 cm from rest. From the familiar expression for constantly accelerated motion,

$$s = v_0 t + \tfrac{1}{2} g t^2$$

we can write

$$s_1 = v_0 \, (\Delta t_1) + \tfrac{1}{2} g (\Delta t_1)^2$$

$$s_2 = v_0 \, (\Delta t_2) + \tfrac{1}{2} g \, (\Delta t_2)^2$$

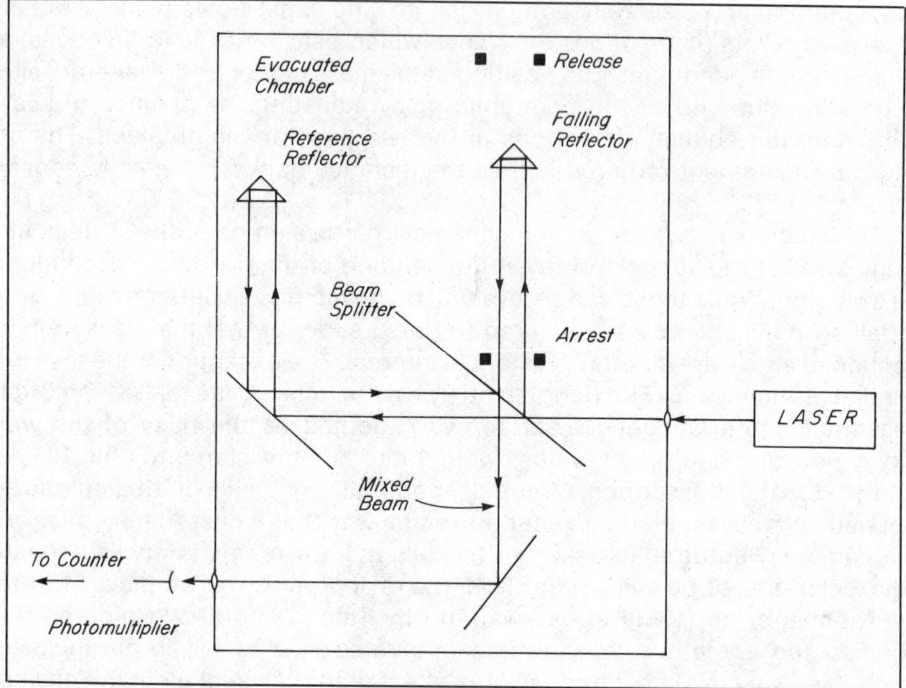

Figure 10.1 Schematic diagram of the Hammond-Faller absolute gravity instrument.

where s_1 and s_2 are the vertical distances traveled in the two intervals. Eliminating v_0,

$$g = \frac{2\left(s_2 - s_1 \dfrac{\Delta t_2}{\Delta t_1}\right)}{\Delta t_2{}^2 - \Delta t_1 \Delta t_2}$$

or, in terms of the wavelength of the laser light, λ, and the fringes counted, m_1 and m_2, in the two time intervals,

$$g = \frac{\lambda\left(m_2 - m_1 \dfrac{\Delta t_2}{\Delta t_1}\right)}{\Delta t_2^2 - \Delta t_1 \Delta t_2}$$

The drops are carried out in an evacuated chamber, and a drop and recovery of the cube can be made in rather less than one minute. The chief source of error lies in seismic motions at the site, as these alter the position of the reference cube with respect to the falling cube. Hammond and Faller (1971) have transported the apparatus to the sites of previous absolute determinations, including those at Teddington, England and Sevres, France. They estimate the standard error of one of their observations, based on 50 drops, to be approximately 0.050 mgal. At this level of precision, correction for the tidal variation of gravity (Chapter 14) is necessary, to reduce all observations to a common time. In addition, a small correction is applied for the difference in g between the mid-point of the free-fall path and the base of the apparatus. Table 10.1 gives a comparison of their measurements with some of the older determinations. It is

TABLE 10.1 ABSOLUTE MEASUREMENTS OF GRAVITY

Location of Absolute Measurement	Observer	Adjusted Value (referred to a common site) gals
National Bureau of Standards, Washington, D.C.	Tate (1968)	980.10477
"	Hammond and Faller (1971)	980.104234
National Physical Laboratory, Teddington, England	Cook (1967)	
	Cook and Hammond (1969)	981.18181
"	Hammond and Faller (1971)	981.18186
Bureau International des Poids et Mesurs, Sevres, France	Sakuma (1971)	980.925949
"	Hammond and Faller (1971)	980.925960

apparent that the world gravity network can now be placed on a much firmer absolute basis than was formerly possible. Almost as important, the relative instruments to be discussed in the next section can now be calibrated with much greater precision, through readings taken at absolute sites.

10.2 RELATIVE MEASUREMENTS

The determination of the differences in gravity between points on the earth was first accomplished by comparing the periods of physical pendulums at the different points. Provided all conditions, including the length of the equivalent simple pendulum, remain constant, the ratio of the values of g at two stations is equal to the inverse ratio of the squares of the periods. The accuracy of relative measurements with the pendulum thus depends on the accuracy of timing and the constancy of the conditions. We note, however, that the difference in gravity is obtained in absolute units, and there is no question of calibration, as we shall find in the case of the static gravimeter.

Up to about 1930, the pendulum was the only instrument available for relative gravity measurements, even for local surveys made in connection with geophysical prospecting (Gay, 1940). Following the development of static gravimeters, interest in pendulum measurement became restricted to providing base stations for the calibration of these gravimeters. However, with the growing number of truly absolute determinations, it appears that even this application of relative pendulum measurements will disappear.

A static gravimeter is simply a mass suspended on a spring, or some form of vertical seismograph, arranged so that detectable changes in the system will occur when it is set up at places where g is different. While the proposal for such an instrument goes back at least to 1849, the development of a portable instrument capable of measuring differences to 0.1 or 0.01 mgal came in much more recent years, and chiefly as a result of developments in geophysical prospecting.

The Worden gravimeter (Fig. 10.2) gives an example of some of the principles employed in a modern instrument. A system of quartz rods and fibers is constructed, so that the suspension is equivalent to a vertical seismograph. The main spring is arranged so that its restoring force is very small, which is

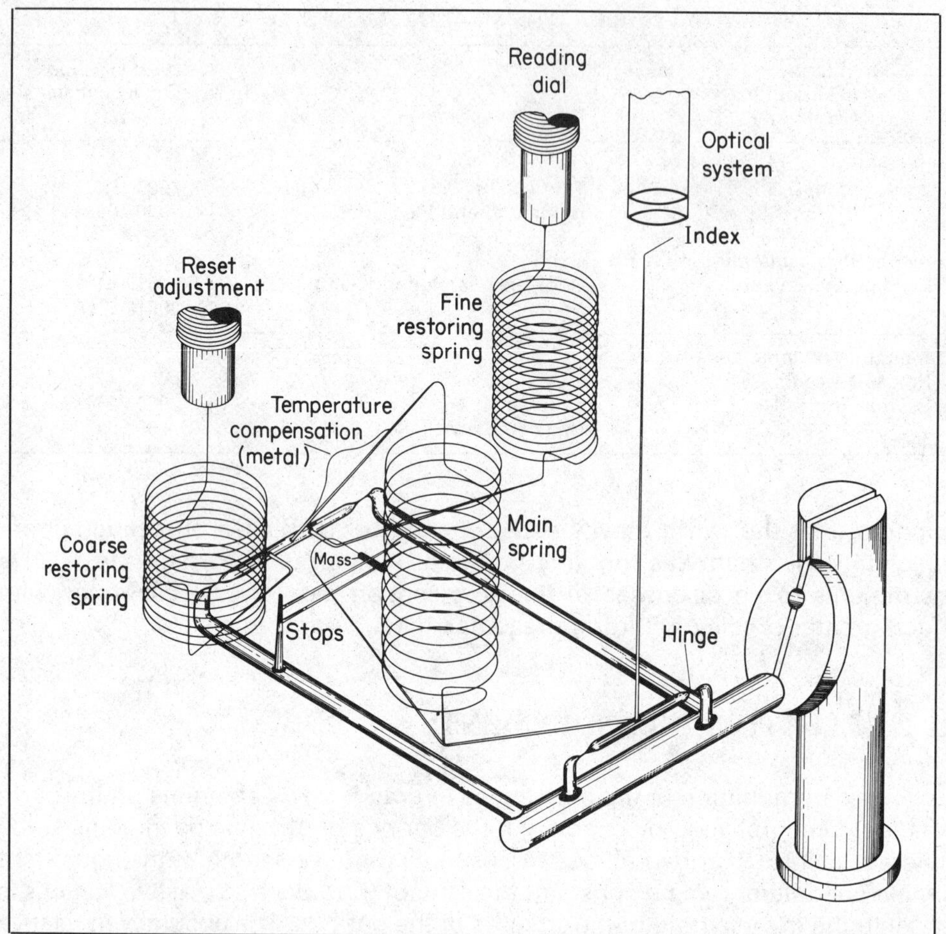

Figure 10.2 Working parts of the Worden gravimeter. All parts of the suspension are of fused quartz, except the metal strip used to provide temperature compensation.

equivalent to making the period very long. The position of the mass would depend on its weight, and therefore on the local value of g, but in fact it is always brought to a standard position, as indicated by the position of the index in the eyepiece, by adjusting dials connected to the restoring springs. Differences in the readings of these dials between stations are found to be proportional to the difference in gravity, although the constant of proportionality itself varies over different parts of the instrument's range. The reason for having two restoring springs is that the dial connected to the coarse spring may be left fixed, except when large differences are to be measured, allowing small differences to be measured with greater precision.

The instrument illustrates a number of the problems which must be overcome in gravimeter design and operation. Changes in temperature will obviously change the balance position, and therefore the reading. Some gravimeters are maintained at constant temperature by thermostatically controlled heating circuits, but in the Worden system, the expansion of a single metal part is arranged (by trial and error) to compensate for the other temperature effects. In addition, however, all gravimeters exhibit a slow change with time in the balance position, when kept at one station, due to the aging of the elastic materials of which the system is constructed. This has led to a standard method of determining gravity differences, in which observations are made in closed circuits, with repeated readings at base stations. The gradual change in reading, or drift, is then distributed over time so that its effect on all readings can be estimated.

Finally, it is clear that measurements of the differences in gravity will be obtained in quite arbitrary units: divisions on a graduated dial, for example. The relation between these units and milligals can only be determined by reading the instrument at a number of points where g is known, through absolute or relative pendulum measurements. Because an instrument cannot be assumed to have a completely linear response, the network of known points must cover the entire range of gravity over which the gravimeter is to be used; in other words, the gravimeter becomes an instrument of interpolation.

Because of the requirements for an absolute datum, for calibration stations, and for effective drift control in gravimeters, some systematic approach is necessary before the gravitational field of the earth can be determined from surface measurements. This has been formalized by the establishment of a world gravity network, defined by the International Association of Geodesy, and described by Woollard and Rose (1963). The network includes points at which absolute determinations have been made, and also a world-wide distribution of first order bases, mostly at airports, between which gravity differences have been determined by rapid direct flights. Regional and national gravity surveys are then connected to one or more of the first order bases. In different continents lines of calibration stations have been recognized, so that all gravimeters used for the world-wide network will have consistent calibrations.

Instruments suitable for recording very small changes in gravity with time at a fixed point are important for earth tide measurements (Chapter 14) and for studies of secular changes in gravity. While the best conventional gravimeters can be modified for this purpose, it appears that greatly improved long-term stability can be achieved with instruments operating at liquid helium temperatures and employing the phenomenon of superconductivity (Prothero and Goodkind, 1972). In their apparatus, a superconducting sphere is held in the magnetic field of two superconducting coils. The persistent currents in the superconducting coils remain extremely constant with time, and the sphere is held locked to the constant magnetic field of these currents. Variations in gravity tend to change the position of the sphere, but this is opposed by an electrostatic force applied through capacitor plates above and below the sphere. The voltage required to keep the sphere in a fixed position is the parameter which is recorded. Prothero and Goodkind give examples of recordings which indicate resolution to a very few microgals (1 microgral = 10^{-6} cm sec^{-2}), and long-term drift of less than 50 microgals over a time interval of one month.

10.3 MEASUREMENTS FROM MOVING VEHICLES

Since g is an acceleration, the problem of its measurement from a vehicle that is accelerating relative to the earth raises fundamental problems. On the other hand, the incentive to measure gravity on ships, and possibly in the air, is very great. Vening Meinesz (1929) recognized the desirability of measuring gravity at sea, since without such measurements only a small portion of the earth's field would become known, and he was able to modify the only instrument available to him—the pendulum—for observations in submarines. As analyzed by Vening Meinesz, the movements of a support include rotations and accelerations, the latter both horizontal and vertical. Rotations can be reduced by gimbal mounting, but a horizontal acceleration, $\ddot{y}$, in the plane of swing of a pendulum of equivalent length l, enters directly into the equation of motion

$$\ddot{\theta}_1 + \frac{g}{l}\,\theta_1 = -\frac{\ddot{y}}{l} \qquad 10.3.1$$

where θ_1 is the angular displacement of the pendulum. However, if two pendulums of the same equivalent length l are swung on the same support, the difference of their equations of motion yields

$$(\ddot{\theta}_1 - \ddot{\theta}_2) + \frac{g(\theta_1 - \theta_2)}{l} = 0 \qquad 10.3.2$$

showing that the difference of amplitude $(\theta_1 - \theta_2)$ behaves as the amplitude of a fictitious pendulum, free of horizontal accelerations. Vening Meinesz devised a pendulum apparatus, in which the difference in amplitude was optically measured and photographically recorded.

Vertical accelerations, in principle, cannot be separated from g, and their effect can only be reduced if they tend to average to zero over the period of an observation. The mean value of z is

$$\bar{\ddot{z}} = \frac{1}{t}\,(\dot{z}_t - \dot{z}_0) \qquad 10.3.3$$

where $\dot{z}_0$ and $\dot{z}_t$ are the vertical velocities of the support at the beginning and end, respectively, of the observation period t. In Vening Meinesz's case, the period of the fictitious pendulum was determined from swings lasting 30 to 60 minutes; t was therefore of the order of 2000 seconds, the vertical velocity of the submarine could be kept below 2 cm/sec, and $\bar{\ddot{z}}$ was less than 1 mgal. Equation 10.3.3 is of fundamental importance if measurements from an aircraft are contemplated. If gravity measurements corresponding to a short section of flight path are to be obtained, t must be short, and $\dot{z}$ will be required to an accuracy of a fraction of 1 cm/sec if $\bar{\ddot{z}}$ is to be measured to 1 mgal.

The above effects of $\ddot{y}$ and $\ddot{z}$ are of the first order. For any instrument which responds to the total gravity vector, there is an important second order effect, pointed out by Browne (1937). If the horizontal accelerations $\ddot{x}$ and $\ddot{y}$ have periods much longer than the natural period of the instrument, the quantity measured will be the resultant $g[1 + (\ddot{x}^2 + \ddot{y}^2)/2g^2]$, which is systematically larger than g.

The Vening Meinesz pendulums continued to be used until fairly recent years, but two gravimeters have been developed which may be used on surface ships, and these have largely supplanted the pendulums for gravity measurements at sea. The original Graf gravimeter (Graf and Schulze, 1961) employed a heavily damped beam, whose displacement was continuously recorded; as the instrument was normally used on a gyroscopically controlled platform, the mean position of the beam, over an interval of time, indicated the relative value of gravity, without the necessity of the Browne correction. The Lacoste-Romberg instrument (Lacoste, 1959) employs gimbal mounting, so that the system responds to the total vector. Accelerometers measure the horizontal accelerations and the Browne correction is computed in the instrument and applied to the reading. The Lacoste-Romberg instrument has been used on surface ships, without gyroscopically controlled platforms, and has been tested in an aircraft (Nettleton, Lacoste, and Harrison, 1960). Recent modifications have reduced the differences between the Graf and Lacoste instruments: in the Graf gravimeter, the beam is now returned to a standard position and the restoring force continuously recorded, while the Lacoste meter is used on a stabilized platform whenever possible.

The above considerations of the problems of measurement on a moving support have dealt with local or instrumental effects. There is a still more general consideration, known as the *Eötvös effect,* that an observer moving relative to the earth experiences a different centripetal acceleration, and therefore a different g. If ω is the angular velocity of the earth on its axis, and $d\omega$ the change in the observer's angular velocity due to an east-west velocity component v, the change in acceleration is

$$da = 2r\omega\ d\omega = 2\omega v \tag{10.3.4}$$

where r is the distance from the axis. At geocentric latitude ϕ, the effect on gravity is

$$dg = da \cos \phi \tag{10.3.5}$$

In middle latitudes, an east-west velocity of 1 mile per hour corresponds to a change in g of 5 mgals. Obviously the course and speed of any moving vehicle must be well determined during gravity measurements, if the latter are to be accurate to 1 mgal or better. At aircraft speeds, even motion in the north-south direction produces a change in g, although this is considerably smaller than the effect noted above. It is now generally agreed that the greatest limitation to gravity measurements from moving vehicles is the lack of navigation systems which could be used anywhere on earth to give a continuous record of the vehicle's velocity.

Problems

1. Derive an expression for the potential at a point on one of the principal axes of a triaxial ellipsoid of uniform density.

2. Prove that for the physical pendulum, for any point of support A, there is a complementary point B on the opposite side of the mass center, such that the periods when the pendulum is swung from A and B are equal, and that the distance AB is the length of the equivalent simple pendulum.

3. Show that g may be determined if a small object is projected upward past two slits, separated by a vertical distance d, and the times T_1 and T_2 for the object to return past each slit are measured. Show also that the expression is unaltered by the addition of a viscous retarding force, such as would be produced by residual air in the chamber.

4. To what accuracy must the period of a pendulum be measured, if differences in gravity are to be determined to 0.1 mgal?

5. In the construction of gravimeters, it is usually desirable to make the period of vertical oscillations very long. Show that it is possible to suspend a boom from a spring, such that the period is infinite, provided the tension in the spring is proportional to its actual length. (A spring with this property must be specially wound; it is known as a zero-length spring.)

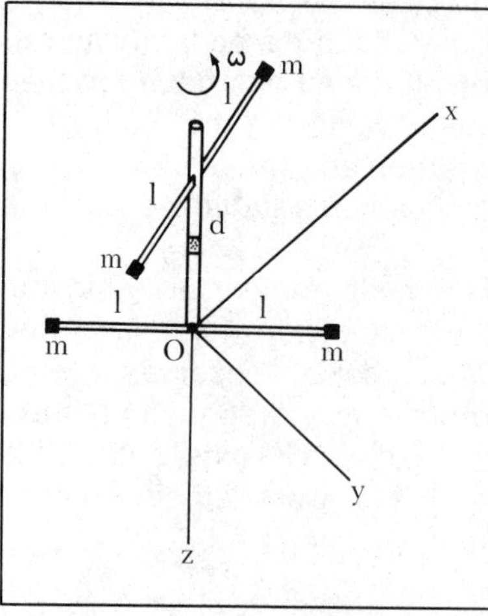

6. Some of the difficulties of measuring g from moving vehicles are removed if the gradient of gravity is measured. Show that if the element in the sketch is rotated about the z-axis at angular velocity ω, a difference in torque on the two perpendicular arms results, depending upon the gravity gradients $\partial^2U/\partial x\partial z$ and $\partial^2U/\partial y\partial z$. Write the expression for the time-dependence of this torque in terms of the mass m and the lengths l and d. (It may be measured by means of a strain transducer placed in the shaft.)

7. Gravity anomalies reaching 200 mgals are believed to be associated with the mares of the moon. If the mass excess (mascon) is in the form of an extensive sheet close to the moon's surface, what maximum gravity gradient (in Eötvös units) would be experienced by a lunar satellite orbiting at a height of 30 km? (A gradient of 1 Eötvös unit is equivalent to 1×10^{-9} cm/sec^2 per cm.)

BIBLIOGRAPHY

Browne, B. C. (1937) The measurement of gravity at sea. *Mon. Nat. A. S., Geophys. Suppl.*, **4**, 271.

Clark, J. S. (1940) An absolute determination of the acceleration due to gravity. *Phil. Trans. Roy. Soc. Lond. A*, **238**, 65.

Cook, A. H. (1967) A new determination of the acceleration due to gravity at the National Physical Laboratory, England, *Phil. Trans. Roy. Soc. Lond. A*, **261**, 211.

Cook, A. H. (1968) Report on absolute measurements of gravity. *Trav. de l'Assoc. Int. de Géodésie*, Tome **23**, 271.

Cook, A. H., and Hammond, J. A. (1969) The acceleration due to gravity at the National Physical Laboratory. *Metrologia*, **5**, 141.

Faller, J. E. (1965) Results of an absolute determination of the acceleration of gravity. *J. Geophys. Res.*, **70**, 4035.

Gay, Malcolm W. (1940) Relative gravity measurements using precision pendulum equipment. *Geophysics*, **5**, 176.

Graf, Anton, and Schulze, Reinhard (1961) Improvements in the Sea Gravimeter Gss2. *J. Geophys. Res.*, **66**, 1813.

Hammond, J. A., and Faller, J. E. (1967) Laser-interferometer system for the determination of the acceleration of gravity. *IEEE, J. Quantum Electron.*, **3**, 597–602.

Hammond, J. A., and Faller, J. E. (1971) Results of absolute gravity determinations at a number of different sites. *J. Geophys. Res.*, **76**, 7850–7854.

Heyl, Paul R., and Cook, Guy S. (1936) The value of gravity at Washington. *J. Res. Nat. Bur. Stand.*, **17**, 805.

Kater, H. (1818) An account of experiments for determining the length of the pendulum vibrating seconds in the latitude of London. *Phil. Trans. Roy. Soc. Lond.*, **108**, 33.

Kuhnen, F., and Furtwängler, P. (1906) Bestimmung der Absoluten Grösse der Schwerkraft zu Potsdam. *Veröff. Press. Geodät Inst.*, **27**, 397 *pp.*

Lacoste, Lucien (1959) Surface ship gravity measurements on the Texas A and M College Ship, the "Hidalgo." *Geophysics*, **24**, 309.

Nettleton, L. L., Lacoste, L., and Harrison, J. C. (1960) Tests of an airborne gravity meter. *Geophysics*, **25**, 181.

Preston-Thomas, H., Turnbull, L. G., Green, E., Dauphinee, T. M., and Kalra, S. N.

(1960) An absolute measurement of the acceleration due to gravity at Ottowa. *Can. J. Phys.*, **38**, 824.

Prothero, W. A., Jr., and Goodkind, J. M. (1972) Earth-tide measurements with a superconducting gravimeter. *J. Geophys. Res.*, **77**, 926–937.

Sakuma, A. (1971) Recent developments in the absolute measurement of gravity. In *Proceedings of the International Conference on Precision Measurement and Fundamental Constants,* ed. D. N. Langenberg and B. N. Taylor. Nat. Bur. Stand. Spec. Publ. 343.

Tate, D. R. (1968) Acceleration due to gravity at the National Bureau of Standards. *J. Res. Nat. Bur. Stand.*, 72C, 1.

Thulin, A. (1961) *Trav. Mem. Bur. Int. Poids et Mes.*, 22, A1.

Vening Meinesz, F. A. (1929) *Theory and practice of gravity measurements at sea.* Delft.

Volet, C. (1952) Gravitation. Mesure de l'accéleration due à la pesanteur, au Pavillon de Breteuil. *C. R. Acad. Sci. Paris,* **235**, 442.

Woollard, G. P., and Rose, J. C. (1963) International gravity measurements. *Soc. Explor. Geophysicists,* Spec. Publ. Tulsa, Oklahoma.

11 THE FIGURE OF THE EARTH

11.1 GEODESY AND GEOPHYSICS

We noted in Chapter 9 that mean sea level, unperturbed by winds or tides, is an equipotential surface of the earth's gravitational field. In this chapter we pursue the determination of the form of this sea level surface from the field, but it is desirable first to establish the importance of a knowledge of this form.

The geodesist has traditionally fixed the location of points on the earth's surface in terms of latitude, longitude, and height above sea level. To locate these points absolutely in space, a knowledge of the form of the sea-level surface is necessary. The geophysicist is primarily interested in whether or not the figure of the sea-level surface is an equilibrium figure for a rotating fluid mass, for if it is not, long-term strength within the earth is required to support the difference.

Because of the relative complexity of the gravitational field, due to inhomogeneities in density in the outer part of the earth, the equipotential surfaces of the field do not have a simple mathematical form. It has therefore been the custom to approximate the sea level surface with a uniform surface enclosing the same volume, and then to determine the departures of the actual sea-level equipotential from this. Let us begin by defining the actual sea-level surface, which is known as the *geoid,* as that closed surface which corresponds with mean sea level over the oceans, and with the extension of the same equipotential beneath the continents (this extension is often described in terms of the heights to which water would rise in narrow canals cut through the continents). We now visualize an ideal earth with no lateral variations in density, and with an external form given by equilibrium between gravitational and rotational forces. On such an earth, gravity at sea level would vary smoothly from equator to poles, and the equipotentials would be regular surfaces. The surface corresponding to the geoid in this case is known as the *spheroid.* In the case of the actual earth, we think of the spheroid as that regular figure (actually very nearly an ellipsoid of revolution) which gives the best approximation to the geoid.

The approximate form of the spheroid can be determined by measuring the length of a degree of latitude, in different latitudes. By the eighteenth century, it was known that the spheroid was oblate, with a polar radius shorter than the equatorial radius by about one part in 300. About this time Clairaut (1743) showed that the form can be determined from the variation of gravity with latitude, and we shall follow his development. We shall then discuss the departure of the actual figures from an equilibrium form, and the implications for the internal strength of the earth. Finally, we shall indicate how both gravity measurements on earth and observations on artificial satellites permit the actual geoid to be built upon the spheroid.

11.2 CLAIRAUT'S THEOREM

As noted above, this theorem deals with the smooth variation in g from equator to poles. We take the spheroid to surround all of the mass of the earth, and consider the potential for an observer, at an external point P, who rotates with the earth at an angular velocity ω about the z-axis. The origin is at the center of mass (Fig. 11.1) and primed quantities refer to the position of P. Considering gravitation and centripetal acceleration, we may write

$$U = G\int \frac{dm}{\rho} + \tfrac{1}{2}(x'^2 + y'^2)\omega^2 \qquad 11.2.1$$

where dm is an element of mass and the integration is over the entire earth.

To obtain expressions valid to the desired order of small quantities, it is usual to expand U into a series. This may be done in two ways, which lead to identical results. In the first, the term $1/\rho$ is expanded, and the resulting series is integrated term by term; in the second, U is immediately written in terms of the known solutions of Laplace's equation in spherical coordinates, which are spherical harmonics. We shall trace both approaches.

The quantity $1/\rho$ may be written

$$\frac{1}{\rho} = \frac{1}{r'}\left[1 - 2\frac{r}{r'}\cos\gamma + \frac{r^2}{r'^2}\right]^{-1/2} \qquad 11.2.2$$

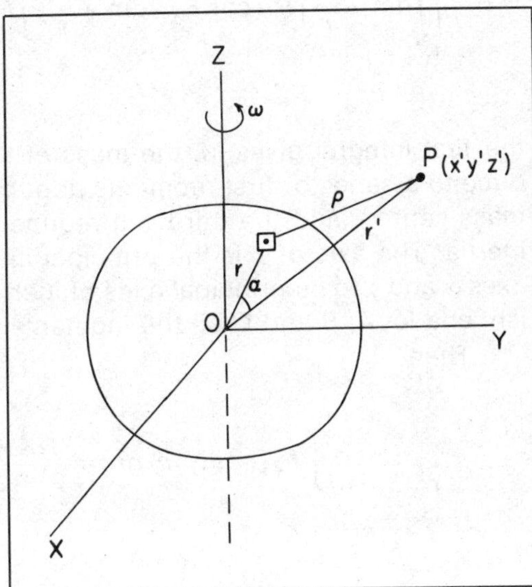

Figure 11.1 Notation for the development of Clairaut's theorem.

Provided $\left|2\left(\frac{r}{r'}\right)\cos\gamma - r^2/r'^2\right| < 1$, the bracketed quantity may be expanded by the binominal theorem to give

$$\frac{1}{\rho}=\frac{1}{r'}\left[P_0(\cos\gamma)+\frac{r}{r'}P_1(\cos\gamma)+\left(\frac{r}{r'}\right)^2P_2(\cos\gamma)+\cdots\right] \qquad 11.2.3$$

where

$$P_0(\cos\gamma)=1$$
$$P_1(\cos\gamma)=\cos\gamma$$
$$P_2(\cos\gamma)=\tfrac{3}{2}(\cos^2\gamma-\tfrac{1}{3})$$
$$\vdots$$
$$P_n(\cos\gamma)=\sum_{k=0}^{n/2}\frac{1\cdot 3\cdots(2n-2k-1)(-1)^k\cos^{n-2k}\gamma}{2^k k!(n-2k)!}$$

The functions $P_n(\cos\gamma)$ are Legendre polynomials; for the present, they are defined only by the general term above.

With the angle ϕ as latitude (not co-latitude θ) and λ as longitude, we may write

$$\cos\gamma=\frac{xx'+yy'+zz'}{rr'}=\cos\phi\cos\phi'\cos(\lambda-\lambda')+\sin\phi\sin\phi' \qquad 11.2.4$$

In the expression for U:

$$U=\frac{G}{r'}\left[\int dm+\frac{1}{r'}\int P_1(\cos\gamma)\,r\,dm+\frac{1}{r'^2}\int P_2(\cos\gamma)\,r^2dm\cdots\right] \qquad 11.2.5$$
$$+\tfrac{1}{2}(x'^2+y'^2)\omega^2,$$

the first integral gives M, the mass of the earth, the second will evidently reduce to a series of first moments about O, and therefore vanish (since O is the mass center), and the third will reduce to a sum of moments and products of inertia. The axis of z is the principal axis of the greatest inertia; let us select axes x and y to be principal axes of inertia also, so that products of inertia vanish, and let A, B and C be the moments of inertia about x, y and z respectively.

Then

$$\int P_2(\cos\gamma)r^2\,dm=\frac{3}{2}\left(\frac{A+B}{2}-C\right)(\sin^2\phi'-\tfrac{1}{3})$$

$$11.2.6$$

$$+\tfrac{3}{4}(B-A)\cos^2\phi'\cos 2\lambda'$$

Dropping all higher terms in the expansion, we may write

$$U = \frac{MG}{r}\left[1 + \frac{K}{2r^2}(1 - 3\sin^2\phi) + \frac{3(B-A)}{4Mr^2}\cos^2\phi\cos 2\lambda + \frac{\omega^2 r^3}{2MG}\cos^2\phi\right] \tag{11.2.7}$$

where $K = \left(C - \frac{A+B}{2}\right)\Big/M$; since the integration has been completed, primes have been dropped.

Equation 11.2.7 indicates the relation between a longitude term in U, and the difference in equatorial moments of inertia, $B-A$; however, if this difference is taken to be zero, as is appropriate to the order of small quantities considered,

$$U = \frac{MG}{r}\left[1 + \frac{K}{2r^2}(1 - 3\sin^2\phi) + \frac{\omega^2 r^3}{2MG}\cos^2\phi\right] \tag{11.2.8}$$

We now introduce the condition that the external surface is an equipotential, $U = U_0$, and that P lies on it. Then the shape of this surface is defined by

$$r = \frac{MG}{U_0}\left[1 + \frac{K}{2a^2}(1 - 3\sin^2\phi) + \frac{\omega^2 a^3}{2MG}\cos^2\phi\right] \tag{11.2.9}$$

where the equatorial radius a has been substituted for r in the (small) second and third terms.

The term $\omega^2 a^3/MG$ will be denoted by m, where

$$m = \frac{a\omega^2}{MG/a^2} = \frac{\text{centripetal acceleration at equator}}{\text{attraction at equator}}$$

Rearranging, we have

$$r = \frac{MG}{U_0}\left(1 + \frac{K}{2a^2} + \frac{m}{2}\right)\left[1 - \left(\frac{3K}{2a^2} + \frac{m}{2}\right)\sin^2\phi\right] \tag{11.2.10}$$

which is of the form

$$r = a(1 - f\sin^2\phi)$$

The quantity f is evidently given by

$$\frac{\text{equatorial radius} - \text{polar radius}}{\text{equatorial radius}}$$

and is known as the *flattening* of the spheroid.

Therefore,

$$f = \frac{3K}{2a^2} + \frac{m}{2} \tag{11.2.11}$$

Gravity at any distance r is given to the first order by

$$g = -\,\partial U/\partial r$$

or

$$g = \frac{MG}{r^2}\left[1 + \frac{3K}{2r^2}(1 - 3\sin^2\phi) - m\cos^2\phi\right] \qquad 11.2.12$$

and gravity on the spheroid, γ_0, is obtained by substituting r from Equation 11.2.10 into Equation 11.2.12, to give

$$\gamma_0 = \frac{U_0^2}{MG}\left(1 + \frac{K}{2a^2} - 2m\right)\left[1 + \left(2m - \frac{3K}{2a^2}\right)\sin^2\phi\right] \qquad 11.2.13$$

This is of the form

$$\gamma_0 = \gamma_{\text{equator}}\,[1 + B_2\sin^2\phi] \qquad 11.2.14$$

where

$$B_2 = 2m - \frac{3K}{2a^2} \qquad 11.2.15$$

is recognized as the proportional difference between equatorial and polar gravity. A comparison of Equations 11.2.11 and 11.2.15 shows that

$$B_2 = \frac{5}{2}m - f \qquad 11.2.16$$

In this equation, which is the expression of Clairaut's theorem to the first order in f, all quantities have a straightforward physical significance, and the flattening f is seen to be directly related to the term B_2 in the expression for γ_0 as a function of latitude.

For practical applications in geodesy, Equation 11.2.16 must be modified by the inclusion of higher order terms. It should be noted, however, that as far as the first order variations from equator to pole are concerned, the only assumption has been that the external surface (i.e., sea level) be an equipotential. Hydrostatic equilibrium in the interior is not required; it is important to emphasize this, as Clairaut also considered the implications of hydrostatic equilibrium, which we consider below, and it is sometimes assumed that Equation 11.2.16 requires it.

11.3 EXPANSION OF THE POTENTIAL BY SPHERICAL HARMONICS

In the treatment followed in the previous section, the potential of an element of the earth's mass was first developed as an infinite series of terms, each of which was then integrated over the whole earth. The method had the advantage of showing the physical significance of succeeding terms, but it is more convenient in practice to immediately write the potential in terms of spherical harmonics. An introduction to these is given in Appendix A.

The potential of the earth's field, without the rotation term, is written as

$$U = \frac{MG}{r}\left[\sum_{l=0}^{\infty}\left(\frac{a}{r}\right)^l \sum_{m=0}^{l} P_l^m(\cos\theta)\{C_{lm}\cos m\lambda + S_{lm}\sin m\lambda\}\right] \qquad 11.3.1$$

where C_{lm} and S_{lm} are the coefficients which give the contributions of the various spherical harmonic terms. Alternatively, it may be written in terms of cos $m\lambda$ alone with a phase angle:

$$U = \frac{MG}{r}\left[1 - \sum_{l=2}^{\infty} J_{l,0}\left(\frac{a}{r}\right)^l P_l(\cos\theta) + \sum_{l=2}^{\infty}\left(\frac{a}{r}\right)^l \sum_{m=0}^{l} J_{l,m} P_l^m(\cos\theta)\cos m(\lambda - \lambda_{l,m})\right] \qquad 11.3.2$$

where the constants $J_{l,m}$ and $\lambda_{l,m}$ are now the two constants associated with each harmonic. As we saw in the previous section, there is no term in $l = 1$. In practice, there is usually a difference between the numerical values of even the zonal coefficients $C_{l,0}$ and $J_{l,0}$, because the signs are changed to make $J_{2,0}$ a positive quantity, and also because the expression for the potential given in 11.3.1 is used with *normalized* forms of the functions $P_l^m(\cos\theta)$ (Appendix A). The form of the expression given in Equation 11.3.2 was introduced by Merson and King-Hele (1958), the coefficients being designated J in honor of Sir Harold Jeffreys. Since $P_2^0(\cos\theta) = \frac{1}{2}(3\cos^2\theta - 1)$, a comparison of Equations 11.2.8 and 11.3.2 shows that

$$J_{2,0} = \frac{C - A}{Ma^2}$$

The quantity $J_{2,0}$ is known as the *dynamical form factor* of the earth; it is closely related to the flattening f.

Great contributions have been made in recent years to the determination of the harmonics by observations on the dynamics of artificial satellites. The effect on the orbit is most pronounced in the case of $J_{2,0}$, and we can see this most clearly by considering the dynamics of a satellite in a field containing only the constant potential and the $J_{2,0}$ term. Following King-Hele (1958), we slightly rewrite a truncated Equation 11.3.2 as

$$U = ga\left[\frac{a}{r} + J_{2,0}\frac{3a^2}{2r^2}\left(\frac{1}{3} - \cos^2\theta\right)\right] \qquad 11.3.3$$

where g is the equatorial value of gravity on the earth's surface and a is the earth's radius. There is no term in ω because, after launch, the satellite does not partake of the earth's rotation. To simplify the development, terms of the order of J^2 will be neglected and the satellite's orbit will be assumed to be nearly circular. In order to appreciate the formulation of the problem, it is useful to predict the type of motion that will result. A satellite moving in the gravitational field of a spherically symmetrical earth is well known to follow an elliptical path in a fixed plane, according to Kepler's laws. The difference in the present case is the presence of the latitude-dependent term in the potential, arising from the earth's oblateness. Forces in the θ-direction, arising from the attraction of the additional matter in the earth's equatorial bulge, produce torques on the satellite as it moves in its orbit. The situation is similar to that of a gyroscope acted on by a torque normal to its spin axis. The latter precesses about a direction fixed in space. In the case of the satellite, the normal to its instantaneous orbital plane precesses at a constant inclination to the earth's axis. The motion is therefore no longer strictly planar; however, because the rate of precession is much slower than the angular speed of the satellite, the concept of an instantaneous orbital plane is still useful.

With this background, let us consider the specific situation illustrated in Figure 11.2. Axes *x, y, z* are taken to be fixed in space, with origin *O* at the earth's center, the *x-y* plane in the equatorial plane of the earth, and the *z*-axis coincident with the earth's axis of rotation (taken as fixed in direction). The orbit of the satellite is inclined at an angle α to the *x-y* plane. Initially, the instantaneous plane is taken to intersect the equatorial plane along the *x*-axis. At the instant portrayed in the figure, precession has rotated the plane of orbit through the angle Ω so that the intersection with the equatorial plane is given by the moving axis Ox'. On the instantaneous plane of orbit, the position of the satellite can be given by one angle; we conveniently choose the angle *AOS*, which we denote ψ and measure from *A*, the highest point of the orbit above the equatorial plane. The point *A* lies in the moving plane zOy'. We shall also use, for the instantaneous position of the satellite, spherical polar coordinates (r, θ, λ) measured with respect to the fixed axes, with θ the co-latitude and λ the azimuth reckoned from the plane *zOx*.

Trigonometric relations between all of the angular quantities can be established as:

$$\begin{aligned} \cot\theta &= \tan\alpha\,\sin(\lambda+\Omega) \\ \cos\psi &= \operatorname{cosec}\alpha\,\cos\theta \\ \sin\psi &= -\sin\theta\,\cos(\lambda+\Omega) \end{aligned} \qquad 11.3.4$$

The equations of motion of the satellite are obtained by equating the accelerations in the directions of increasing (r, θ, λ) to the forces, or gradients of the potential, in these directions. Components of acceleration in spherical co-

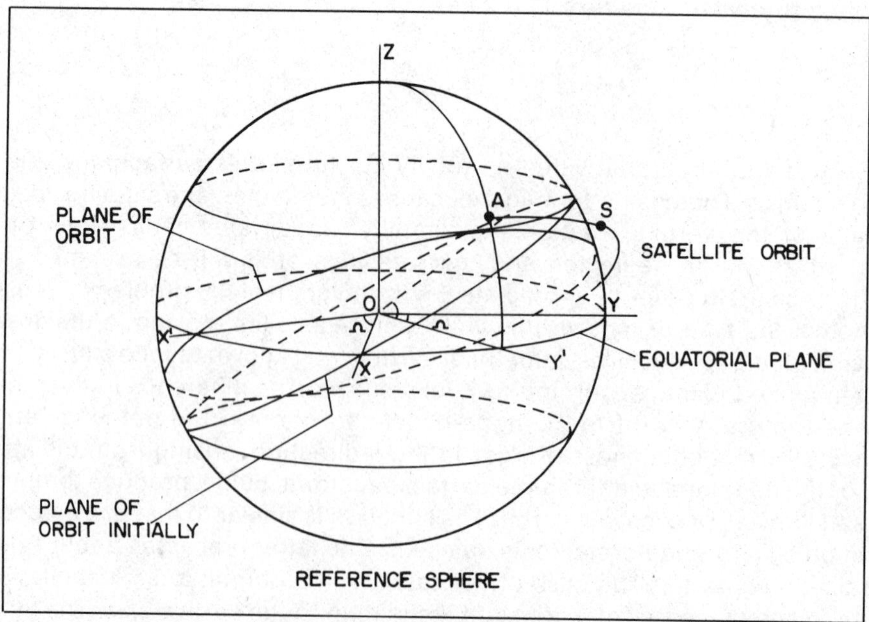

Figure 11.2 A satellite S in orbit in a rotating plane (after King-Hele).

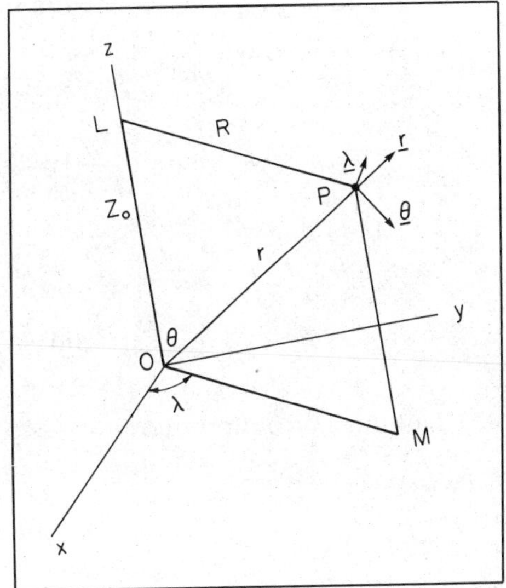

Figure 11.3 Quantities used in defining acceleration in spherical coordinates.

ordinates are derived in many texts on dynamics, but because they may appear unfamiliar, the derivation is reviewed here with the aid of Figure 11.3.

Consider first the motion in the plane $z = z_0$, for which (R, λ) are the polar coordinates. Plane dynamics establishes that the component accelerations for a point with Cartesian accelerations $\frac{d^2x}{dt^2}, \frac{d^2y}{dt^2}$ are

$$\frac{d^2R}{dt^2} - R\left(\frac{d\lambda}{dt}\right)^2 \qquad \text{(along } LP\text{: increasing radius plus instantaneous centripetal)}$$

$$R\,\frac{d^2\lambda}{dt^2} + 2\frac{dR}{dt}\frac{d\lambda}{dt} = \frac{1}{R}\frac{d}{dt}\left(R^2\,\frac{d\lambda}{dt}\right) \qquad (\perp \text{ plane } \lambda = \text{const.: tangential plus Coriolis})$$

The second of these is the λ-component we require. Now consider the accelerations in the plane λ = constant. We have determined $\left[\frac{d^2R}{dt^2} - R\left(\frac{d\lambda}{dt}\right)^2\right]$ along LP, and there will be an acceleration $\frac{d^2z}{dt^2}$ along MP. By the plane polar transformation just employed, the accelerations $\frac{d^2R}{dt^2}$ and $\frac{d^2z}{dt^2}$ in Cartesian coordinates would be equivalent to $\frac{d^2r}{dt^2} - r\left(\frac{d\theta}{dt}\right)^2$ along OP and $\frac{1}{r}\frac{d}{dt}\left(r^2\,\frac{d\theta}{dt}\right)$ perpendicular to OP. The additional acceleration $-R\left(\frac{d\lambda}{dt}\right)^2$ may be resolved into $-R\left(\frac{d\lambda}{dt}\right)^2 \sin\theta$ along OP and $-R\left(\frac{d\lambda}{dt}\right)^2 \cos\theta$ perpendicular to OP. Combining all contributions, and using $R = r\sin\theta$, we have for the required components:

$$\frac{d^2r}{dt^2} - r\left(\frac{d\theta}{dt}\right)^2 - r\sin^2\theta\left(\frac{d\lambda}{dt}\right)^2 \qquad (r)$$

$$\frac{1}{r}\frac{d}{dt}\left(r^2\,\frac{d\theta}{dt}\right) - r\sin\theta\cos\theta\left(\frac{d\lambda}{dt}\right)^2 \qquad (\theta)$$

$$\frac{1}{r\sin\theta}\frac{d}{dt}\left(r^2\sin^2\theta\,\frac{d\lambda}{dt}\right) \qquad (\lambda)$$

Reference to Figure 11.3 will also confirm that the potential gradients in the required directions are $\frac{\partial U}{\partial r}$, $\frac{1}{r}\frac{\partial U}{\partial \theta}$, and $\frac{1}{r \sin \theta}\frac{\partial U}{\partial \lambda}$. The equations of motion are thus

$$\left.\begin{aligned} \frac{d^2}{dt^2}(r) - r\left(\frac{d\theta}{dt}\right)^2 - r \sin^2 \theta\left(\frac{d\lambda}{dt}\right)^2 &= -g\frac{a^2}{r^2} - \frac{3J_{2,0}\, ga^4}{2r^4}(1 - 3\cos^2\theta) \\ \frac{1}{r}\frac{d}{dt}\left(r^2\frac{d\theta}{dt}\right) - r\sin\theta\cos\theta\left(\frac{d\lambda}{dt}\right)^2 &= \frac{6J_{2,0}\, ga^4}{2r^4}\sin\theta\cos\theta \\ \frac{d}{dt}\left(r^2\sin^2\theta\frac{d\lambda}{dt}\right) &= 0 \end{aligned}\right\} \quad 11.3.5$$

The third equation shows that

$$r^2 \sin^2 \theta \frac{d\lambda}{dt} = \text{constant} = p \quad 11.3.6$$

where p is the angular momentum of the satellite about the O_z-axis. Elementary conditions for a circular orbit of radius r show also that

$$p^2 = r \cos^2 \alpha g a^2 \quad 11.3.7$$

We return to the second equation of 11.3.5, and, on the basis of 11.3.6, replace d/dt by $\frac{p}{r^2 \sin^2 \theta}\, d/d\lambda$. Then

$$\frac{d^2}{d\lambda^2}(\cot\theta) + \cot\theta = \frac{-6J_{2,0}\, ga^4}{2rp^2}\sin^3\theta\cos\theta \quad 11.3.8$$

If the plane of the orbit precesses, as supposed, Ω will vary with time, but more slowly than ψ. In that case we could write

$$\frac{d\Omega}{d\psi} = \epsilon \quad 11.3.9$$

where ϵ is a small constant (of the same order as $J_{2,0}$). If this were so, direct manipulation of Equations 11.3.4 would give

$$\frac{d^2}{d\lambda^2}(\cot\theta) + \cot\theta = -2\epsilon \sec^3\alpha \sin^3\theta\cos\theta \quad 11.3.10$$

Comparison of Equations 11.3.8 and 11.3.10 shows that Ω indeed varies slowly with ψ, provided

$$\epsilon = \frac{3J_{2,0}a^2\cos\alpha}{2r^2} \quad 11.3.11$$

The rate at which the orbital plane precesses is then

$$\frac{d\Omega}{dt} = \frac{3J_{2,0}a^2\cos\alpha}{2r^2}\frac{d\psi}{dt} = \frac{3J_{2,0}a^2\cos\alpha}{2r^2}\omega' \quad 11.3.12$$

where ω' is the angular velocity of satellite in its orbital plane. If we accept the value of ω' for a satellite in the field of a spherical earth:

$$\omega' = \frac{(ga^2)^{1/2}}{r^{3/2}}$$

we obtain finally

$$\frac{d\Omega}{dt} = \frac{3}{2} J_{2,0} \cos \alpha \left(\frac{a}{r}\right)^2 \sqrt{g} \frac{a}{r^{1.5}} \qquad 11.3.13$$

which indicates that the rate of precession varies directly as $J_{2,0}$, and is controlled also by cos α and r. For example, with the value of $J_{2,0}$ of 0.001, corresponding to the earth, the rate of rotation of the plane of a satellite with $\alpha = 20°$, at a height of 400 nautical miles, is about 6.5° per day. Since $\frac{d\Omega}{dt}$ can be measured to 0.1 per cent or better, it is apparent that Equation 11.3.13 provides a very sensitive method for the determination of $J_{2,0}$. For this reason the value of the flattening of the spheroid, f, became better determined within one year of the launching of the first artificial satellite in 1957 than it had ever been in pre-satellite days. As will be shown in the next section, there is international agreement on the value of $J_{2,0}$ to five significant figures, individual determinations agreeing to within a few places in the sixth figure. It is one of the best determined physical parameters of the earth.

The behavior of the orbital path in its plane is governed by the first equation of 11.3.5. When the corresponding equation is evaluated for the case of an elliptical orbit, it is found that the ellipse revolves in its plane, at a rate which is also dependent upon $J_{2,0}$.

The treatment of the dynamics for the case of higher-order terms in the potential is considerably more complex (Kaula, 1966; Caputo, 1967; Lundquist and Veis, 1966) and will not be given here.

Precise tracking of a satellite to determine perturbations of the orbit can yield higher harmonics in the gravitational field. Notice, however, a fundamental difference between zonal harmonics (for which $m = 0$ and the contributions are independent of longitude) and the tesseral harmonics (which are longitude-dependent). Rotation of the earth beneath the satellite does not affect the perturbing influence of the former, and the perturbations build up with each successive pass of the satellite. These can then lead to long-term, observable changes in the orbit, as we saw above for $J_{2,0}$. In contrast, the effect of tesseral harmonics is usually averaged out as the earth rotates beneath the orbit. The only exception is if the satellite orbit is such that the satellite follows the same tracks above the earth on successive days; such satellites are said to be *resonant* (King-Hele, 1972). In the case of a polar orbit, the condition is equivalent to the satellite making an integral number of circuits per sidereal day. Since the period of most available satellites is of the order of 1.5 to 2 hours, resonance with 9 to 15 equally spaced tracks around the earth is possible. The orbits of these satellites are sensitive to perturbations caused by tesseral harmonics of the same order number, as Figure 11.4 suggests.

A limitation to the determination of harmonics of increasing degree (i.e., decreasing wavelength on the earth's surface) is provided by the fact that the effects are attenuated more and more rapidly with height, so that low-level

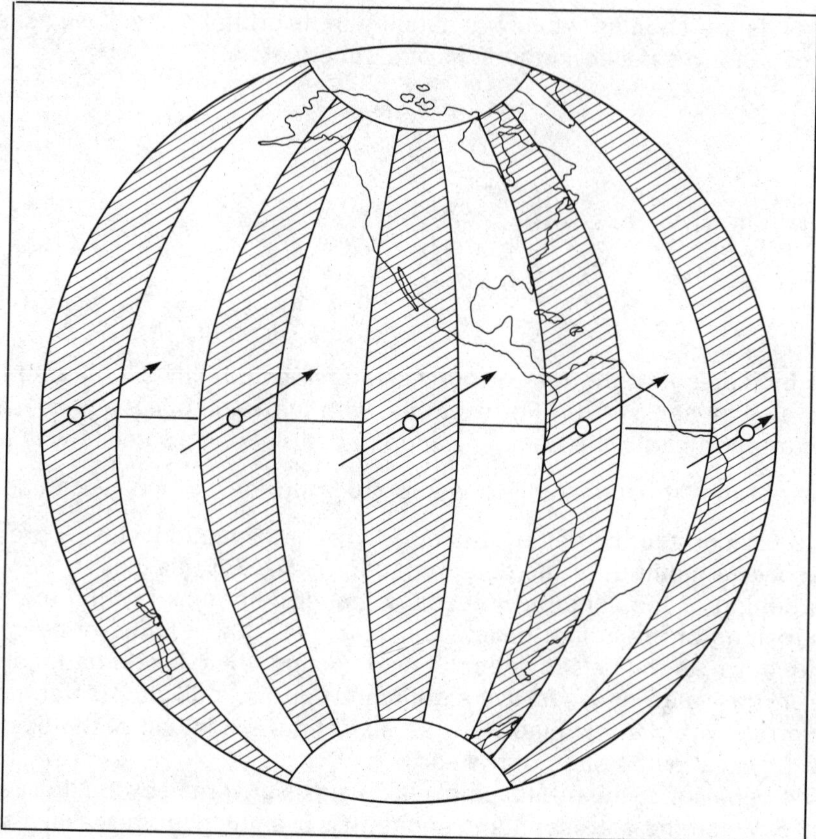

Figure 11.4 Example of the path of a satellite resonant to a ninth-order meridional harmonic in the gravitational field. The satellite must retrace precisely its paths over the earth on future passes.

satellites are required to sample them. Two problems arise: low-level satellites experience atmospheric drag, and they are difficult to track continuously with a limited number of ground stations. The most promising approach appears to be in the use of high-altitude satellites for the tracking of low-level ones. Schwarz (1972), in a simulated recovery of an assumed gravity field, concluded that a satellite at an altitude of 200 km, which could be tracked from very high geostationary satellites, could resolve wavelengths as short as 200 km, corresponding to degree 90.

The available spherical harmonic analyses based on satellite observations alone do not yet approach this figure. In fact, most solutions incorporate information from earth-surface gravity measurements to provide control for the higher degree harmonics. Gaposchkin (1974), by including resonant satellite and surface gravity information, has given the expansion of the field complete to degree and order 18, together with some higher terms. His results will be given below pictorially in Figures 11.6 and 11.7. The status of recent similar determinations has been summarized by Rapp (1974).

11.4 GRAVITY FORMULAE

An expression for gravity on the spheroid as a function of latitude, with the form of the expression given by theory as in Equation 11.2.13, and the parameters determined by fitting observations of gravity to it, is known as a gravity formula. It will be apparent from the discussion above that the adoption of the constant B_2 is equivalent to establishing the flattening.

The International Gravity Formula, adopted by the International Union of Geodesy and Geophysics in 1930, was

$$\gamma_0 = 978.049(1 + 0.0052884 \sin^2 \phi - 0.0000059 \sin^2 2\phi) \qquad 11.4.1$$

where ϕ is now geographic latitude.

In this expression, 978.049 gal is the value at the equator, derived from the available observations which were, of course, on the standard defined by the measured value of g at Potsdam, 981.274 gal. The flattening corresponding to the coefficient 0.0052884 is 1/297. The coefficient of $\sin^2 2\phi$ was not determined from observations; rather, the value adopted is that given by theory for an ellipsoid of revolution.

A gravity formula provides a reference against which measured values of g may be compared. Departures of the latter from the smoothed variation thus provide evidence of lateral density variations in the earth. For this purpose, the International Gravity Formula served well for almost 40 years, but it became evident that both of the values 978.049 and 0.0052884 required revision. The probable correction to the Potsdam standard has been discussed in Chapter 10, while the value of the coefficient $J_{2,0}$, as determined from satellite orbits, corresponds to a flattening of 1/298.247. For the derivation of a new international formula, therefore, the approach has been to adopt the values MG, $J_{2,0}$, and the equatorial radius, and to compute values of gravity from these (Caputo, 1967). The internationally adopted constants of the "Geodetic Reference System 1967" (Bull. Geod. *86*, 1967, p. 367) are:

$$a_{\text{equatorial}} = 6{,}378{,}160 \text{ m}$$

$$MG = 398{,}603 \times 10^9 \text{ m}^3/\text{sec}^2$$

$$J_{2,0} = 10{,}827 \times 10^{-7}$$

A question may be raised at this point regarding the mass of the earth's atmosphere, which is largely within the orbits of satellites, but external to observers on the earth's surface. It appears, however, that to the accuracy quoted above, the effect of the atmosphere is negligible on values of MG or $J_{2,0}$ determined from satellite orbits, and these values may be adopted for determining gravity on the spheroid. The most convenient expression for computing γ_0 on the new standard is:

$$\gamma_0 = 978.03185\,[1 + 0.005278895 \sin^2 \phi + 0.000023462 \sin^4 \phi]$$

11.5 THE FIGURE OF A HYDROSTATIC EARTH

It is of the greatest interest to determine whether the figure of the actual earth, as given by the flattening of the spheroid, corresponds to the equilibrium form of a rotating fluid mass. If it does not, long-term strength in the interior is implied, to maintain the departure from equilibrium.

Intuition suggests that an ellipsoid of revolution is a possible form for the external surface, but this is true only for a fluid of uniform density (Darwin, 1899). The equilibrium form of a fluid whose density increases toward the center was considered by Clairaut and treated in detail by Darwin (1899) and de Sitter (1924). It was found that such a variation in density introduces the next higher harmonic in form, corresponding to a departure from the ellipsoid, but for any reasonable density distribution in the earth this departure is very small. Secondly, the flattening f was found to be related to the precessional constant H (related to the period of precession, Chapter 1, and given by $(C-A)/C$, Appendix B) and to the constant $C_{2,0}$ or $J_{2,0}$ in the spherical harmonic expansion for the potential. This relation suggested that the flattening of an earth in hydrostatic equilibrium could be calculated, but $C_{2,0}$ was not well determined in the pre-satellite era, and the calculations of the hydrostatic flattening therefore contained approximations, which made the comparison with actual flattening uncertain. Caputo (1965) showed that, for hydrostatic equilibrium, $C_{2,0}$ and H must follow a linear relation (Fig. 11.5) while the known parameters for the actual earth, based on satellite observations and the precession, do not plot on this line. This indicates immediately that the earth is not in hydrostatic equilibrium, but the adoption of the most representative equivalent hydrostatic flattening is less straightforward. Caputo suggests holding the ratio $C_{2,0}/H$ fixed at the value measured for the actual earth; the corresponding hydrostatic would be 1/299.49. Comparing this with the adopted value for the actual flattening, 1/298.247, we see that the real earth has a greater flattening, by about one part in 300 of f, than an earth in hydrostatic equilibrium. The difference could arise from a lateral variation in mass distribution of very low harmonic degree, or could indicate that the earth has inherited and preserved a hydrostatic form appropriate to a larger value of ω. Recent investigations, however, suggest that the latter explanation is unlikely (Section 13.2).

Hydrostatic equilibrium within the earth would imply that surfaces of equal density are also equipotentials. The flattening of these surfaces decreases toward the center. Bullard (1948) obtained the value 1/390 for the flattening of the core boundary, on the hydrostatic theory.

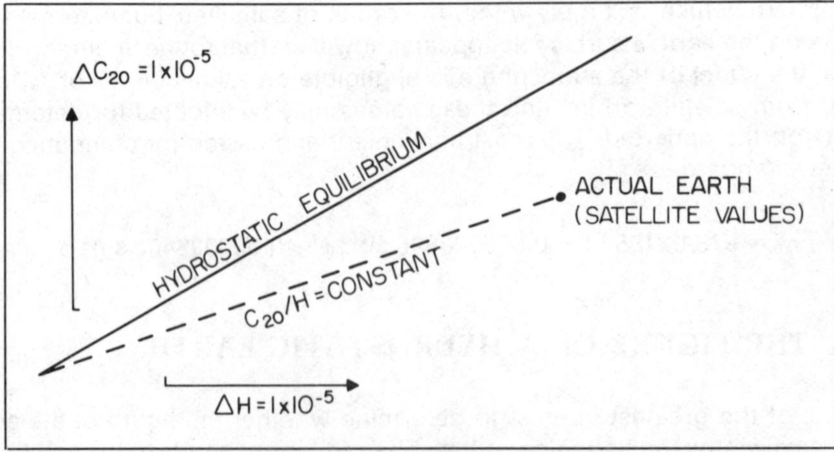

Figure 11.5 Relations between the parameters $C_{2,0}$ and H for an earth in hydrostatic equilibrium (solid line) and the actual earth. The intersection of the solid and broken lines gives an estimate of the equivalent hydrostatic quantities for the earth. (After Caputo)

11.6 THE GEOID

Inhomogeneities in the distribution of mass within the earth produce warpings of the actual sea-level equipotential surface, or geoid, from the regular form of the spheroid. The same inhomogeneities produce departures of the measured value of g, at sea level, from the value predicted by formula; the aim is then to use the variations in g to determine the warpings of the geoid.

Since the geoid is an equipotential surface of the actual field, the effect of a local mass excess, which adds potential U, is to produce an outward *warp* of the geoid through a distance N from the spheroid, where

$$gN = U \tag{11.6.1}$$

and g is the local value.

A number of problems complicate the simple picture. The effect of mass in the earth's crust above sea level has already been neglected, in the derivation of the spheroid; we shall postpone once more the consideration of it, and consider all mass to lie within the geoid. Secondary, the warping N at any point is, of course, controlled by the total effect of mass excesses and deficiencies over the earth. Stokes (1849) obtained an integral expression for N at any point, in terms of the gravity anomalies, or departures of g from γ_0, taken over the surface of the whole earth. The gravity anomaly, Δg, is given by $(g-\gamma_0)$, where g is the measured value at sea level, and γ_0 is the value given by formula for the same latitude. We note that Δg arises from two sources. One is the attraction of anomalous masses, whose effect can be written $-\partial U/\partial r$ if U is the anomalous potential. The second is the effect of additional height N, since g relates to a point on the geoid, and γ_0 to a point on the spheroid. We shall establish in Section 12.1 that the height effect can be written

$$\frac{-2g}{r} N = \frac{-2U}{r} \tag{11.6.2}$$

$$\therefore \; \Delta g = \frac{-\partial U}{\partial r} - \frac{2U}{r} \tag{11.6.3}$$

Stokes, working from Equation 11.6.3, obtained an expression for U, and therefore N, at a point, in terms of Δg at all points over the globe. For N, his expression was

$$N = \frac{a}{2\pi g} \int \Delta g f(\psi) \, d\sigma \tag{11.6.4}$$

where ψ is the angular distance between the points of observation at the element of surface $d\sigma$,

$$f(\psi) = [\tfrac{1}{2} \operatorname{cosec} \psi/2 - 1 - \cos \psi] + 3[1 - \cos \psi = 2 \sin \psi/2$$
$$- \cos \psi \log_e(\sin \psi/2 + \sin^2 \psi/2)]$$

and a and g are taken to be mean values.

In theory, N at any point is determined uniquely if Δg is measured, even though the form of the disturbing masses is not known. The great difficulty with the application of Stokes' formula has been the slow convergence of the weighting function $f(\psi)$. It turns out that gravity anomalies from parts of the surface remote from the station have an appreciable effect, and until recently there have been few places on earth where N could be evaluated without uncertainties from unsurveyed areas.

It has proven to be more fruitful to proceed from the spherical harmonic expansion of the field. We let the actual potential be written in the form introduced in Equation 11.3.1:

$$U_1 = \frac{MG}{r} \sum_{l=0}^{\infty} \left(\frac{a}{r}\right)^l \sum_{m=0}^{l} P_l^m(\cos\theta)[C'_{lm} \cos m\lambda + S'_{lm} \sin m\lambda] \qquad 11.6.5$$

and the normal potential U_0 by a similar expression, in which the constants are C_{lm} ($C_{lm} = 0$ if $m \neq 0$) and $S_{lm} = 0$. The anomalous potential is $U = U_1 - U_0$. We now imagine Δg to be expanded in a similar series with coefficients $\Delta g_{C_{lm}}$ and $\Delta g_{S_{lm}}$; and using Equation 11.6.3 evaluated at some mean radius a, we obtain the following relation between corresponding coefficients:

$$\begin{Bmatrix} \Delta g_{C_{lm}} \\ \Delta g_{S_{lm}} \end{Bmatrix} = (l-1)\bar{g} \begin{Bmatrix} C'_{lm} - C_{lm} \\ S'_{lm} \end{Bmatrix} \qquad 11.6.6$$

where $\bar{g} = \frac{MG}{a^2}$ is a mean value of g.

From Equation 11.6.1 we have, for the constants in an analogous expansion for N,

$$g \begin{Bmatrix} N_{C_{lm}} \\ N_{S_{lm}} \end{Bmatrix} = \frac{MG}{a} \begin{Bmatrix} C'_{lm} - C_{lm} \\ S'_{lm} \end{Bmatrix} \qquad 11.6.7$$

and if g is replaced by $\bar{g}$,

$$\begin{Bmatrix} N_{C_{lm}} \\ N_{S_{lm}} \end{Bmatrix} \frac{1}{a} = \begin{Bmatrix} C'_{lm} - C_{lm} \\ S'_{lm} - S_{lm} \end{Bmatrix} \qquad 11.6.8$$

Finally,

$$\begin{Bmatrix} N_{C_{lm}} \\ N_{S_{lm}} \end{Bmatrix} = \frac{a}{(l-1)\bar{g}} \begin{Bmatrix} \Delta g_{C_{lm}} \\ \Delta g_{S_{lm}} \end{Bmatrix} = a \begin{Bmatrix} C'_{lm} - C_{lm} \\ S'_{lm} - S_{lm} \end{Bmatrix} \qquad 11.6.9$$

within the limitations involved in the substitution of mean values.

The warpings of the geoid can thus be synthesized from expansions of either the gravity anomalies or the anomalous potential, the latter being given by observations on satellites. Geoids have been constructed on the basis both of surface measurements of gravity (Kaula, 1966) and of satellite observations (Guier and Newton, 1965; Gaposchkin, 1966; Gaposchkin and Lambeck, 1970; Gaposchkin, 1974). Figure 11.7 exhibits the geoid based upon the spherical harmonic analysis given in the last reference. It indicates that the undulations, or departures from the spheroid, range up to about 100 meters, for example in the depression south of India.

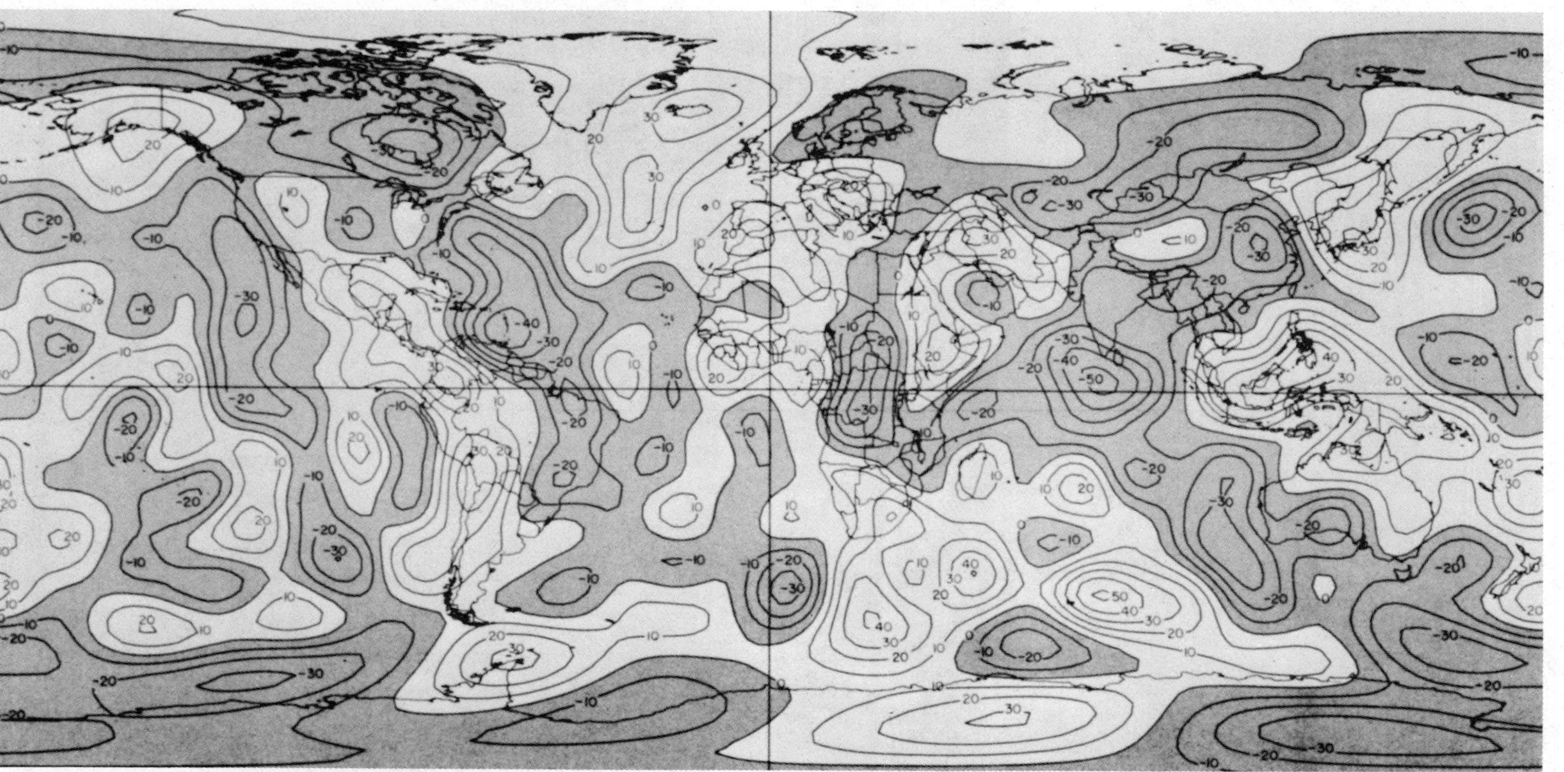

Figure 11.6 Gravity anomaly in milligals relative to ellipsoid of f = 1/298.256, based on satellite and terrestrial data; harmonics to eighteenth degree. Courtesy of E. M. Gaposchkin.

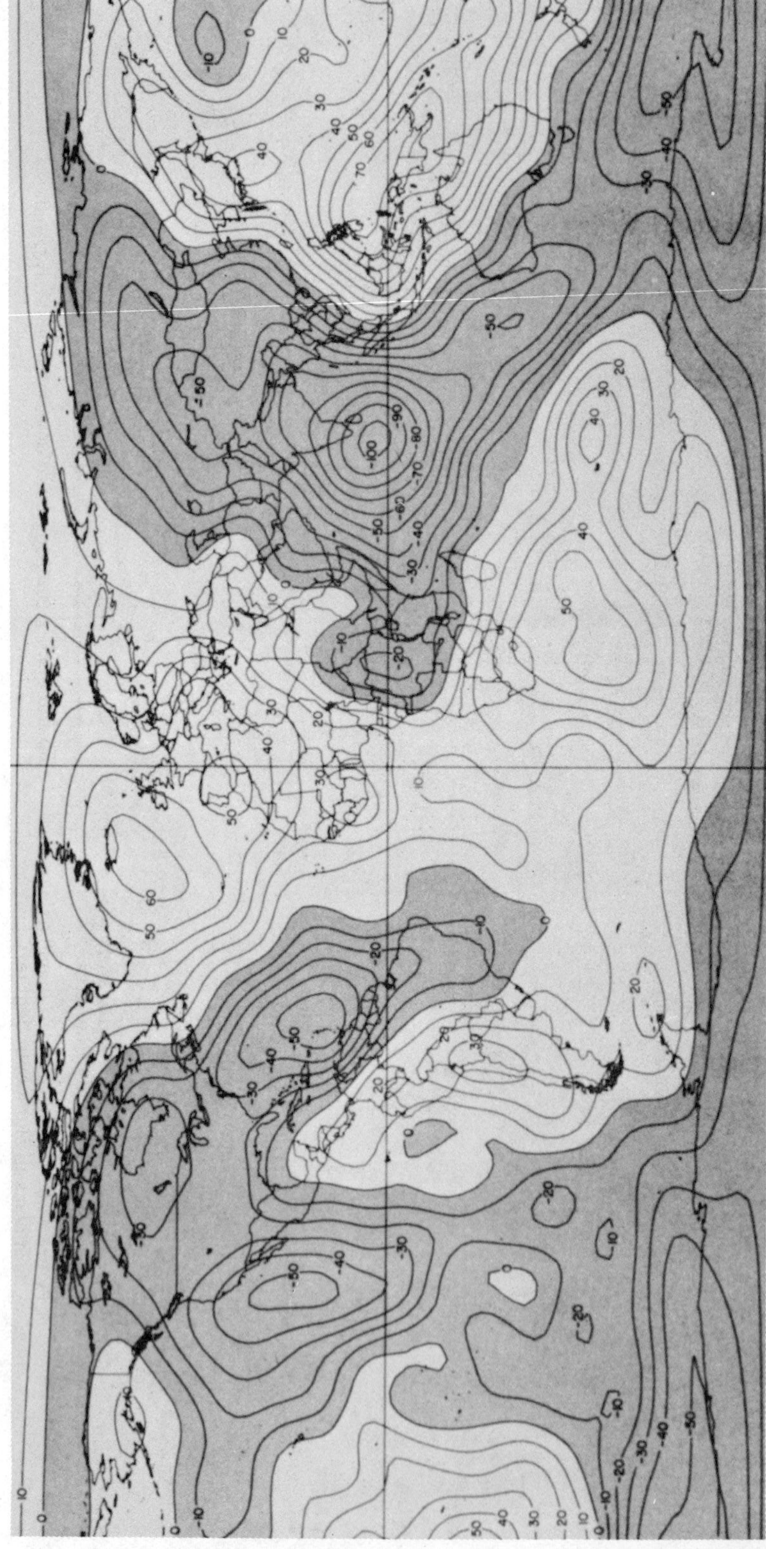

Figure 11.7 Geoid height in meters, relative to ellipsoid of f = 1/298.256, based on satellite and terrestrial data; harmonics to eighteenth degree. Courtesy of E. M. Gaposchkin.

From the geophysicist's point of view, a map of the geoid is essentially a pictorial representation of the low harmonics of the gravitational field, Equation 11.6.9 indicating that higher harmonics are suppressed as $1/l-1$. This can be seen by comparing the geoid of Figure 11.7 with the gravity field itself, as shown in Figure 11.6, computed from the same spherical harmonic expansion. Shorter wavelength features are much more prominent on the latter. The conclusion to be drawn from Figures 11.6 and 11.7 is that there are important widespread inhomogeneities in mass distribution within the earth. Looking ahead somewhat, we may indicate that these mass distributions may be taken as evidence of strength in the earth, if they are elastically supported, or as evidence of no strength, if they result from differences in density in a convecting fluid.

For the geodesist, the geoid is an important reference surface. The simple technique of spirit leveling to determine elevations measures differences in height above the geoid, so that a knowledge of the form of the latter is necessary to fix points in space. Also, the actual direction of gravity at any point is normal to the geoid. The direction of the vertical, or plumb line, at any point on earth is therefore deflected from normal by an angle equal to the slope of the geoid at that point. This *deflection of the vertical* is a critical quantity in any measurements which are made relative to the local plumb line direction. A familiar case is the determination of position by measuring, with a theodolite, the angle between the local vertical and the direction to some star. The computed position on earth will be in error by an amount proportional to the local deflection.

Problems

1. Show that, to the order of 1 part in 300, the International Gravity Formula (Equation 11.4.1) corresponds to a spheroid of flattening 1/297, and that potential is constant on this spheroid (using the results of Section 11.2).

2. If an aircraft flew at a constant height above sea level over a region where the geoid had undulations of wavelength λ measured along the track, would the plane suffer vertical accelerations? What would the order of magnitude and sign of these be, if the gravity anomalies responsible for the undulations had amplitudes Δg?

3. Evaluate Equation 11.3.13 for the moon as a satellite of the earth, and compare your result with the known motion of the moon. Is there an analogous motion for the earth as a satellite of the sun?

4. Very often in geodesy, the deflection of the vertical is desired, rather than the geoidal undulation N. (The deflection at any point is the angle between the normal to the geoid and the normal to the spheroid.) Derive expressions for the deflections in the north-south and east-west vertical planes, in terms of the derivatives of N with respect to latitude ϕ and longitude λ; then find expressions for these analogous to Equations 11.6.4 and 11.6.9.

BIBLIOGRAPHY

Bullard, E. C. (1948) The figure of the earth. *Mon. Not. R. A. S., Geophys. Suppl.*, **5**, 186.
Caputo, Michele (1965) The minimum strength of the earth. *J. Geophys. Res.*, **70**, 955.
Caputo, Michele (1967) *The gravity field of the earth from classical and modern methods.* Academic Press, New York.
Clairaut, H. C. (1943) *Théorie de la figure de la terre.* Paris.
Darwin, Sir G. H. (1910) *Scientific Papers.* Cambridge Univ. Press. Cambridge.

de Sitter, W. (1924) On the flattening and the constitution of the earth. *Bull. Astron. Inst. Netherlands*, **55**, 97.

Gaposchkin, E. M. (1966) Tesseral harmonic coefficients and station coordinates from the dynamic method. In *Smithsonian Astrophysical Observatory, Sp. Report 200,* ed. C. A. Lundquist and G. Veis.

Gaposchkin, E. M. (1974) Earth's gravity field to the eighteenth degree and geocentric coordinates for the 104 stations from satellite and terrestrial data. *J. Geophys. Res.*, **79**, 5377–5411.

Gaposchkin, E. M., and Lambeck, K. (1970) 1969 Smithsonian Standard Earth II. *Smithsonian Astrophysical Observatory, Sp. Report 315.*

Guier, W. H., and Newton, R. R. (1965) The earth's gravity field as deduced from the Doppler tracking of five satellites. *J. Geophys. Res.*, **70**, 4613.

Kaula, W. M. (1966) Tests and combinations of satellite determination of the gravity field with gravimetry. *J. Geophys. Res.*, **71**, 5303.

King-Hele, D. G. (1958) The effect of the earth's oblateness on the orbit of a near satellite. *Proc. Roy. Soc. Lond. A,* **247**, 49.

King-Hele, D. G. (1972) Heavenly harmony and earthly harmonics. *Quart. J. R. A. S.*, **13**, 374–395.

Kozai, Y. (1964) New determination of zonal harmonic coefficients in the earth's gravitational potential. Space research. *Proc. 5th Int. Space Sci. Symp., Florence,* North Holland Publ., Amsterdam.

Lundquist, C. A., and Veis, G. (Editors) (1966) *Smithsonian Astrophysical Observatory, Special Report 200.*

Merson, R. H., and King-Hele, D. G. (1958) Use of artificial satellites to explore the earth's gravitational field: results from Sputnik 2 (1957β). *Nature,* **182**, 640–641.

NASA (1969) *Report of a Study at Williamstown, Mass.* Cambridge, Mass.

Rapp, R. H. (1974) Current estimates of mean earth ellipsoid parameters. *Geophys. Res. Letters,* **1**, 35–38.

Schwarcz, Charles R. (1972) Refinement of the gravity field by satellite-to-satellite Doppler tracking. In *The use of artificial satellites for geodesy,* ed. Soren W. Henriksen, Armando Mancini, and Bernard Chovitz. American Geophysical Union, Monograph 15.

Stokes, G. G. (1849) On the variation of gravity and the surface of the earth. *Trans. Camb. Phil. Soc.*, **8**, 672.

REDUCTION AND INTERPRETATION OF GRAVITY ANOMALIES 12

12.1 CORRECTIONS FOR ELEVATION

In all of the discussions of Chapter 11, measured values of g at sea level were assumed to be available. If g is measured on the land surface, above sea level, it must be corrected for elevation, before it is compared to γ_0. The effect of changing distance from the center of a spherical earth could be obtained immediately. For, if

$$g = \frac{GM}{r^2} \tag{12.1.1}$$

then

$$\partial g / \partial r = \frac{-2GM}{r^3} = -2g/r \tag{12.1.2}$$

At sea level, the gradient is equivalent to -0.3086 mgal/m; the effect of the earth's ellipticity is usually negligible. This change in g is one that could be observed in "free air" (i.e., with no other masses involved) and the correction, $0.3086h$, where h is the height in meters, is known as the free-air correction. If g_0 is the observed value of gravity, the quantity

$$\Delta g_{F.A.} = (g_0 + 0.3086h \times 10^{-3}) - \gamma_0 \tag{12.1.3}$$

is known as the *free-air anomaly,* where g_0 and γ_0 refer to the same latitude.

However, in comparing points on the land surface with points at sea level, there is an effect which is not included in the free-air correction. Bouguer (1749) realized that, at the point P (Fig. 12.1), there would be an additional downward attraction due to the attraction of mass above sea level, and that, contrary to the free-air effect, this would tend to increase g there. If the material is approximated by an infinite horizontal slab, of thickness h and mean density ρ, the attraction of it (Section 12.4) in gals is

$$\Delta g = 200\pi G\rho h \tag{12.1.4}$$

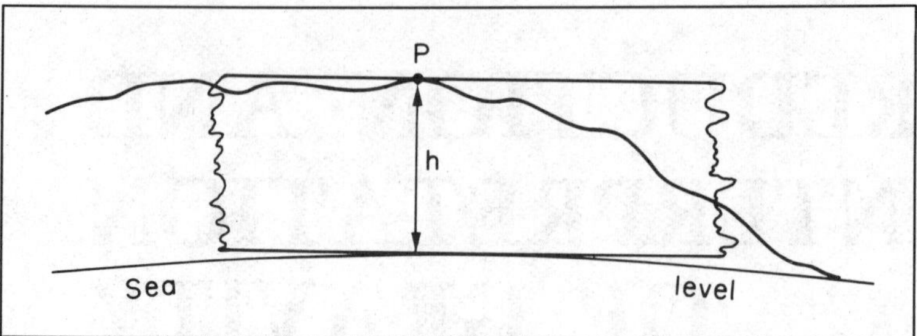

Figure 12.1 Bouguer's approximation of mass above sea level by a slab.

When this effect is combined with the free-air effect, we obtain the *Bouguer anomaly*, Δg_B, where

$$\Delta g_B = [g_0 + (0.3086 \times 10^{-3} - 200\pi G\rho)h] - \gamma_0 \qquad 12.1.5$$

The terrain surrounding a station may appear to make the slab approximation unjustified; however, it is possible to make an additional correction for this effect, which rarely exceeds 1 mgal outside of mountainous areas. Also, the term which was derived in Equation 11.6.2 to correct for the departure of the geoid from the spheroid is often omitted in the calculation of Bouguer anomalies for non-geodetic purposes.

In general, free-air anomalies have been used in geodesy for the determination of the spheroid and geoid, and Bouguer anomalies have been used for the study of internal mass distributions. The free-air anomaly, in the calculation of which something appears to be omitted, can go to large positive values in elevated regions. But its values are, in fact, very nearly those that would be observed if the mass of the topography were condensed onto the geoid, and this permits the geoidal undulations due to the topography to be calculated, without violation of the assumption that there is no mass outside of it.

The Bouguer anomalies characteristically trend toward very large negative values in mountainous areas (Fig. 12.2), suggesting that in allowing for the attraction of mass above sea level, too much has been subtracted.

In the case of gravity measurements made at sea, free-air and Bouguer anomalies require a slightly different definition. Observations made on surface ships require no free-air correction; those made in submarines are corrected upward to sea level by subtraction of the free-air term and also by addition of twice the Bouguer term for the layer of water above the submarine. The difference of the corrected observations and γ_0 gives the free-air anomaly. The Bouguer anomaly for sea stations is obtained by adding a term which effectively replaces sea water by normal crustal rock through the depth of ocean under the station; the anomaly tends toward large positive values over the deep ocean.

12.2 ISOSTASY

One point to be noticed in an examination of Figures 11.6 and 11.7 is that there is no very obvious correlation between the gravity field and the distribution of continents and oceans. If the continents represented simply additional mass on the surface of the earth, one would expect the field to be systematically positive over them. Apparently regions which stand high above sea level are somehow compensated by mass deficiencies beneath. This idea is very old and, as in the case of continental drift, it is difficult now to assign credit for it. The quantitative study dates from the middle of the nineteenth century, when the geodetic survey of India had been extended into the foothills of the Himalayas. This survey included the determination of the positions of stations by astronomical observations, which, as mentioned in Section 11.1, are subject to local variations in the direction of the vertical, and by triangulation, whose results are independent of the local vertical direction. As might be expected, stations closest to the mountains exhibited the greatest difference between astronomical and geodetic position. (A difference of 100 seconds of arc in latitude or longitude is considered to be very large.) In 1855, Pratt published a study of these differences, in which he included direct calculations of the effect to be expected at each station from the attraction of the observed mass of the Himalayas. He pointed out that differences were everywhere smaller than those predicted by this calculation. Very soon, Airy (1855) offered an explanation (Fig. 12.3) in terms of a light "crust," whose lower boundary reflected the topographic height. It must be remembered that this was many years before the recognition of the crust seismologically, and was a remarkable prediction for its day. Pratt (1859) published a second paper, in which he offered an alternative explanation: that the density above a certain level in the earth varies inversely as the topographic height. In a sense, he visualized mountains being produced by vertical expansion of the near-surface material. Either mechanism provided a "root" or region of mass deficiency beneath mountains and could be adjusted to explain quantitatively the Indian observations. Notice that neither Pratt nor Airy had access to measurements of the intensity of gravity; they worked entirely with the direction of the field.

Relations between the parameters for either mechanism, shown in Figure 12.3, may be derived by expressing the fact that the mass per unit area in any column extending to the "depth of compensation" is to be the same as for a column at sea level. For the Airy case with topographic height h above sea level,

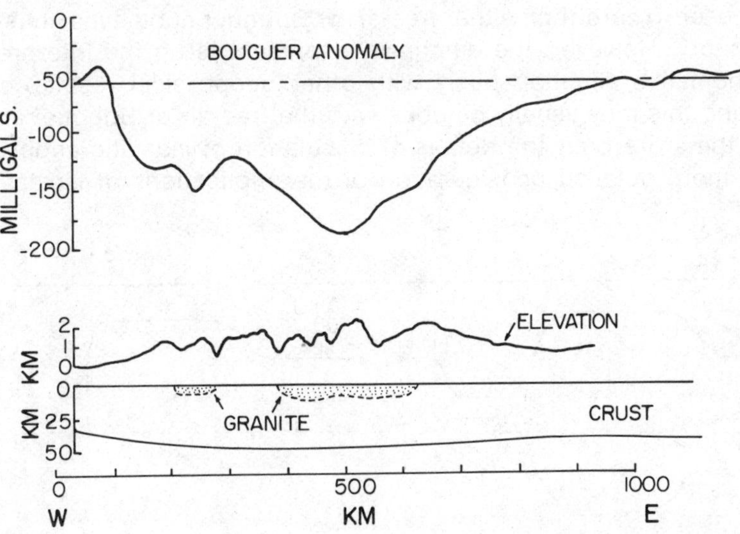

Figure 12.2 Negative Bouguer anomaly observed across the Canadian Cordillera, showing the inverse relation between anomaly and elevation. The interpretation shown attributes most of the mass deficiency to crustal thickening, with an additional influence of low-density granite bodies. (After Garland and Tanner)

$$\rho_c (H + h + y) = \rho_c H + \rho_s y$$

where y is the additional crustal thickness or "root." Thus

$$y = h \frac{\rho_c}{\rho_s - \rho_c}$$

so that the base of the crust forms a magnified mirror image of the topography. For the Pratt mechanism

$$\rho (H + h) = \rho_n H$$

or

$$\rho = \rho_n \frac{H}{H + h}$$

giving the density ρ of any column of height h.

Neither Pratt nor Airy used the word *"isostasy,"* which was introduced by Dutton (1889) to describe the condition of the compensation of the topography, and a state of hydrostatic stress below a certain depth. The tendency of the Bouguer anomaly to go to large negative values in elevated regions is evidence of compensation by concealed mass deficiencies. If the effects of these deficiencies were added to the Bouguer anomaly, and if compensation were nearly complete, a new anomaly close to zero should be obtained. For compensation by either the Pratt or Airy mechanism, the effect on gravity of the roots of any given topography may be calculated; when this effect is added to the Bouguer anomaly, the so-called isostatic anomaly is obtained. The isostatic anomaly is not unique, but depends on the parameters (density, depth of compensation) of the model chosen.

A great deal of work has been done on the testing of isostasy through the calculation of isostatic anomalies (Heiskanen and Vening Meinesz, 1958). It was found that, with a suitable choice of parameters, isostatic anomalies remained closer to zero than either free-air or Bouguer anomalies, thus supporting the theory. However, the emphasis at present is on the interpretation of gravity anomalies simultaneously with other geophysical results, especially seismic, and this may usually be done with the free-air or Bouguer anomalies. We shall therefore omit the details of calculation of isostatic anomalies, and defer the more detailed consideration of the implications of isostasy for the

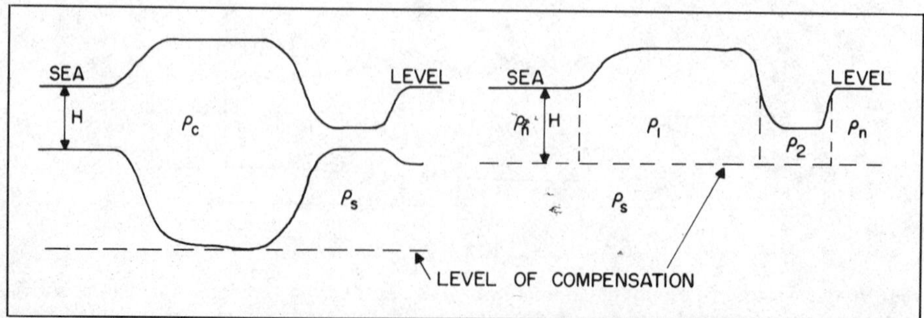

Figure 12.3 Views of isostatic compensation proposed by Airy (*Left*) and Pratt. In Airy's model, ρ_c is the constant crustal density; in Pratt's, ρ_n is the normal density of the crust.

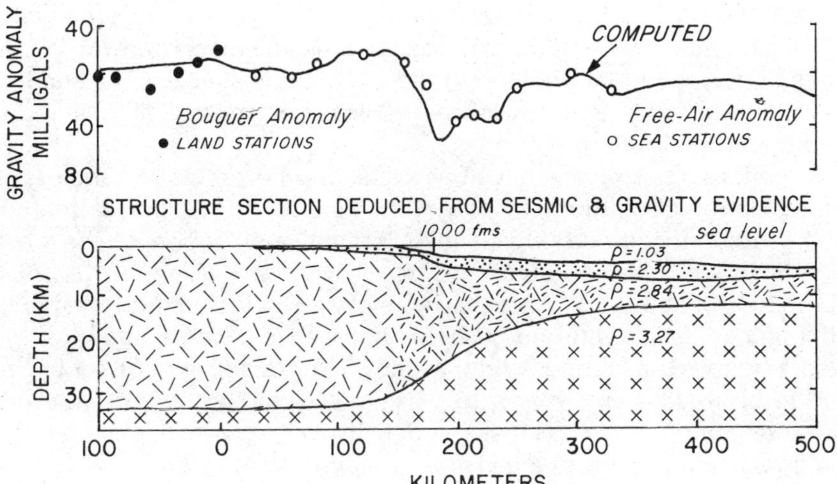

Figure 12.4 Gravity anomaly profile across the eastern coast of the United States obtained by combining Bouguer anomalies for land stations and free-air anomalies for sea stations. The ocean is included as part of the structure; its mass deficiency is nearly compensated by the thinning of the crust. After Worzel and Shurbet (1955).

rheological nature of the earth to the following chapter. The important point to bear in mind is that, although recent developments have made any single model of isostatic compensation appear oversimplified, the principle that the earth over broad areas tends toward a state of isostasy remains very real, and this principle must be provided for in any theory of tectonics.

We conclude this section with a further look at the significance of the free-air and Bouguer anomalies. If a topographic feature, such as a plateau, is fairly broad and is compensated, the free-air anomaly over it will be only moderately positive. This is because the negative effect of the roots will be almost as great as the positive effect of mass above sea level. Very large free-air anomalies, over land or sea, indicate significant departures from isostatic equilibrium. Secondly, it was mentioned above that the Bouguer anomalies become very large and positive over the oceans. This is a consequence of the method of computing the Bouguer anomaly for sea stations and the fact that the apparent mass deficiency in the oceans is largely compensated for by the difference in thickness of continental and oceanic crust. The free-air anomaly at sea stations remains much closer to zero, except in the presence of anomalies due to local structures, and it is preferred by most workers for the interpretation of profiles over the oceans. For land stations, the Bouguer anomaly remains (on the average) close to zero for points at low elevations. In profiles crossing continental margins (Fig. 12.4), Bouguer anomalies on land have actually been connected to free-air anomalies at sea. This is permissible provided the water layer beneath oceanic stations is included as part of the structure whose effect is calculated to explain the profile.

12.3 INTERPRETATION

The complete physical interpretation of the anomalous gravitational field would involve the deducing of a mass distribution which would uniquely produce the observed field. It follows from the characteristics of potential fields that such a unique interpretation is not possible. One may, however, place useful limits on the form and position of the anomalous masses. The second part of the interpretation is then to attach a physical or geological meaning to the inferred mass distribution.

Two general approaches have been used in interpretation. One of these is known as the direct method, in spite of the acknowledged lack of uniqueness, while the other technique, which involves the comparison of the observed anomaly field with the computed effects of hypothetical models, is called indirect. In the direct method, some parameters are selected, and the best mass distribution consistent with the observed anomalies is computed directly from the field. For example, suppose that, within an area small enough for the earth's surface to be treated as a plane, it is desired to find a solution in terms of a surface distribution of excess mass of density $\sigma(x, y)$ over a plane at depth h. The field very close to this plane is given (Section 12.4) by $\Delta g_{x,y,h} = 2\pi G\sigma(x, y)$, so that if $\Delta g_{x,y,h}$ were known, $\sigma(x, y)$ could be obtained directly. It turns out that the field at depth, $\Delta g_{x,y,h}$, can be computed from the surface field, so that a "direct" interpretation is possible. A method of downward continuation of the field that is straightforward, at least in principle, is that of Tsuboi (1938). Suppose that there exists, at depth h, a plane distribution σ that is a harmonic function of x:

$$\sigma = \sigma_0 \cos px \tag{12.3.1}$$

Then

$$\Delta g_h = 2\pi G\sigma_0 \cos px \tag{12.3.2}$$

and the field on the earth's surface is found to be:

$$\Delta g_0 = 2\pi G\sigma_0 e^{-ph} \cos px = e^{-ph}\Delta g_h \tag{12.3.3}$$

If any gravity profile of general form, but with variation in the x-direction only, is resolved into Fourier components, each component may be projected downward, to give the corresponding component of Δg_h, through multiplication by e^{+ph}. From these components, σ may then be synthesized. The calculation may be extended to gravity variations in two directions, by performing a double Fourier analysis. A term proportional to cos px cos qy is projected downward through multiplication by $e^{+(p^2+q^2)^{1/2}}$. Tsuboi's method is, in fact, the result for a plane earth that is analogous to an inward projection of spherical harmonics.

To investigate the spherical case, suppose that, at a depth z beneath the earth's surface, and therefore at radius $a - z$, there is an excess mass distribution $\Sigma\, \sigma_l S_l$, where S_l is a surface spherical harmonic of degree l. The potential at radius r, where $r > a - z$ (MacRobert, 1947; also Appendix A) is

$$\Delta U = 4\pi G \sum_{l=0}^{\infty} \frac{1}{2l+1} \frac{(a-z)^{l+2}}{r^{l+1}} \sigma_l S_l \tag{12.3.4}$$

and the anomaly in gravity at the earth's surface is

$$\Delta g = 4\pi G \sum_{l=0}^{\infty} \frac{l+1}{2l+1} \left(\frac{a-z}{a}\right)^{l+2} \sigma_l S_l \tag{12.3.5}$$

If the anomaly field is expressed in spherical harmonics, as

$$\Delta g = \Sigma\, g_l S_l \tag{12.3.6}$$

the comparison of corresponding terms in Equations 12.3.5 and 12.3.6 yields

$$\sigma_l = \frac{1}{4\pi G}\frac{2l+1}{l+1}\left(\frac{a}{a-z}\right)^{l+2} g_l \tag{12.3.7}$$

The unknown coefficients, σ_l, of the mass distribution may therefore be obtained from the expansion of Δg. These examples of downward projection of the field show clearly that limits may be placed on the acceptable solutions. For, if a solution is sought at too great a depth, the excess mass density that is inferred oscillates so widely as to be physically unreasonable, because of the factors e^{+ph} or $(a/a-z)^{l+2}$. Obviously a solution which requires negative rock densities must be discarded, but in practice the range of reasonable rock densities puts even more stringent limits on the possible solutions.

12.4 CALCULATION OF ANOMALIES FOR SIMPLE MASS DISTRIBUTIONS

The indirect method of interpretation requires that anomalies be calculated for certain simple forms, for comparison with observed anomaly fields. Normally, the comparison is made with the information displayed in profile form, and we take the earth's surface as plane.

For a general mass distribution (Fig. 12.5) the gravity anomaly, which is the vertical component of the gravitational attraction of the excess mass, is given by

$$\Delta g = G\Delta\rho \int_V \frac{\sin\phi\, dV}{R^2} \tag{12.4.1}$$

where $\Delta\rho$ is the excess density, the integration is over the volume of the body, and all physical quantities are in c.g.s. units. The ease of integration depends upon the form of the distribution. In the case of a sphere of radius R, the anomaly is immediately found to be

$$\Delta g = \tfrac{4}{3}\pi R^3 G\Delta\rho \frac{z}{(x^2+z^2)^{3/2}} \tag{12.4.2}$$

where z is the depth to center, and x is measured from a point above the center.

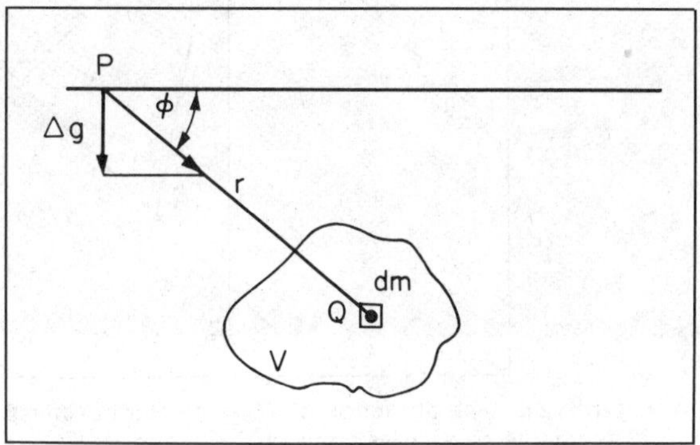

Figure 12.5 Quantities involved in the calculation of gravity anomalies.

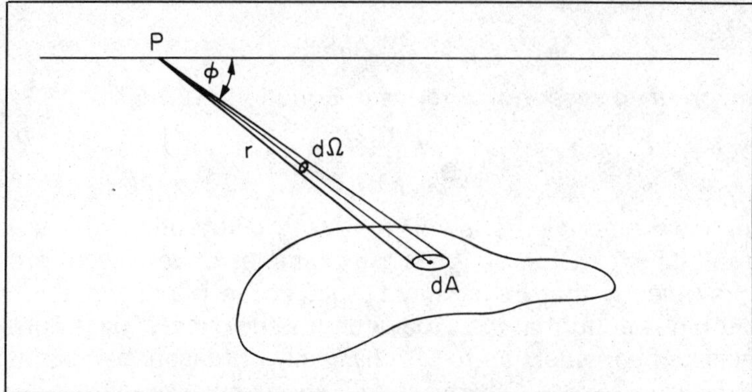

Figure 12.6 Attraction of a flat sheet of mass.

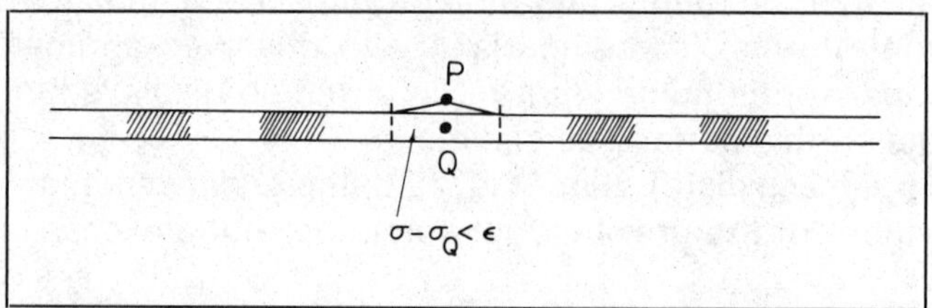

Figure 12.7 Attraction of a sheet with variable surface density of mass. As P approaches the sheet, the solid angle subtended by the circle drawn about Q approaches 2π.

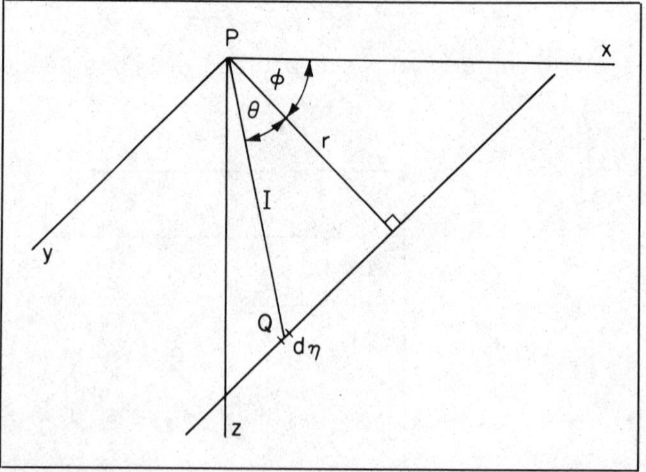

Figure 12.8 The attraction of a line element of mass is obtained by integrating the contribution of segments as θ varies between $-\pi/2$ and $\pi/2$.

For structures whose vertical dimension is small compared to their depth and horizontal extent, a flat sheet approximation is often useful. The attraction due to a sheet of excess density, $\Delta\sigma$ (Fig. 12.6), is

$$\Delta g = G\Delta\sigma \int \frac{\sin\phi\, dA}{r^2} \qquad 12.4.3$$

$$= G \cdot \Delta\sigma \cdot \Omega$$

where Ω is the solid angle subtended by the sheet. If the sheet is of great horizontal extent

$$\Delta g = 2\pi G\Delta\sigma \qquad 12.4.4$$

independent of the distance of the point P. An infinite horizontal slab of thickness h and excess volume density $\Delta\rho$ may therefore be replaced by an assemblage of sheets of excess surface density $\Delta\rho\Delta h$ so that

$$\Delta g = 2\pi G\Delta\rho \cdot h, \qquad 12.4.5$$

as found by Bouguer.

The effects of three-dimensional forms of irregular shape may be computed numerically, by dividing the bodies into plane sheets, each of which is approximated by a polygon. Talwani and Ewing (1960) have given a method adaptable for high-speed computation, which may be used in this case. Alternatively, the body may be approximated by rectangular prisms, the attraction of which is given by Nagy (1966) in a form suitable for computation.

To return to the plane distribution of mass, let us suppose that the surface density is not constant, but is a continuous function of horizontal coordinates x and y. As the observer approaches the sheet, at a point Q (Fig. 12.7) where the density is $\sigma(x, y)$, the limiting value of the field is

$$\Delta g(x, y) = 2\pi G\sigma(x, y) \qquad 12.4.6$$

This may be established by considering a small circle about Q, within which σ does not vary from the central value by more than a small prescribed amount ϵ. In the limit, the angle subtended by the circle, at the observer, is 2π, and Δg has the value shown. This result is the basis for the methods of direct interpretation discussed in the previous section.

Many geological structures are elongated in one direction (the strike) and have a uniform cross-section along the strike. Their effect can be obtained by superimposing the attractions of infinite line elements. For a line parallel to the y axis carrying excess mass m per unit length (Fig. 12.8) the anomaly in gravity is

$$\Delta g = \frac{2Gm \sin\phi}{r^2} \qquad 12.4.7$$

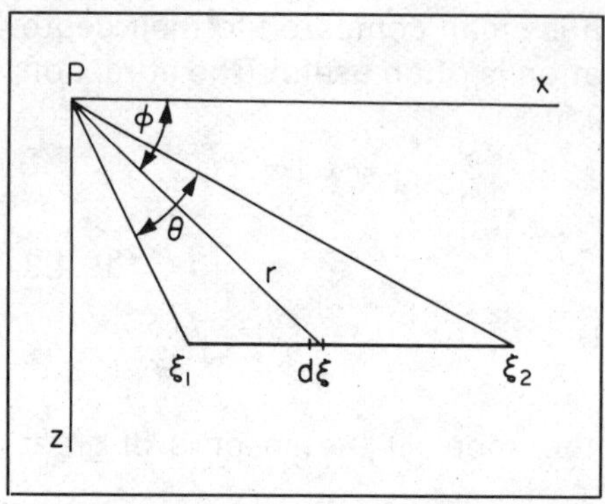

Figure 12.9 Attraction of a two-dimensional sheet, of infinite extent normal to the page.

The effect of two-dimensional forms of finite cross section are then obtained by integration in the x-z plane. For the cylinder of radius R

$$\Delta g = 2\pi G R^2 \Delta\rho\left(\frac{z}{x^2 + z^2}\right) \qquad 12.4.8$$

where z is the depth to the axis and x is measured from a point over the axis. A plane sheet of infinite extent in the y-direction (Fig. 12.9) produces an anomaly

$$\Delta g = 2G\Delta\sigma \int_{\xi_1}^{\xi_2} \frac{\sin\phi \, d\xi}{r} = 2G \cdot \Delta\sigma \cdot \theta \qquad 12.4.9$$

where θ is the plane angle subtended by the trace of the sheet.

Once again, it is hardly worth going to more complicated integral expres-

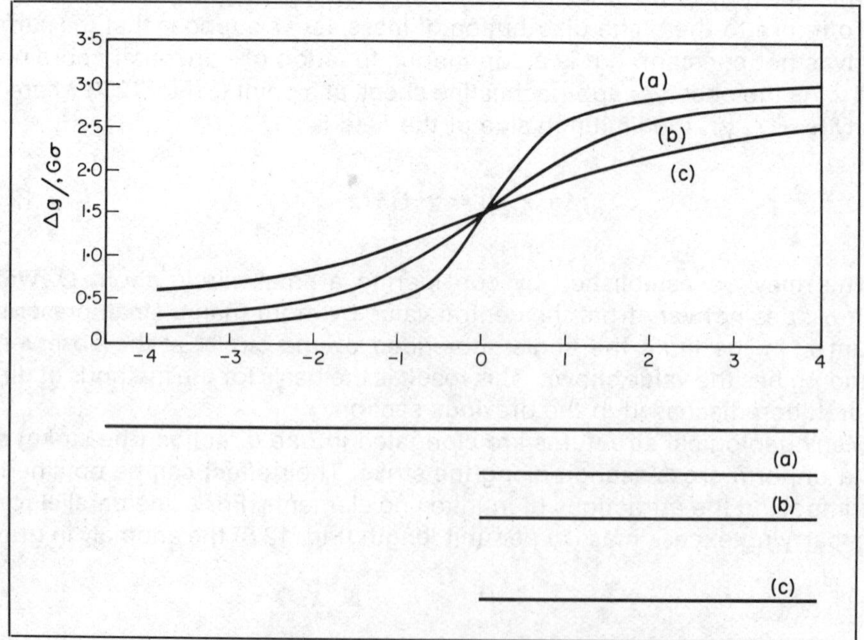

Figure 12.10 Anomaly profile over the edge of a sheet of mass, for three depths of the sheet. Note the reduction in maximum gradient (immediately over the edge) with depth.

sions, as numerical calculations can usually be done by combining simple shapes.

Model calculations can be used to demonstrate a most important characteristic of potential field anomalies (Fig. 12.10). As the source of the anomaly is made deeper, features of the anomaly profile smear out, slopes become less steep, peaks become smaller in amplitude and broader. In fact, for the same excess mass, the area under the anomaly profile remains constant, but short-wavelength effects are filtered out. Depth, as Equation 12.3.3 shows, acts as a low-pass filter on the anomaly field.

The fundamental ambiguity remains, for even though good agreement between the computed effect of a model and the observed anomaly is obtained, there is no guarantee that quite a different model will not produce as good an agreement.

Problems

1. The downward continuation method illustrated in Section 12.3 depends upon the attraction of a sheet of mass in which the surface density varies as cos px. Confirm the expression for this attraction, introduced in Equation 12.3.3.

*2. If there were warpings of the core-mantle interface which could be represented by a 10th degree spherical harmonic, what amplitude of gravity anomaly would be observed at the earth's surface for each 1 km of warp?

3. Show from potential theory that the attraction of a buried horizontal cylinder is identical to that of a massive line element along its axis; then write the expression for the gravity anomaly due to such a cylinder.

4. A deep sedimentary basin consists largely of shale. At the surface, the shales have a density of 2.2 gm/cm^3, but the density increases exponentially with depth, so that at the bottom of the section (5 km deep), their density becomes 2.7 gm/cm^3, the normal crustal density of the region. What gravity anomaly would be observed over the center of the basin, assuming it is extensive in area? What depth would have been estimated for it, if a constant density of 2.2 gm/cm^3 had been assumed?

5. A circular plug of basic volcanic rock, of diameter 10 km, produces a positive Bouguer gravity anomaly of 30 mgals. The density of the basic rock is 3.0 gm/cm^3, and it intrudes rock of density 2.8 gm/cm^3. Assuming that the Bouguer anomalies are accurate to within 1 mgal, what is the best estimate of the depth extent of the plug, and what is the uncertainty of this estimate? (It will, of course, be necessary to derive and investigate the behavior of the expression for the attraction of a fine vertical cylinder.)

6. It has been noted that any gravity anomaly could be due to one of an infinite number of forms. Show, however, that the total excess mass involved is always the same, and that for an isolated body, this mass can be determined by integrating the anomaly to the limits of its perceptibility over the earth's surface (cf. Equation 9.2.7).

7. Find an explicit expression for the maximum gravity gradient over the edge of a semi-infinite sheet of mass (Figure 12.10) in terms of the depth of the sheet.

BIBLIOGRAPHY

Airy, G. B. (1855) On the computations of the effect of the attraction of the mountain masses as disturbing the apparent astronomical latitude of stations in geodetic surveys. *Phil. Trans. Roy. Soc. Lond.*, **145**, 101.

Bouguer, Pierre (1749) *La figure de la terre.* Paris.

Dutton, C. E. (1889) On some of the greater problems of physical geology. *Bull. Phil. Soc., Washington,* **11**, 51.

Heiskanen, W. A., and Vening Meinesz, F. A. (1958) *The earth and its gravity field.* McGraw-Hill, New York.

MacRobert, T. M. (1947) *Spherical harmonics* (2nd Ed.) Methuen, London.

Nagy, Dezsö (1966) The gravitational attraction of a right rectangular prism. *Geophysics,* **31**, 362.

Pratt, J. H. (1855) On the attraction of the Himalaya Mountains and of the elevated regions beyond upon the plumb-line in India. *Phil. Trans. Roy. Soc. Lond.,* **145**, 53.

Pratt, J. H. (1859) On the deflection of the plumb-line in India caused the attraction of the Himalaya mountains and the elevated regions beyond; and its modification by the compensating effect of a deficiency of matter below the mountain mass. *Phil. Trans. Roy. Soc. Lond.,* **149**, 745–778.

Talwani, Manik, and Ewing, Maurice (1960) Rapid computation of gravitational attraction of three-dimensional bodies of arbitrary shape. *Geophysics,* **25**, 203.

Tsuboi, Chuji (1938) *Proc. Imp. Acad. Tokyo,* **14**, 170.

Worzel, J. L., and Shurbet, G. L. (1955) Gravity interpretations from standard oceanic and continental crustal sections. In *Crust of the earth,* ed. A. Poldervaart. Geol. Soc. Amer. Spec. Paper 62.

GRAVITY AND THE INTERIOR OF THE EARTH 13

13.1 THE MECHANISM OF ISOSTATIC COMPENSATION

The fact that isostatic gravity anomalies—calculated on either the hypothesis of Airy or that of Pratt, or on some modification of them—are systematically closer to zero than are Bouguer anomalies shows that some mechanism of isostatic compensation is operative. If there is any relationship between density and seismic velocity, we should expect the mechanism of compensation to be apparent on seismological profiles of the crust and upper mantle. When the idea of the seismological crust became established, the Airy mechanism became widely favored, although a great many earlier studies had concentrated on tests of the Pratt model. If the base of the Airy crust is associated with the Mohorovičić discontinuity, the latter should show the inverse relationship with topography described in Section 12.2. Reasonable density factors for the crust and uppermost mantle are approximately 2.8 and 3.2 gm/cm^3 respectively, so that the multiplying factor for warpings of the discontinuity would be 2.8/(3.2 − 2.8) or about 7. The compilation of seismically determined crustal sections shown in Figure 13.1 for oceans and continents shows that, in a very general way, the relationship is maintained. Most mountainous areas are characterized by a thick crust and most deep oceans by a thin one.

The general relationship between seismic *P*-wave velocity and density was mentioned in Section 7.3: velocity increases with density. More specifically, the relationship is suggested by the curves in Figure 13.2, based on laboratory measurements by different workers. If this empirical relationship is accepted, it means that a density may be assigned to any region in which the seismic velocity has been measured. Worzel and Shurbet (1955) have compared a number of continental and oceanic profiles and have suggested standard density columns (Fig. 13.3) which are in isostatic balance, for oceanic areas of average water depth, and for continental areas at sea level.

The departures from ideal Airy conditions become apparent when the results of seismic crustal studies are examined in more detail. For example, Figure 13.4 shows the velocity layering and crustal thicknesses observed for a number of profiles in the western United States (Steinhart and Meyer, 1961). We observe immediately that elevated areas, such as Arizona and Nevada, do not have as great a crustal thickness as other areas of lesser elevation. Furthermore,

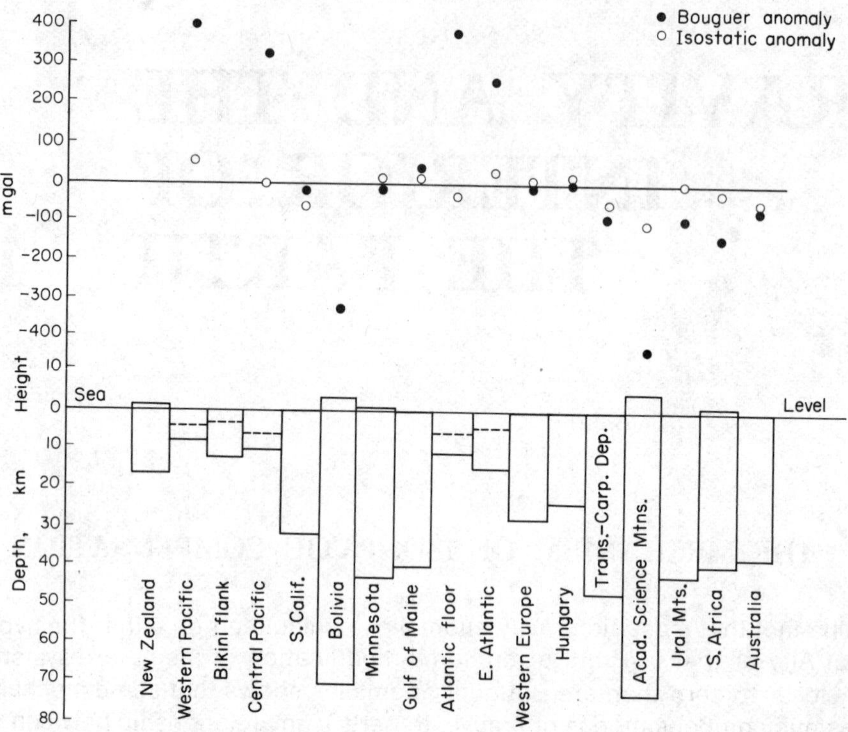

Figure 13.1 Relation between Bouguer anomaly, isostatic anomaly and seismologically determined crustal thickness, for regions in different continents and oceans.

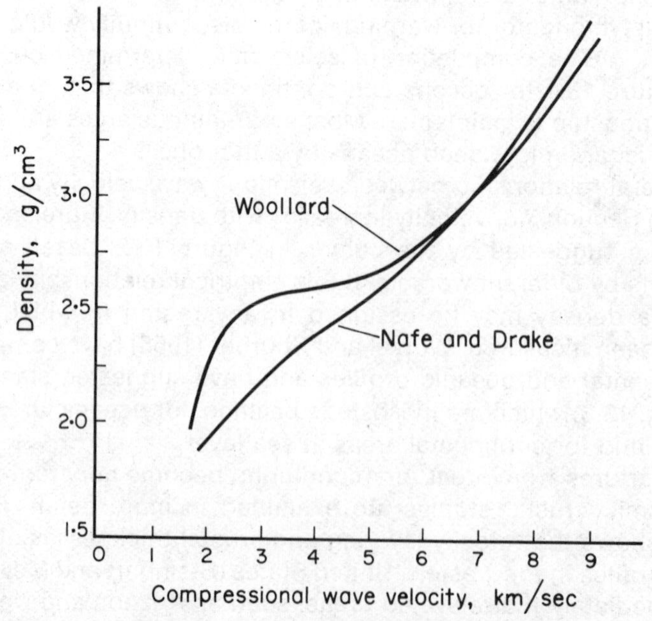

Figure 13.2 Proposed relationship between density and P-wave velocity.

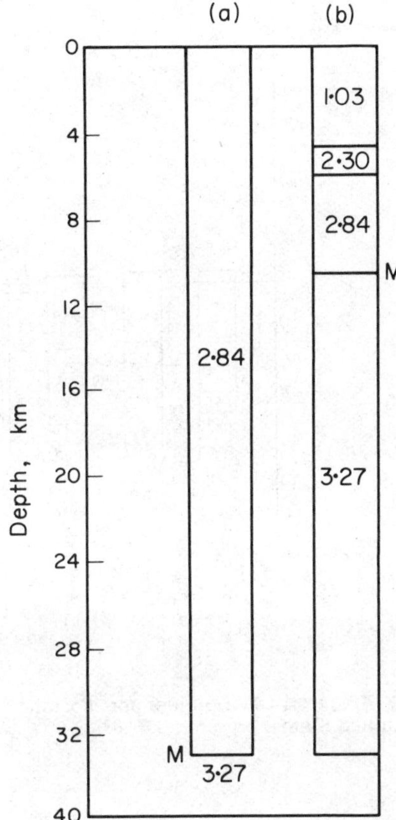

Figure 13.3 Standard columns for continental (*Left*) and oceanic sections. (After Worzel and Shurbet)

if we associate densities with the velocities by means of the curve of Figure 13.2, we may calculate the mass excess, relative to the standard section, down to the Mohorovičić discontinuity. For the Nevada section, with a mean elevation of 1830 meters, it is 3.2×10^5 gm/cm^2. On the other hand, the isostatic gravity anomaly, which is slightly negative, suggests a mass excess of -0.2×10^5 gm/cm^2 (using an infinite slab approximation). The only possible conclusion is that while the topography is nearly compensated, much of this compensation is located in the upper mantle. To some extent this is borne out by the known variations in upper mantle *P*-wave velocity, for although individual profiles in Figure 13.4 show upper mantle velocities greater than 8.0 km/sec, the generalized map, Figure 7.7, suggests that western North America is a region of systematically low upper mantle velocity. We are thus brought to a Pratt-type compensation, with differences in density in the upper mantle contributing to the support of topography in the crust above. It is by no means clear how this situation would arise, but it is possible that, during mountain building, lower density crustal material becomes mixed with the upper mantle, producing the lesser density there (Cook, 1962).

Further evidence of the fact that isostatic compensation prevails, to a good first approximation, is obtained from the detailed analysis of the relation between the Bouguer anomaly and height (Woollard, 1969). For perfect compensation of material above sea level over a broad area, the Bouguer anomaly Δg_B

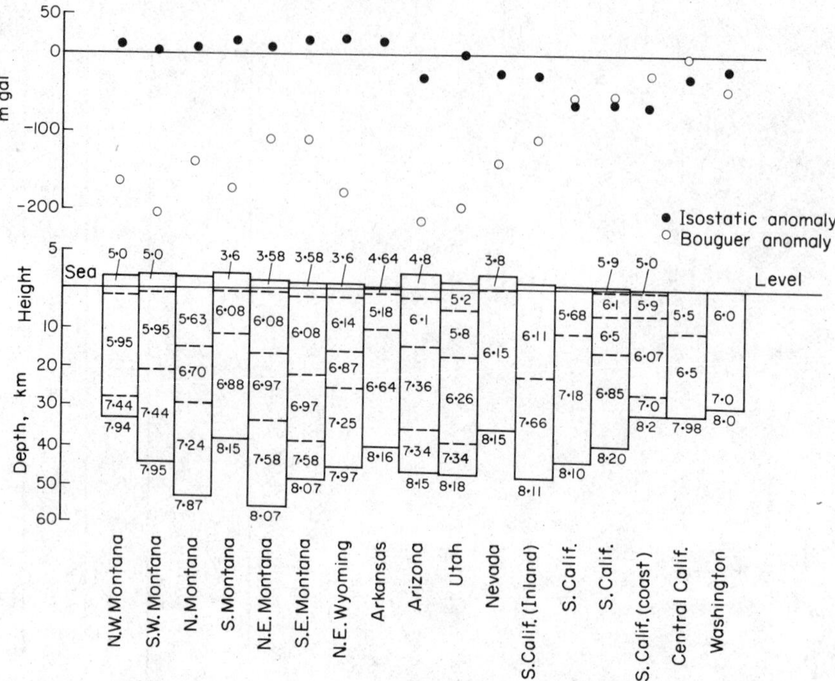

Figure 13.4 Bouguer and isostatic anomalies, and seismic velocity sections, from the western United States.

is negative by just the amount of the Bouguer correction, $2\pi G h \rho_c$, where ρ_c is the crustal density. In a given region, therefore,

$$\Delta g_B = \text{constant} - 2\pi G h \rho_c \qquad 13.1.1$$

Again for perfect compensation, the free-air anomaly is independent of elevation, and the mean free-air anomaly of the region must be the constant of Equation 13.1.1. Thus

$$\Delta g_B = \overline{\Delta g}_{F.A.} - 2\pi G h \rho_c \qquad 13.1.2$$

If we take $\rho_c = 2.67$ gm/cm³, as is usually assumed in making the reductions (ρ_c is here the density of material above sea level, for which 2.67 gm/cm³ is probably a low, but reasonable, estimate; the density shown on Fig. 13.3 is for the crust as a whole) the coefficient of h is -0.1118 mgal/m. Woollard (1969) investigated the relation between Bouguer anomalies, averaged over 1° or 3° "squares," and height, for several continental areas. In general, the relation is one to be expected on the basis of the above model, but different areas may be characterized by different values of the intercept, indicative of abnormal crustal density, or by different gradients, showing departure from perfect compensation of the topography. For the United States, using 3° squares, the relation which fits the observations between heights of 200 and 1700 meters is

$$\Delta g_B = -6 - 0.1005h$$

indicating that the overall crustal density is close to normal, and compensation is rather well established.

13.2 STRESSES DUE TO UNCOMPENSATED LOADS

Lateral variations in the distribution of mass within the earth represent departures from hydrostatic equilibrium which must be supported by non-hydrostatic stresses. If the outer part of the earth is taken to be an elastic solid, the stresses due to certain simple mass distributions may be calculated. We do not rule out the possibility that the masses are supported by, or even result from, some other process such as viscous flow, and this will be considered in the next section.

For the two-dimensional elastic case, we follow Jeffreys (1962) and consider a flat earth, on which a load at the surface ($z = 0$) is placed, with all quantities independent of y. The equilibrium conditions are

$$\frac{\partial p_{ij}}{\partial x_j} = 0$$

or

$$\frac{\partial p_{31}}{\partial x} + \frac{\partial p_{33}}{\partial z} = 0 \qquad \frac{\partial p_{11}}{\partial x} + \frac{\partial p_{31}}{\partial z} = 0 \tag{13.2.1}$$

It follows that a function χ can be found such that

$$p_{11} = \frac{\partial^2 \chi}{\partial z^2}; \qquad p_{13} = -\frac{\partial^2 \chi}{\partial x \partial z}; \qquad p_{33} = \frac{\partial^2 \chi}{\partial x^2} \tag{13.2.2}$$

To satisfy the equations relating stress components to the strains, it may be shown that χ must satisfy

$$\left(\frac{\partial^2}{\partial x^2} + \frac{\partial^2}{\partial z^2}\right)^2 \chi = 0 \tag{13.2.3}$$

For a prescribed load, p_{33} will be given at the surface, and p_{13} will be zero there. The solution of a problem then involves the finding of a function χ, satisfying Equation 13.2.3, subject to these boundary conditions.

As a simple case, we take a surface density of mass varying harmonically, with amplitude σ, so that

$$p_{33} = g\sigma \cos kx, \text{ at } \quad z = 0 \tag{13.2.4}$$

It is easily found that

$$\chi = (g\sigma/k^2)(1 + kz)e^{-kz} \cos kx \tag{13.2.5}$$

so that

$$\left.\begin{aligned} p_{11} &= g\sigma(1-kz)e^{-kz}\cos kx \\ p_{13} &= -g\sigma z e^{-kz}\sin kx \\ p_{33} &= g\sigma(1+kz)e^{-kz}\cos kx \end{aligned}\right\} \quad 13.2.6$$

Having obtained the components of stress as a function of z and x, we must adopt some characteristic of them to specify the non-hydrostatic conditions to which the material is subjected. This is sometimes taken to be the maximum shear stress, and sometimes the stress-difference, or difference between greatest and least normal stress. In either case, the components must be related to the principal axes of stress through the properties of the stress ellipsoid (Section 2.2). The greatest and least principal stresses, P and Q, are obtained by rotating the axes in the $x-z$ plane until these correspond to the geometric axes of the ellipsoid. They are

$$\left.\begin{aligned} P+Q &= p_{11}+p_{33} \\ PQ &= p_{11}p_{33}-p_{13}^2 \end{aligned}\right\} \quad 13.2.7$$

from which we find

$$\left.\begin{aligned} P &= g\sigma e^{-kz}\cos kx + g\sigma kz e^{-kz} \\ Q &= g\sigma e^{-kz}\cos kx - g\sigma kz e^{-kz} \end{aligned}\right\} \quad 13.2.8$$

The stress difference is then $2g\sigma kze^{-kz}$, which reaches a maximum value of $2g\sigma/e$ at a depth $z=$ (wavelength of loading)$/2\pi$. As a first approximation, it may be said that the maximum stress difference is equal to two-thirds of the normal stress produced by the load. The stress difference is independent of x, but the orientation of the stress ellipsoid is not (Fig. 13.5).

For a specific example, we consider a pattern of gravity anomalies with a harmonic form along the x-axis and extended along the y-axis, with an amplitude of 50 mgals and a wavelength of 200 km. Such anomalies would be large (100 mgals peak-to-peak) but not exceptional. The equivalent σ is $50\times10^{-3}/2\pi G$ or 1.2×10^5 gm/cm^2 and the maximum stress difference is about 0.9×10^8

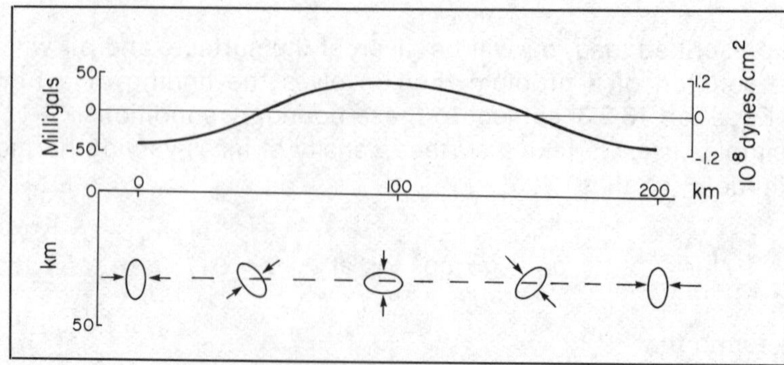

Figure 13.5 Orientation of the stress ellipsoid at the depth of maximum stress-difference beneath a harmonically-varying load.

dynes/cm^2. This occurs at a depth of about 30 km. The gravitational field thus shows that stress differences of between 1×10^8 and 1×10^9 dynes/cm^2 exist in the crust and uppermost mantle. This produces no difficulty in the case of short-wavelength anomalies, because it is known from laboratory measurements that crustal rocks can support stress differences of at least 1×10^9 dynes/cm^2. There is a problem in the support of broad, uncompensated loads if the strength of the earth decreases with depth. As the above analysis shows, the depth at which maximum stress difference occurs is proportional to the wavelength of loading. For the low harmonics in the gravitational field, such as those represented by the undulations of the geoid (Fig. 11.7) this depth would be several hundreds of kilometers. Jeffreys investigated the stress distribution for a model of the mantle once held: that of strength extending in depth to the limit of deep-focus earthquakes. He assumed elastic support above 600 km and no strength below it, and concluded that stress differences of 3.3×10^8 dynes/cm^2 exist down to the 600 km depth. His arguments have been extended by McKenzie (1967) to the more current view of a lithosphere composed of the crust and uppermost mantle floating upon denser, fluid material. For a lithosphere 100 km thick, the maximum stresses found by McKenzie, for wavelengths up to 300 km, are not greatly different from those given by the simple theory, because the maximum stress-difference implied by the latter would occur within the strong layer. Longer wavelengths are shown to require impossibly large stresses in the lithosphere if no support from below is provided.

The limiting case of long-wavelength departure from hydrostatic conditions is the flattening, discussed in Section 11.5. The calculation of stress due to loads imposed on an elastic sphere is not simple (Jeffreys, 1943) and obviously the plane-earth approximation should not be valid. However, a remarkable result is that for a uniform sphere, the maximum stress-difference for loads represented by low spherical harmonics is approximately equal to $g\sigma$, the load amplitude, and therefore not very different from the value obtained above for the plane earth. We may apply this fact to estimate the stress due to the external, non-hydrostatic form. The flattening of the spheroid causes the equatorial radius to exceed the polar radius by about 21 km. A variation in the flattening of 1 part in 300 is equivalent to an additional layer reaching a thickness of 0.72 km at the equator, and for a density 3.5 gm/cm^2, $g\sigma$ would be about 25×10^6 dynes/cm.^{2} For a uniform sphere, this would be approximately the maximum stress-difference, which would occur near the center. Caputo (1965) gives a value of about 70×10^6 dynes/cm^2 for the stress-difference in the mantle, if no strength is assumed in the core. Unless motions in the core or mantle support the departure from hydrostatic form, this figure represents the "minimum strength of the earth."

In particular, if the non-hydrostatic form represented an inheritance from an earlier period, when the rotation rate was higher, the implication would be that this stress had been supported by the earth for hundreds of millions of years. This would place severe limitations on the types of motions which would be possible in the mantle. However, by a very simple and direct argument, Goldreich and Toomre (1969) showed that the hypothesis of the earth's maintaining a figure from an earlier time of faster spin is unlikely. We have seen (Section 11.3) that the principal moments of inertia A, B and C are related by

$$J_{2,0} = \frac{C - \frac{1}{2}(B + A)}{Ma^2}$$

An analoguous relationship, which we do not derive, is

$$J_{2,2} = -\frac{(B-A)}{4Ma^2}$$

Both $J_{2,0}$ and $J_{2,2}$ are well determined by satellite observations, so that all differences between A, B and C for the actual earth are known. The difference between B and A is completely non-hydrostatic, and the non-hydrostatic portion of C can be computed from the non-hydrostatic flattening. Denoting by primes the non-hydrostatic contributions, Goldreich and Toomre give

$$C' - B' = 6.9 \times 10^{-6}\ Ma^2$$

$$B' - A' = 7.2 \times 10^{-6}\ Ma^2$$

These are the differences in principal moments of inertia that would be observed if the earth's rotation on its axis were suddenly stopped, and the hydrostatic bulge due to the present rotation, but nothing else, were allowed to disappear. The remarkable fact is that $B' - A'$ is larger than $C' - B'$; in other words, inhomogeneities in mass distribution relative to two axes in the equatorial plane are as important as any feature related to the rotation axis. Goldreich and Toomre suggest that all non-hydrostatic contributions to the moments of inertia, and therefore to the figure, are best explained by such inhomogeneities in density. There is no need to invoke a minimum strength for the earth, if these inhomogeneities arise from mantle convection currents; conversely, if there is no evidence for such minimum strength or rigidity, much of the objection to deep mantle convection disappears. This explanation of the non-hydrostatic form is therefore of very great importance to geodynamics (Chapter 27).

13.3 EXPLANATIONS OF THE LOW HARMONICS

The ability of the mantle to withstand stress-differences due to superimposed loads would appear to argue against flow in the mantle, but the ironic situation is that the gravity anomalies themselves may be evidence of density differences associated with flow. This is certainly one of the leading paradoxes of geophysics.

The type of flow which is most likely to be involved concerns convection currents, the mechanism of which we defer to Chapter 25. However, we may note here that thermal convection involves systematic differences in density between rising and sinking currents, and a worldwide pattern of such currents could produce the observed features of the gravitational field. For a temperature perturbation of the form $f(r)S_l(\theta, \lambda)$, where $f(r)$ is a function of radius and $S_l(\theta, \lambda)$ is a surface spherical harmonic of degree l, the disturbance in density would be $-\alpha\rho_0 f(r)s_l(\theta, \lambda)$, where α is the volume coefficient of thermal expansion and ρ_0 is the undisturbed density. An anomalous potential U would arise, with U satisfying Poisson's equation

$$\nabla^2 U = 4\pi G\alpha\rho_0 f(r)S_l(\theta, \lambda) \qquad 13.3.1$$

in the mantle, and

$$\nabla^2 U = 0$$

outside.

Gravity anomalies will undoubtedly be associated with U, but a difficulty arises because of uncertainties in the boundary conditions. If the convection currents warp the outer surface, or any surface of discontinuity in density above them, additional terms in the gravity field, of the same spherical harmonic degree, may well obscure the anomalies directly due to thermal effects. For example, a rising convection current is associated with higher temperatures and lower density, but because of uplift it may produce a positive rather than negative anomaly, as was found by Pekeris (1935) in one model of convection.

Later numerical studies of the gravitational field (McKenzie, 1968) have verified this: when the upper boundary is free, positive rather than negative anomalies are associated with rising convection currents. On the other hand, if the upper boundary is perfectly rigid, there is no warping of the boundary and the gravity field reflects only the difference in density within the convecting fluid. In this case, positive anomalies will be found over sinking currents and negative anomalies over rising ones. Runcorn (1964, 1967) proposed determining the flow pattern directly from the gravitational field. His analysis appears to attribute the major part of the anomalies to density differences in the convecting system, as would be appropriate with a rigid upper boundary. The difficulty is that the upper boundary condition is not known and it may not be everywhere the same. In terms of plate tectonics, convection, if it exists, is restricted to the mantle beneath the base of the lithosphere, which would provide the upper boundary. Where the lithosphere is thick, the boundary could be expected to resist deformation by convection more readily than where it is thin. We shall return to a more specific comparison of the gravitational field with elements of plate tectonics after some alternative hypotheses are examined.

Terms corresponding to low harmonics in the field might also be caused by long-wavelength warping of some surface of discontinuity, say at the base of the crust, or possibly even as deep as the core-mantle boundary. In this case, the relation between spherical harmonic terms in the expansion for gravity (g_l) and those in the equivalent surface distribution of mass (σ_l) is given by Equation 12.3.7:

$$\sigma_l = \frac{1}{4\pi G}\,\frac{2l+1}{l+1}\left(\frac{a}{a-z}\right)^{l+2} g_l \qquad 12.3.7$$

If the boundary in question separates material of density ρ_1 (above) and ρ_2, this surface distribution of mass is

$$\sigma_l = (\rho_2 - \rho_1)\; N_l \qquad 13.3.2$$

where N_l represents the upward warping of the boundary, of degree l. In addition to producing gravity anomalies at the earth's surface, this distribution of mass produces a stress distribution beneath itself, in a manner analogous to the distribution on a plane earth, considered in the previous section. We saw there that the maximum stress difference was proportional to the amplitude of the load (actually equal to $2/e$ times this amplitude). Higbie and Stacey (1970) have considered the spherical case, in which the maximum stress difference is also proportional to the load amplitude (closer to $\frac{1}{2}$ this amplitude), and have sought the depth z for the boundary under the condition that the stress-difference be independent of the degree of harmonic term. The stress-difference will be proportional to

$$\sigma_l g_z = \frac{g_z}{4\pi G} \frac{2l+1}{l+1} \left(\frac{a}{a-z}\right)^{l+2} g_l \qquad 13.3.3$$

where g_z is the value of g at depth z. If values for g_l are taken from the satellite-determined expansion of the field (Higbie and Stacey actually worked with terms in the potential, for which the relation is only slightly different), a relation between stress difference and degree l can be computed for any given depth z. Higbie and Stacey found that the maximum stress difference was most nearly independent of l for a depth of 655 km, and that at this depth it differed by less than a factor of 2 from 15×10^6 dynes cm^{-2} over the first 14 harmonic terms. For other values of z, for example 2000 km, much greater stresses, up to 1000×10^6 dynes cm^{-2}, would be required. They consider such large variations in stress to be unreasonable and suggest the depth of about 600 km, in the upper mantle, to be the most probable for the source of the low harmonics in the field. Their results really reflect the fact, mentioned in Section 12.3, that if the source of a given harmonic is sought at too great a depth, unreasonable mass distributions are required. For harmonics above degree 8, the source does appear to be in the upper mantle, but gravity alone is incapable of establishing a minimum depth for the cause. The very low harmonics could in fact have a source as deep as the core-mantle boundary (Garland, 1957; Hide and Horai, 1968; Hide and Malin, 1971). The latter authors suggest that harmonics up to order 4 could be produced by reasonable (of the order of 1 km) warpings in the core boundary. Furthermore, they show a remarkable correlation with the same harmonics in the expansion of the earth's magnetic field. This correlation might be expected if the magnetic field results from core motions (Chapter 17) which distort, or are influenced by, the core-mantle boundary. However, the surprising fact is that corresponding features of the magnetic field are displaced to the west, relative to the gravitational field, by an average of 160° in longitude. The geoidal positive which trends from the North Atlantic to south Africa (Fig. 11.7), for example, has its counterpart in a magnetic positive trend over the Pacific Ocean. At present, all explanations for this resemblance between the fields are highly tentative, and we shall postpone further consideration of them until Chapter 17.

13.4 GRAVITY ANOMALIES AND STRUCTURES

When we turn to gravity anomalies of smaller extent, we find we are dealing with the effects of variation in density closer to the earth's surface. In many cases, these involve juxtapositions of rocks of known type, so that interpretation depends upon some knowledge of the average densities of rocks. For igneous rocks, porosity is usually low, and the rock density is a weighted average of the densities of the constituent minerals; but for sedimentary rocks, porosity is a controlling factor and the density of a given rock type will vary with depth of burial. It is possible only to list ranges of density for representative rocks, as follows.

	ρ, gm/cm^3
Shale	2.00–2.65
Limestone	2.25–2.80
Sandstone	2.20–2.70
Granite	2.65–2.75
Diorite	2.70–2.95
Gabbro	2.85–3.10
Dunite	3.20–3.30

Gravity measurements have been used to study a great many types of geological structure, ranging in scale down to the dimensions of ore bodies. It is possible here to discuss only a few examples, which appear to have significance on a global scale.

Some of the most striking types of gravity anomalies are the long, narrow strips of negative anomaly (Fig. 13.6) discovered by Vening Meinesz, and known to be associated with oceanic trenches and island arcs. Since these regions are essentially at sea level, the effect is evident on all types of anomaly. They must result from fairly shallow strips of mass deficiency, and Vening Meinesz (1954) concluded that they were expressions of a symmetrical downbuckle of a light crust into the denser mantle. More recently it has become apparent that interpretation of gravity anomalies is not independent of the structure determined seismologically, because of the relation between density and velocity (Section 13.1). This relation has been used, for example, in the study of the Tonga Trench (Talwani, Worzel, and Ewing, 1961) to produce a combined interpretation which attributes much of the negative anomaly to a thick accumulation of sedimentary material at the top of the oceanic crust. Still more recently, Hatherton (1969) has called attention to the broad positive anomaly which is always associated with the negative strip, and has interpreted this (Fig. 13.7) on the assumption of a dense plate of the lithosphere, descending several hundred kilometers into the mantle. The assumption that a descending plate is of greater density than the mantle into which it is dragged rests on the further assumption that the plate is cold.

Positive anomalies associated with down-going slabs have also been observed over the Aleutians (Grow, 1973) and over the Chile trench west of South America (Grow and Bowin, 1975). Grow and Bowin analyzed the free-air anomalies on an east-west profile extending 1000 km, from west of the trench to east of the Andes (Fig. 13.8). Their section is based on seismic data, where available, and the densities in the slab are estimated for relatively cold lithospheric material at pressures corresponding to the depths to which it has been pushed. Note that while the density within the slab is variable, partly because of inferred phase changes in some minerals, at every depth the slab is denser than normal mantle at the same depth. The primary effect of the slab is to produce a broad gravity positive, of about 100 mgal, over the main body of the Andes, in spite of the root of greatly thickened crust. The large oscillations in free-air anomaly at 0 and 600 km on the profile are due to the effects of the trench itself and of the edge of the root, respectively. Seaward of the deep-sea trenches, there are often positive free-air anomalies also. Watts and Talwani

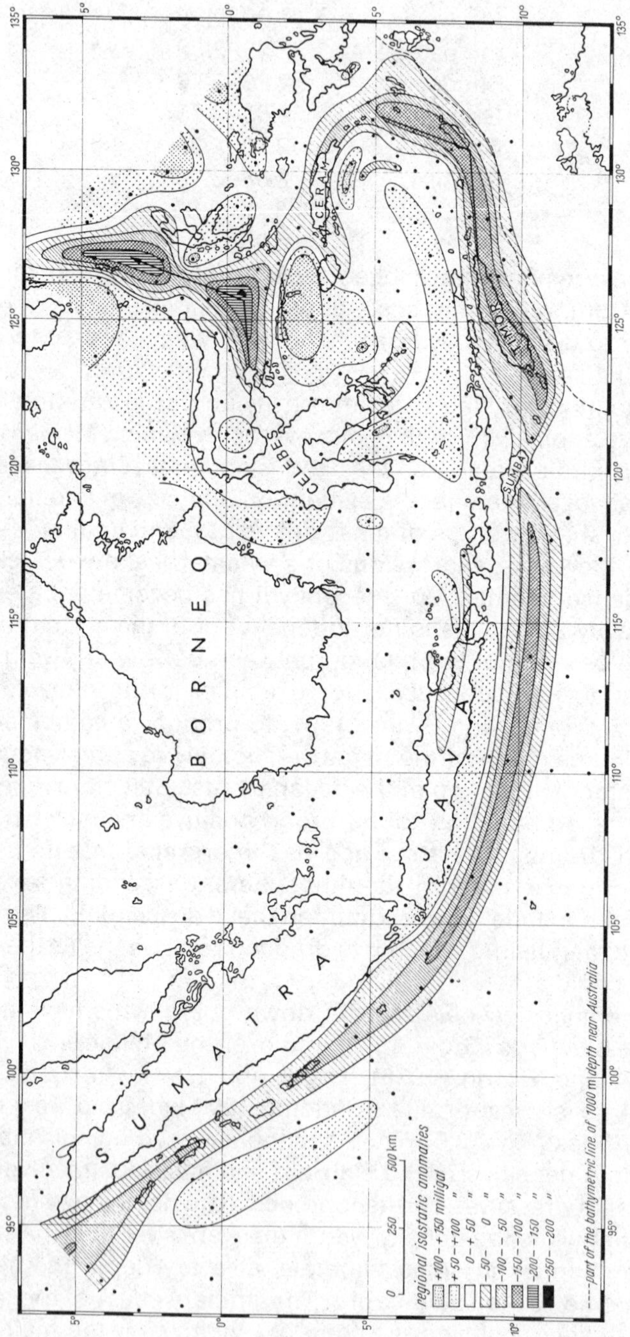

Figure 13.6 Isostatic anomaly map of the East Indies. Small circles show gravity stations. (After Vening Meinesz)

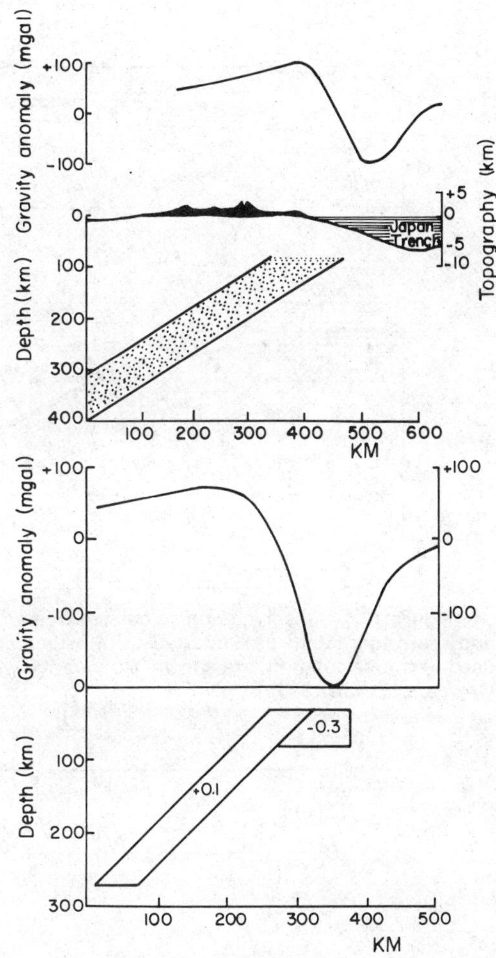

Figure 13.7 *Top,* observed gravity anomaly across the Japan Trench, in relation to the island arc and seismic zone; *Bottom,* a possible interpretation in terms of anomalous densities. (After Hatherton)

(1974) show that these are best explained by near-surface features, including possible warping of the oceanic lithosphere.

For the mid-ocean ridges, Talwani (1970) has summarized what gravity information is available. Most ridges show a broad free-air positive anomaly, reaching a maximum near the ridge axis. Upon this broad anomaly are superimposed many short-wavelength variations, which appear to be related to local ocean floor topography. The amplitude of the broad anomaly is variable; over the Mid-Atlantic Ridge north of 30°N, which has been studied in detail, the anomaly is well developed. The southern part of the mid-Atlantic Ridge and the East Pacific Rise have much smaller free-air anomalies. In all cases, however, the positive is very much less than it would be if the ridge were uncompensated; for example, the Mid-Atlantic Ridge would produce a free-air anomaly of 200 mgals in that case. The question which remains concerns the mechanism of compensation. One explanation places the deficiency in mass close to the earth's surface. An example is shown in Figure 13.9, after Talwani, LePichon, and Ewing (1965). Here the interpretation is based on seismic control (where available), and it attributes a large measure of the compensation to a volume of anomalously low-density material (3.15 gm/cm^3) in the lower crust and upper mantle. In their interpretation, the base of this material was adjusted

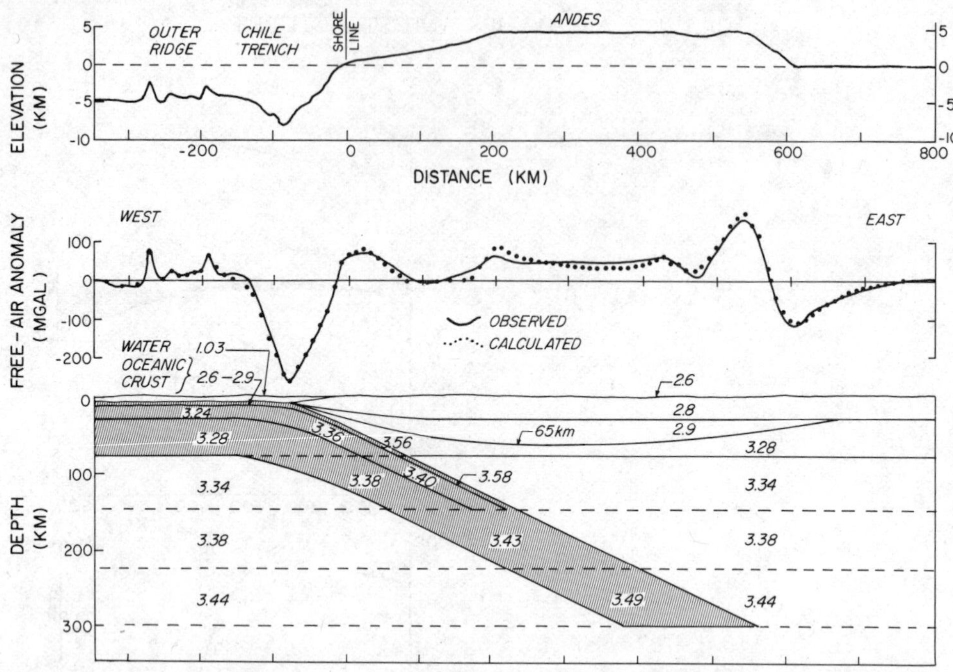

Figure 13.8 Observed and calculated free-air anomalies on a profile across the Chile Trench and the Andes. Densities, inferred from estimates of temperature and pressure in the slab and used in the calculation, are shown. Boundaries within the crust are based on seismic data. (After Grow and Bowin 1975).

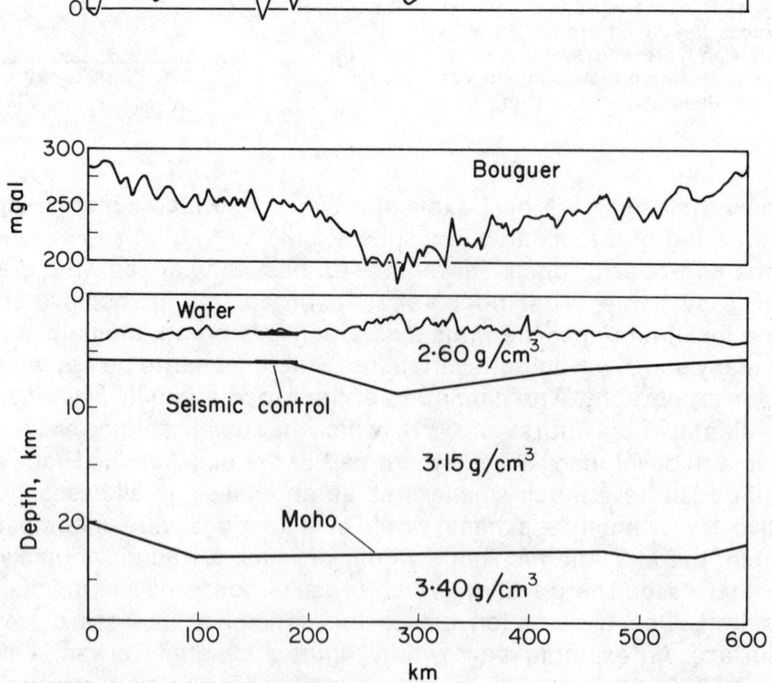

Figure 13.9 Free-air and Bouguer gravity anomalies across the Mid-Atlantic Ridge. Seismic control was available only where shown; the remainder of the section is inferred from gravity. Note that the free-air anomaly reflects the attraction of local bottom structure, but not that of the ridge as a whole. (After Talwani, LePichon and Ewing)

to fit the gravity observations. Other authors (Morgan, 1965; Keen and Tramontini, 1970) have suggested that the mass deficiency lies in a region of moderately reduced density (0.04 gm/cm³ less than normal) extending to depths of 200 to 400 km. This explanation would be consistent with the density deficiency associated with a rising convection current. The choice between the two models must await more detailed gravity and seismic studies of ridge areas. In the meantime, the observation remains valid, that ridges are characterized by small, but positive-tending, broad free-air anomalies.

Within continental areas, we may make a very general division between gravity anomalies related to sedimentary basins, and those related to densities in the crystalline rocks of the shields. As the table above indicates, sedimentary rocks are normally less dense than the average rocks of the continents, and deep sedimentary basins produce relative negative effects in the Bouguer anomaly field. From the point of view of large-scale tectonics, the most important application of this effect would be to estimate the depth of some of the deepest basins. As basins are very broad compared to their depth, this should be possible by application of the Bouguer slab formula, but a difficulty arises because of the increase with depth of the density of sedimentary rocks. The difference in density between the basin rocks and the underlying rocks of the crust becomes small and uncertain, so that any depth determined tends to be a minimum one.

Within the continental shields, striking negative anomalies are frequently observed over masses of granite (Fig. 13.10). When these were first observed they were occasion for surprise, as it was thought that the upper part of the continental crust was typically granitic. But the density of a true granite is close to 2.67 gm/cm³, and it now appears that the mean density of the igneous and metamorphic rocks, even of the upper part of the continental crust, is close to 2.75 gm/cm³. A further consequence of the negative anomalies over granite batholiths is that the granite could not have been produced by differentiation in place, for the settling out of denser minerals would leave a layer of mass excess whose effect, if it were present, would have largely obscured the negative anomaly.

At the other end of the range of igneous rocks are the masses of ultrabasic rocks, exposed at the earth's surface in various places. Because of their high density, they produce large positive gravity anomalies. It is important to analyze these to determine if any of the masses are in fact "windows" to the mantle, since the upper mantle is believed to be of similar composition. Harrison (1956) studied the anomalies over the ultrabasic rocks of Cyprus (Fig. 13.11) and his results are somewhat typical of what has been found elsewhere. The masses appear to be thin in comparison with the thickness of the crust, and while it is impossible to rule out the presence of a small-diameter "feeder," the exposures in general cannot be windows through to the mantle.

Apart from the direct effects of varying rock densities, there do appear to be negative anomalies over continental areas which result from a former depression of the regions under the load of ice sheets. Examples are found in Fennoscandia (Honkasalo, 1959) and northern Canada (Innes, 1960), in both cases in areas which are believed to have been centers of Pleistocene glaciation. If an ice load on the crust depresses it by an amount h, and if this occurs through the outflow of mantle material of density ρ_n, over a broad area the magnitude of the resulting negative anomaly would be

$$\Delta g = 2\pi G \rho_n h \qquad 13.4.1$$

If partial uplift has occurred since the retreat of the ice, the observed Δg permits the evaluation of the remaining uplift h required before equilibrium is restored. We note that the expression involves ρ_n, the upper mantle density, and not the difference in density between crust and mantle. For both Fennoscandia and Canada, remaining uplifts of about 200 meters have been estimated, if complete equilibrium is to be restored. We shall return to this question in Chapter 27, when the mantle viscosity is discussed.

Some of the features mentioned above, including slabs, ridges and glaciated areas, produce anomalies which are sufficiently broad for them to be seen on global maps of the gravity field. The increased resolution of the gravity field itself, Figure 11.6, makes it preferable to the geoid, Figure 11.7, for this comparison. The geoid, as we noted, is dominated by the very lowest harmonics. Kaula (1969, 1972) has in fact categorized the global field in terms of major type areas related to the tectonics. His 1972 study is based on a map of free-air gravity referred to the hydrostatic figure and therefore slightly differ-

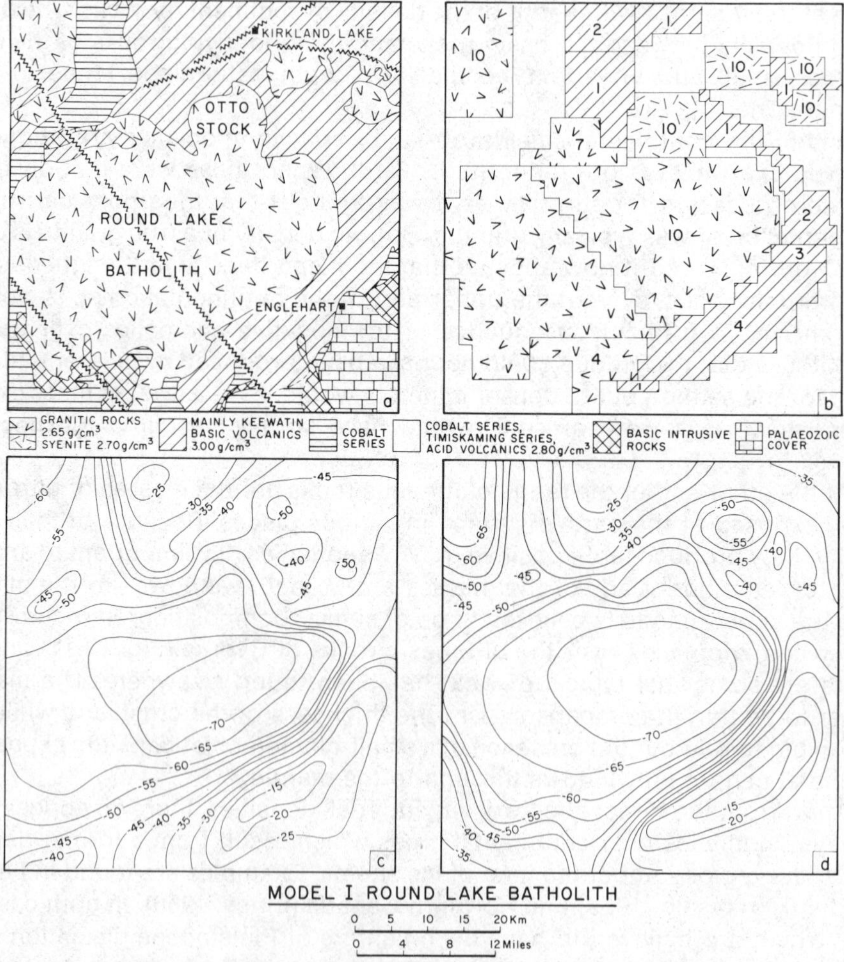

Figure 13.10 Observed Bouguer anomalies (c) over a granite batholith in the Canadian Precambrian shield (a). The contours shown in (d) are for the computed field of the assemblage of prisms shown in (b), where the figures indicate thickness of the prisms in kilometers. (Courtesy of R. A. Gibb and J. van Boeckel, and the Earth Physics Branch, Dept. of Energy, Mines and Resources, Canada)

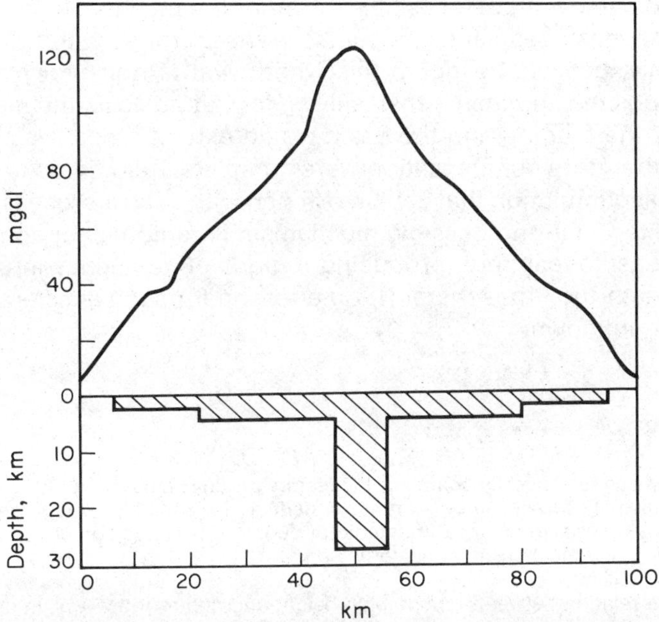

Figure 13.11 Bouguer gravity anomaly observed over Cyprus, and inferred shape of dense ultrabasic rocks. (After Harrison)

ent from Figure 11.6, but the main features remain obvious. Briefly, the map is characterized by about 30 regions, of area at least 10^6 km^2, over which the anomaly maintains the same sign. In agreement with our above discussion of structures, positive anomalies are indicated over mid-ocean ridges or rises, and island arcs associated with down-going slabs. Negative anomalies are associated with glaciated regions and basins, both deep ocean basins and sedimentary basins on the continents. Of the first type (mid-ocean ridges and rises), examples are seen, on Figure 11.6, over the northern part of the Mid-Atlantic Ridge and the East Pacific Rise. Positive anomalies associated with island arcs or subduction zones are apparent over the Philippines area and over the Andes. Areas of oceanic volcanism not associated with ridges, such as Hawaii, also tend to give positive anomalies. Negative features probably related to the most recent period of glaciation are evident over Hudson Bay in Canada, and around the coast of Antarctica. Kaula points out, however, that the extent and magnitude of the Antarctic features are such that some origin additional to glaciation is suggested, and it may share some of the characteristics of the basins. Negative features associated with oceanic basins may be seen on each side of the Mid-Atlantic Ridge and the East Pacific Rise and in a number of other places. The most striking anomaly on the map is over the Indian Ocean, just south of India. The reason for the large amplitude of this particular anomaly is not known. It is separated from a more moderate negative trend over the Himalayas to the north; the latter feature has been attributed partly to crustal thickening during continental collision and partly to the removal of mass by rapid erosion of the young Himalayas. A good example of a negative anomaly over a continental basin is that associated with the Congo basin in central Africa.

The correlation of the global gravity field with large scale tectonic features is therefore largely what would be expected on the basis of individual associations. Perhaps the chief unexplained feature is the precise cause of the negative anomaly associated with deep basins, or rather the mechanism of basin formation itself. The mass deficiency presumably arises from a withdrawal of mantle material below, but what triggers this withdrawal is not known. It is not the weight of deposited sediments themselves, for in that case, the mantle outflow would follow the loading and the basin would exhibit a positive anomaly until outflow and therefore compensation were complete. Also, there is a limit to the thickness of accumulation that can be reached with such a process (Problem 3). Kaula suggested that one possible mechanism is regional horizontal accelerations of the aesthenosphere, producing regions of deficiency in upper mantle material beneath the lithosphere. But the reason for such accelerations, if they exist, is itself unknown.

Problems

1. Does an isostatically compensated feature produce a free-air gravity anomaly? If you believe the answer to be affirmative, deduce the anomaly for topography represented by a surface spherical harmonic of degree l, which is completely compensated by the Airy mechanism at depth h.

2. The relation between topographic height and depth of the Airy crust was obtained by considering prisms with vertical sides. But suppose a long section of otherwise compensated crust (or lithosphere) becomes decoupled from its surroundings, by parallel faults dipping inward at 45°. Will the equilibrium level of the region between the faults be different from that of the surroundings? Examine the case of faults dipping outward at 45° also.

3. When the isostatic compensation of crustal topography by the Airy mechanism was considered, the difference in density between crust and substratum was involved. But if a superficial load, of ice or sediment, is simply added on top of a crust whose thickness remains constant, and compensation is achieved by the outflow of an equal mass of mantle material, only the densities of load and mantle are required. Consider the deposition of sediments of density 2.0 gm/cm^3 in sea water of density 1.1 gm/cm^3. If the density of the mantle is 3.3 gm/cm^3, what thickness of sedimentary rock could accumulate beneath a sea originally 1 km deep? How does your result compare with known thicknesses of shallow-water sedimentary formations?

BIBLIOGRAPHY

Cook, Kenneth L. (1962) The problem of the mantle-crust mix: lateral inhomogeneity in the uppermost part of the earth's mantle. *Advances in Geophysics,* **9**, 295.

Garland, G. D. (1957) The figure of the earth's core and the non-dipole field. *J. Geophys. Res.,* **62**, 486–487.

Gibb, R. A., and van Boeckel, J. (1970) Three-dimensional gravity interpretation of the Round Lake batholith, Northeastern Ontario. *Can. J. Earth Sci.,* **7**, 156.

Goldreich, Peter, and Toomre, Alar (1969) Some remarks on polar wandering. *J. Geophys. Res.,* **74**, 2555–2567.

Grow, J. A. (1973) Crustal and upper mantle structure of the central Aleutian arch. *Bull. Geol. Soc. Amer.,* **84**, 2169–2192.

Grow, J. A., and Bowin, C. O. (1975) Evidence for high-density crust and mantle beneath the Chile trench due to the descending lithosphere. *J. Geophys. Res.,* **80**, 1449–1458.

Harrison, J. C. (1956) An interpretation of gravity anomalies in the eastern Mediterranean. *Phil. Trans. Roy Soc. Lond. A,* **248**, 283.

Hatherton, Trevor (1969) Gravity and seismicity of asymmetric active regions. *Nature,* **221**, 353.

Hide, R., and Horai, K-I. (1968) On the topography of the core-mantle interface. *Phys. Earth Plan. Interiors,* **1**, 305–308.

Hide, R., and Malin, S. R. C. (1971) Novel correlations between global features of the earth's gravitational and magnetic fields: further statistical considerations. *Nature Phys. Sci.,* **230**, 63.

Higbie, J., and Stacey, F. D. (1970) Depth of density variations responsible for features of the satellite geoid. *Phys. Earth Plan. Interiors,* **4**, 145–148.

Honkasalo, Tauno (1959) *Geofisica,* **7**, 2.

Innes, M. J. S. (1960) Gravity and isostasy in Northern Ontario and Manitoba. *Pub. Dom. Observatory Ottowa,* **21**, 261.

Jeffreys, H. (1943) The determination of the earth's gravitational field (second paper). *Mon. Nat. R. A. S., Geophys. Suppl.,* **5**, 55.

Jeffreys, H. (1962) *The earth* (4th ed.). Cambridge Univ. Press, Cambridge.

Kaula, W. M. (1969) A tectonic classification of the main features of the earth's gravitational field. J. Geophys. Res., **74**, 4807.

Kaula, W. M. (1972) Global activity and tectonics. In *The nature of the solid earth,* ed. E. C. Robertson. McGraw-Hill, New York.

McKenzie, D. P. (1967) Heat flow and gravity anomalies. *J. Geophys. Res.,* **72**, 6261.

McKenzie, D. P. (1968) The influence of the boundary conditions and rotation on convection in the earth's mantle. *Geophys. J. R. A. S.,* **15**, 457–500.

Morgan, W. Jason (1965) Gravity anomalies and convection currents. *J. Geophys. Res.,* **70**, 6175 and 6189.

Pekeris, C. L. (1935) Thermal convection in the interior of the earth. *Mon. Nat. R.A.S., Geophys. Suppl.,* **3**, 343.

Runcorn, S. K. (1964) Satellite gravity measurements and a laminar viscous flow model of the earth's mantle. *J. Geophys. Res.,* **69**, 4389.

Runcorn, S. K. (1967) Flow in the mantle inferred from the low degree harmonics of the geopotential. *Geophys. J.,* **14**, 375.

Steinhart, John S., and Meyer, Robert P. (1961) Explosion studies of continental structure. *Carnegie Inst. Washington Publ. 622.*

Talwani, M., LePichon, X., and Ewing, M. (1965) Crustal structure of the mid-ocean ridges. *J. Geophys. Res.,* **70**, 341.

Talwani, Manik, Worzel, J. Lamar, and Ewing, Maurice (1961) Gravity anomalies and crustal section across the Tonga Trench. *J. Geophys. Res.,* **66**, 1255.

Talwani, M. (1970) Gravity. In *The sea,* vol. 4, part 1, ed. Arthur E. Maxwell. Wiley-Interscience, New York.

Vening Meinesz, F. A. (1955) Plastic buckling of the earth's crust: The origin of geosynclines. *Geol. Soc. Amer. Spec. Paper,* **62**, 319.

Watts, A. B., and Talwani, M. (1974) Gravity anomalies seaward of deep-sea trenches and their tectonic implications. *Geophys. J. R. A. S.,* **36**, 57–90.

Woollard, George P. (1969) Regional variations in gravity. In *The earth's crust and upper mantle,* ed. Pembroke J. Hart. *Amer. Geophys. Un. Monograph 13.*

Worzel, J. L., and Shurbet, G. L. (1955) Gravity anomalies at continental margins. *Proc. Nat. Acad. Sci.,* **41**, 458.

14 EARTH TIDES

14.1 THE TIDE-RAISING POTENTIALS

We have, until now, discussed only those features of the gravitational field which result from the attraction of the earth, or parts of it. The field, however, includes the attractions of the sun and moon, which vary with time, and this produces an effect on the earth. The earth responds to these attractions in a way that depends upon its rigidity, so that measurement of the variation of gravity with time provides further information on the mechanical properties of the interior. Since the critical part of the analysis consists in comparing the time variations on the actual earth with those to be expected on a completely rigid earth, we shall outline first the development of the accelerations produced by sun and moon that would be observed on a rigid earth, and then consider the effect of yielding.

It is convenient to consider first the effect of the moon (Fig. 14.1), and assume that the earth and moon revolve about their common mass center, without rotation of the earth on its own axis. We take O and M to be the centers of mass of the earth and moon respectively; at these points, the mutual gravitational attraction precisely provides the centripetal acceleration; at other points, such as P, it does not. At P, we may write for the moon's tidal potential, or potential of the difference between attraction and centripetal acceleration,

$$U_m = GM\left(\frac{1}{\rho} - \frac{r\cos\theta}{R^2} + C\right) \tag{14.1.1}$$

where M is the mass of the moon and C is a constant. If C is chosen to make

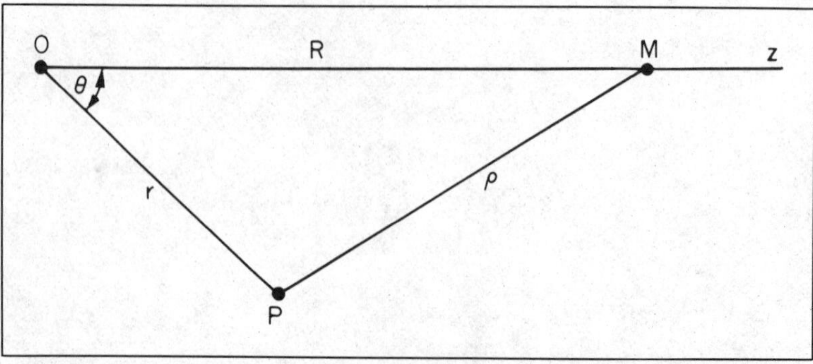

Figure 14.1 Notation for expression of the tidal attraction of the moon.

U_m vanish at O, we have

$$U_m = GM\left(\frac{1}{\rho} - \frac{1}{R} - \frac{r \cos \theta}{R^2}\right) \qquad 14.1.2$$

We now expand $1/\rho$ in terms of r, R and the Legendre polynomials, as was done in Chapter 11. This leads to

$$U_m = \frac{GMr^2}{R^3}\left[P_2(\cos \theta) + \frac{r}{R} P_3(\cos \theta) \ldots\right] \qquad 14.1.3$$

The first term in this expansion is dominant, since r/R is about 1/60. We note in passing the importance of the second degree harmonic, which shows at once the presence of two high tides at any time. It is usual to rewrite Equation 14.1.3 slightly, in terms of the moon's mean distance c, to give

$$U_m = G(r)\left[\left(\frac{c}{R}\right)^3 (\cos 2\theta + \tfrac{1}{3}) + \frac{1}{6}\frac{r}{c}\left(\frac{c}{R}\right)^4 (5 \cos 3\theta + 3 \cos \theta)\right] \qquad 14.1.4$$

where

$$G(r) = \tfrac{3}{4}GM\left(\frac{r^2}{c^3}\right)$$

The expression for the tidal potential of the sun follows in the same way; in terms of the sun's mean distance c_s and its mass S,

$$U_S = G_S(r)\left[\left(\frac{c_S}{R_s}\right)^3 (\cos 2\theta + \tfrac{1}{3}) + \frac{1}{6}\left(\frac{r}{c_s}\right)\left(\frac{c_S}{R_s}\right)^4 (5 \cos 3\theta + 3 \cos \theta)\right]$$

14.1.5

where

$$G_s(r) = \tfrac{3}{4}GS(r^2/c_s^3)$$

The effect of the earth's rotation is to cause the observer to move through a potential field which varies with time. With the geometry of Figure 14.2, we have, for the moon,

$$\cos \theta = \sin \phi \sin \delta + \cos \delta \cos \phi \cos (t - 180^\circ) \qquad 14.1.6$$

where δ is the declination, ϕ is the geocentric latitude of P, and t is the moon's hour angle. Finally,

$$U_m = G(r)\left(\frac{c}{R}\right)^3 [3(\tfrac{1}{3} - \sin^2 \delta)(\tfrac{1}{3} - \sin^2 \phi) - \sin 2\phi \sin 2\delta \cos t \qquad 14.1.7$$

$$+ \cos^2 \phi \cos^2 \delta \cos 2t]$$

where only the leading term of Equation 14.1.4 has been retained. A similar expression for U_s holds.

The significant point of our analysis is the manner in which the terms in Equation 14.1.7 vary with time, and with the latitude of P. The first term is independent of t, but varies with δ and R, and leads to tides of long period. In the second term, the factor $\cos t$ leads to a period of one lunar day, while the third

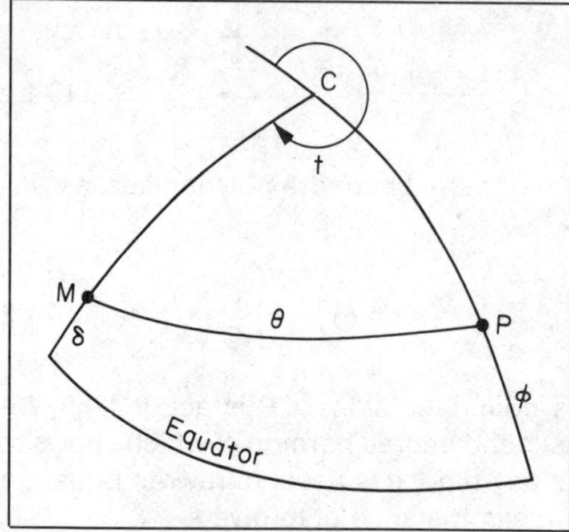

Figure 14.2 Quantities required for tidal calculations on a rotating earth. C is the pole of the celestial sphere, M the moon, P the point on earth. The moon is located by hour angle t, and declination δ; ϕ is the geocentric latitude of P.

term has a period of one-half lunar day. The diurnal tides are antisymmetric about the equator, while the other two are symmetric.

When the tide-producing potential of the sun is considered in addition, we obtain the following principal tides, with the periods shown:

Symbol	Name	Period (hr)
M_2	Principal lunar	12.42
S_2	Principal solar	12.00
N_2	Lunar ellipticity	12.66
K_2	Lunisolar	11.97
K_1	Lunisolar	23.93
O_1	Lunar declination	25.82
P_1	Solar declination	24.07

On a completely rigid earth, the tide-producing potentials would produce changes in g, given directly by the term $-\partial U/\partial r$. The variation with latitude of the amplitude, in microgals, of the various tidal components, is shown in Figure 14.3.

14.2 THE YIELDING OF THE EARTH

The earth as a whole yields to the tide-raising potentials (to maximum amplitudes of about 10 cm), but in contrast to the treatment of oceanic tides, it is possible to neglect dynamic effects and use an equilibrium theory. This implies that, for an additional potential U at any point, the increased surface elevation will be U/g. Love (1911) showed that with this assumption and for a tidal potential of second degree, both the surface displacement and the potential due to the mass in the deformation are proportional to the tidal potential. The total disturbed potential at any point can be written

$$U = U_0 + U_t + U' - gu \qquad 14.2.1$$

where U_0 is the undisturbed potential due to attraction and rotation, U_t is the tidal potential due to sun and moon, U' is the potential due to the redistributed

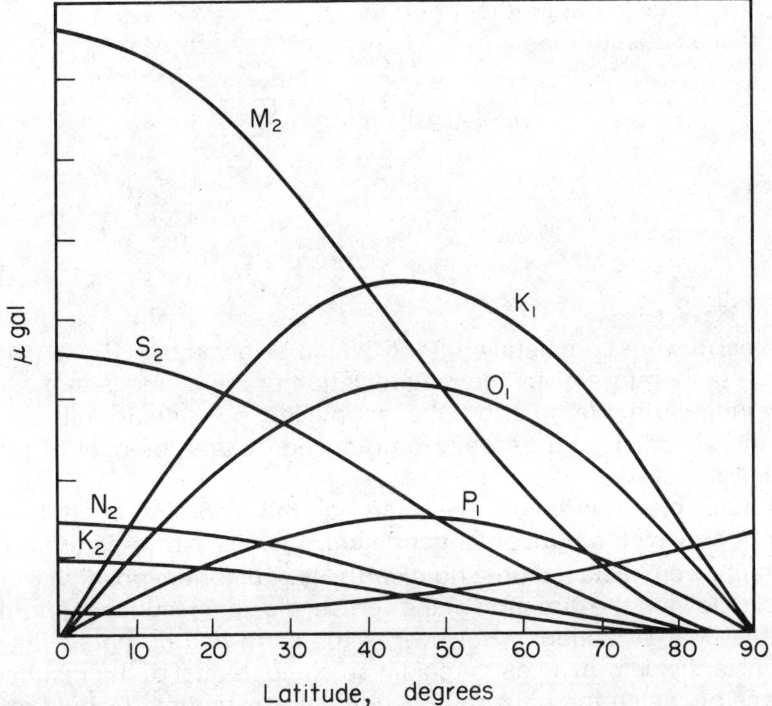

Figure 14.3 Variation with latitude, on a rigid earth, of the principal tidal components.

mass in the tidal deformations, and u is the radial displacement at the point. Because of the proportionality mentioned above, we may write

$$u = h\ (U_t/g) \qquad U' = kU_t \tag{14.2.2}$$

where h and k are constants known as Love's numbers. Then

$$U = U_0 + U_t(1 + k - h) \tag{14.2.3}$$

The tidal component in gravity is

$$g_t = -\frac{\partial U_t}{\partial r} - \frac{\partial U'}{\partial r} + u\,\frac{\partial g}{\partial r} \tag{14.2.4}$$

Since U_t depends on r^2 (Equation 14.1.4) we may write

$$\frac{\partial U_t}{\partial r} = +\frac{2U_t}{r}$$

The term $\partial U'/\partial r$ is the attraction of a mass coating which is distributed over the earth according to a second degree harmonic; it may be written

$$\frac{\partial U'}{\partial r} = -\frac{3}{r}\,U'$$

Also, $\partial g/\partial r$ is the vertical gradient of g, or $-2g/r$.

With these substitutions

$$g_t = -[2 + 2h + 3k]\frac{U_t}{r} \qquad 14.2.5$$

or

$$g_t = -\left[1 + h - \frac{3}{2}k\right]\frac{\partial U_t}{\partial r}$$

If the earth were completely rigid, g_t would be given by $-\partial U_t/\partial r$. The factor $\delta = 1 + h - \frac{3}{2}k$ therefore gives the magnification of the changes in g due to the yielding of the earth. We note that the magnification results from the displacement of the observer, and tends to be reduced by the potential of the redistributed mass.

Earth tides may also be observed through the recording of changes in the direction of gravity. If a horizontal pendulum, such as a long-period horizontal seismograph, were installed on a rigid earth, its equilibrium position would vary with time, following the direction of the vertical (i.e., the resultant of normal gravity and the tidal attractions). However, if the earth had no rigidity its surface would constantly deform so as to remain an equipotential of the resultant field. The surface on which the horizontal pendulum was mounted would always be normal to the vertical, and the pendulum would show no displacement relative to its support. The effect of yielding is thus to reduce the amplitude of a horizontal pendulum, and an analysis similar to that above shows that the reduction is by the factor $\lambda = (1 + k - h)$. A combination of gravimeter and horizontal pendulum measurements thus permits h and k to be determined individually.

14.3 RESULTS OF OBSERVATIONS

Techniques for observing earth tides and a discussion of the results obtained have been given by Melchior (1966). Because long series of observations are required, both horizontal pendulums and recording gravimeters must be installed in constant-temperature environments, free from local disturbances, and, if possible, remote from the broader influence of oceanic tides. The horizontal pendulums that have been employed are not unlike long-period horizontal seismographs. For the recording of changes in the magnitude of g, instruments similar to portable gravimeters, but with increased sensitivity, have been employed. The total range in g resulting from the tides is about 0.3 mgal (Fig. 14.4), but as the interest is in the direction of fairly small differences from the theoretical values for a model earth, changes in g as small as one microgal must be recorded. For this purpose the superconducting gravimeter (Chapter 10) holds great promise. It is from the recording gravimeters installed originally for earth tide measurements that some of the finest records of the free oscillations of the earth have been obtained.

Whether the deflection of a horizontal pendulum or the change in g is measured, the instrumental response consists of the resultant of a number of tidal components of different periods. It is important, for comparison with the computed effect of earth models, to make this comparison for each component

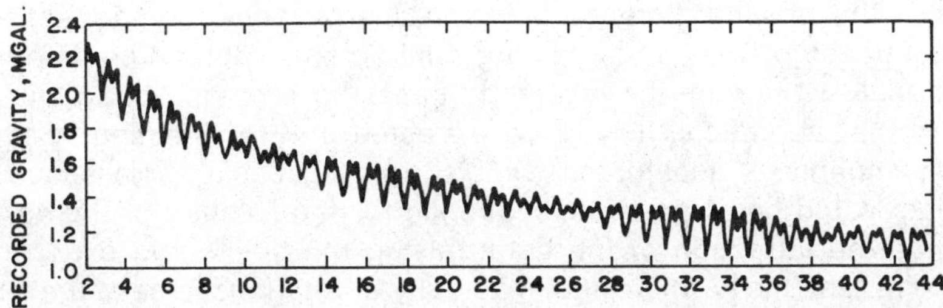

Figure 14.4 Gravimeter record of earth tide observed at Glendora, Calif., showing modulation, over a month, of the characteristic double-peaked curve. The horizontal axis shows days. (After J. C. Harrison et al.)

individually and this implies accurate separation of the frequencies in the record. This separation is complicated by the fact that the peaks occur as two closely spaced groups, around the periods of 12 and 24 hours. Melchior (1966) has described appropriate methods of harmonic analysis, while Harrison et al. (1963) have employed specially shaped numerical filters, centered on the theoretical frequencies, for convolution with the digitized records.

Calculations of Love's numbers for model earths were first made by Takeuchi (1950). Assuming distributions of rigidity and density in the mantle similar to those we have seen in Chapter 3, Takeuchi investigated, in particular, the effect of rigidity in the core. The effect of increasing the modulus of rigidity of the core from 0 to 10^{11} dynes/cm^2 was to reduce h from 0.60 to 0.53 and k from about 0.29 to 0.24.

Unfortunately, there is not yet complete agreement on the best observational determinations of h and k. For the tide M_2, which has been the most thoroughly studied, the gravimetric factor $(1+h-\frac{3}{2}k)$ has been found to range from 1.19 to 1.14; the horizontal pendulum factor, $(1+k-h)$ has been found to be more consistent, with a value of about 0.71. Adoption of the lower value of the gravimetric factor leads to values of h and k of 0.58 and 0.29 respectively, and there is support for these values from a connection with the period of the Chandler wobble. In combination with Takeuchi's model calculations, they would favor a core rigidity significantly less than 1×10^{11} dynes/cm^2. The difficulty, not yet resolved, is that the lower value of δ is observed at stations in Asia (e.g., Moscow, Kiev, Tashkent, Kyoto), but not at stations in Europe, Africa, or America. Until the reason for this difference is explained, the deduced values of h and k must be accepted with caution.

A second effect of a liquid core, still harder to observe, is a theoretical dependence of Love's numbers upon frequency. This effect results from the fact that the equilibrium theory, applicable to the solid earth, is not valid for the core, since free oscillations in the core have periods comparable to the tidal periods. Analyses by Jeffreys and Vicente (1957) and Molodensky (1961) have shown that the gravimetric factor should reach a minimum for periods very close to 24 hours (the K_1 component). A difference of 0.02 to 0.04 in the factor could be expected between the tidal components K_1 and O_1. Similarly, the horizontal pendulum factor should reach a maximum, by 0.04 to 0.06, for the K_1 tide. The results which have been obtained to date are still rather scattered, but it may be said that horizontal pendulum stations virtually always give differences of the correct sign and approximately correct magnitude. Many gravimeter stations do also, but contradictory results have been reported for some stations.

Finally, the phase difference between any observed component and the theoretical phase of the corresponding tidal potential should be determinable from the analysis. In fact, the reported phases have ranged between ±5° with respect to the theoretical. It is to be expected that the yielding of the earth under the influence of tidal forces would show a lag, because of viscous effects. Such a lag would be of great consequence for the rotation of the earth (Fig. 14.5), since the attraction of the body raising the tide would exert a couple tending to reduce the earth's angular velocity of rotation; some of the energy of rotation would be dissipated in overcoming the viscosity, and angular momentum would be transferred to the orbit of the tide-raising body. Slichter (1960) has pointed out that the observed phase difference for any component gives the phase relation of the resultant of the direct effect (which is in phase with the theoretical tide) and the effect of the distortion. For the gravimetric factor, the phase difference of the latter component alone, which is of interest here, would be six times that observed, while for horizontal pendulum observations, it would be about twice that observed, and of opposite sign (Fig. 14.6). Darwin (1910) showed that a phase lag in the deformation as small as 1° could account for the known angular deceleration of the earth. Since part of the deceleration may be attributed to oceanic tidal friction on the bed of shallow seas (Jeffreys, 1920; Munk and MacDonald, 1960), this figure is an upper limit. The observed viscous phase lag to be expected in gravimetric measurements is thus only $\frac{1}{6}^\circ$, and we must conclude that the scatter in observed values is still far too large for this to be detected.

All of the above applications—comparisons with models, variation with frequency, and phase relations—show the need for further earth tide measurements of high standards. Long series of observations, at stations as far as possible from the oceans, are required. It has been suggested, for example in Britain

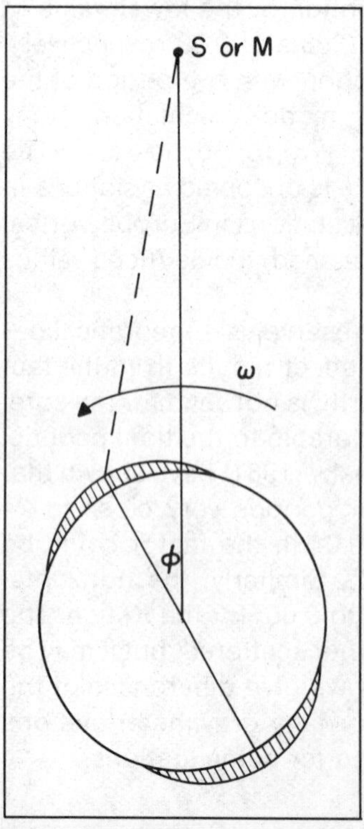

Figure 14.5 Effect of phase lag on the attraction of the tidal bulges.

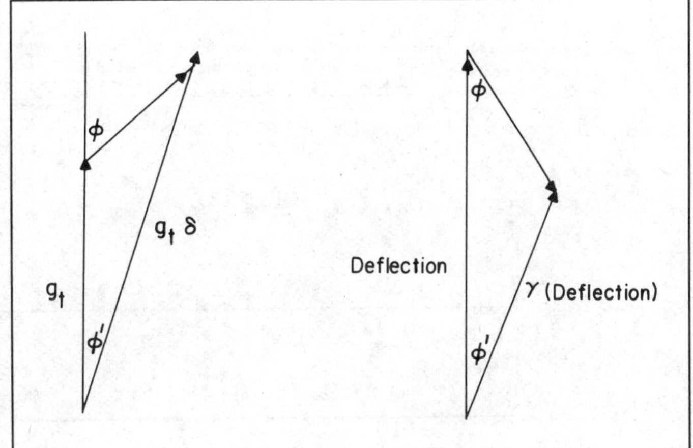

Figure 14.6 Vector diagrams illustrating the relation between the observed phase difference (ϕ') and the true phase difference (ϕ) of the deformation for gravimetric (*Left*) and horizontal pendulum observations. The vectors "g_t" and "Deflection" give the theoretical rigid-earth response.

(Tomaschek, 1954) and Japan (Nishimura, 1950), that the tidal response as measured by δ or λ shows some relationship to the regional structure, but this interpretation is not conclusive.

The truth is that the effect of the much larger ocean tides on the response of the solid earth has been underestimated in the past. Oceanic tides alter the observed tidal variation in gravity in three ways: by the loading of the sea floor through the redistributed water mass, by the further distortion of the earth through the attraction of this water mass, and by the direct effect on gravity at a station of this attraction. The calculation of a correction for the oceanic tides is not yet practicable, not so much because of theoretical difficulties as because of a lack of information on the ocean tides in the open seas. However, two types of experiment on earth tides serve to show the effect: long profiles of earth tide stations across continents, and earth tide measurements at a geographic pole of the earth.

As an example of the first, Kuo, Jackens, Ewing, and White (1970) have published observations on the factor δ and the phase, for the M_2 and O_1 tides, on a coast-to-coast profile across North America. Figure 14.7 summarizes their results. Very noticeable is an approach to almost constant values once a distance of 1000 km or less from the coast is reached. The effect near the coast is in reasonable agreement with the estimated effect of the ocean tides, based on the available information for the Atlantic and Pacific, except for the O_1 component of δ near the Pacific Coast. The profile crosses several major geological provinces, but the authors conclude that "there is no observable correlation between tidal gravity parameters and the regional geology." It may even be the case that the greatest value of earth tide measurements on long profiles is in refining our knowledge of ocean tides, through comparison of the observed and computed perturbations near the coast.

Measurements of the tidal variation of gravity at the south pole have been reported by Jackson and Slichter (1974). Their observations cover almost one year of recording and were made with a digitally-recording tidal gravimeter, counting to 0.1 microgal. According to the predicted variation of amplitude with latitude (Fig. 14.3), all of the principal tidal components should vanish at the geographic poles. But Jackson and Slichter observed these same components, at levels well above the background noise. They concluded that ocean tides, at the same frequency, were altering the value of gravity at the pole, through the three mechanisms mentioned above. The largest amplitude observed, for K_1, was 0.6 microgal. The authors show that even if only the direct attraction of the

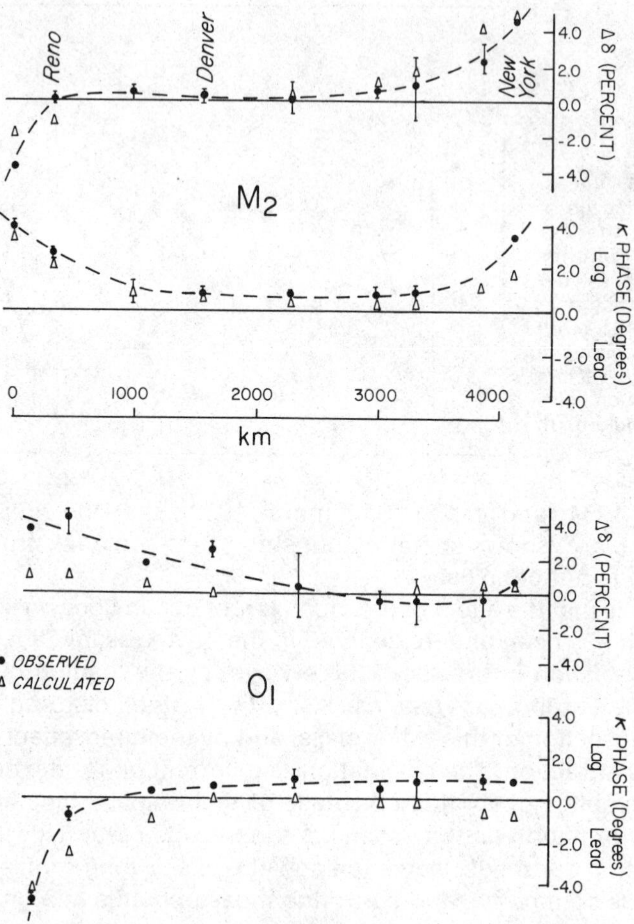

Figure 14.7 Variation in the gravimetric factor δ and the phase angle, observed for two tidal components on a profile across the continental United States. Calculated points are based on the estimated effect of oceanic tides. After Kuo et al (1970).

sea water were operative, a global tide over the oceans, reaching a maximum amplitude of 26 cm, could produce the effect. Apart from demonstrating the importance of the ocean tides, this experiment furnished an important negative result. The noise background in the recordings was extremely low for periods down to two hours. Jackson and Slichter estimate that a tidal component with amplitude as low as 0.005 microgal would have been detected in the computed spectra if it had been consistent in phase over the period of the observations. It had been estimated (Busse, 1974) that a free oscillation in the inner core could produce a signal with a period of about 7 hours, but if this exists, its amplitude at the south pole must be less than 0.005 microgal or it must be re-excited in random phase relationship over a one-year interval.

There is an additional area of common interest between earth tides and seismology: the possibility of the triggering of earthquakes by tidal stresses. While this suggestion goes back many years, positive evidence of correlations has been very difficult to obtain. The maximum value of the stress in the crust induced by earth tides is only about 10^{-2} or 10^{-3} of the stress drop associated with moderate earthquakes, so that the influence of the tides would be limited to triggering, in regions already under tectonic stress approaching the failure point. Recent revival of interest in the possible association is perhaps due to observations on the moon (Lammlein et al., 1974) which show characteristic times of occurrence of moonquakes. However, the very large diurnal variation of thermal stress on the moon makes the situation quite different from that on earth. Positive evidence of a tide-related effect in the case of earthquake swarms has now been found by Klein (1976) for several regions of the earth. The best examples are near active rifts: Iceland, the Mid-Atlantic Ridge, and the Imperial Valley–Gulf of California area. Klein showed a correlation both in time of occurrence and in the direction of the tidally induced stress. For the first, swarms were found to occur preferentially when the semidiurnal component of tidal gravity was increasing upward. The orientation of the tidal stress system is computed from the fact that the direction of maximum tension at any time is along the resultant line of the moon's and the sun's attraction. Furthermore, if the tidal stress is to increase the shear stress on, for example, a right-lateral, east-west striking fault, the direction of maximum extension must be in the northeast quadrant. Klein found (Fig. 14.8), for the rift-associated areas, that swarms did indeed occur preferentially when the direction of maximum extension was in the quadrant predicted to increase shear stress on the right-lateral faults of the San Andreas system, and to promote spreading on the associated minor spreading centers. The pattern indicated on Figure 14.8 as the expected one, if earthquakes occurred randomly in time, results from the fact that east-west tidal extension is operable for the maximum period of time each day. Areas not associated with rifts gave more variable results: aftershocks of large earthquakes in the Aleutian trench area showed a positive correlation, but a swarm in Japan did not.

14.4 TIDAL FRICTION IN THE PAST

Even though measurements of the phase lag in earth tides are still inconclusive, the angular deceleration of the earth is very real, as was noted in Chapter 1. The uncertainty is in apportioning the cause between lunar tides in the solid earth, and lunar tidal friction at the bed of the oceans. In earlier times, the moon must have been closer to the earth, and this poses a problem for theories of its origin. Slichter (1963) pointed out that 2×10^9 years ago the moon must have been very close; tidal forces would have been extremely great, for, as Equation 14.1.4 shows, the height of the tide varies as $1/r^3$. Evidence of major tidal deformation, or melting, of the surfaces of the earth or moon could be expected. On the other hand, very recent evidence, based on the dating of lunar samples (Albee et al, 1970), suggests that the age of the moon is 4.5×10^9 years. It is thus unclear how to account for the first 2.5×10^9 years of the moon's history. An extreme hypothesis would be that the moon was captured by the earth, and went into orbit, only after the end of Precambrian time. A second is that the present value of the tidal frictional torque (5 to 9×10^{23} dyne cm) is greater than its average value in the past, and that the projection backward in time of the moon's distance is not reliable. However, Runcorn (1964, 1967) has applied measurements on corals to suggest that the average value of the torque since Devonian time (370×10^6 years ago) has been remarkably close to its present value. A middle Devonian coral was shown by Wells (1963) and Scrutton

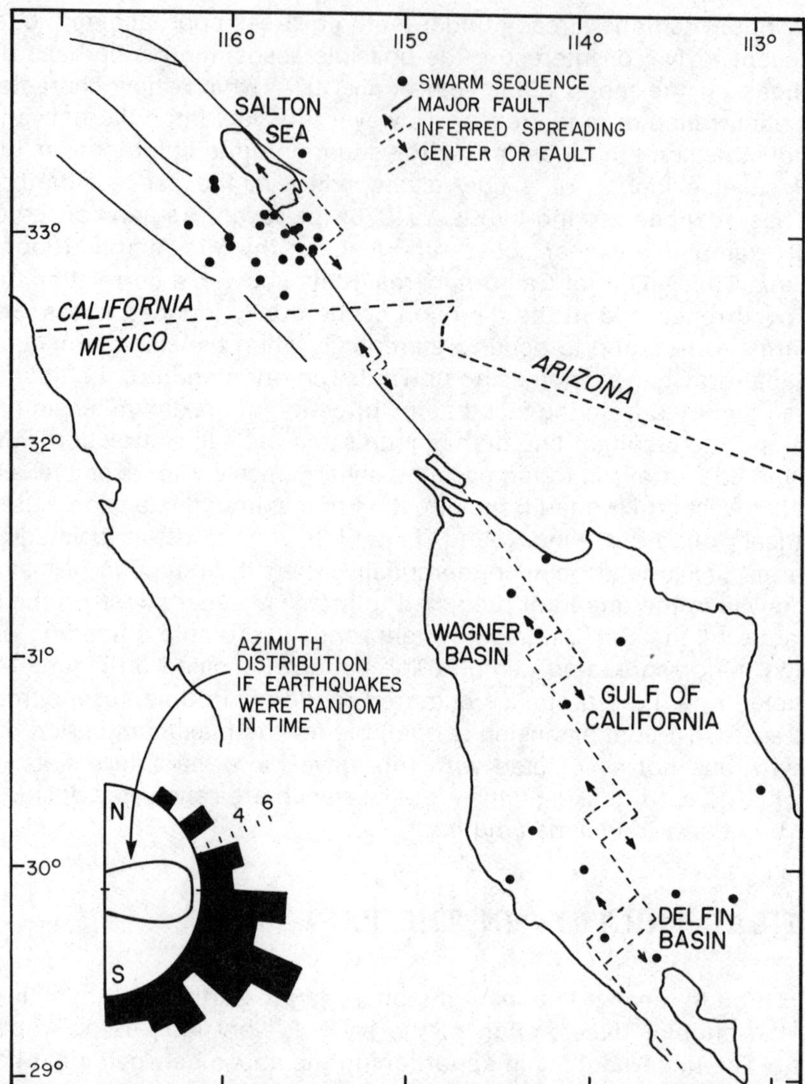

Figure 14.8 Location of earthquake swarms relative to the tectonics of the Gulf of California–Salton Sea area, and, inset, their frequency of occurrence as a function of the azimuth of maximum tidal extension at the time. After Klein (1976).

(1964) to have daily growth bands, the pattern of which displayed modulations that were apparently annual and monthly (the latter resulting from the tidal depth of water at the place where the coral grew). Counts by Wells gave the length of the year in the Devonian as 400 ± 7 solar days, and this is in accord with the present rate of angular deceleration. However, a change in the earth's angular velocity of rotation could result from a change in the earth's axial moment of inertia, such as would arise from internal redistribution of mass, and it is necessary to use the length of the month to separate the effects of moment of inertia and of tidal friction. Following Runcorn, we apply Kepler's third law to the earth-moon system:

$$\frac{r^3}{T^2} = \frac{G(M+m)}{(2\pi)^2} \tag{14.4.1}$$

where M and m are the masses of the earth and moon, r is the mean distance, and T is the length of the sideral month. The orbital angular momentum of the moon at any time is given by

$$\frac{L}{2m}=\frac{\pi r^2}{T} \qquad 14.4.2$$

Therefore

$$[G^2(M+m)^2m^3]T=2\pi L^3 \qquad 14.4.3$$

A similar relationship must hold for the earth-sun system; if N is the earth's orbital angular moment and M_s the mass of the sun,

$$G^2M_s^2M^3Y=2\pi N^3 \qquad 14.4.4$$

where Y is the length of the sidereal year. A relationship between the two orbital angular momenta that holds at any time is then obtained as

$$\frac{L^3}{N^3}=\frac{m^3}{M_s^2M}\left(1+\frac{m}{M}\right)^2\frac{T}{y} \qquad 14.4.5$$

Since the effect of the solar tide on the earth is much less than that of the lunar tide, and since N is now of the order of 10^6L, the value of N may be taken as constant at its present value. Writing Equation 14.4.5 for the present (subscript 0) and for a time in the past, and taking ratios, we obtain

$$\left(\frac{L_0}{L}\right)^3=\frac{T_0}{Y_0}\cdot\frac{Y}{T} \qquad 14.4.6$$

Taking Wells' value of 399 sidereal days for Y, and Scrutton's of 28.4 sidereal days in the month, Runcorn obtained

$$\frac{L_0}{L}=1.016\pm0.003$$

For the orbital angular momentum of the moon to have changed by this amount in 370×10^6 years, the mean retarding torque must have been 3.9×10^{23} dyne-cm—slightly less than the present value.

Turning to the question of the earth's moment of inertia, we may write for the changes in spin and orbital angular momenta:

$$\frac{2\pi I}{t}-\frac{2\pi I_0}{t_0}=(L_0-L)+(N_0-N) \qquad 14.4.7$$

where I is the earth's axial moment of inertia, t is the length of the sidereal day, and the subscript 0 again refers to present values. The first term on the right of Equation 14.4.7 is dominant; Runcorn obtained $I=0.99I_0$, with uncertainty in the third decimal place, depending upon the value of (N_0-N).

While the specific figures for the lengths of the month and year used above undoubtedly require refinement, the analysis did indicate the importance of

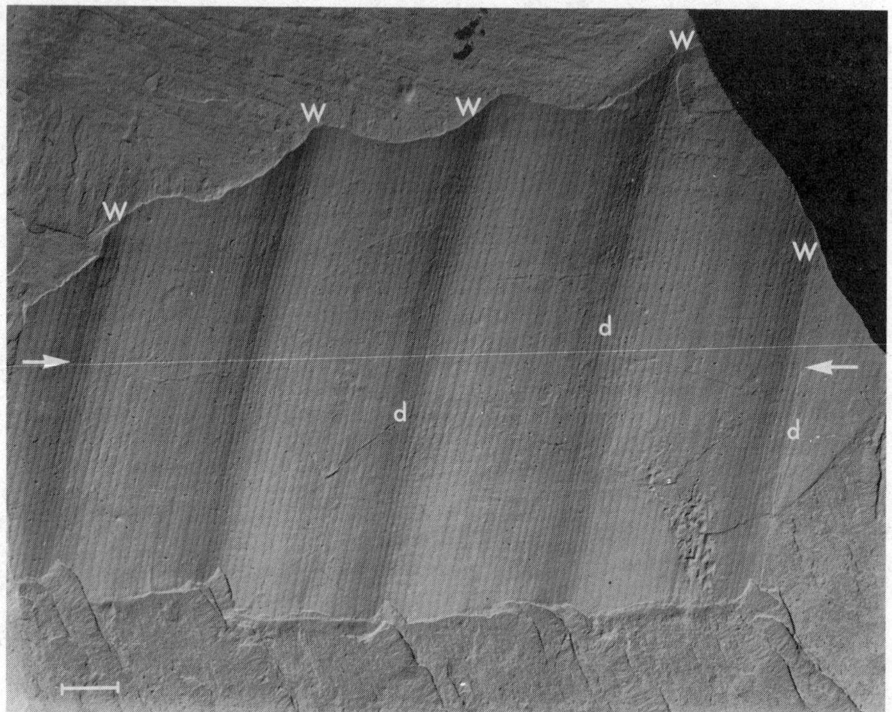

Figure 14.9 Banding in a Middle Devonian Cephalopod. Periods of slow growth (W) are interpreted as winter, so that 4 years are represented between the white arrows; disturbances (d) are also more frequent in winter. Finer banding is believed to be fortnightly. Counting over the 4 years gives 26.5 lunar semi-months per year from the Middle Devonian. Direction of growth was from left to right; line at lower left indicates 1 cm. (Courtesy of G. Pannella).

paleontological estimates of these quantities. The tentative conclusions were, first, that lunar tidal friction had not varied by an order of magnitude since the Devonian period, and second, that the earth's axial moment of inertia had remained constant within about one per cent. The first result showed that the difficulty pointed out by Slichter was still a very real one in the consideration of the history of the moon; the second result was evidence against certain theories of the earth's behavior, such as a major change in radius.

The rather fascinating connection between fossil growth bands (Fig. 14.9) and the history of the earth's rotation stimulated a great deal of further research, both biological and geophysical (Rosenberg and Runcorn, 1975). First, the reality of tidally controlled modulation on corals and bivalves appears to be demonstrated by the study of modern material, including species grown under artificial conditions of varying water depth. Secondary, similar bands have been found on a much earlier form of life, stromatolites, which can be traced back into Precambrian time (Pannella, 1975). While it is recognized that the daily banding in particular on stromatolites is much more difficult to interpret than are the bands on corals and bivalves, Mohr (1975) estimated the year at a time 2000 million years ago to contain between 800 and 900 days and the month to contain about 26 days. The presence of a monthly banding at that time confirms the fact that the moon was in orbit about the earth, ruling out the theory of later capture. The inferred very close approach of the moon to the earth is pushed further back into Precambrian time, and if there were a tidally induced episode of cracking and melting on either the earth or moon its traces have been eradicated. One point of agreement is that it is dangerous to extrapolate the present rate of deceleration of the earth back into geological time; if the major part of the slowing is due to tidal forces in shallow seas, a different distribution of the world's oceans would lead to a very different rate.

Problems

1. Deduce the expression $\lambda = (1+k-h)$ for the reduction in amplitude of a horizontal pendulum on a yielding earth (Section 14.2).

2. In Section 14.3, the possibility of the oscillation of the inner core about its equilibrium position was mentioned. To estimate the period, it is necessary to investigate the restoring force for a small displacement δ, and show that the force is proportional to δ. Slichter (1961) wrote, "the gravitational restoring force is precisely the attraction between two spheres with centers separated by δ, one having the size of the inner core, but of different density $\rho_0 - \rho_1$ and the other of density ρ_1 and radius δ." Here ρ_0 is the density of the inner core and ρ_1 the (constant) density of the liquid outer core.

(a) Confirm the above statement and show that the gravitational restoring force is proportional to δ.

(b) Even if the outer core has zero viscosity, there is an effect due to the mass of fluid set in motion when the inner core moves. It is shown in standard works on hydrodynamics that this can be accounted for (in an infinite fluid) by taking the "effective mass" of the inner core to be its true mass plus one half of the mass of fluid displaced.

Taking account only of the two above effects (gravitation and fluid inertia), estimate the period of oscillation of the inner core, for $(\rho_0 - \rho_1) = 0.3$ gm cm^{-3}, $\rho_1 = 11.0$ gm cm^{-3}. How might such an oscillation be excited and how might it be observed at the earth's surface? (In reality, other forces, viscous and electromagnetic, play a role in determining the period.)

BIBLIOGRAPHY

Albee, A. L., Burnett, D. S., Chodos, A. A., Eugster, O. J., Huneke, J. C., Papanastassiou, D. A., Podosek, F. A., Price Russ II, G., Sanz, H. G., Tera, F., and Wasserburg, G. J. (1970) Ages, irradiation history and chemical composition of lunar rocks from the Sea of Tranquillity. *Science,* **167**, 463.

Busse, F. (1974) On the free oscillations of the earth's inner core. *J. Geophys. Res.,* **79**, 755–757.

Darwin, Sir G. H. (1910) *Scientific papers.* Cambridge Univ. Press, Cambridge.

Harrison, J. C., Ness, N. F., Longman, I. M., Forbes, R. F. S., Kraut, E. A., and Slichter, L. B. (1963) Earth tide observations made during the International Geophysical Year. *J. Geophys. Res.,* **5**, 1497.

Jackson, B. V., and Slichter, L. B. (1974) The residual daily earth tides at the South Pole. *J. Geophys. Res.,* **79**, 1711–1715.

Jeffreys, H. (1920) Tidal friction in shallow seas. *Phil. Trans. Roy. Soc. Lond. A,* **221**, 239.

Jeffreys, H., and Vicente, R. O. (1957) The theory of nutation and the variation of latitude. *Mon. Nat. R.A.S.,* **117**, 142 and 162.

Klein, Fred W. (1976) Earthquake swarms and the semidiurnal solid Earth tide. *Geophys. J. R. A. S.,* **45**, 245–295.

Kuo, J. T., Jackens, R. C., Ewing, M., and White, G. (1970) Transcontinental tidal gravity profile across the United States. *Science,* **168**, 968–971.

Lammlein, D., Latham, G. V., Dorman, J., Nakamura, Y., and Ewing, M. (1974) Lunar seismicity, structure and tectonics. *Rev. Geophys. Space Phys.,* **12**, 1–21.

Love, A. E. H. (1911) *Some problems in geodynamics.* Cambridge Univ. Press, Cambridge.

Melchior, P. J. (1966) *The earth tides.* Pergamon Press, New York and Oxford.

Mohr, R. E. (1975) Measured periodicities of the Biwabik (Precambrian) stromatolites and their geophysical significance. In *Growth rhythms and the history of the earth's rotation,* ed. G. D. Rosenberg and S. K. Runcorn. John Wiley and Sons, London.

Molodensky, M. S. (1961) Quatrième symp. int. sur les marèes terrestres. *Comm. de l'Observatoire Royale de Belgique No.* 188, 25.

Munk, W. H., and MacDonald, G. J. F. (1960) *The rotation of the earth.* Cambridge Univ. Press, Cambridge.

Nishimura, E. (1950) On earth tides. *Trans. Amer. Geophys. Un.,* **31**, No. 3, 357.

Pannella, G. (1972) Palaeontological evidence on the Earth's rotational history since Early Precambrian. *Astrophys. Space Sci.,* **16**, 212–237.

Rosenberg, G. D., and Runcorn, S. K. (eds.) (1975) *Growth rhythms and the history of the earth's rotation.* John Wiley and Sons, London.

Runcorn, S. K. (1964) Changes in the earth's moment of inertia. *Nature,* **204**, 823.

Scrutton, C. T. (1964) Periodicity in Devonian coral growth. *Palaeontology,* **1**, 552.

Slichter, L. B. (1960) In: Comptes rendus des réunions de la commission permanente des mareés terrestres à l'assembleé générale de'Helsinki. *Bull. d'Inform. Mareés Terrestres,* **21**, 368.

Slichter, L. G. (1963) Secular effects of tidal friction upon the earth's rotation. *J. Geophys. Res.,* **68**, 4281.

Takeuchi, H. (1950) On the earth tide of the compressible earth of variable density and elasticity. *Trans. Amer. Geophys. Un.,* **31**, 651.

Tomaschek, R. (1954) Variations of the total vector of gravity at Winsford (Cheshire) Part 1. General results and maritime load influences. *Mon. Nat. R.A.S., Geophys. Suppl.,* **6**, No. 9, 540.

Wells, J. W. (1963) Coral growth and geochronometry. *Nature,* **197**, 948.

GEOMAGNETISM: SCOPE AND FUNDAMENTAL CONCEPTS 15

15.1 THE MAGNETISM OF THE EARTH

The study of geomagnetism occupies a place in the physics of the globe which is rather different from that of the other disciplines, for it cannot be divorced from considerations of effects in space. Indeed, the different applications of the study are so varied that it is well to take an overall look, before concentrating on any particular phase.

Ever since the main magnetic field of the earth was discovered, at least as early as the eleventh century in China (Needham, 1962), man has been fascinated by the fact that the earth is a magnet. While the methods of analysis of the magnetic field bear some relation to those we have discussed for the gravitational field—the expansion in spherical harmonics, for example—there is a fundamental difference in approach, for magnetism, unlike gravity, is not an inherent property of all bodies of the universe. The search for the cause of the earth's magnetic field stimulated the early researches of such men as William Gilbert and Edmund Halley, in the centuries before a complete description of the field was available. Apart from the purely scientific interest, there is the fact that throughout the nine centuries since the discovery of the field, its direction has remained a most useful tool for navigation.

We shall discuss theories of origin of the field in succeeding chapters, where we shall also find that its changes with time are of almost as great importance as the field itself. Some of the changes result from internal effects and in themselves provide direct information on the interior of the earth. Some are external in origin, but all provide information on an important parameter of the material, the electrical conductivity. In addition, the external variations are related to other phenomena, such as sunspots and the aurora, which have in themselves fascinated man since their discovery. The necessity of recording the time changes in the magnetic field led to international cooperation in the establishment of magnetic observations, which, in fact, predated international cooperation in seismology.

Local distortions of the magnetic field may be analyzed in terms of structure, in a manner analogous to the study of gravity anomalies. There is, however, a most important additional source of information, in that rocks may become permanently magnetized at the time of their formation. From measurements of this magnetization on samples, the past history of the field may be deduced, with a completeness that is not possible for other geophysical variables. It is this study of the past history of the field, and of the samples in it, which has thrown new light on the question of relative motions of the earth's crust.

15.2 FUNDAMENTAL EQUATIONS

The consideration of the various phenomena sketched above requires the analysis of magnetic fields in a variety of physical situations: the fields of steady and alternating electric currents, secondary fields arising from electromagnetic induction in a conductor, and the fields of permanently magnetized ferromagnetic material. In all cases, of course, the magnetic field arises from the motion of electric charge (including electron spin and orbital motion), and *Maxwell's equations* apply. In electromagnetic units, these are

$$\begin{aligned} \nabla \times \underline{E} &= -\dot{\underline{B}} \\ \nabla \times \underline{H} &= 4\pi \underline{C} + \dot{\underline{D}} \\ \nabla \cdot \underline{B} &= 0 \\ \nabla \cdot D &= 4\pi\rho \end{aligned} \qquad 15.2.1$$

where $\underline{E}$ and $\underline{H}$ are the electric and magnetic field intensities, $\underline{D}$ and $\underline{B}$ are the electric and magnetic inductions, $\underline{C}$ is the conduction current density, and ρ is the space-charge density. The vectors are connected by the relations

$$\underline{D} = \epsilon \underline{E}, \qquad \underline{B} = \mu \underline{H}, \qquad \underline{C} = \sigma \underline{E} \qquad 15.2.2$$

where ϵ, μ, σ are the permittivity, permeability, and electrical conductivity respectively.

While the form used in Equations 15.2.1 is the most compact for the expression of Maxwell's equations, it is important (and helpful to memory) to recognize the physical significance of each. The first equation is simply the statement of Faraday's law, that a time-varying magnetic field produces an electromotive force. The second contains, in the first term on the right, the statement of Ampere's law, that an electric current produces lines of magnetic field around itself, and in the second term on the right, Maxwell's famous *displacement current* which arises from a time-varying electric field, and which may contribute to the magnetic field. Equation three is the statement that magnetic single poles do not exist, and equation four asserts that the lines of electric field begin and end on electric charges.

In all of the problems we discuss, even those involving oscillatory fields of relatively short periods, the term involving $\dot{\underline{D}}$ will be negligible. Within a conductor, $\dot{\underline{D}}$ is normally much smaller than $4\pi\underline{C}$. In fact, if L and T are characteristic length and time, $\dot{\underline{D}}$ is of order $\epsilon\dot{\underline{E}} = \epsilon\underline{E}/T = \epsilon\mu\underline{H}L/T^2$ and $\nabla \times \underline{H}$ is of order $\underline{H}/L$; $\underline{D}$ is therefore negligible if

$$T \gg (\epsilon\mu)^{1/2}L \tag{15.2.3}$$

But the electromagnetic wave velocity is given by $(\epsilon\mu)^{-1/2}$ so that the condition expressed in Equation 15.2.3 is that characteristic times be great compared to the time of travel of an electromagnetic wave across the region. For the entire earth this is about 0.03 second, and in all of the problems we are to consider, the condition will be satisfied. In non-conducting regions, $\underline{C} = 0$ also, and $\nabla \times H = 0$,

$$\therefore \underline{H} = \pm \nabla U \tag{15.2.4}$$

where U is a scalar, and we apply the methods of potential theory. A difference between the potential function as applied in gravitational and in magnetic problems becomes apparent only when the effects of magnetized structures are considered.

There is, first, the difference that like poles repel, so that if the potential U of a north pole of strength m is written as $+m/r$, the field (i.e., force on a unit north pole) is correctly given as $-\partial U/\partial r$. In the gravitational case, we had the field given by $+\partial U/\partial r$. Secondly, all magnetized bodies are composed of assemblages of dipoles, and the potential will be obtained by integration of the potential of a single dipole over the body. The potential of a dipole aligned in the i-direction is given by

$$\begin{aligned} \delta U &= m\left(\frac{1}{r_1} - \frac{1}{r_2}\right) \\ &= -M\frac{\partial}{\partial i}\left(\frac{1}{r}\right) \end{aligned} \tag{15.2.5}$$

where r_1 and r_2 are the distances of the observer from north and south poles, $M = m \cdot dl$ is the moment of the dipole and $\partial/\partial i$ signifies the directional derivative. When this expression is integrated to give the potential of a fine body, the result is the directional derivative of the gravitational potential of the same body.

The situation regarding units in geophysics is not completely straightforward. While c.g.s. units are widely adopted, with the electromagnetic system (emu) being used for most problems, rationalized M.K.S. units will be found in some papers dealing with electromagnetic induction and the SI system (Appendix E) is now insisted upon by some journals. In the following chapters, the units will be emu unless otherwise specified. It will be recalled that this system is constructed from the expression for the force between unit magnetic poles, with magnetic permeability taken as dimensionless, and equal to unity for free space. Electric current is defined in terms of the force on a unit pole, the electromagnetic unit current being equivalent to 10 amperes. The fields H and B are measured in oersted and gauss, respectively, but in free space the distinction is not always observed. A widely used unit for magnetic fields in geophysics is the gamma (γ), which is 10^{-5} oersted.

The electromagnetic unit of resistance is equivalent to 10^{-9} ohms, so that 1 emu of resistivity is equal to 10^{-9} ohm cm. This conversion factor is very frequently needed, because tabulated values of properties are usually quoted in the latter units or in ohm m. The reciprocal property, conductivity, must be in em units when inserted in fundamental relations, but is often quoted in reciprocal ohm cm or reciprocal ohm m for convenience.

BIBLIOGRAPHY

Chapman, S., and Bartels, J. (1940) *Geomagnetism.* Clarendon Press, Oxford.

Needham, J. (1962) *Science and civilization in China; Vol. IV.* Cambridge Univ. Press, Cambridge.

MEASUREMENT OF THE MAGNETIC FIELD 16

16.1 THE PROBLEM

The earth's main field at the surface is of the order of 0.5 gauss. This is not a large magnetic field when compared, for example, to the fields of artificial sources in industrialized areas. For geophysical purposes it is necessary to measure this, certainly to one part in 10^3, and at times to one part in 10^5. As in the case of gravity measurements, it is desirable to measure the field, at some places, in absolute terms, and then to cover the surface of the earth with relative measurements. However, the distinction between an absolute and a relative measurement is not so clear-cut as in the case of gravity.

The progression of magnetic-field measuring instruments from those available to Gauss to those now available reflects much of the progress of physics over the past century. We shall trace this progress, without attempting to discuss every instrument which has ever been applied to the problem.

16.2 THE METHOD OF GAUSS

A magnet of moment M suspended horizontally by a fiber is directed into the magnetic meridian by a couple MH, where H is the horizontal intensity. If the magnet is displaced, it oscillates about the meridian with a period T, where

$$T = 2\pi\sqrt{\frac{I}{MH}} \qquad 16.2.1$$

Here I is the moment of inertia of the suspension about its axis, and the elastic restoring torque of the fiber is neglected. The measurement of T in this experiment thus determines the product MH.

During the late eighteenth and early nineteenth centuries, a great number of relative measurements of H were made by timing the period of oscillations in this way. As many of these were made on long expeditions, the only way of ensuring that the moment M had not changed was to measure the period at a base station at infrequent times. The procedure was thus similar to the relative measurement of gravity with pendulums. Gauss (1839) recognized the importance of modifying the experiment to yield a measurement of field in terms of fundamental quantities, independent of changes in the property of the magnet. He proposed that the same magnet, of moment M, be brought up to an auxiliary magnet, suspended as a compass, in such a way that its axis is perpendicular to the axis of the auxiliary magnet. The latter comes to equilibrium under torques due to H and to the magnet M. If its axis then makes an angle θ with the meridian, we may equate torques to give

$$\frac{H}{M}=\frac{2}{r^3}\cdot\frac{1}{\sin\theta} \qquad 16.2.2$$

where r is the distance between the centers of the two magnets. It is apparent that each magnet is here treated as a dipole, and if this is not justified, there will be a correction for the size of the magnets. With this reservation, Equations 16.2.1 and 16.2.2 can be solved for M and H, so that the horizontal intensity is determined absolutely; time, length and angle are the only measured quantities.

Until very recently, most of the absolute stations of the world were established by the above method. In addition to H, the declination was determined by astronomical observations and the inclination by dipping needle, yielding the complete specification of the vector field (Whitham, 1960).

16.3 SATURATION INDUCTION MAGNETOMETERS

A number of magnetometers were developed which made use of electric circuits, using galvanometers or the magnetron tube as detectors. However, the real increase in our capability to measure the field under a variety of conditions came with the development of the saturation-core or *fluxgate* magnetometer.

Two identical cores (Fig. 16.1) of high permeability material are placed parallel to each other, and wound with series-opposing primary coils. A second winding surrounds the two primary windings. The primaries are connected to an a.c. source and the cores are driven through saturation, in alternate directions, twice per cycle. At the moment of saturation, each system suffers a decrease in its self-inductance, producing an irregularity in the primary waveform. However, if the two primaries are balanced, no voltage is induced in the secondary coil. The existence of an external field along the axis of the cores produces an imbalance by bringing one core to saturation sooner, and a voltage pulse is produced in the secondary twice per primary cycle. For small fields, the amplitude of this pulse is proportional to the field, and the device is therefore a magnetometer, capable of measuring the field component along the axis of the cores. The secondary pulses can be shaped, amplified, rectified, and recorded; to avoid non-linearity, magnetic fields of the order of the earth's field are measured by first nullifying a large, known fraction with an additional coil wound around the detector.

The flux-gate magnetometer can be used in a variety of ways in terrestrial magnetism. In its original application as an airborne magnetometer (Wyckoff, 1948) it was continuously oriented, by means of two auxiliary detectors and a servo-motor arrangement, into the direction of the total field, and it measured scalar changes in this field. Serson and Hannaford (1956) adapted it for component measurements on the ground by mounting the detector on the optical system of a transit, and Serson, Mack, and Whitham (1957) invented a three-

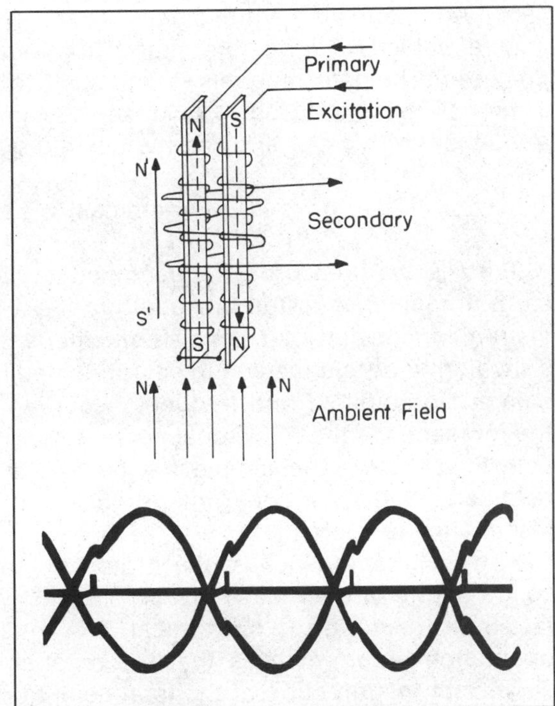

Figure 16.1 Element of a flux-gate magnetometer. The traces at bottom show the voltage across each primary coil, and the induced secondary voltage (after Wyckoff).

component airborne measuring system by mounting three mutually perpendicular detectors on a gyroscopically-oriented platform. When operated at a magnetic observatory, the flux-gate magnetometer is a useful instrument for measuring time changes in the field with periods as short as a few seconds.

16.4 THE PROTON PRECESSION MAGNETOMETER

Precession of nuclei around a magnetic field direction is a well-known phenomenon in nuclear physics. Its application to the measurement of magnetic fields depends upon the precise measurement of the frequency of this precession.

We visualize a sample, say of water, containing a large number of protons. Normally, the dipole moments of the protons are in a random orientation. If a strong magnetic field is applied to the sample, for example by means of a coil wound around the container, these dipoles become aligned in the direction of the field. When the polarizing field is suddenly removed, the spinning nuclear magnets precess for a short time around the direction of the earth's ambient field, before interactions bring the system to its random state again. If the spin angular momentum of the proton is L and its magnetic moment μ, the angular velocity of precession in a field F is

$$\omega = \frac{\mu F}{L}$$

The quantity μ/L is known as the gyromagnetic ratio of the proton; it has been measured in the laboratory to six significant figures (Driscoll and Bender,

1958). An unknown magnetic field can therefore be measured if the precession angular velocity can be determined. It is found that the coherent motion of the precessing nuclear magnets is sufficient to induce a detectable signal in the same coil that was used to polarize the sample, and the measurement of the frequency of this signal gives the desired quantity. The working relation is

$$F \text{ (gammas)} = 23.4874f$$

where f is the measured frequency in Hz. It is evident that, over most of the earth, frequencies in the low audio range will be encountered. In essence the instrument consists of a sample container surrounded by a coil, with electronic switching to permit the coil to be connected in turn to a d.c. supply or to a high-gain audio amplifier and frequency counter. The frequency is counted during the few seconds that the signal persists; after the signal has decayed exponentially toward zero, the sample may be repolarized. The proton precession magnetometer therefore does not produce a continuous record, but a series of discrete measurements.

Two characteristics of the proton precession magnetometer are the absolute nature of the measurement, since only a frequency is required, and the fact that orientation is not critical. The total field is measured, and the only orientation requirement is that the polarizing field make a sufficiently great angle with the direction of the total field. For this reason, the magnetometer is widely used on rockets and other space vehicles.

16.5 ALKALI VAPOR MAGNETOMETERS

These instruments are essentially masers or *optical pumps,* whose frequency of self-oscillation is a function of the ambient magnetic field. In the rubidium magnetometer (Fig. 16.2) circularly polarized light from a rubidium lamp is passed through a cell containing rubidium vapor to a photocell. Absorption of photons from the light excites atoms of rubidium 87 to a higher level, as shown on the energy diagram of Figure 16.3. This diagram illustrates the sub-levels into which the energy levels are split, according to the Zeeman effect, by the ambient magnetic field. Excited atoms return to the ground state, but, because of the circular polarization of the light, to a higher m sub-level, in which they are trapped. This trapping, and the consequent elimination of transitions, makes the cell completely transparent, and the current output of the photocell is at a maximum. This current is amplified, shifted in phase, and passed through

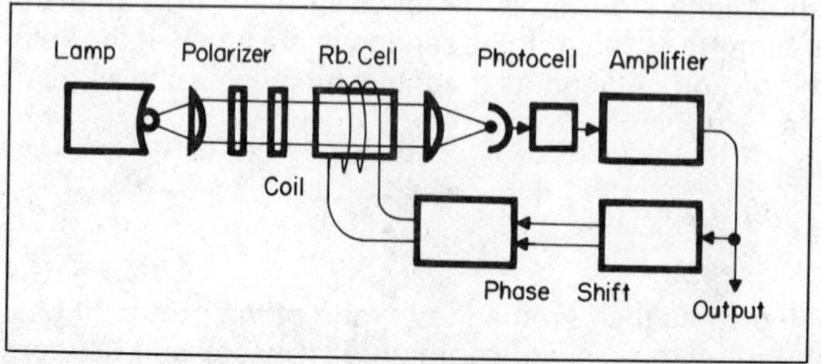

Figure 16.2 Principal components of the rubidium vapor magnetometer.

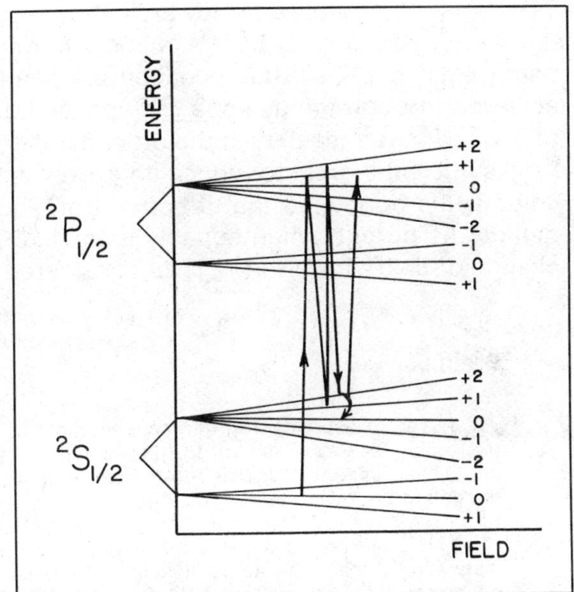

Figure 16.3 Energy levels in rubidium as a function of magnetic field strength. The transformations involved in an operation cycle of the magnetometer are shown; they are separated for clarity, but all occur at the same value of ambient field. Trapping occurs in the $m = +2$ sub-level of the lower state.

a coil surrounding the cell (Fig. 16.2). The magnetic field of this coil is sufficient to redistribute the population of atoms in the sub-levels. Transitions are again possible, and the process is repeated. The complete system oscillates at a frequency, known as the Larmor frequency, which is determined by the difference in energy levels, and therefore by the ambient field.

As in the case of the proton precession magnetometer, the total field is measured, and orientation is not critical. The output frequency in the earth's field, however, is much higher, being 200 to 300 kHz. With this higher frequency, counting over comparable times leads to a more accurate determination of frequency. Indeed, the alkali vapor magnetometer is the most precise instrument available for measuring the earth's magnetic field.

16.6 INDUCTION MAGNETOMETERS

Various types of "earth inductors" in which coils were flipped or rotated have been used to measure the static field, but we consider here the problem of measuring short-period time variations in the field. As we shall see, there is considerable interest in measuring time variations in the components of the field with periods of one second or less, and with an amplitude as small as 0.01 gamma. This is below the level of measurement of the flux-gate magnetometer, and while the alkali vapor magnetometer can be used, it does not measure a component. For these reasons, induction coils have been widely used in the study of these micropulsations, as they are called.

The design of the coil requires some care. For a field rate-of-change of 1×10^{-2} gammas per second, the output in volts of a coil of N turns and area of cross-section A cm² is

$$E = NA \times 10^{-7} \times 10^{-8} \text{ volts}$$

Taking 1×10^{-7} volts as the lower limit of usable signal in this frequency range, it is seen that NA must be 1×10^8 cm^2. For any reasonable dimensions, N can hardly exceed 10^5, so that an effective area of 10^3 cm^2 is required. This is often achieved through the use of a rod-shaped core, a few centimeters in diameter, of alloy with permeability of the order of 1000, and having a length-to-diameter ratio sufficiently great to reduce demagnetization effects (Selzer, 1957). Such a coil may be buried, so that it is free from induction effects produced by wind motion. Its output, when suitably amplified, provides a record of the rate-of-change of field, down to the level considered above.

Problems

1. The measurement of the magnetic field by the method of Gauss requires the expression for H/M (Equation 16.1.2) when the magnet of moment M is used to deflect an auxiliary magnet. What is the full expression for H/M if neither magnet is short enough to be treated as a dipole?

2. It is desired to measure the vertical and horizontal components of the earth's magnetic field, but only a total-field magnetometer is available. Show that the components may be measured, provided known bias fields, in the directions of Z and H, are applied successively to the instrument, and readings are taken with and without these fields.

3. Helmholtz coils are well known to produce a region of uniform field by the elimination, at the center, of the first and second derivative of field along the axis of the coils. Design a system employing a second pair of coils, to cancel the next higher derivative, and so produce a larger region of uniform field.

BIBLIOGRAPHY

Driscoll, R. L., and Bender, P. L. (1958) Proton gyromagnetic ratio. *Phys. Rev. Letters,* **1**, 413.

Selzer, E. (1957) La méthode "Barre-Fluxmètre" d'enregistrement des variations magnétiques rapides. *Ann. Int. Geophysical Year,* **4**, 287.

Serson, P. H., and Hannaford, W. L. W. (1956) A portable electrical magnetometer. *Can. J. Technol.,* **34**, 232.

Serson, P. H., Mack, S. Z., and Whitham, K. (1959) A three-component airborne magnetometer. *Pub. Dom. Obs. Ottawa,* **19**, No. 2.

Whitham, K. (1960) Measurement of the geomagnetic elements. In *Methods and techniques in geophysics,* ed. S. K. Runcorn. Interscience, New York.

Wyckoff, R. D. (1948) The Gulf airborne magnetometer. *Geophysics,* **13**, 182.

THE MAIN FIELD 17

17.1 GENERAL DESCRIPTION

The analysis of the earth's magnetic field bears some resemblance to the study of the gravitational field, but there are important differences. Both are vector fields defined over an almost spherical surface, but because the direction of the gravitational field departs only slightly from the radial direction, its vector characteristic is not so apparent. Secondly, the time variations in the magnetic field are both greater in relative magnitude, and more complex in frequency structure, than are the tidal variations in gravity. Finally, there is the point that we are compelled to enquire into the cause of the earth's magnetism.

The time variations of the magnetic field will be discussed in following sections. For the present, we assume that measurements of the intensity and direction of the field have been made over the earth's surface, and that on the basis of repeated measurements at certain stations, these have been reduced to a single instant of time, or *epoch,* in such a way that variations with periods less than a few years are averaged out. These reduced measurements define the *main field* at the time chosen. The main field changes with time, but rather slowly, and we shall show that it is the result of sources within the earth, free of external effects.

Because of the vector nature of the field, its description at any point on the earth requires the specification of three quantities. The usual definition of these components or angles (some of which, such as declination, have come into everyday usage) is shown in Figure 17.1. The global representation of the field is then achieved by the mapping of one of these elements. The declination, as the quantity of interest to navigators, and also the easiest to measure, is that for which the oldest charts are available. But the general nature of the field is shown more clearly by maps of the dip and total intensity (Fig. 17.2a, b, c). The progress made in the mapping of the fields since the nineteenth century has been reviewed in Zmuda (1971). Before about 1950, three-component measurements were limited to land stations, and to observations made on the cruises of a very few specially-designed non-magnetic ships. A great advance was made when three-component airborne instruments became available and a number of long flights over the oceans were made to fill gaps in the coverage. The field remained poorly defined in the polar regions until the satellite era, beginning in 1957. Satellites permit the sampling of the field on a uniform basis over the whole globe, but the measurements have been limited to total field, rather than component, determinations. Nevertheless, the satellite observations, when combined with all available three-component measurements, permit the first truly global portrayal of the geomagnetic field.

The definition of the poles requires a little care. The two (in principle)

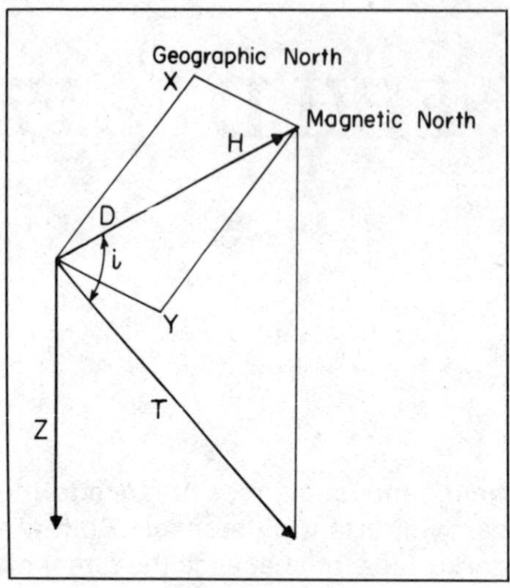

Figure 17.1 Relationship of the total intensity T, horizontal component H, vertical component Z, declination D, and inclination i.

points on the earth's surface where the dip reaches ±90° are known as the dip poles. For various reasons, these points are not actually as significant as is sometimes implied. Because of local anomalies, there may in fact be several points in the general polar region where the dip is vertical. Also, as far as the effect of the earth as a magnetized body in space is concerned, the poles as seen from a distance are more significant and these we shall define below. Particular care must be taken not to confuse the "poles" with the magnetic poles of a bar magnet, for example. The field intensity does not by any means vary with distance from the poles on the surface of the earth according to an inverse square law.

17.2 MATHEMATICAL DESCRIPTION

Modern work on geomagnetism is essentially based on the analysis of Gauss (1839), in which the main field was expanded into a series of spherical harmonics. The important feature of this analysis is that, for each harmonic, it is possible to assign an internal or external source.

The main field, as we have defined it, is only slowly varying with time. If it is due to electric current flow, these currents are obviously not flowing across the earth's surface; at the surface, the field may be derived from a scalar potential U. We investigate the behavior of U over the surface of the earth, assumed to be a sphere of radius a. In spherical coordinates (r, θ, λ) the equation satisfied by U (Appendix A) is

$$\frac{\partial}{\partial r}\left(r^2 \frac{\partial U}{\partial r}\right) + \frac{1}{\sin^2 \theta} \frac{\partial}{\partial \theta}\left(\sin \theta \frac{\partial U}{\partial \theta}\right) + \frac{1}{\sin^2 \theta} \frac{\partial^2 U}{\partial \lambda^2} = 0 \qquad 17.2.1$$

In the important case of axial symmetry, which is appropriate here, the solution may be written as series of Legendre polynomials, but a different dependence

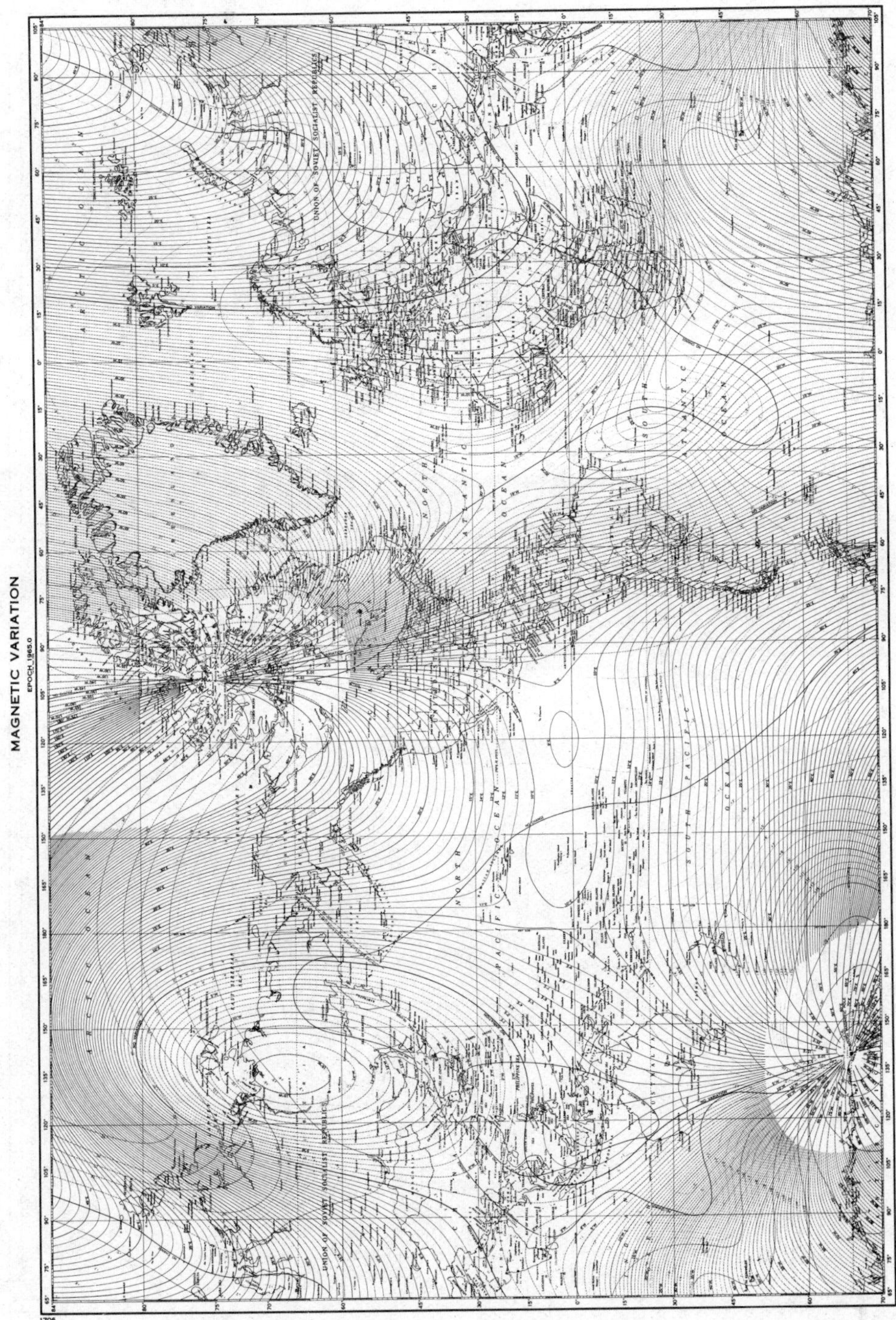

Figure 17.2(a) World map of magnetic declination. The solid contours are at intervals of 1°; broken contours give the change in degrees per year. (Reproduced with permission from U.S. Naval Oceanographic Office, Chart 1706, 8th Ed., Feb. 1966)

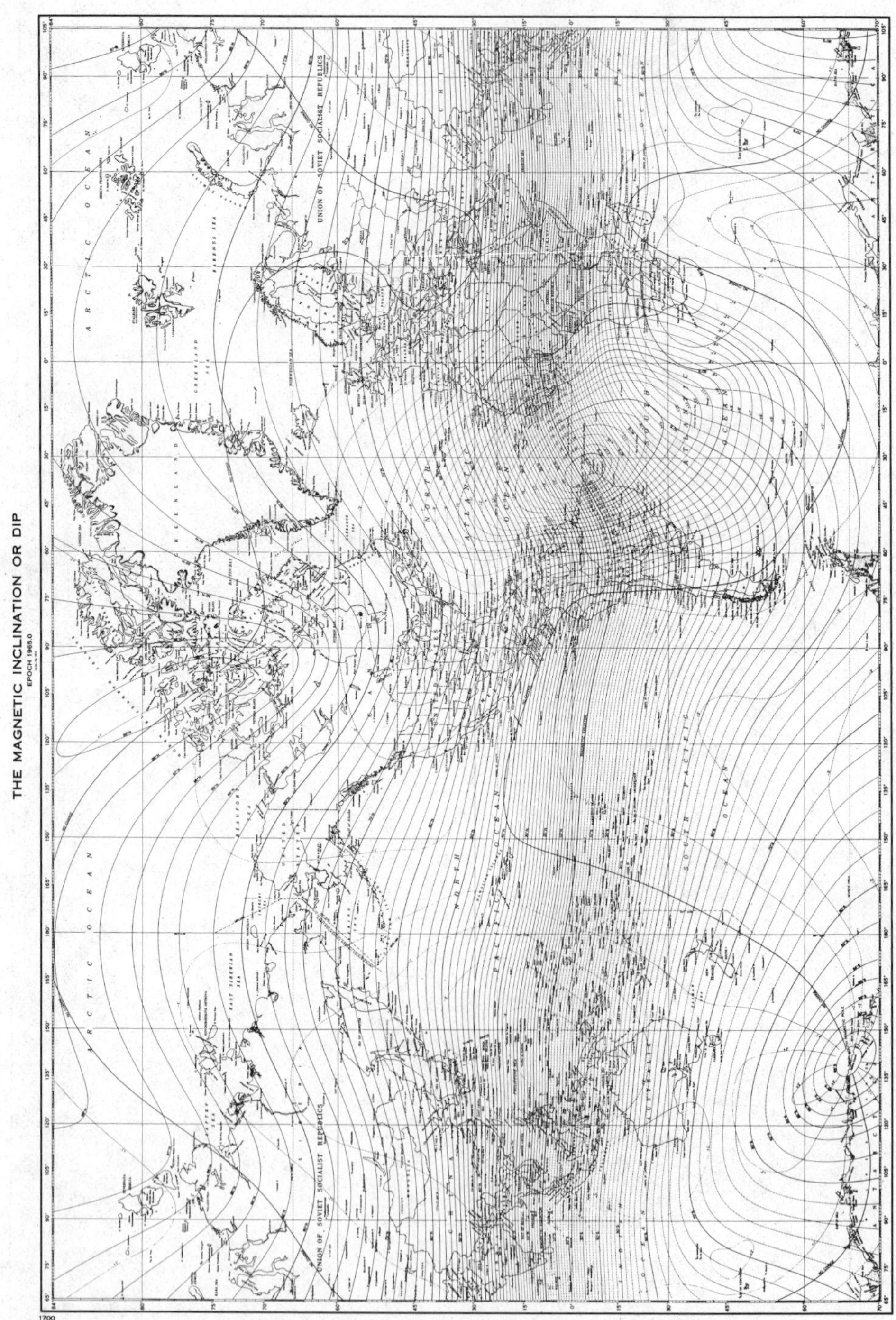

Figure 17.2(b) World map of inclination or dip. The solid contours are at intervals of 2°; broken contours give the change in minutes per year. (Reproduction with permission from U.S. Naval Oceanographic Office, Chart 1700, 8th Ed., Feb. 1966)

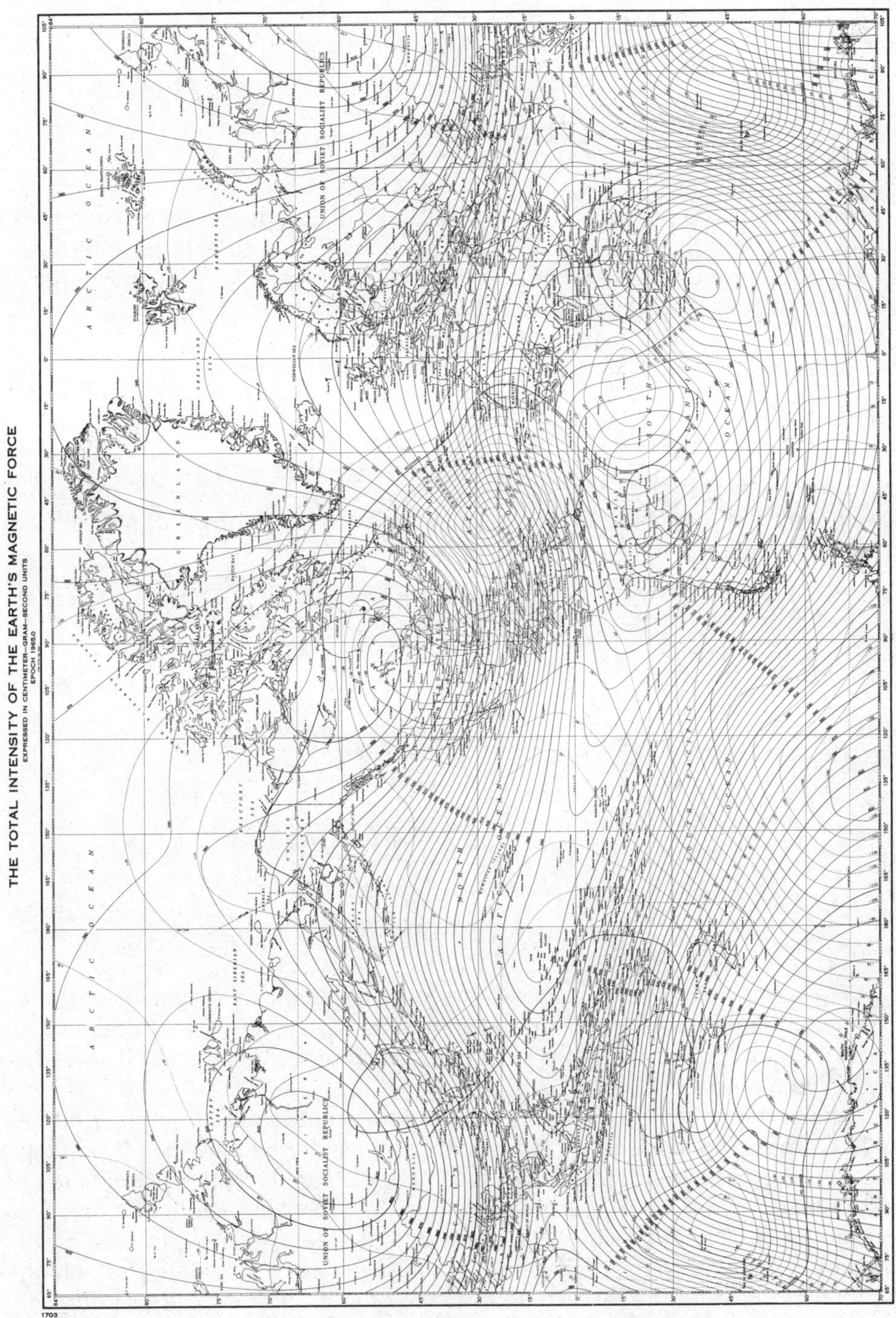

Figure 17.2(c) World map of total intensity. The solid contours are at intervals of 0.010 gauss; broken contours give the change in gammas per year. (Reproduced with permission from U.S. Naval Oceanographic Office, Chart 1703, 8th Ed., Feb. 1966)

upon r is involved, according to whether the source of the field is internal or external to the sphere. In particular:

$$\left.\begin{aligned} U_e &= a \sum E_l \left(\frac{r}{a}\right)^l P_l(\cos\theta) \\ U_i &= a \sum I_l \left(\frac{a}{r}\right)^{l+1} P_l(\cos\theta) \end{aligned}\right\} \qquad 17.2.2$$

Here, U_e is the potential, inside or on the sphere, due to external sources, while U_i is the potential, outside or on the sphere, due to external sources. E_l and I_l are amplitude factors or *coefficients* which specify the contribution of external and internal sources to the lth harmonic term; the equations are written so that they have the dimensions of magnetic field.

The total potential on the earth's surface is therefore

$$U = a \sum \left[I_l \left(\frac{a}{r}\right)^{l+1} + E_l \left(\frac{r}{a}\right)^l \right] P_l(\cos\theta) \qquad 17.2.3$$

Components of the field (X northward; Y eastward; Z downward) are obtained by differentiation as

$$\begin{aligned} X &= \frac{1}{a}\frac{\partial U}{\partial \theta} = \sum \left[E_l \left(\frac{r}{a}\right)^l + I_l \left(\frac{a}{r}\right)^{l+1} \right] \frac{\partial P_l(\cos\theta)}{\partial \theta} \\ Y &= 0 \qquad\qquad 17.2.4 \\ Z &= \frac{\partial U}{\partial r} = \sum \left[l E_l \left(\frac{r}{a}\right)^{l-1} - (l+1) I_l \left(\frac{a}{r}\right)^{l+2} \right] P_l(\cos\theta) \end{aligned}$$

Let us consider the behavior of the terms involving $P_l(\cos\theta)$:

$$X_1 = -[E_1 + I_1] \sin\theta$$

$$Z_1 = [E_1 - 2I_1] \cos\theta$$

The terms in brackets can be obtained by fitting observed values of X and Z at different latitudes to a sine or cosine function. Two equations in the two unknowns E_1 and I_1 are therefore obtained, and the equations can be solved for these two amplitude factors.

The extreme cases of an entirely external source and an entirely internal source are shown in Figure 17.3. Notice that the external influence which gives rise to a $P_1(\cos\theta)$ distribution over the sphere is a uniform field, while the corresponding internal source is a central dipole. If measurements of X only were available, no distinction could be made. It is the behavior of Z in relation to X which provides for the separation. The analysis can be extended to the higher harmonics, yielding the relative importance of internal and external sources.

Gauss did not, in fact, limit himself to zonal spherical harmonics, but determined the coefficients in the potential for the 24 terms up to $l = 4$, $m = 4$. His analysis stands as a triumph of geophysical investigation, the more so since it was done with no mechanical aids to computation. The extent of the measurements of the field available to Gauss in 1839 was perhaps greater than is usually appreciated. In 1837 Sabine had published a global chart of the total in-

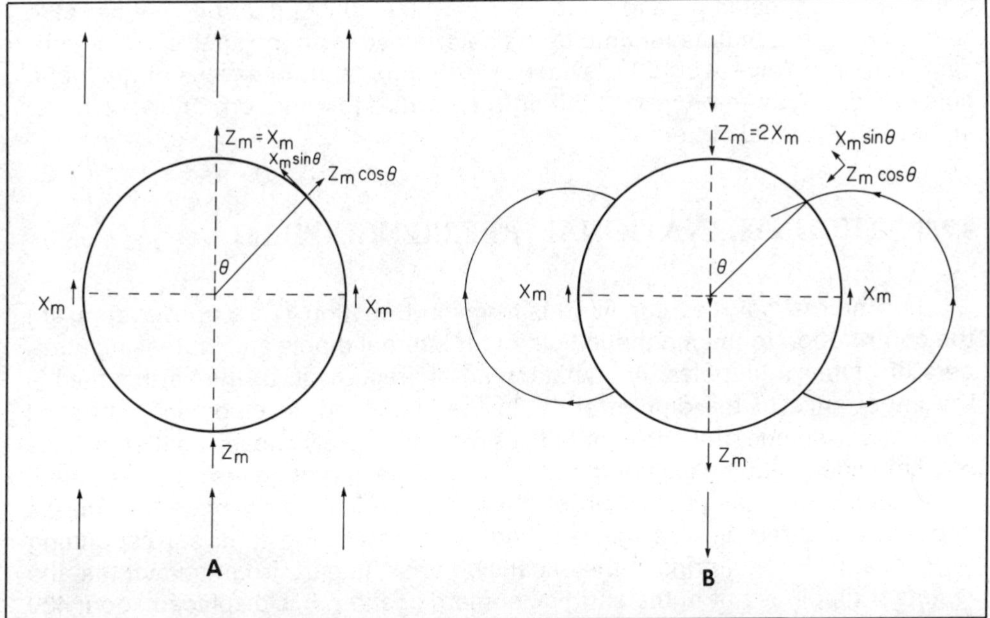

Figure 17.3 Variation of the vertical and horizontal magnetic components over the earth, for (left) a uniform external source and (right) an internal dipole. Subscript m indicates maximum values; notice that (A) and (B) yield the same variation of X over the surface, but that the behavior of Z is different.

tensity, and older charts of the declination and inclination were available. Of course, there were great regions over which the chart contours were uncertain and they were not reduced to a common point in time. Gauss picked off the values at 12 points on each of 7 parallels of latitude, using these as the starting points in the determination of coefficients.

The dipole nature of the field was immediately made apparent when these coefficients were determined, the coefficient of the axial dipole term alone being more than five times greater than that of the next most important term. Carrying out an analysis such as that traced above, Gauss concluded that *all* of the field had an internal origin. Some of the analyses made in the interval since his work have proposed that a small proportion of the field owes its origin to external sources, but the current view is that all of the "main field" is internal in origin. An apparent contribution from external sources may very easily arise if the basic observations used in the analysis are incorrectly reduced to free them from time variations (Chapter 18).

To return to Gauss, he made a number of remarkable suggestions for further investigations, to be carried out when more adequate observations were available. The first was whether the field was completely derivable from a potential, as was assumed in the expansion. If it is not, then around a closed contour on the earth's surface

$$\oint H \cdot ds \neq 0$$

Gauss pointed out that this would be the case if electric currents flowed across the earth's surface. Again, some later investigations suggested the presence of a non-potential portion of the field, but Bartels' (1941) explanation, that they also result from incorrect reduction to epoch, has usually been accepted.

Gauss also suggested that, in future analyses, the constant term P_0 not be set at zero *a priori,* but that its magnitude be determined from the analysis. A significant non-zero value would be evidence, of course, for an excess of magnetic poles of one type (monopoles) within the earth. Most modern analyses have, in fact, set P_0 equal to zero.

17.3 THE INTERNATIONAL REFERENCE FIELD

For internal sources, the various terms in Equation 17.2.3 above represent the contribution to the magnetic field of an internal dipole and increasing numbers of higher multipoles. An outstanding characteristic of the earth's field is the importance of the dipole terms. In the simplified development with axial symmetry assumed, the coordinate axes were effectively chosen so that the axis of z coincided with the axis of the resultant dipole. If other axes are prescribed, with z along the axis of rotation of the earth, the field is no longer symmetric about z and three dipoles are required to represent the term corresponding to $P_1(\cos\,\theta)$. Higher terms in the expansion arise, in part, from the fact that the resultant dipole is not at the earth's geometric center. It is displaced about 400 km north.

The points where the axis of the resultant dipole, extended, intersects the earth's surface are known as the geomagnetic poles. It is these points which are more significant than the dip poles. The north geomagnetic pole is in northwestern Greenland, whereas the north dip pole is near Bathurst Island in the western Canadian arctic.

As in the case of the gravitational field, it is useful to have an internationally accepted reference to provide a background for the isolation of more local magnetic anomalies. The International Association of Geomagnetism and Aeronomy has adopted, for epoch 1965.0, a field, entirely internal in origin, defined in terms of the values, to $l = m = 8$, of spherical harmonic terms (Zmuda, 1971). It was the relative complexity of the magnetic field, as compared to the gravitational, that prevented the establishment of the international reference magnetic field until forty years after the adoption of the International Gravity Formula. The magnetic field description had to await the extension of measurements over the oceans, initially by shipborne and later by airborne magnetometers, and also the availability of high-speed computers for the analysis.

For some applications it is convenient to relate points on the surface of the earth to a system of coordinates based on the magnetic dipole. Geomagnetic latitude is measured from the geomagnetic equatorial plane, taken normal to the dipole axis; geomagnetic longitude is measured eastward from a great circle which passes through both geographic and geomagnetic poles. Nomograms and maps (Matsushita and Campbell, 1967, Appendix 1) are available to determine the geomagnetic position of a station. (See also Fig. 19.5.)

17.4 THE CAUSE OF THE MAIN FIELD

The problem of finding the cause of the earth's magnetism is now better defined. We seek an internal source, and may be content to find a source for the dipole component only. We note that the dipole is approximately aligned with

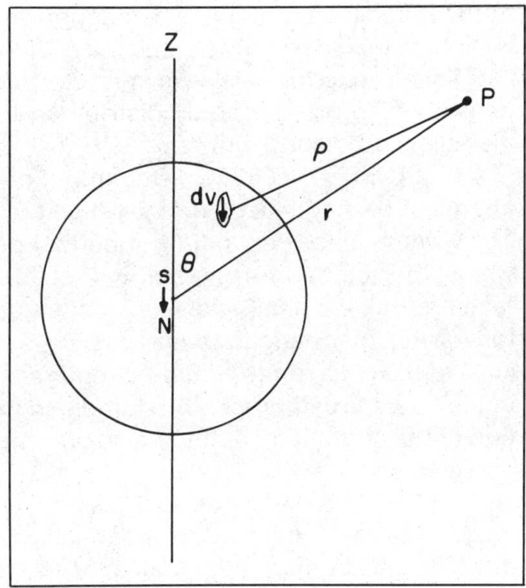

Figure 17.4 The magnetic effect of a uniformly magnetized sphere is equivalent to that of a central dipole.

the earth's axis of rotation, which suggests that rotation may be involved in the source.

It is important to realize that the predominance of the dipole term does not, in itself, suggest a source near the center of the earth, for a uniformly magnetized sphere possesses an external field identical to that of a central dipole. To show this, consider the sphere of Figure 17.4, uniformly magnetized to intensity J in the negative z-direction. The potential at P due to an element dv is

$$\delta U_p = \frac{\partial}{\partial z}\left(\frac{1}{\rho}\right) J\,dv \qquad 17.4.1$$

When the effect of the whole sphere is considered, the order of z-differentiation and integration may be interchanged, but the integration is then equivalent to that for the gravitation potential, which we know leads to a value identical to that of a point mass at the center. Thus

$$\left.\begin{aligned} U_p &= J\frac{\partial}{\partial z}\left(\frac{V}{r}\right) \\ \text{or} \qquad & \\ U_p &= -M\frac{\cos\theta}{r^2} \end{aligned}\right\} \qquad 17.4.2$$

in which V is the volume of the sphere and M is its magnetic moment. The final expression is identical to that of a central dipole of moment M aligned on the z-axis.

It is instructive at this point to look at the magnetic field inside the magnetized sphere. The above procedure can be applied and the first of Equations 17.4.2 shows that the magnetic potential is the same as the z-component of gravitational force exerted by a body of unit density, except that J replaces G. We know that this force component at radius r inside a sphere is $-\frac{4}{3}Gr\cos\theta$, so the magnetic potential is $-\frac{4}{3}\pi Jr\cos\theta$ or $-\frac{4}{3}\pi Jz$. The internal magnetic field is therefore entirely in the z-direction, uniform, equal to $\frac{4}{3}\pi J$, and opposite in sense to the magnetization. It is a demagnetizing field tending to reduce the magnetization of the sphere itself, and we shall have cause to refer to it in Chapter 21.

To return to the earth, we must consider possible causes of the earth's magnetism which are uniformly distributed throughout the volume. Most of these models can be shown to fail.

FERROMAGNETISM. One might consider the main field to be the resultant effect of magnetized rocks, which are known to exist at the earth's surface. The volume involved would not be that of the earth, but of a spherical shell in which the temperature is below the Curie point for magnetic minerals. This temperature, we shall show, is reached at a depth of about 20 km. To produce the earth's magnetic moment (M) of 92×10^{24} c.g.s. units, an intensity in the shell of 8.0 c.g.s. units would be required. This is much greater than is observed in the usual crustal rock. In addition, in this theory there would be no explanation of the alignment with the axis of rotation, nor would the secular change be provided for.

VARIOUS ROTATION THEORIES. These usually involve either the gyromagnetic effect, which is the magnetic polarization, observed in the laboratory, of rotating ferromagnetic bodies, or the rotation of separated electric charges in the earth. The first fails by several orders of magnitude (Barnett, 1933), and the second because the associated electric field that would be expected is not observed.

Blackett (1947) proposed a new theory, in which magnetization would be a fundamental property of all mass in rotation, with the resultant magnetic moment related linearly to angular momentum through a proportionality constant containing $G^{1/2}/2c$. At the time this was proposed, it appeared to predict correctly the relative magnetic moments of the sun and earth in terms of their angular momenta. However, with a revised value for the sun's magnetic moment, this agreement disappeared. Blackett himself (1952), in a classic negative experiment, showed that a rotating mass does not acquire a magnetic moment of the required magnitude. In so doing, he incidentally led the way to greatly increased activity in the study of rock magnetism, by bringing the astatic magnetometer to a new standard of sensitivity.

There is one experiment by which it may be determined whether the field arises from a deep source, or from the earth's whole body. That is to measure the change in strength of some field component with depth *inside* the earth. (It follows from our discussions in potential theory, Section 9.2, that external measurements cannot make the distinction.) If the cause is actually a central dipole, field strength should increase with depth according to an inverse-cube law of distance from the center. If the source is distributed throughout the earth, the horizontal component of field strength should decrease with depth, much as gravitational attraction decreases inside a uniformly dense sphere. Measurements (Runcorn, Benson, Moore, and Griffiths, 1951) suggest that the cause in fact is deep-seated. We are thus led to consider the role of the earth's core in producing the main field. Theories involving the core have included a thermoelectric effect at the core-mantle boundary (Runcorn, 1956) and a self-sustaining dynamo within the fluid core. The latter we shall investigate in some detail.

17.5 DYNAMO THEORIES

Dynamo theories of the earth's magnetic field suppose the existence of motions in the fluid, electrically conducting core. The motion of material in the field is supposed to induce electrical currents which maintain the field.

An analogue, which is simple in principle but difficult to construct in practice, is the Faraday *disc generator,* Figure 17.5. If a conducting disc of radius a is rotated in an uniform field H_0, parallel to the axis of rotation, an emf is set up between the periphery of the disc and the axis. Current will flow through the single-turn coil if brush connections are made as shown, and this current will be equal to $\frac{1}{2}H_0va/R$, where v is the peripheral velocity of the disc and R is the total resistance of the circuit. This current will produce a field, along the axis, of strength H, where

$$H = \frac{2\pi I}{a} = \frac{\pi H_0 v}{R}$$

(Strictly, this is the value at the axis; for simplicity we take it as constant over the area of the disc.)

If $v = R/\pi$, an initial field will just be sustained; if $v > R/\pi$, the condition with no current is unstable, and any small initial field will be amplified, provided the agent driving the disc can maintain the angular velocity against the opposing electromagnetic torque. On the other hand, if the agent can supply only a constant torque, G say, a rather interesting behavior results (Bullard, 1955). The opposing electromagnetic torque is given by

$$IH \int_0^a r\,dr = \tfrac{1}{2} IHa^2$$

Then if C is the moment of inertia of the disc about the axis of rotation, the disc must obey

$$C\dot{\omega} = G - \tfrac{1}{2} IHa^2 = G - \pi a I^2$$

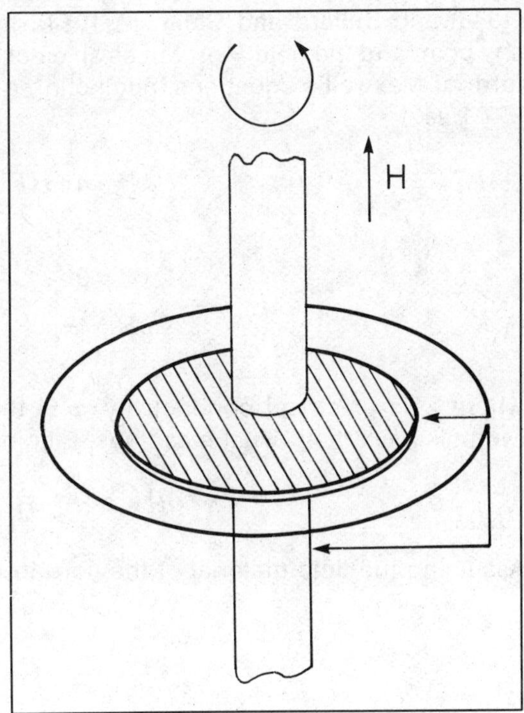

Figure 17.5 The Faraday disc generator.

The only steady state ($\dot{\omega} = 0$) is when

$$I = I_c = \sqrt{\frac{G}{\pi a}}$$

If the disc is started with too high an angular velocity, the field H will initially be amplified, but the disc will slow down until the current falls below the critical value, I_c, when it will again speed up. A rather asymmetric but repetitive pattern of field variation results; Bullard, however, showed that neither the current nor the field actually reverses in sign.

The energy relations in the disc generator are instructive also. In the steady state, with $\omega = R/\pi a$, the rate of input of energy by the driving agent is

$$G\omega = GR/\pi a = I_c^2 R$$

The energy expended is just that required to maintain the Joule heating in the resistive circuit. A self-sustaining dynamo has been constructed and operated by Lowes and Wilkinson (1963). In their model, two rotors of ferromagnetic, conducting metal are free to spin about perpendicular axes. The rotors are in electrical contact with a surrounding block of the same material. It was found that, at sufficiently high speed velocities, a flow of current was maintained in the system.

In the extension to the earth's core, the same principles are involved. A self-sustaining dynamo depends upon the existence of suitable motions, low-resistance circuits, and an energy source for maintaining the motion. The source of the original small field, required to begin the generation process, has not been considered a serious problem; presumably it could have been located outside of the earth, in the solar wind (Chapter 19), for example. The problem is a difficult one of magnetohydrodynamics, investigated originally by Elsasser (1946) and Bullard and Gellman (1954). Interaction between the velocity v at any point and the field $\underline{H}$ produces an electric field $\underline{v} \times \underline{H}$, so that the appropriate form of Maxwell's equations, neglecting displacement currents and assuming $\mu = 1$, is

$$\begin{aligned} \nabla \times \underline{H} &= 4\pi\sigma(\underline{E} + \underline{v} \times \underline{H}) \\ \nabla \times \underline{E} &= -\dot{\underline{H}} \\ \nabla \cdot \underline{H} &= 0 \\ \nabla \cdot \underline{E} &= 4\pi\rho c^2 \end{aligned} \qquad 17.5.1$$

where σ is electrical conductivity, ρ is the density of charge, and the other symbols have their usual meanings. Elimination of $\underline{E}$ yields

$$\nabla \times \nabla \times \underline{H} = 4\pi\sigma[-\dot{\underline{H}} + \nabla \times (\underline{v} \times \underline{H})] \qquad 17.5.2$$

Assuming the fluid material of the core to be incompressible,

$$\nabla \cdot \underline{v} = 0 \qquad 17.5.3$$

The problem is then to find a solution of 17.5.3 which, when inserted in 17.5.2, yields a field $\underline{H}$ of the required characteristics. At the core boundary, all components of $\underline{H}$ and of tangential $\underline{E}$ must be continuous, and the normal component of current density must be zero. Outside the core, the field $\underline{H}$ must be equivalent to that of a central dipole.

It is instructive to recast Equation 17.5.2 by means of the vector identity

$$\nabla \times \nabla \times \underline{H} = \nabla(\nabla \cdot \underline{H}) - \nabla^2 \underline{H} \tag{17.5.4}$$

and the fact that $\nabla \cdot \underline{H} = 0$. For then,

$$\dot{\underline{H}} = \nabla \times (\underline{v} \times \underline{H}) + \frac{1}{4\pi\sigma} \nabla^2 \underline{H} \tag{17.5.5}$$

If motion stops, $\underline{v} = 0$, the first term on the right vanishes, and the equation becomes the diffusion equation. Its solution represents a field $\underline{H}$ which decays exponentially with time, at a rate controlled by $\frac{1}{4\pi\sigma}$. The *decay time* of any field, if generation were to cease, therefore increases with σ.

No completely general solution of the equation is yet available. Most of the approaches have involved the expansion of the vector fields $\underline{v}$ and $\underline{H}$ into sums of elementary *toroidal* (i.e., having no radial component) and *poloidal* vectors analogous to the torsional and poloidal oscillations discussed in Chapter 8, with the investigation of the behavior of the first few terms. Some useful rules have emerged from this analysis. In particular, it is known that a purely toroidal motion cannot yield a solution, nor can a field which is symmetric about an axis be supported (Cowling, 1934).

Bullard and Gellman investigated in detail a combination of T_1 and S_2^2 motion (the notation corresponds to that of Chapter 8). Such a combination would, in fact, be expected if the core were in a simple mode of convection, with a pattern involving four convection cells (Fig. 17.6). The poloidal motion S_2^2 represents the convection itself, while the toroidal motion T_1, which is motion relative to the main body of the rotating earth, results from the conservation of angular momentum (hence reduction in angular velocity) of material raised by the convection. Figure 17.6 outlines the successive interactions, starting with the T_1 motion in the S_1 (dipole) field. The current resulting from this interaction produces a T_2 field, which is completely contained within the core. Interaction of this field with the S_2^2 motion generates currents whose fields interact once again with the T_1 motion. The field arising from the currents so generated interact again with the S_2^2 motion to produce a toroidal T_1 current. This current produces the dipole S_1 field, which is thereby maintained.

Bullard and Gellman were forced, in the days before high-speed computers were available, to choose an extremely simple type of motion, and it was perhaps not completely unexpected when Gibson and Roberts (1969) showed that their solution did not converge. More recent work (Gubbins, 1973, 1974) has emphasized smaller-scale, more complicated motions, but the problem remains a very difficult one. In the first place, the solution of Equation 17.5.2 alone, with prescribed motions, is not adequate, for it neglects the force exerted on the moving fluid by the magnetic field itself. This force (the Lorentz force) is proportional to $\underline{J} \times \underline{H}$, where $\underline{J}$ is the electric current density, and (in the limit of high electric conductivity) it restricts relative motion between the fluid and the

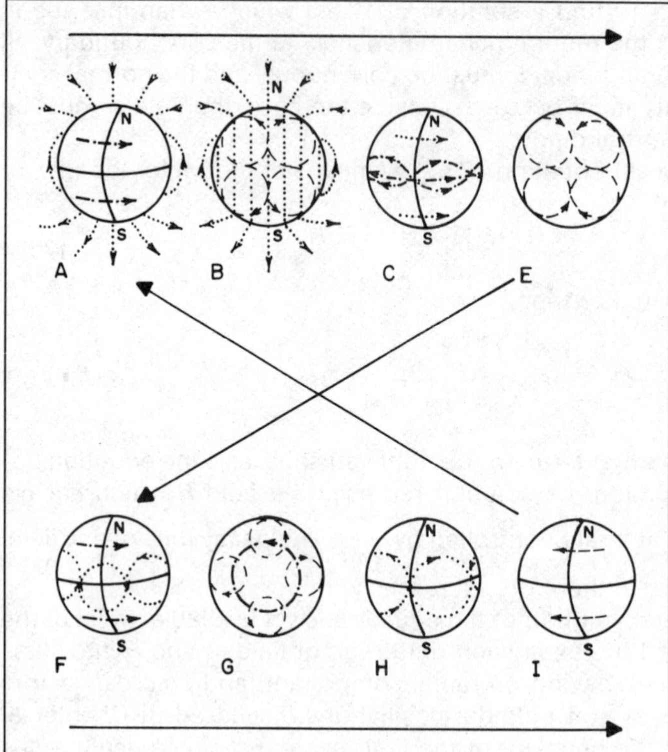

Figure 17.6 Bullard's model of a self-sustaining dynamo in the core. Diagrams *A, C, F, H,* and *I* are external views; *B* is a meridian section; *E* and *G* are equatorial sections. Dotted lines represent magnetic flux; broken lines, electric current; and dot-dash lines, fluid motion.

field. Neglect of it is equivalent, in the case of the Faraday disc generator, to assuming that the driving agent can maintain a constant angular velocity, as opposed to assuming that the driving torque is constant. Dynamos which have been proposed with the Lorentz force neglected are known as kinematic dynamos, while a solution of the complete problem would yield a true hydromagnetic dynamo.

The importance of small-scale motion in kinematic dynamos can be demonstrated in several ways. G. O. Roberts (1970) showed that a two-dimensional periodic pattern consisting of fluid motion in spirals could produce a mean field in a direction perpendicular to an existing field, with amplification under suitable conditions. This was a most important development, for, applied to the earth, it meant that core-wide convection did not have to be invoked; a set of smaller convection cells could possibly form a dipole field. It is even possible to construct kinematic dynamos from completely turbulent motion, as could be represented by a series of eddies, provided the turbulence is not completely random ("isotropic") in orientation. In this case, the motion and the field produced cannot be traced in detail, but space- and time-averaged quantities are considered. A result which emerges, and which is contrary to ordinary experience, is that a mean current density $\underline{J}$ can be produced, in the same direction as the mean field; that is,

$$\underline{J} = \alpha \underline{H}$$

where α can be positive or negative depending on the sense of rotation in the eddies. Now if some mean differential toroidal motion in the core, as well as a system of eddies, is permitted, a dipole field can be converted into a toroidal field and back into a dipole field (Fig. 17.7) as in the Bullard-Gellman dynamo. The first part of the diagram shows winding of the lines of force by the mean toroidal motion, as a result of the hydromagnetic locking of the fluid to the field. The second stage is the production of a toroidal current by the "α-effect," and this current can sustain a dipole field. Parker (1955) first proposed a dynamo of this type, arguing that α has different signs in the two hemispheres because of Coriolis effects, analogous to those in the atmosphere and oceans. An example of a small-cell kinematic dynamo in a sphere is shown in Figure 17.8, from the work of Gubbins (1973). Fluid motion in the cells has symmetry about the vertical axis of the diagram. An initial field, from right to left, is distorted, and reconnects to form the T_2^1 field shown at left. If the direction normal to the page is taken to be parallel to the earth's axis, the latter field can be seen as a contribution to the geomagnetic dipole field. By numerical calculation at a grid of points, Gubbins established the convergence of this solution.

A close approach to a hydromagnetic dynamo has been obtained by Busse (1975). He considered first the convection of fluid in a rotating region between coaxial cylinders and showed that the pattern would consist of a number of small cylindrical cells. The kinematic problem was then solved for these cells, and finally the effect of the Lorentz force was estimated and superimposed on the motion. Busse showed that with a suitable choice of parameters an external dipole field could be supported, with reasonable velocities in the core. He argues that because of rotational effects his assumed cylindrical geometry is an approximation of the actual motions in the earth's core. A consequence of this is that heat should be more efficiently transferred to the core-mantle boundary in equatorial than in polar regions; the temperature of the boundary, or even the core's contribution to heat flow at the earth's surface, should have a latitudinal dependence, but these effects cannot yet be observed.

A complication is provided by paleomagnetism (Chapter 21), which strongly suggests that the polarity of the main field has reversed, with a rather complicated periodicity, through geological time. It is difficult to visualize how a single dynamo mechanism could so reverse, but a series of coupled dynamos could.

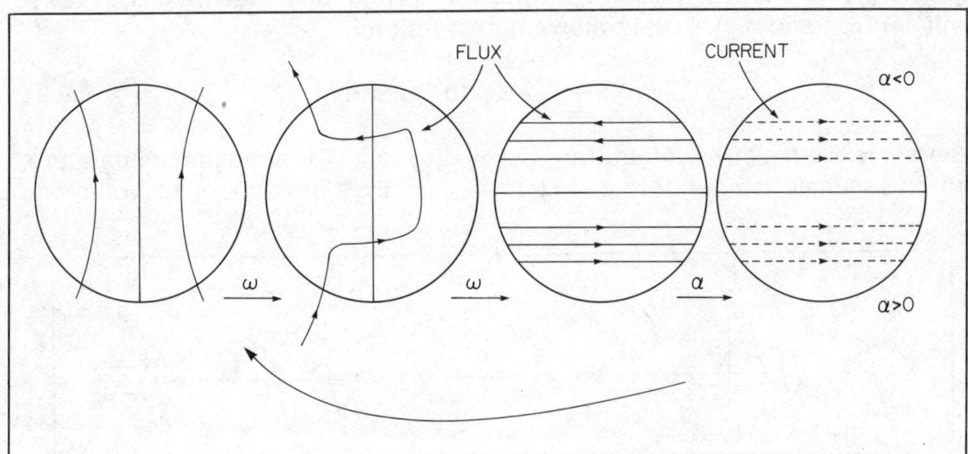

Figure 17.7 Maintenance of a dipole field by differential core rotation ("ω") and the creation, by turbulence, of current parallel to the distorted field lines ("α"). After Gubbins (1974).

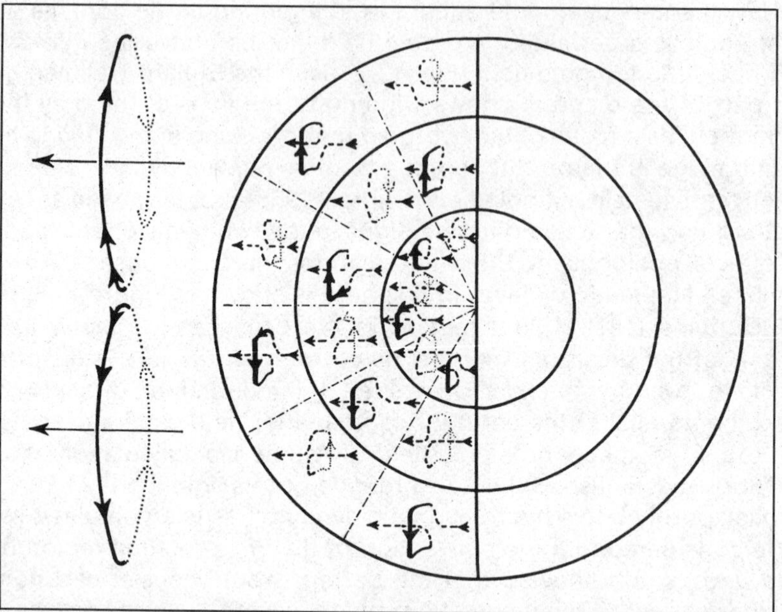

Figure 17.8 A possible dynamo based upon small cells. An initial field, shown by light arrows at left, is distorted by fluid motions. Heavy field lines are in front of the plane of the page. Reconnection gives the field shown at left as a double toroid. From D. Gubbins (1973); courtesy of the author and the Royal Society of London.

This was considered in detail by Rikitake (1958) and Allan (1958). Allan's example of the reversing external field (Fig. 17.9) does indeed resemble the record of polarity of the earth's field. The implication is that there is not one, but two or more competing systems of interactions in the core, any one of which becomes predominant for a certain period of time.

The toroidal magnetic field T_2, which is mentioned above, is an important feature of dynamo theories. Although it cannot be directly detected at the earth's surface, it probably plays an important role both in the secular change of the magnetic field and in the coupling of the mantle to the core, as we shall see presently.

The electrical conductivity of the core remains an uncertain quantity. Bullard (1948) adopted, as an extrapolation of the measured values of resistivity for iron and nickel, the following expression:

$$\rho = (40 + 0.22T)\ 10^{-6} \text{ ohm cm}$$

where ρ is the resistivity of an iron-nickel alloy, and T is absolute temperature. For an assumed temperature of 10,000° C, this gives $\rho = 2200 \times 10^{-6}$ ohm cm,

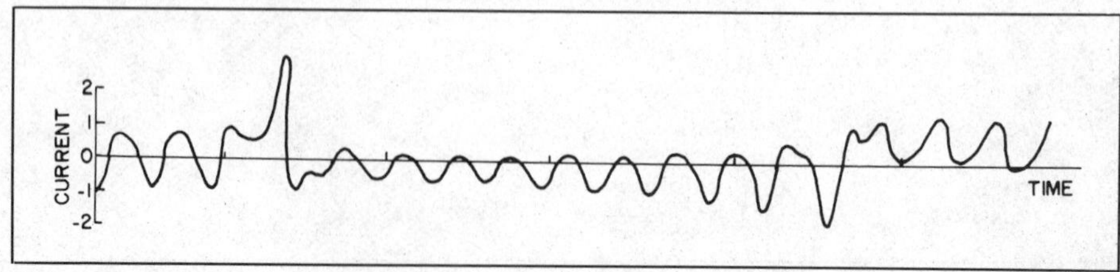

Figure 17.9 Computed current in two coupled disc generators, showing reversals. (After Allan)

at atmospheric pressure. The effect of compression was assumed to be given by

$$\frac{\rho_p}{\rho_0} = \left(\frac{D_0}{D_p}\right)^{3.6}$$

where D is density, and the subscripts refer to the compressed and uncompressed states. The effect of increasing pressure consists of two opposing terms: the dominant one, corresponding to a decrease in resistivity, results from an increase in order in the lattice; a smaller, contrary influence is the decrease in electron energy with decreasing volume. In the outer core ρ_p/ρ_0 is thus about 0.47, and this led to $\rho = 0.47 \times 2200 \times 10^{-6}$ ohm cm, or about 1×10^6 emu. This value has often been quoted. Modern views on the temperature within the core (Chapter 25) suggest a lower temperature than 10,000°C, but the effect of this may be offset by the presence of silicates and other non-metallic elements within the core, so that modern estimates are within an order of magnitude of Bullard's value.

Gardiner and Stacey (1971) reviewed the information available on the melting point resistivity of pure iron, and on the extrapolation to temperatures and pressures representative of the outer core and near the inner core boundary. They showed that the value 3×10^{-4} ohm cm was as good an estimate for both regions as was possible on the knowledge available. Johnston and Strens (1973) investigated the effect, on the melting-point resistivity, of the alloying of an iron-nickel mix with sulfur and carbon. Since modern ideas on the core do favor alloying (Chapter 26), they propose increasing Gardiner and Stacey's estimate of resistivity by about 30 per cent, corresponding to a content of 15 per cent sulfur. A value of ρ in the outer core of about 4×10^{-4} ohm cm (4×10^5 emu) therefore remains as a reasonable estimate. There are, in principle, theoretical limits on the geophysically acceptable value of core resistivity, but the range they permit is rather broad. If the resistivity were as high as 1×10^{-2} ohm cm (1×10^7 emu), say, then unreasonably high velocities would be required for the generation term of Equation 17.5.5 to predominate over the diffusion term. On the other hand, because of an established relationship between electrical and thermal conductivity, the assumption of a very low resistivity would imply a very high thermal conductivity and an unreasonably high flow of heat from the core, in comparison with the known flux at the earth's surface (Chapter 24). Gardiner and Stacey estimate 1×10^{-4} ohm cm |1×10^5 emu) as the probable lower limit of electrical resistivity on this basis.

In conclusion, we may say that the outstanding problems remaining with the dynamo theory are the source of energy for the motions, and the question of whether the combination of velocity and conductivity is sufficient to maintain the field. The latter condition is often expressed by considering the value of the dimensionless combination of parameters $4\pi\sigma Lv$ (where L is a characteristic length), which is known as the magnetic Reynold's number. In other words, it is the value of this product which is significant in maintaining a given field; smaller values of σ could be tolerated if v were great, and vice versa.

Energy sources both within and external to the core have been proposed to drive the motions. Bullard and Gellman considered thermal convection in the fluid core. This would require a source of heat, presumably a concentration of radioactivity, whose presence cannot be established. Convection could be maintained by other than thermal processes, for example, if cooler and denser iron from the mantle were continuing to drop into the core from the base of the mantle (Urey, 1952; Runcorn, 1965). But the possibility of energy being transferred from the outer part of the earth to the core cannot be excluded. Malkus (1963) considered the influence of the earth's precession, which is controlled by the ratio $(C - A)/C$ (Chapter 1). Since the ratio of principal moments of inertia for the core is not quite equal to that for the earth as a whole, the core is driven at an "incorrect" precession velocity. Whether this could maintain the motions necessary for a self-sustaining dynamo depends upon details of the core-mantle coupling that are not yet established. Rochester et al. (1975) conclude that it is unlikely.

Problems

1. What are the geomagnetic coordinates of a station at the geographic position (ϕ, λ) if the geomagnetic pole is at (ϕ_p, λ_p)?

2. A "non-potential" portion of the earth's magnetic field would be produced by electric currents flowing across the surface, from the interior to the atmosphere. Attempts have been made to evaluate this current by measuring $\oint F \cdot ds$ around a closed contour on the surface, and applying Ampere's law. Assuming that field components are measurable to 1 gamma, and directions to 1 minute, design an experiment to evaluate the integral, and estimate the limit of detectability of earth-air current, in amps/cm^2, for the contour you select.

3. "A rotating sphere with a radial variation of electric charge density produces an axial dipole magnetic field." Is this a correct statement? Estimate the order of magnitude of electric field that would exist in the earth, if separated charges were the source of the earth's magnetic field.

4. Find the relation between the magnetic moment of a magnetized shell, the intensity of magnetization, and the thickness of the shell, and confirm the statement (Section 17.4) that an intensity of 8.0 c.g.s. units would be required in the shell above the Curie temperature isotherm to produce the earth's field by permanent magnetization.

5. In connection with the dynamo theory, it was pointed out that the sustaining of a field depends upon the relative magnitudes of the generation term and the diffusion term of Equation 17.5.2. Show that this can be expressed by the "magnetic Reynold's number" $4\pi\sigma Lv$, where L is a characteristic length and v is velocity. Is the combination of quantities dimensionless?

BIBLIOGRAPHY

Allan, D. W. (1958) Reversals of the Earth's magnetic field. *Nature,* **181**, 469.

Barnett, S. J. (1933) Gyromagnetic effects: History, theory and experiments. *Physica,* **13**, 241.

Bartels, J. (1939) Some problems of terrestrial magnetism. In *Terrestrial magnetism and electricity,* ed. J. A. Fleming. McGraw-Hill, New York (reprinted 1949, Dover Publ., Inc.).

Blackett, P. M. S. (1947) The magnetic field of massive rotating bodies. *Nature,* **159**, 658.

Blackett, P. M. S. (1952) A negative experiment relating to magnetism and the earth's rotation. *Phil. Trans. Roy Soc. A,* **245**, 309.

Bullard, E. C. (1948) On the secular change of the earth's magnetic field. *Month. Not. R.A.S., Geophys. Suppl.,* **5**, 248.

Bullard, E. C. (1955) The stability of a homopolar dynamo. *Proc. Camb. Phil. Soc.,* **51**, 744–760.

Bullard, E. C., and Gellman, H. (1954) Homogeneous dynamos and terrestrial magnetism. *Phil. Trans. Roy. Soc. Lond. A,* **247**, 213.

Busse, F. H. (1975) A model of the geodynamo. *Geophys. J. R. A. S.,* **42**, 437–459.

Cowling, T. G. (1934) The magnetic field of sunspots. *Mon. Not. Roy. Ast. Soc.,* **94**, 39–48.

Elsasser, W. M. (1946) Induction effects in terrestrial magnetism. *Phys. Rev.,* **69**, 106 and **70**, 202.

Gardiner, R. B., and Stacey, F. D. (1971) Electrical resistivity of the core. *Phys. Earth Plan. Int.*, **4**, 406–410.

Gauss, C. F. (1839) *Allgemeine Theorie des Erdmagnetismus.* Leipzig. (Also in Gauss, Werke, **5**, Göttingen, 1877.)

Gauss, C. F. (1841) Intensitas vis magneticae terrestris ad mensuram absolutam revocata. *Kon. Akad. Wiss. Göttingen.* **8.** (Also in Gauss, Werke, **5.** Göttingen, 1877.)

Gibson, R. D., and Roberts, P. H. (1969) The Bullard-Gellman dynamo. In *Applications of modern physics to the earth and planetary interiors,* ed. S. K. Runcorn. Interscience, New York.

Gubbins, D. (1973) Numerical solutions of the kinematic dynamo problem. *Phil. Trans. Roy. Soc. Lond. A,* **274,** 493–521.

Gubbins, D. (1974) Theories of the geomagnetic and solar dynamos. *Rev. Geophys. Space Phys.*, **12**, 137–151.

Johnston, M. J. S., and Strens, R. G. J. (1973) Electrical conductivity of molten Fe-Ni-S-C core mix. *Phys. Earth Plan. Int.*, **7**, 217–218.

Lowes, F. J., and Wilkinson, I. (1963) Geomagnetic dynamo: a laboratory model. *Nature,* **198**, 1158.

Malkus, W. V. R. (1963) Precessional torques as the cause of geomagnetism. *J. Geophys. Res.*, **68**, 2871.

Matsushita, S., and Campbell, W. H. (1967) *Physics of geomagnetic phenomena.* Academic Press, New York.

Parker, E. N. (1955) Hydromagnetic dynamo models. *Astrophys. J.*, **122**, 293–314.

Rikitake, T. (1958) Oscillations of a system of disc dynamos. *Proc. Camb. Phil. Soc.*, **54**, 89.

Roberts, G. O. (1970) Spatially periodic dynamos. *Phil. Trans. Roy. Soc. Lond. A,* **266**, 535–558.

Rochester, M. G., Jacobs, J. A., Smylie, D. E., and Chong, K. F. (1975) Can precession power the geomagnetic dynamo? *Geophys. J.*, **43**, 661–678.

Runcorn, S. K. (1956) The magnetism of the earth's body. In *Handbuch der Physik,* Vol. 47, ed. J. Bartels. Springer, Berlin.

Runcorn, S. K. (1965) Changes in the convection pattern in the earth's mantle and continental drift. *Phil. Trans. Roy. Soc. Lond. A,* **258**, 228.

Runcorn, S. K., Benson, A. C., Moore, A. F., and Griffiths, D. H. (1951) Measurements of the variation with depth of the main geomagnetic field. *Phil. Trans. Roy. Soc. Lond. A,* **244**, 113.

Sabine, Edward (1837) Report on the variations of the magnetic intensity observed at different points of the earth's surface. *Brit. Assoc. Advance. Sci.*, **6**, 1–37.

Urey, H. (1952) *The origin of the earth and planets.* Oxford Univ. Press, Oxford.

Zmuda, A. J. (ed.) (1971) *The world magnetic survey.* Int. Assoc. Geomag. Aeronomy, Bull. 28.

18 TIME CHANGES IN THE MAGNETIC FIELD: THE SECULAR VARIATION

18.1 DETECTION OF TIME CHANGES

The realization that the elements of the magnetic field at points on the earth's surface change with time over a period of years dates back to early records of magnetic declination at Paris, London, and a few other cities. The curve showing the change in declination at London from 1540, for example, has often been reproduced. While the earlier records refer only to the direction of the field, measurements of the field intensity are now available for about 130 years, and these also show that very significant changes, with periods of the order of a century, are a characteristic of the geomagnetic field. These changes constitute the secular variation.

We will consider in this chapter the analysis of the observations which are available for historical time, noticing in passing that the secular change can also be studied by the methods of paleomagnetism, using the uppermost sedimentary layers (usually post-glacial) and prehistoric artifacts (archeomagnetism). While in principle the secular variations could be traced by repeating measurements of the field at selected points, say once each year, the situation is more complicated. As we shall show in the next chapter, the field is subject to a wide spectrum of shorter period variations, whose effect must first be removed from the repeat measurements. The variations of greatest amplitude are those of daily period, and of storms lasting over a few days. To isolate the secular variation it is necessary to average the field elements over many days, so that year-to-year trends may be detected. For this purpose, the records provided by magnetic observatories are ideal.

18.2 THE MAGNETIC OBSERVATORY

In a manner similar to the seismological observatory, the magnetic observatory produces daily continuous records, usually photographic, of a selection of the magnetic elements. As indicated in Figure 18.1, each of the elements *D, H,* and *Z* may be continuously recorded by means of mirrors attached to

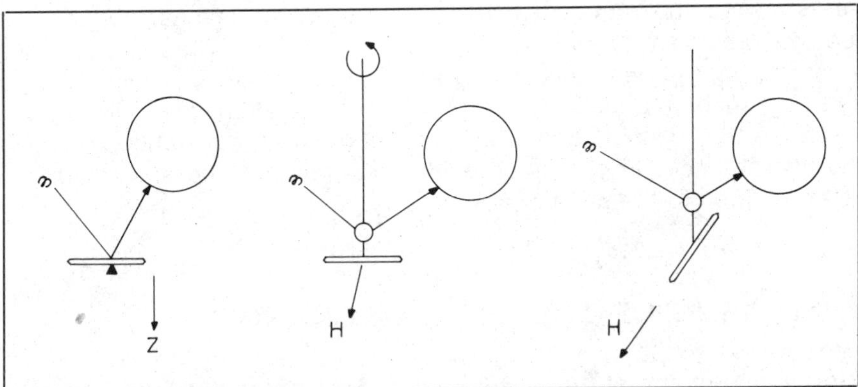

Figure 18.1 Schematic representation of variographs to record (from left) changes in *Z, H,* and *D.* Light is reflected to a recording drum from mirrors carried on a balanced magnetic needle, a needle perpendicular to the meridian and a needle in the meridian respectively.

the suspensions of suitably oriented magnetic needles. Drum speeds are normally adjusted to give one rotation each 24 hours. Such a system produces standard magnetograms of the type shown in Figure 18.2, in which variations with periods as short as a few minutes are detectable. Shorter period variations are filtered by the inertia of the suspension, and if it is desired to record these, systems such as described in Chapter 16, together with higher recording speeds, are required.

For a magnetogram to be useful, the observatory must fulfill four basic requirements: freedom from artificial disturbance, accuracy of time marks on the record, determination of the absolute values corresponding to the base line on the record of each element, and knowledge of the scale factor for each element. Base-line values are established through frequent absolute measurements at the observatory, while the scale factors are determined by applying known artificial fields to the instruments and measuring the record deflection.

Magnetic observatories record and publish a number of parameters, including mean values of the elements for hourly or other periods, and various

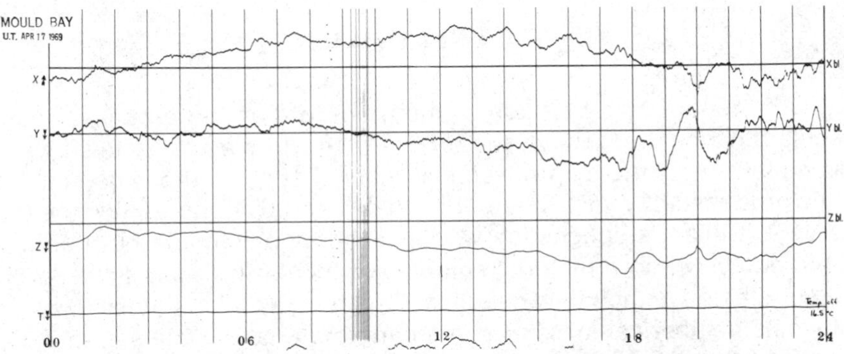

Figure 18.2 A magnetogram for a moderately disturbed day. Hours are shown at the bottom. Components *X* and *Y,* rather than *H* and *D,* are recorded. Between eight and nine hours, the recorders were calibrated for sensitivity by the application of known fields provided by Helmholtz coils. Between nine and ten hours, extra time lines show the instants of absolute measurements made to determine the baseline values. The prominent event at about 20 hours corresponds to decreases of 70, 200, and 40 gammas in *X, Y,* and *Z* respectively. (Courtesy of P. H. Serson and the Earth Physics Branch, Dept. of Energy, Mines and Resources, Canada)

indices showing the range of activity. As far as the secular change is concerned, it is apparent that the change in average value of a magnetic element from year to year is best determined at an observatory. However, it has been found that, except in anomalous areas, the daily variations are uniform over distances of several hundred kilometers, so that other repeat stations can be used to establish the secular change, provided they are corrected for diurnal variation by reference to the nearest observatory.

18.3 MAPPING OF THE SECULAR CHANGE

World-wide charts of the *secular change* are normally plotted for selected elements, as a rate of change in gammas or degrees per year. These charts can in fact be produced in a number of ways: by direct contouring of rates established at magnetic observatories or repeat stations; by determining the rate of change from charts of the element itself, drawn for two different epochs; or, most elegantly, by determining the rate of change of each spherical harmonic of the field, through analyses at two epochs and synthesis of the harmonics. Examples of the secular change for a particular element, for two epochs, are shown in Figure 18.3.

Certain characteristics are immediately evident. The earth's surface is broken down into regions, of approximately continental size but not directly correlated with the continents, within which the field increases or decreases. Also, the fact that the highest rates of change shown, 150 gammas per year, persist in the same sense over periods of decades, shows that total field changes as great as 10,000 gammas must be expected. As this can amount to 20 per cent of the actual field at certain points, and therefore be far greater than any non-internal portion of the field, the secular change must have an internal origin. We are therefore faced with an internal parameter of the earth which can change greatly in a few decades. A very close examination of Figure 18.3 will reveal some more subtle characteristics, which we shall investigate in Section 18.5.

18.4 THE CAUSE OF THE SECULAR CHANGE

Bullard (1948) drew attention to the fluid core as the probable source of the secular change, because of the relatively short time scale. He showed that the main features of the map of secular change, such as those on Figure 18.3, could be produced by a distribution of variable vertical dipoles over the surface of the core. Bullard's suggestion was that the dipoles could be produced by turbulent eddies of core material spinning in the earth's main field. As eddies start up or die down, their field changes. To estimate the quantities involved, we may imagine a spherical mass of conducting material of radius a, rotating in a uniform field F, and surrounded by an insulator. The calculation of the magnetic moment produced, even in this case, is more involved than may appear (Bullard, 1949), but we may use a physical argument to anticipate the limiting case of high conductivity or high angular velocity. In these cases, by the laws of magnetohydrodynamics (Chapter 19), the resultant field must be zero inside the sphere (since field lines are "frozen out" of the spinning material), which

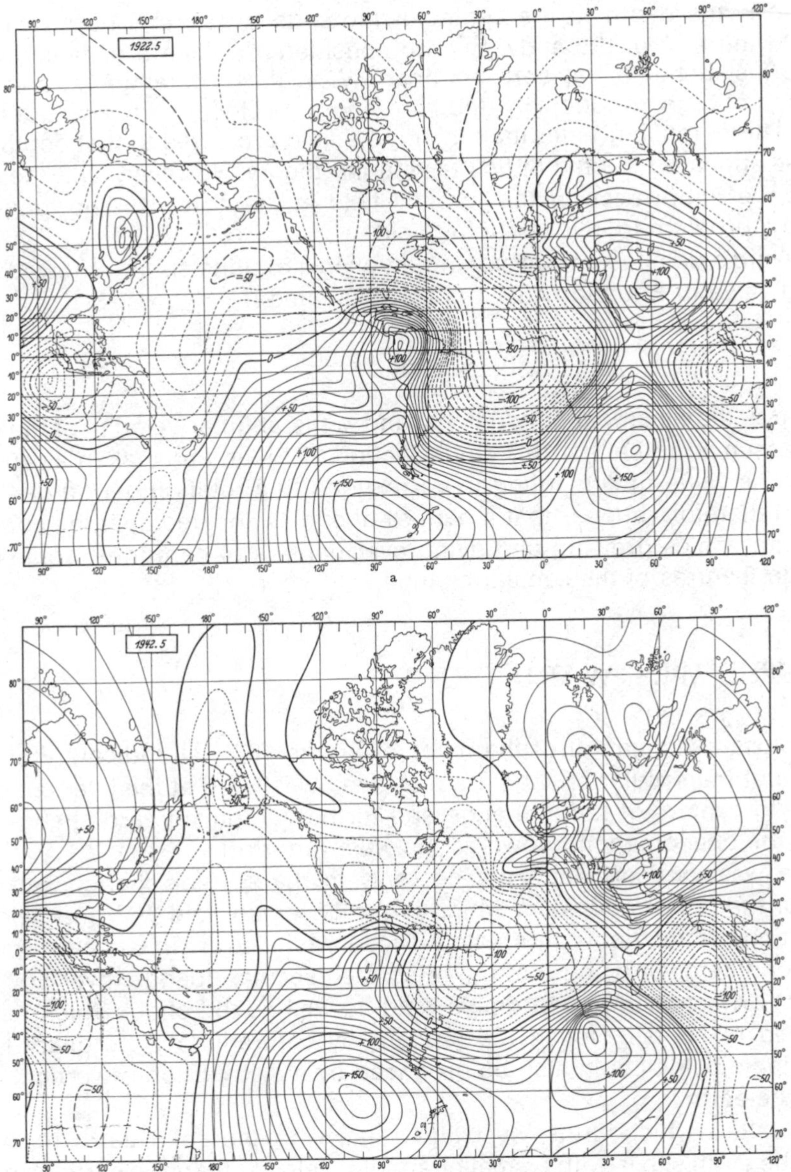

Figure 18.3 The secular change in the earth's magnetic field, for 1922.5 and 1942.5. Contours give the rate of change of Z in gammas per year. (After Vestine, illustration courtesy of S. K. Runcorn and Springer-Verlag)

requires the sphere to produce a magnetic moment M, where

$$M = \tfrac{1}{2}Fa^3 \qquad 18.4.1$$

with M directed opposite to F. For a numerical example, we take the area of decrease in horizontal intensity located west of Africa, where the field had been reduced by 0.08 gauss in 100 years. The value of F required to produce a sufficient moment is then

$$F = 2 \times 0.08 \times \frac{d^3}{a^3} \qquad 18.4.2$$

where d is the depth to the center of the sphere. There is no direct way to estimate the size of the eddy. Bullard considered a sphere of radius 400 km, located just beneath the core boundary. With these parameters, and the assumption of high conductivity and high angular velocity, F turned out to be 90 gauss. This is much greater than the dipole field projected to that depth, and at first this appeared to be a difficulty with the theory. However, with the realization of the importance of the toroidal field in the core, it is no longer considered a problem to invoke such a field.

Bullard's detailed analysis of induction in a rotating sphere showed that the limiting value of Equation 18.4.1 is approximated if

$$2\pi\sigma\omega a^2 > 50 \qquad 18.4.3$$

where σ is the electrical conductivity and ω the angular velocity of the sphere. If σ is assumed to be 1×10^{-6} emu (Section 17.5), and a is taken as 400 km as before, the value of ω is found to correspond to a rotation period of 40 years. It would therefore appear that a pattern of eddies, located near the core boundary, with eddies growing and dying over periods of centuries, could explain the main features of the secular change.

18.5 THE WESTWARD DRIFT

A close examination of the maps of Figure 18.3 will indicate that certain features of the secular change are displaced westward at the later epoch. This applies to centers of rapid change and to the contour of zero change, and is part of the phenomenon known as the westward drift.

The westward drift appears not only in the secular change, but also in maps of the *non-dipole field,* which is the observed field with the field of the best-fitting central dipole subtracted. These two fields are in fact closely related, the chief difference being that the integrated secular change, over any period, includes any change in the dipole field as well as the effect which has been attributed to eddies. Notice the (apparently contradictory) corollary that the "non-dipole field" could also be produced by a distribution of vertical dipoles within the earth.

A feature which drifts westward relative to the solid earth must rotate about the earth's axis with a smaller angular velocity than the earth as a whole. If the non-dipole field arises from motions in the outermost layer of the core, the implication is that it is this layer which has the smaller angular velocity. This is consistent with the discussion in Section 17.5 in which the toroidal motion T_1 was introduced. It is therefore of very considerable interest to determine the rate of westward drift with more precision than can be done by a simple inspection of maps.

Bullard et al. (1950) considered profiles of both the secular change and non-dipole field around parallels of latitude. A computer program was designed to minimize the residuals between corresponding profiles for different epochs, for various rates of drift. It was found that the best rate varied with latitude, but the average over all latitudes, for the non-dipole field, was 0.18° per year. It has since been found (Whitham, 1958) that there are very considerable areas of the earth's surface for which this figure is not applicable, the drift apparently being much smaller. However, it does appear that, in the relatively short time of

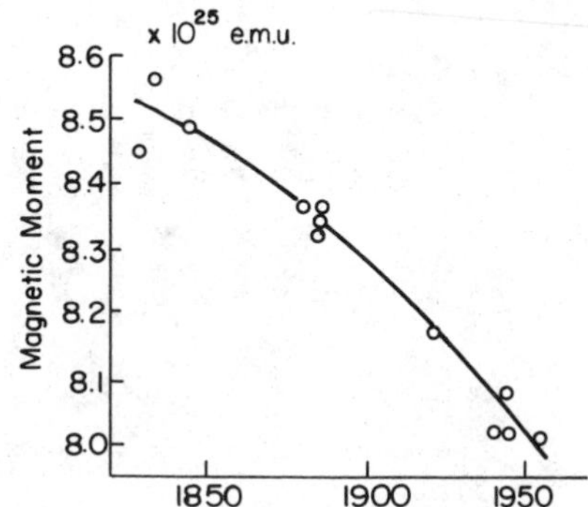

Figure 18.4 The change in strength of the central dipole since 1840. (Based on spherical harmonic coefficients tabulated by Vestine, 1967)

2000 to 5000 years, the outer layer of the core could drift through 360° relative to the rest of the earth. Averaged over times of this order, the field at any point on earth should be very nearly that of a central dipole, a fact which is of importance in paleomagnetism. The westward drift also provides information on the mechanism of coupling between the mantle and core (Chapter 26).

18.6 THE CHANGE OF THE DIPOLE FIELD

The spherical harmonic analysis of the main field, as discussed in Section 17.2, yields, among others, the harmonics corresponding to the central dipole. When analyses for different epochs are available, the change in moment of this dipole may be traced. As we have noted, the first analysis was that of Gauss, in 1839, but a reasonable distribution of stations was not available until 1885. Even in this century, it is difficult to make comparisons, as the later analyses are based on a mapping of the field that is much more complete than the earlier ones. With this reservation in mind, we may examine the trend in strength of the dipole, as shown in Figure 18.4. The most prominent trend is a decrease over the past 100 years, possibly toward the next change in polarity or reversal.

Problem

1. What is the skin-depth (Section 20.1) in the earth's core for time-changes of period 100 years? If the secular change is the result of eddies in the outer core, what can you conclude about the location of these eddies?

BIBLIOGRAPHY

Bullard, E. C. (1948) On the secular change of the earth's magnetic field. *Mont. Not. R.A.S., Geophys. Suppl.*, **5**, 248.

Bullard, E. C. (1949) Electromagnetic induction in a rotating sphere. *Proc. Roy. Soc. A*, **199**, 413.

Bullard, E. C., Freedman, C., Gellman, H., and Nixon, J. (1950) The westward drift of the earth's magnetic field. *Phil. Trans. Roy. Soc. A*, **243**, 67.

Vestine, E. H. (1967) Main geomagnetic field. In *Physics of geomagnetic phenomena*, ed. S. Matsushita and Wallace H. Campbell. Academic Press, New York.

Whitham, K. (1958) The relationships between the secular change and the non-dipole fields. *Can. J. Phys.*, **36**, 1372.

19 THE EXTERNAL MAGNETIC FIELD

19.1 SOLAR-TERRESTRIAL RELATIONSHIPS

No branch of geophysics has shown a more dramatic increase in the past few years, and in particular since the International Geophysical Year, than the study of the external magnetic field, and of the influence of the sun upon it. This is indicated diagrammatically in Figure 19.1, which compares the classical view of the earth's dipole field in space with modern ideas of our earth, surrounded by a "magnetosphere" and bathed in a "solar wind." The only question is the extent to which it is possible, or proper, to consider this branch of geophysics in a work devoted to the solid earth.

There are at least two important reasons for the solid-earth geophysicist to consider these external, largely solar-controlled, phenomena. All of the broad spectrum of time variations in the geomagnetic field, of period shorter than the most rapid secular change, have their primary cause outside of the earth. In

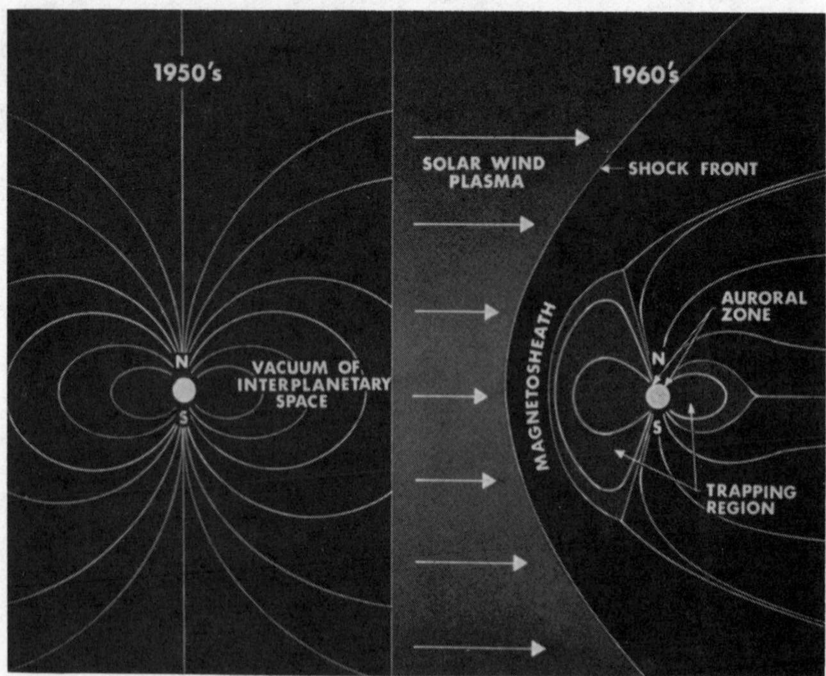

Figure 19.1 Old (*Left*) and new ideas on the external magnetic field. (Courtesy of B. J. O'Brien and Academic Press)

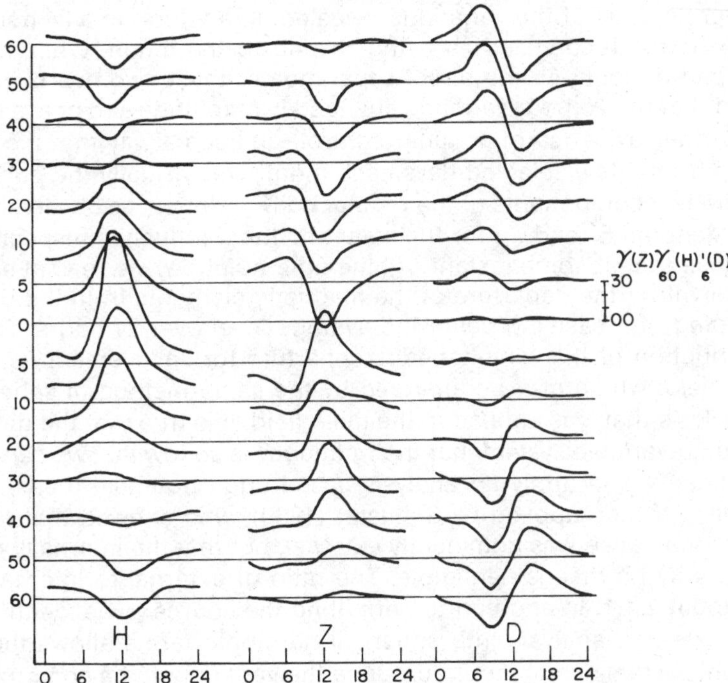

Figure 19.2 Diurnal variation in the magnetic elements as a function of latitude. Horizontal scale is local time. (After Matsushita)

one sense, these time variations are a nuisance in the study of the internal field, for measurements of the magnetic elements must be corrected for them. For this purpose, it is necessary to have a knowledge of the characteristic periods and amplitudes of the different types of variations. But there is an even more important reason, which is that every variation in the field induces electric currents inside the earth. The changes measured at the earth's surface are the resultant of the fields of the external sources and of the internal currents. From analyses of these changes, the distribution of electrical conductivity within the earth can be estimated. This problem will form the basis of Chapter 20; in the present chapter we discuss the causes of the principal magnetic variations of external origin.

19.2 THE DIURNAL VARIATION

The magnetogram of Figure 18.2 suggests that there may be a regular cyclic change in the magnetic elements, of period about 24 hours, upon which more irregular disturbances are superimposed. Soon after magnetic observatory records became available in the ninteenth century, it was found that if one averaged the records for a number of days, free from irregular disturbance, a smooth daily variation curve was indeed obtained, and this curve was found to be characteristic of the particular observatory. The amplitude of the variation was found to depend upon the magnetic latitude of the observatory; for observatories at the same latitude, the variations were similar, provided they were plotted against local time (Fig. 19.2). The latter observation therefore suggested the sun as a controlling factor.

Fourier analysis of the variations revealed, in addition to a 24 hour period, one of 25 hours, recognized as being related to the moon. While the earlier analyses had difficulty in the direct separation of these two nearby periods, it was found that the 25 hour variation changed in amplitude and phase through a month—further evidence of its lunar control—in such a way that the averages of records from a few selected days each month left virtually the pure 24 hour period. The two components of the regular daily variation, as observed on quiet days, are denoted S_q and L_q, for the solar and lunar influence respectively.

Suppose that at some instant of time on a quiet day we read at each magnetic observatory the departure of the magnetic elements from the baseline of the recording, the baseline being the average level over 24 hours. The worldwide distribution of this simultaneous departure for any element represents a magnetic field, which may be analyzed by the same method of spherical harmonic analysis that was applied to the main field. It is true that the departure is part of a time-variable system, but the frequency is so low that we can treat it as quasi-stationary and apply an analysis based upon potential theory. This approach was, in fact, applied by Schuster (1889), who showed that for all low harmonics the cause was dominantly external, but that the internally produced portion was by no means negligible. The ratio of external to internal effect is actually about 2.4:1. In addition to permitting the source to be identified as internal or external, analysis into spherical harmonic terms allows the equivalent current system, which could produce the variation, to be constructed.

The potential of the magnetic field at the earth's surface is closely related to the current density of the equivalent current system, which we imagine to be a flowing spherical shell which surrounds the earth. It is convenient to introduce a current function J, such that the (closed) lines of J on the shell represent the directions of flow, and such that the flow between lines J_0 and $J_0 + dJ$ is dJ units of current. The current density is therefore dJ/ds where ds is the spacing between lines. The magnetic potential of a current flowing around a closed element of the shell is identical to that of a distribution of magnetic dipoles over the shell; in the case of the strip between current lines J_0 and $J_0 + dJ$, the equivalent magnetic moment per unit area would be dJ over the cap of the shell bounded by the J_0 contour. If all current elements are replaced by equivalent distributions, the density of dipoles at any point will be equal to J at the point, apart from a constant. The function J can then be expanded as a series of spherical harmonics $J_1(\theta, \phi)$, and the problem of determining the potential on the surface of the earth is identical to that considered in Appendix A, of finding the gravitational potential due to a harmonic distribution of mass over a sphere.

If r is the radius of the sphere on which the potential is defined, and a is the radius of the shell in which currents flow, we have

$$\left.\begin{aligned} U &= -\frac{4\pi}{10}\sum_{l=0}^{\infty}\left(\frac{l+1}{2l+1}\right)J_l(\theta,\phi)\left(\frac{r}{a}\right)^l, && \text{for} \quad r < a \\ U &= \frac{4\pi}{10}\sum_{l=0}^{\infty}\left(\frac{l}{2l+1}\right)J_l(\theta,\phi)\left(\frac{a}{r}\right)^{l+1}, && \text{for} \quad r > a \end{aligned}\right\} \qquad 19.2.1$$

The factor 10 arises from the fact that J is normally plotted in amperes rather than electromagnetic units. For the converse problem of determining the current function, we simply write

$$\left.\begin{aligned} J_l(\theta, \phi) &= -\frac{10}{4\pi}\frac{2l+1}{l+1}\left(\frac{a}{r}\right)^l U_l, \quad &\text{for} \quad r < a \\ J_l(\theta, \phi) &= \frac{10}{4\pi}\frac{2l+1}{l}\left(\frac{r}{a}\right)^{l+1} U_l, \quad &\text{for} \quad r > a \end{aligned}\right\} \qquad 19.2.2$$

Here, U_l is the lth spherical harmonic in the expansion of the magnetic potential. Reference to the equations of Section 17.2 will indicate that if the coefficients in the expansion of the field are determined, those for the potential are easily found. To determine the equivalent current system for the S_q or L_q variations, the terms in the expansion for the potential corresponding to external sources are used, together with the first equation of 19.2.2. Each term in J is thus determined, and the complete current system can be synthesized by contouring the scalar quantity J. World-wide current systems for both the solar and the lunar variations, based on IGY data, have been given by Matsushita (1967), and an example is shown in Figure 19.3.

A proposal for the source of the S_q and L_q variations was given by Balfour Stewart (1882) in one of the most remarkable forecasts of geophysics. Before the nature of the upper atmosphere was known, even before Schuster's analysis was available, he proposed a dynamo action in conducting upper levels of the atmosphere, resulting from solar and lunar tidal motions. The dynamo theory has gone through many modifications, but it remains as the best explanation of the daily variations. An early difficulty was the absence of a pronounced 12 hour period in the magnetic effects, which would be expected on the basis of tidal motions. The first major advance was made by Chapman (1919), who reasoned that if the conductivity of the upper atmosphere arises from photoionization of the air, it must be dependent on the sun's hour angle, so that the dynamo effect is greatest on the sunlit side of the earth. It is now known that the conductivity in the ionosphere must be taken as a tensor, since current flow in a plasma depends on the direction relative to the magnetic field. Modern calculations on the required dynamo parameters (Matsushita, 1967) show no inconsistencies with what is known from rocket measurements of ionospheric motions or conductivities.

19.3 MAGNETIC DISTURBANCES

The pure daily variation is rarely seen on a single day, as the daily record is distorted by irregular disturbances. These disturbances are characterized by a set of internationally adopted *index figures, K,* which measure the range of magnetic elements in a three hour period. The average value of the indices at a number of observatories, with allowance for the undisturbed daily range at each, is used to provide a universal index K_p. The latter index, on a quasi-logarithmic scale, ranges from 0 for extremely quiet conditions to 9 for a great magnetic disturbance.

Large magnetic disturbances, known as storms, have certain features which have come to be accepted as characteristic, in spite of great irregularities in the detailed behavior of the magnetic elements during the disturbance. The

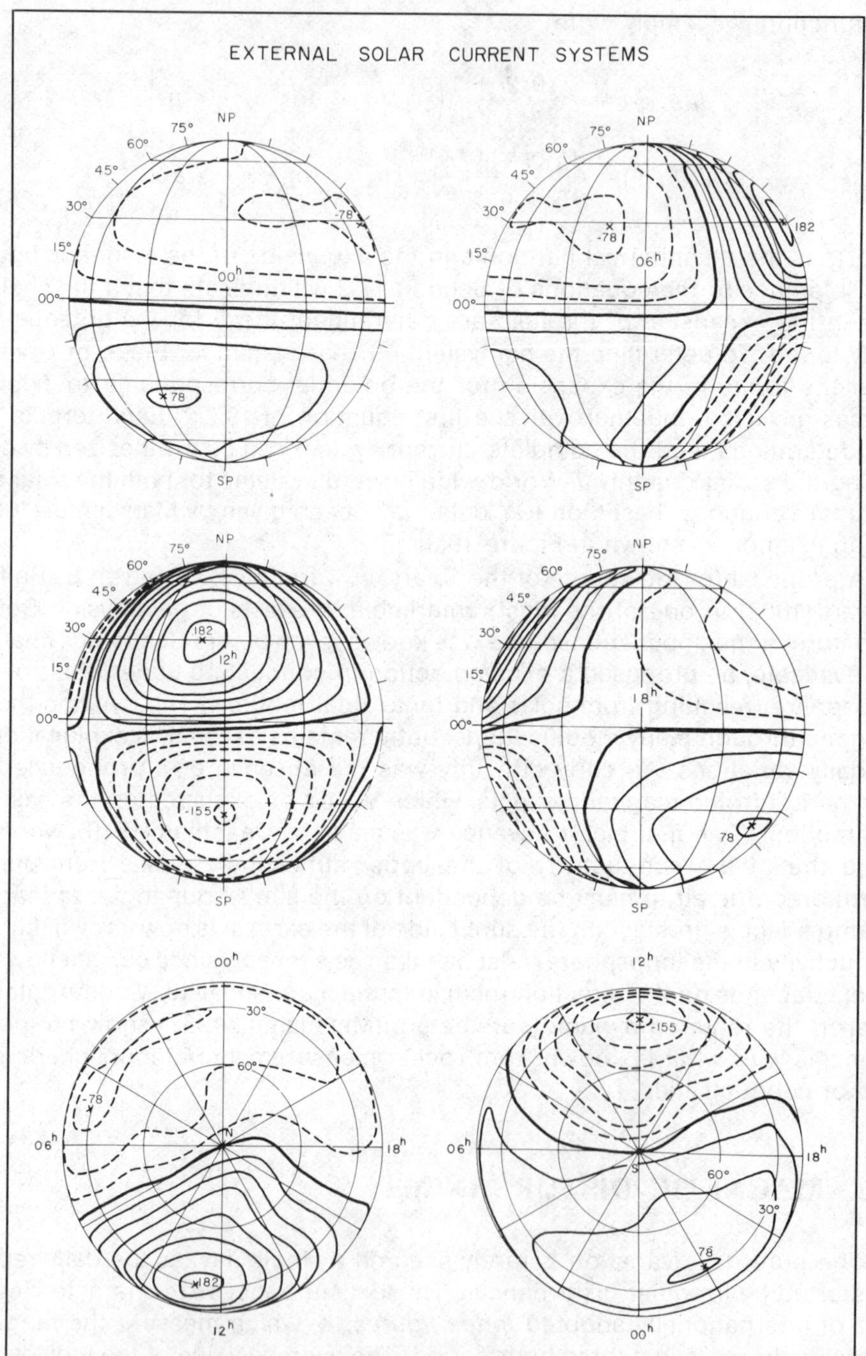

Figure 19.3 Current system for the solar quiet daily variation, averaged over 12 months during the International Geophysical Year. Views of the earth at 0, 6, 12 and 18 hours local time, and from above the north and south geomagnetic poles. The current between adjacent lines is 25×10^3 amperes; solid lines indicate counter-clockwise flow. (Courtesy of S. Matsushita, High Altitude Observatory, Boulder, Colorado)

typical variation of H during a storm is shown in Figure 19.4, which illustrates the sudden commencement, the initial phase in which H is increased, the large decrease in the field, and the gradual recovery. There is ample evidence that storms are solar-controlled: an almost perfect correlation with the cycle of sunspot activity, correlation of individual great storms with observed sunspots (with a time delay), and repetition of storms after a period of 27 days, corresponding to the sun's rotation time on its axis. Birkeland (1908) suggested that the magnetic disturbance on the earth is due to streams of charged particles from the sun. The decrease in H which is characteristic of the main phase of a storm suggests a current circling above the earth in the equatorial plane. It is well known that a charged particle in a magnetic field follows a curved path, and Birkeland and Störmer (1931) investigated in great detail the motions of charged particles to determine if a solar stream could be trapped into such a "ring current." The difficulty with this analysis was that interaction between particles was neglected. Modern theories require conducting material from the sun to arrive, via the solar wind, as a neutral plasma, with charge separation taking place after trapping of particles in the earth's field. To even trace the outline of these it is necessary to consider some principles of magnetohydrodynamics.

19.4 APPLICATION OF MAGNETOHYDRODYNAMICS TO MAGNETIC DISTURBANCES

The mathematical treatment of magnetic fields and particle motion within a plasma of very high electrical conductivity is outlined in Appendix D, but the important results can be described qualitatively here. Probably the single most important property of a plasma is that any magnetic field is permanently locked to the material. This is often expressed by the statement that lines of magnetic field are "frozen" to the plasma. A moving plasma carries any magnetic field

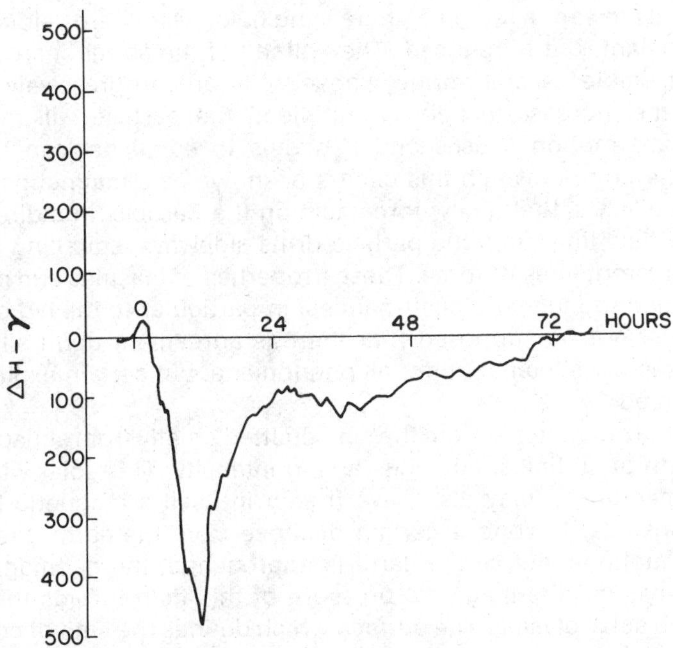

Figure 19.4 A typical magnetic storm. The sudden commencement at 0 hours is followed by a brief period of increased H, after which the field is greatly decreased during the main phase.

lines it may possess with it; conversely, if a plasma has no magnetic field, any external field is permanently frozen out of it. This characteristic arises essentially from Lenz's law, in that any relative motion between field and plasma would induce an electric field whose resulting current flow would oppose the motion. Because of the extremely high conductivity, even an infinitesimally small motion would be immediately opposed by a finite current flow. As a result of that fact that magnetic fields must be "frozen in" or "frozen out" of plasmas, a magnetic field exerts a very real pressure on a mass of ionized gas, and this pressure varies as the square of the field strength.

Small disturbances which may be generated within a plasma are propagated through a mechanism investigated by Alfven (1950). Alfven showed that a new type of wave exists, with the characteristic that propagation is along the lines of magnetic field, with a velocity proportional to the field and inversely proportional to the density of the plasma. Because of the locking of field lines to the fluid, Alfven waves represent both an oscillatory displacement of fluid particles, in directions transverse to the field, and a traveling distortion of the lines of force.

The above characteristics are bulk properties of plasmas. In addition, the analysis of the motion of a single charged particle within a non-uniform magnetic field exhibits some most important results. It is well known from elementary considerations that a charged particle, say an electron, moving with velocity v injected perpendicularly into a *uniform* magnetic field executes circular motion of determined radius r. This is the basis of the mass spectrometer and many other instruments, and arises from the fact that the electrodynamical force (proportional to $\underline{v} \times \underline{H}$) just supplies the centripetal force (proportional to v^2/r). If the particle is injected obliquely to the lines of a uniform field, its path follows a spiral of constant radius. This is because the component of velocity parallel to the field lines remains constant and only the component perpendicular to the field determines the radius of the spiral. However, if the injection is obliquely into a *non-uniform* field, two new effects arise. The first is that if the particle spirals toward a region of increasing field, its forward velocity no longer remains constant, but is reduced. The "pitch" of the spiral decreases, so that the path resembles a coil spring whose coils are progressively more compressed. If the increase in field is sufficient, the particle will reach a point where forward motion ceases and it begins to spiral back in the opposite direction. The point at which this occurs is known as a magnetic mirror point. The second effect is that a new force acts on the particle, in a direction transverse to the field lines, and the particle drifts sideways, spiralling around progressively different lines of force. These properties of plasmas and particles can now be applied to magnetic disturbances, in particular to the typical *magnetic storm.* It must not be supposed that there is agreement upon all aspects, or even a satisfactory explanation for all phenomena, but certain explanations are widely accepted.

Highly ionized material from the sun, emitted chiefly from sunspots, sweeps past the earth at all times, but with varying intensity. This solar wind carries a certain momentum; it may also have frozen in itself a magnetic field carried from the sunspots. Beyond a certain distance from the earth, the solar wind controls all motions, but as the earth is approached, the geomagnetic dipole field increases, and the magnetic pressure of this field forbids the closer approach of the solar plasma. The surface which defines the limit of control by the earth's field is the boundary of the *magnetosphere.* At the onset of a particularly strong burst of solar wind, the lines of force of the earth's field are compressed, as they cannot enter the plasma, and this is reflected at the earth's surface in the sudden commencement and initial phase of a magnetic storm. Clouds of individual particles from the solar wind, of both charge signs, become trapped in the earth's field, and are forced to spiral around lines of this field. Perhaps the greatest difficulty in the model has been the understanding of precisely how particles become trapped. Recent work (Roederer, 1970; McCormac, 1970, 1974) suggests that initial capture takes place on the night side of the earth, in the magnetic tail, with particles subsequently being "pumped" to the daylight side. As we have shown above, these trapped particles, which now constitute the radiation belts, are forced to oscillate between mirror points. At the same time, there is a longitudinal drift, of opposite sense for positive and negative particles, so that the net effect is a single ring current above the equator. We therefore return to the ideas of Birkeland and Störmer for the explanation of the main phase of a magnetic storm, in terms of a *ring current.* However, as we shall show, the drift velocity is a small component of the particles' total motion.

Normally the mirror points are high above the earth, and the trapped particles do not interact appreciably with the earth's atmosphere. If the dipole field is distorted, as it is by the intensified solar wind itself, the mirrors are disrupted. Charged particles may then proceed into the atmosphere, where they are eventually slowed by collisions. In the process, aurorae and other phenomena are produced in the upper atmosphere. One of these phenomena is the great enhancement of electrical conductivity along the aurora zone, along which a current, known as the auroral electrojet, becomes concentrated. The electrojet is now believed to explain certain complications of the magnetic storm. As charged particles escape from the belts, the population in the belts returns to a "normal" level, the ring current decreases, and the magnetic field at the earth's surface returns to its normal value during the recovery phase of the storm.

There is a high degree of *conjugacy* or similarity of conditions at points located in an equivalent way relative to the north or south geomagnetic pole. This is shown by Figure 19.5, in which are plotted the loci of most frequent occurrence of aurorae, as determined by Feldstein (1960) for both polar regions. The auroral electrojets are located approximately, but not precisely, over the curves of most frequent occurrence.

As we have noted, the above outline does not take account of many features on which research is actively in progress, but is intended to provide the solid-earth geophysicist with an appreciation of the main causes of the disturbances he observes at the earth's surface. We conclude this section with examples of the magnitudes of characteristic times for trapped particles of different energies, as quoted by Akasofu and Chapman (1961). Particles of the energy shown are assumed to be injected into the earth's field at a distance of six earth radii, with a pitch angle of $\sin^{-1} 0.1$ in the equatorial plane. The second and third columns refer to motion in the spiral, the fourth to the time between mirror points, and the fifth to the drift through 360° in longitude.

Particle	Radius of gyration, km	Period of gyration, sec.	Period of oscillation, sec	Period of drift
Electron, 1.07×10^3 Kev	3.38	7.48×10^{-7}	0.67	14 min
Proton, 2 Kev	3.38	0.433	400	6.6 days
Proton, 479 Kev	67.6	0.433	20	21 min

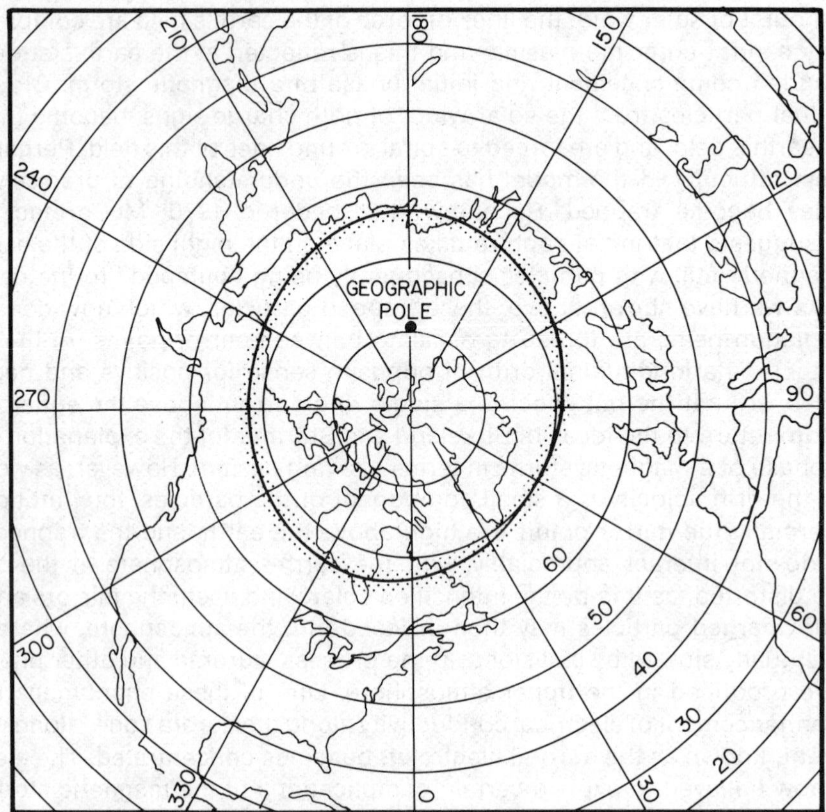

Figure 19.5 Curves of maximum auroral activity (after Feldstein) around north and south geomagnetic poles. The solid lines correspond to an occurrence frequency of approximately 240 nights per year. At the edge of the stippled bands, the activity has fallen to one-half. Note that the maps are plotted in geomagnetic coordinates.

Illustration continued on opposite page.

19.5 THE SPECTRUM OF THE TIME VARIATIONS IN THE EARTH'S FIELD

Although we have not yet considered time variations with period longer than the secular variations, it is convenient at this stage to consider the complete spectrum of time-variations, as shown in Figure 19.6. We have discussed the secular change, the solar and lunar daily variations, and magnetic disturbances. Energy appears at a period of 11 years because of the higher incidence of disturbances at epochs of maximum sunspot activity, at a period of six months because of the greater ability of the earth's field to trap particles when one pole is tipped toward the sun, and at a period of 27 days because of the tendency of magnetic storms to recur after one solar rotation.

A band of periods between one and several hundred seconds represents micropulsations (Troitskaya, 1967), which are believed to be the magnetic effects of hydromagnetic waves trapped in the magnetosphere. Still higher frequencies result from the natural oscillations of the earth-ionosphere cavity as a whole, and from the electromagnetic radiation of lightning discharge.

At the long-period end of the spectrum are shown reversals of the main field, which we shall consider in Chapter 21.

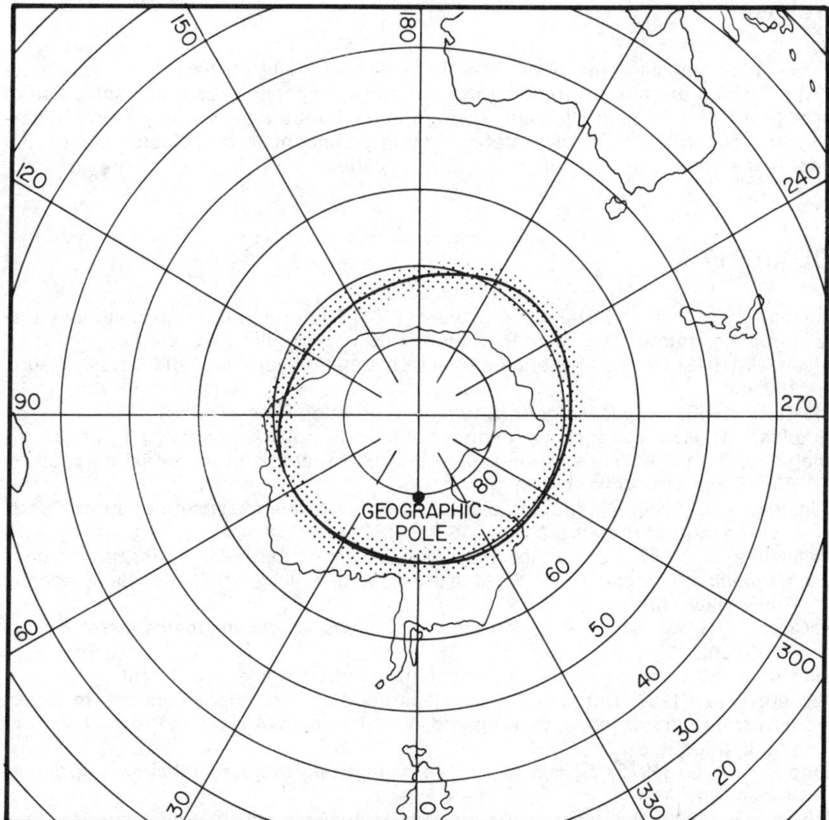

Figure 19.5 *Continued.*

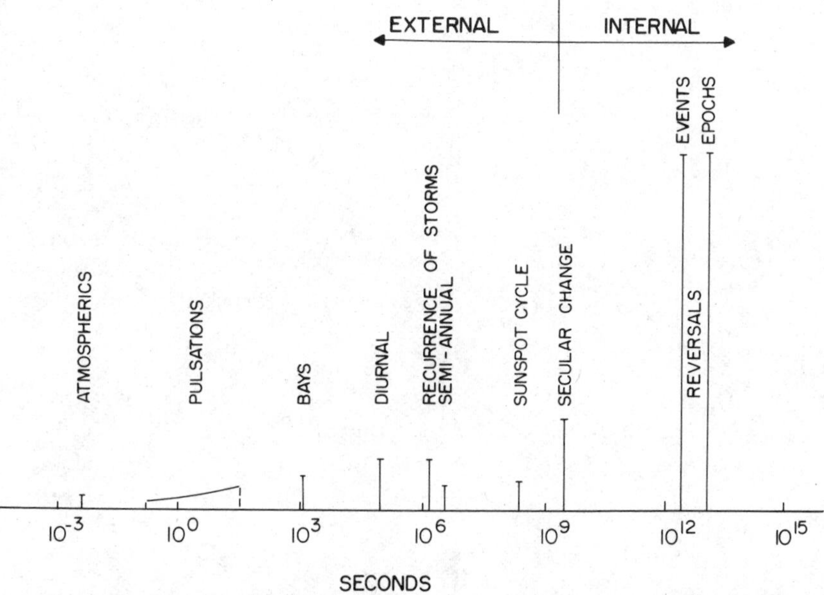

Figure 19.6 The spectrum of changes in the earth's magnetic field; only the most important lines are shown, and the amplitudes of the shorter-period changes are exaggerated relative to the secular change and reversals.

Problem

1. Many transient magnetic effects exhibit a similar behavior at "conjugate points," which are points on the earth's surface at opposite ends of the same line of force of the earth's magnetic field. Taking the field to be that of a dipole only, determine the conjugate point of a station whose geomagnetic coordinates are (ϕ, λ). What is the conjugate point of your present location?

BIBLIOGRAPHY

Akasofu, Syun Ichi, and Chapman, Sydney (1961) The ring current, geomagnetic disturbance, and the Van Allen Radiation belts. *J. Geophys. Res.*, **66**, 1321.

Alfven, H. (1942) On the existence of electromagnetic hydromagnetic waves. *Nature*, **150**, 405.

Alfven, H. (1950) *Cosmical electrodynamics*. Clarendon Press, Oxford.

Birkeland, Kr. (1911) Orages magnétiques et aurores polaires. *Arch. Sci. Phys.*, **32**, 97.

Chapman, S. (1919) The solar and lunar diurnal variations of terrestrial magnetism. *Phil. Trans. Roy. Soc. A*, **218**, 1.

Feldstein, Y. I. (1960) Geographic distribution of aurora and azimuth of auroral arcs. *Investigation of the Aurora, Acad. Sci. USSR*, **4**, 61–78.

Matsushita, S. (1967) Solar quiet and lunar daily variation fields. In *Physics of geomagnetic phenomena*, ed. S. Matsushita and Wallace H. Campbell. Academic Press, New York.

McCormac, B. M. (ed.) (1970) *Particles and fields in the magnetosphere*. Reidel, Dordrecht.

McCormac, B. M. (ed.) (1974) *Magnetospheric physics*. Reidel, Dordrecht.

O'Brien, B. J. (1967) Energetic charged particles in the magnetosphere. In *Solar-terrestrial physics*, ed. J. W. King and W. S. Newman. Academic Press, New York and London.

Roederer, J. G. (1970) *Dynamics of geomagnetically trapped radiation*. Springer-Verlag, New York.

Schuster, A. (1889) The diurnal variation of terrestrial magnetism. *Phil. Trans. Roy Soc. A*, **180**, 467.

Stewart, Balfour (1882) Terrestrial magnetism. *Encyc. Brit.* 9th Ed.

Störmer, C. (1930) Twenty-five years' work on the polar aurora. *Terr. Mag.*, **35**, 193.

Troitskaya, V. A. (1967) Micropulsations and the state of the magnetosphere. In *Solar-terrestrial physics*, ed. J. W. King and W. S. Newman. Academic Press, London and New York.

ELECTROMAGNETIC INDUCTION WITHIN THE EARTH 20

20.1 THE INTERNAL PORTION OF THE TIME VARIATIONS

We have shown (Section 19.2) that a portion of the daily variation, and of time variations of other periods, is known through spherical harmonic analysis to be of internal origin. This portion represents the magnetic fields, observed at the earth's surface, of electric currents induced in the earth by the primary external sources. In this chapter we shall trace the methods that have been employed to estimate the electrical conductivity within the earth from an analysis of these internally produced fields.

Electrical conductivity is one of those basic parameters of the earth which can be studied on a global basis. Its importance lies in the fact that below the crust the value is largely controlled by temperature. The study of the distribution of conductivity therefore assists in establishing the range of temperatures in the deeper parts of the earth.

There is one relationship which is of fundamental importance to all determinations of electrical properties. Any electromagnetic field produced outside of a conductor is attenuated with distance inside the conductor, at a rate which depends upon frequency and conductivity. For a half-space of conductivity σ, a field quantity proportional to $e^{i\omega t}$ (electric or magnetic) at the surface is attenuated in amplitude by $e^{-\alpha z}$ at a depth z, where

$$\alpha = \sqrt{2\pi\sigma\omega}$$

The depth

$$z = \frac{1}{2\pi}\sqrt{\frac{T}{\sigma}} \quad \left(\text{where } T = \frac{2\pi}{\omega}\right)$$

at which the amplitude is reduced to $1/e$ of its surface value, is known as the *skin depth.* In the above relationship, z will be in cm when T is in seconds, and σ in emu. If information is to be obtained on a certain region of the earth from measurements at the surface, it is apparent that the skin depth for the primary field must be sufficiently great for the field disturbance to penetrate to the depth in question. In other words, a spectrum of variations of increasing period will provide information on the conductivity at increasing depths. To some extent the time variations of the earth's field provide this spectrum, but it is unfortunate for conductivity studies that there is not more energy in the spectrum at periods of a few years.

20.2 THE RADIAL VARIATION OF CONDUCTIVITY

When it became known that a significant portion of the daily variation was internal in origin, both Schuster (1908) and Chapman (1919) recognized that within the earth there must be a region of higher conductivity than is typical of crustal rocks. Their solutions were based on uniform conductivity within a sphere of radius smaller than the earth's. Lahiri and Price (1939) extended the treatment to the case of a radial variation of conductivity, and provided the foundation for a global conductivity distribution curve. Their approach was to calculate the ratio of internal field to external field, for a given spherical harmonic and given period, for different assumed distributions of electrical conductivity. As we noted in Section 17.2, analysis of any chosen period of time variation in terms of spherical harmonics gives, for each harmonic, the ratio of the fields due to external and to internal causes. In other words, the response (internally generated field) for a given source geometry and period (externally generated field) is known. Lahiri and Price (Appendix D) showed how the ratio could be computed for a model earth, in which conductivity varied as a simple function of radius; the agreement of the computed ratio with that obtained from the analysis of disturbances then provided a test of the model. The basic model chosen by them consisted of an earth with conductivity decreasing as an inverse power of radius. In the outer 1000 km of the earth, the variation of conductivity with radius is so drastic that a negative power of 30 is required to represent it. But this variation by itself led to extremely low values of conductivity at the earth's surface, values appropriate, perhaps, to the crystalline crustal rocks, but not to an earth largely covered by sea water of high conductivity. (The conductivity of sea water is between 3 and 4 reciprocal ohm m, depending on the salinity.) Lahiri and Price therefore modified the model to consist of a very thin shell of moderately high conductivity, beneath which the conductivity decreased to a very low value, and then increased rapidly (Fig. 20.1). But the rapid increase itself shows that, because of the skin effect, the depth of penetration of a 24-hour period is only about 900 km. The investigators were able to extend this to a depth slightly over 1000 km by using the main phase of magnetic storms, which contain a period of about 48 hours.

The curve of Lahiri and Price has remained as a standard of comparison for the global distribution for many years. Recent workers have proposed modifications, based upon the analysis of variations of other periods. Eckhardt et al. (1963) isolated variations with frequencies of one and two cycles per year. Banks (1969) studied these frequencies, and also those of one, two, and three cycles per 27 days. He suggested that, for all variations up to the period of six months, the variation over the earth is represented by the harmonic P_1^0, which could be produced by fluctuations in the equatorial ring current. His deduced variation of conductivity in the upper mantle is also shown in Figure 20.1; it is characterized by a pronounced step at a depth of 400 km, with only a moderate increase below this depth. Bailey (1970) has demonstrated the applicability of Backus-Gilbert inverse theory (Section 3.3) to the problem of determining a

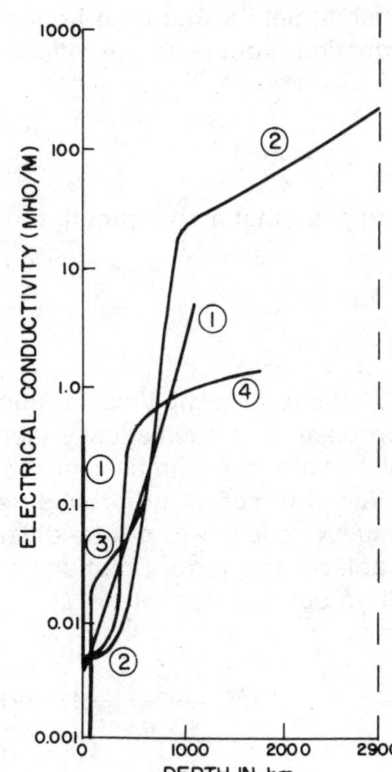

Figure 20.1 Electrical conductivity (in reciprocal ohm-meters) in the earth's mantle, according to (1) Lahiri and Price, (2) McDonald, (3) Cantwell, and (4) Banks. The conductivity in emu is obtained by multiplying the ordinates by 10^{-11}.

continuous distribution of conductivity from a severely limited choice of periods. It is apparent that any estimates for depths greater than 1000 km must still be accepted with great caution.

20.3 LOOKING THROUGH THE MANTLE

Because of the extreme attenuation with depth of the fields of variations in the period range of days, our information on the conductivity of the mantle below 1000 km comes chiefly from consideration of the screening effect of the mantle on the secular change. As we have seen, the secular change has an internal source, and most probably arises from motions in the outer core. The conducting mantle acts as a filter, so that the higher-frequency changes are screened from detection at the earth's surface. If the spectrum of the secular change at the lower boundary of the mantle were known, the conductivity throughout the mantle could be obtained. However, since we know the secular change only after it has been filtered, we may place only an upper bound on the integrated mantle conductivity. Runcorn (1955) drew attention to the possibility of employing "secular change impulses" for this purpose. At certain observatories it had been observed that new trends in the secular change are established in the very short time of three to five years. The mantle filter must therefore pass an impulse of this sharpness.

Although in Appendix D we set up the induction problem for a spherical earth, it is much simpler for the filter problem to follow Runcorn in treating the mantle as a slab bounded by planes $z = 0$ and $z = L$. Let the secular change field at the core boundary be parallel to *y*. Then the relevant Maxwell equation reduces to the *diffusion equation*

$$\frac{\partial^2 H_y}{\partial z^2} - 4\pi\sigma \dot{H}_y = 0 \qquad 20.3.1$$

Suppose that a step-function occurs in the secular change, so that at $z = 0$,

$$H_y = 0 \qquad t < 0$$

$$H_y = 1 \qquad t > 0$$

We require the time-variation of the H_y at the top of the "mantle," $z = L$. The problem is mathematically identical to the determination of temperature in a slab, with given initial and boundary conditions. While the equation is not difficult to solve, the procedure is somewhat lengthy because the equation is usually reduced to a total differential equation through the application of the Laplace transform. Solutions are given in Carslaw and Jaeger (1959), Chapter III; in our notation, at $z = L$,

$$H_y = \frac{2}{\sqrt{\pi}}\left[\left(-1 + \text{erf}\,\frac{\sqrt{\pi\sigma L}}{\sqrt{t}}\right) + \left(1 - \text{erf}\,\frac{3\sqrt{\pi\sigma L}}{\sqrt{t}}\right) + \cdots\right] \qquad 20.3.2$$

where erf is the error function, erf $x = \frac{2}{\pi}\int_0^x e^{-u^2}\,du.$

Figure 20.2, computed from this equation, shows the result of applying a step-function to the base of the "mantle," taken to be 2000 km thick, for two different conductivities. Also plotted is the variation in horizontal intensity at

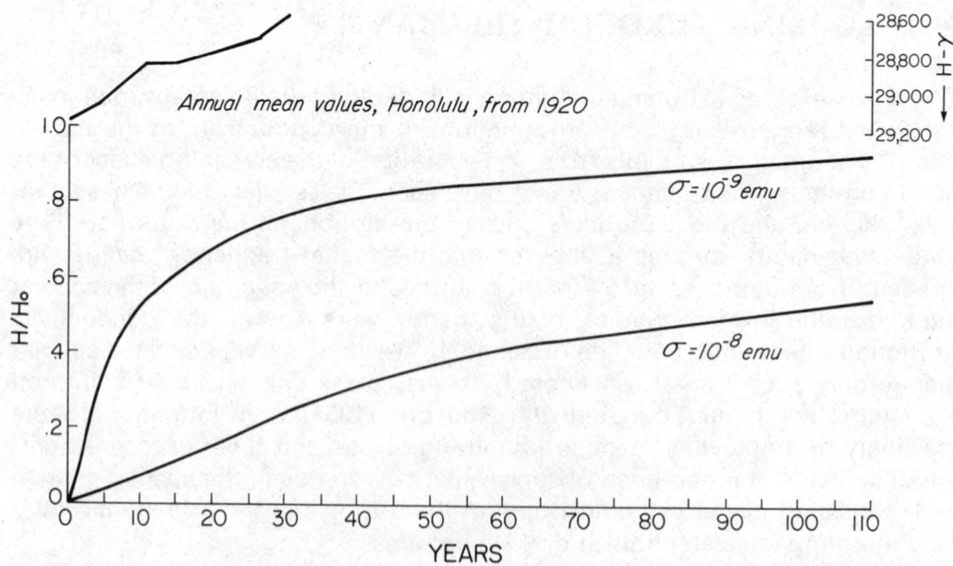

Figure 20.2 The change in field with time after the application of a unit step function at the base of a plane "mantle," for two values of conductivity. At the top is shown an example of a "secular change impulse."

Honolulu from 1920, taken from Walker and O'Dea (1952), who recognized impulses, or abrupt changes in rate of secular changes. A comparison of the curves suggests that the maximum value of mantle conductivity, on the assumption that the impulses are internal in origin, must be close to 10^{-9} emu; otherwise, changes in secular variation rate, even with the physically unreasonable step-function source at the base of the mantle, could never become established in times as short as those observed (for example, the four years between 1930 and 1934). The conclusion could almost have been reached using the expression for skin depth given at the beginning of this chapter. For a slab of conductivity 10^{-9} emu, the skin depth for a period of 10 years is 950 km; even decade changes must penetrate three skin depths to reach the earth's surface and, if the lower mantle conductivity were significantly higher, could never be observed. The extension of this treatment to a spherical earth has been made by McDonald (1957), who investigated secular change impulses in a number of harmonics and whose curve for conductivity, reaching a similar maximum value, is shown in Figure 20.1.

Recently, Alldredge (1977) has questioned whether the "secular change impulses" used in this type of analysis are truly internal in origin. He interprets field variations in the decade period range as manifestations of solar-controlled ring current fluctuations, and suggests that the shortest period change which traverses the entire mantle is 25 years. On this assumption, the lower mantle conductivity could be as high as 10^{-6} emu (10^5 mhos/m), within a factor of 10 of the assumed conductivity of the core. It is not clear how the rather flat lowest portion of Banks' curve could be connected with such a high value.

20.4 LATERAL VARIATIONS IN MANTLE CONDUCTIVITY

We have seen in the previous sections how the radial variation in electrical conductivity, at least in the outer 1000 km of the earth, has been established. As with so many other physical properties of the earth, the second step is to detect lateral departures from this mean curve.

For these more local studies, it is sufficient to treat the earth's surface as a horizontal plane. The problem is to detect variations with horizontal position in the electrical conductivity at some depth. The approach is similar to the analysis of the daily variation, in that the source of magnetic disturbance is external, but for the more local studies a range of time variations with periods decreasing to the micropulsation range is available. The geometry of the sources of these shorter-period variations tends to be complicated. For example, magnetic *substorms* with periods of a few hours appear to arise from currents flowing partly along magnetic field lines, and partly horizontally along the auroral electrojet (Section 19.4).

Some of the geometrical relationships involved in the induction of currents in the earth are suggested in Figure 20.3. In (*A*), a time-varying current above a conductor induces currents in it, whose magnetic effect is the same as that of an image current, of opposite sense to the original. The strength of the image increases with the conductivity σ, becoming equal to the original current as σ approaches infinity. In that case, for points at the surface of the conductor, the vertical component of the magnetic field is annulled, while the horizontal component is doubled. The vanishing of the normal component of field at the sur-

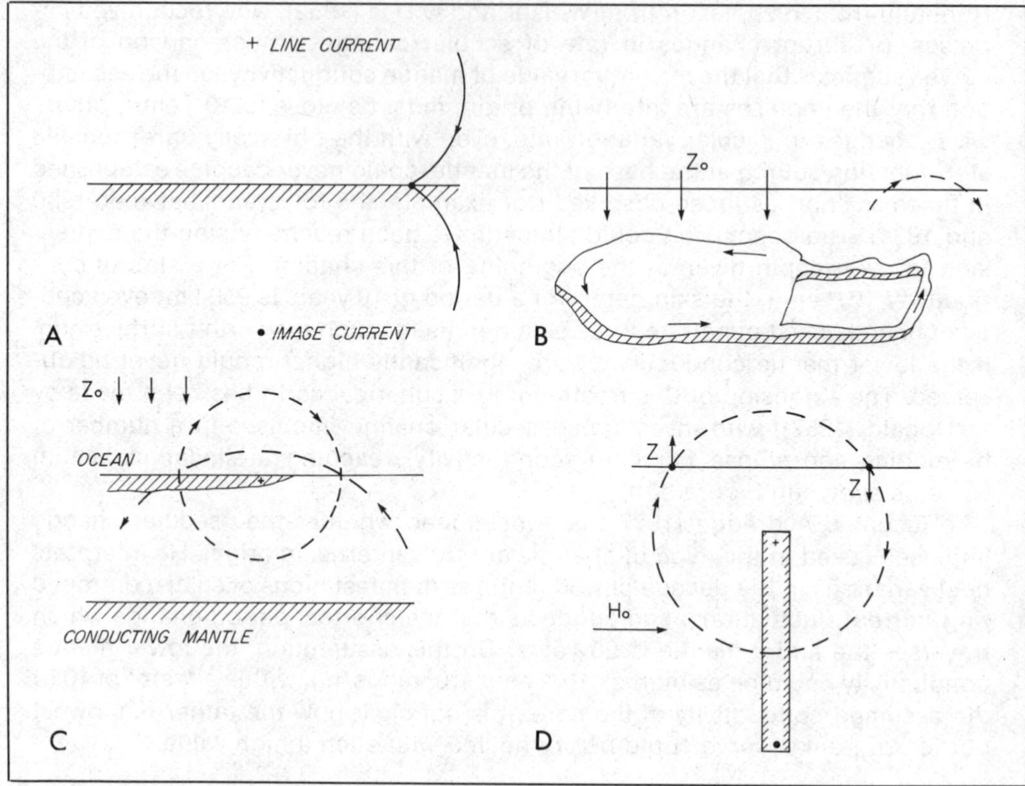

Figure 20.3 Idealized geometry of induction situations in the earth. (*A*) Reduction of the vertical component by the image of an ionosphere source; (*B*) channeling of current, induced in a large region, by a high-conductivity circuit; (*C*) lines of magnetic field around current induced near the edge of an ocean and the image current in the conducting mantle. (*D*) reversal in sign of the vertical magnetic component above a tabular conductor in a horizontal inducing field.

face of a high conductor may be shown analytically from Maxwell's equation,

$$\nabla \times \underline{E} = -\dot{\underline{B}}$$

For the vertical (*z*) component,

$$-\dot{\underline{B}}_z = \left(\frac{\partial E_x}{\partial y} - \frac{\partial E_y}{\partial x}\right)$$

Since electric fields in the conductor are very small, and tangential components of them are continuous across the upper surface, $\dot{\underline{B}}_z$ must approach zero.

It is found everywhere on earth that time variations in the vertical component are smaller than those in horizontal directions. Over a region of very high conductivity, such as the oceans, $\dot{B}_z$ is attenuated even further. However, this does not mean, as has sometimes been implied, that primary vertical fields may be completely neglected in induction studies. The observed vertical component is small only because currents are flowing in the conductor. If these currents leak out through some high-conductivity channel in the crust (Fig. 20.3*B*), they can produce strong local effects including local vertical fields which may be detected. This phenomenon, known as *current channelling*, has become increasingly recognized as important in the study of major crustal structures.

Secondly, any currents induced in near-surface conductors, such as the oceans, will themselves be imaged in the highly conducting deep mantle (Fig. 20.3*C*). A complete analysis of variation of magnetic fields at the earth's surface should therefore include the effect of the external sources, the near-surface induced currents, and the images of both of these in the conducting mantle. In many cases, one or two of these contributions become dominant, and the model is simplified.

In the case of isolated, steeply dipping conductors in the crust (Fig. 20.3*D*), it is true that induction may often be considered as due to a uniform horizontal field H_0. Induced currents flow around the edges of the conductor; of these, the line of current along the top produces the largest secondary magnetic field at the earth's surface. The normally small vertical component is enhanced in one sense to one side of the conductor, and simultaneously in the opposite sense to the other side, "crossing over" through zero immediately over the body.

A number of methods of studying lateral variations in conductivity have been designed with the above considerations in mind. The general aim is to isolate the secondary magnetic field due to induced currents, which has the characteristic that it shows more rapid spatial variation than the primary and may have a larger vertical component. The methods may be classified into those employing a large number of recording magnetometers operating simultaneously in an *array,* those employing statistical studies of relationships between the field components at single stations or small groups of stations, and those involving simultaneous measurements of magnetic and electric fields at the same stations.

Measurements with an array became possible with the development by Gough and Reitzel (1967) of a simple three-component torsion-fiber magnetometer, which could be manufactured in quantity and which could be left unattended for several days of recording. Arrays of these instruments have been operated in America (Porath, Oldenburg, and Gough, 1970; Camfield, Gough, and Porath, 1971), Australia (Gough et al., 1974), South Africa (Gough et al., 1973), and elsewhere. Fourier analysis of the simultaneous records allows the amplitude and phase relationships, at given periods, to be seen across a region (Fig. 20.4). Sometimes, as in the example shown, inspection alone indicates whether an effect is of internal or external origin. In addition, formal separation of the surface field into internal and external contributions is possible.

The application of a Gaussian type of internal-external field separation to a local area is based on the solution to the Neumann problem (Scheube, 1958; Siebert, 1958; Hartmann, 1963), which we have already discussed in connection with gravitational fields. Assuming, as we do in all cases that we meet in this chapter, that the field is derivable from a potential, and taking results from Sections 9.2 and 12.3, we may write the potential of external and internal sources as:

$$\left.\begin{aligned} U_e(x', y', z') &= \frac{1}{2\pi}\iint\limits_{-\infty}^{\infty} \frac{dx\, dy Z_e(x, y, 0)}{[(x'-x)^2+(y'-y)^2+(z'-z)^2]^{1/2}} \qquad \text{for} \quad z \geq 0 \\ U_i(x', y', z') &= -\frac{1}{2\pi}\iint\limits_{-\infty}^{\infty} \frac{dx\, dy Z_i(x, y, 0)}{[(x'-x)^2+(y'-y)^2+(z'-z)^2]^{1/2}} \qquad \text{for} \quad z \leq 0 \end{aligned}\right\} \quad 20.4.1$$

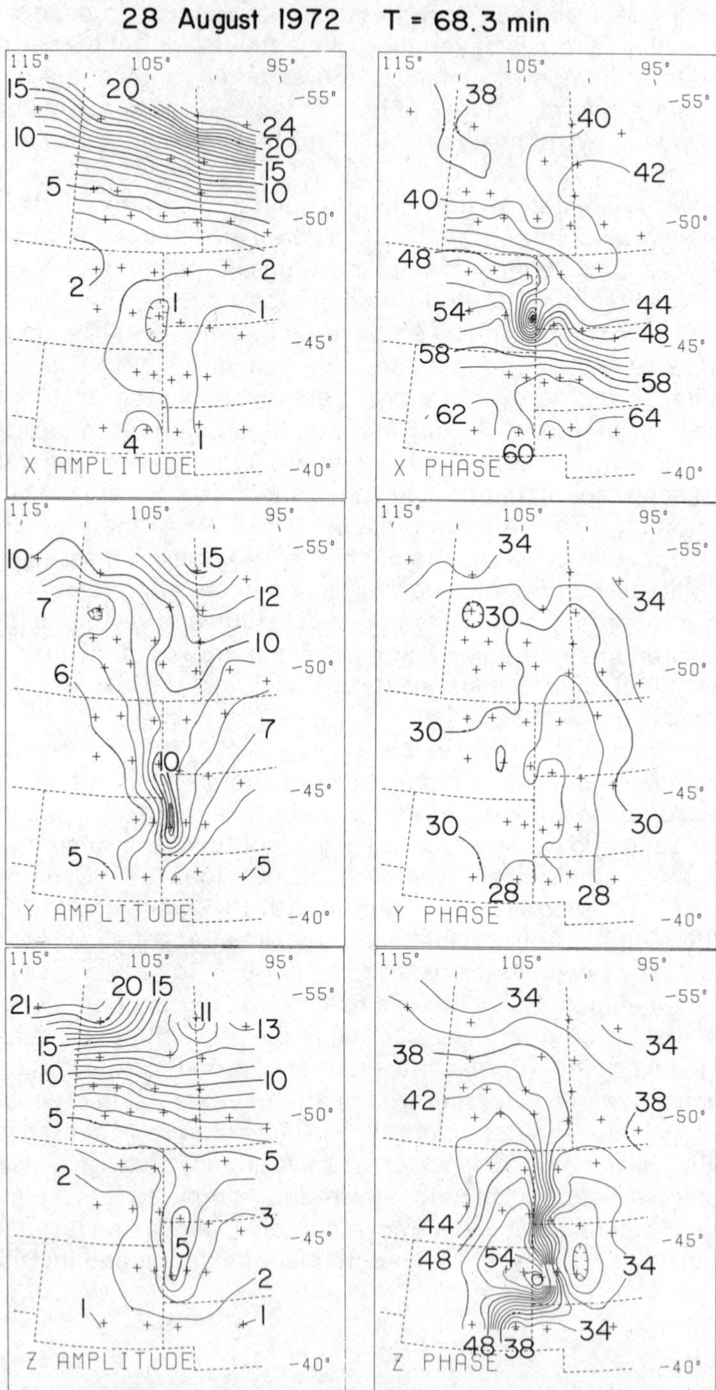

Figure 20.4 An example of results from an array of simultaneously-recording magnetometers, located at the points marked by crosses. For a period of 68.3 minutes, the amplitudes in gammas and the phases in minutes are shown for the three components X, Y, and Z. These observations probably demonstrate a source effect in the northern half of the area and the effect of a sub-surface conductor in the south-central region. (From A. O. Alabi, P. A. Camfield and D. I. Gough 1975; courtesy of the authors)

Here, a set of Cartesian coordinates has been taken, with z directed downward. The potentials at the point (x', y', z') are expressed as surface integrals of an equivalent density, $Z/2\pi$, of magnetic sources spread over the earth's surface. The potentials can also be written in terms of the potential at the earth's surface:

$$\left.\begin{aligned} U_e(x', y', z') &= \frac{z}{2\pi}\iint_{-\infty}^{\infty} \frac{dx\, dy U_e(x, y, 0)}{[(x'-x)^2+(y'-y)^2+(z'-z)^2]^{3/2}} \\ &\qquad\qquad \text{for} \quad z \geq 0 \\ U_i(x', y', z') &= -\frac{z}{2\pi}\iint_{-\infty}^{\infty} \frac{dx\, dy\, U_i(x, y, 0)}{[(x'-x)^2+(y'-y)^2+(z'-z)^2]^{3/2}} \\ &\qquad\qquad \text{for} \quad z \leq 0 \end{aligned}\right\} \quad 20.4.2$$

Differentiation of the first equation of 20.4.2 with respect to z, followed by partial integration with respect to x, leads to

$$Z_e(x', y', 0) = -\frac{1}{2\pi}\iint_{-\infty}^{\infty} X_e(x, y, 0)\, \frac{x'-x}{(y'-y)^2 r}\, dx\, dy \qquad 20.4.3$$

where

$$r = [(x'-x)^2 + (y'-y)^2 + (z'-z)^2]^{1/2}$$

while differentiation of the first equation of 20.4.1 with respect to x gives

$$X_e(x', y', 0) = \frac{1}{2\pi}\iint_{-\infty}^{\infty} Z_e(x, y, 0)\, \frac{x'-x}{r^3}\, dx\, dy \qquad 20.4.4$$

It follows that all field components, both internal and external, may be derived by similar operations on other components. We introduce operators M and N such that

$$\left.\begin{aligned} M_x &= \frac{1}{2\pi}\iint_{-\infty}^{\infty} -\frac{x'-x}{r^3}\, dx\, dy \\ N_x &= \frac{1}{2\pi}\iint_{-\infty}^{\infty} -\frac{x'-x}{(y'-y)^2 r}\, dx\, dy \end{aligned}\right. \quad 20.4.5$$

and similarly for M_y and N_y with x and y interchanged.

Then

$$\begin{aligned} Z_e &= -N_x X_e & X_e &= M_x Z_e \\ Z_i &= N_x X_i & X_i &= -M_x Z_i \\ Z_e &= -N_y Y_e & Y_e &= M_y Z_e \\ Z_i &= N_y Y_i & Y_i &= -M_y Z_i \end{aligned} \qquad 20.4.6$$

which lead to

$$\left.\begin{aligned} Z_e &= \tfrac{1}{2}(Z - N_x Y) \\ Z_i &= \tfrac{1}{2}(Z + N_x Y) \\ X_e &= \tfrac{1}{2}(X - M_x Z) \\ X_i &= \tfrac{1}{2}(X - M_x Z) \\ Y_e &= \tfrac{1}{2}(Y - M_y Z) \\ Y_i &= \tfrac{1}{2}(Y - M_y Z) \end{aligned}\right\} \qquad 20.4.7$$

The right-hand expressions of Equation 20.4.7 are entirely in terms of the total observed field changes ($Z = Z_e + Z_i$ and so forth), and if simultaneous three-component recordings are available over the entire anomalous area, the surface integrals may be evaluated for any given instant of time. The field quantities used are usually taken as differences from a normal background, so that the integration extends only to the limits of the anomaly area. Equations 20.4.7 therefore provide a full separation of the field into internal and external contributions.

The above methods were developed by the geophysicists at Göttingen, in the study of an anomalous area in Germany (Schmucker, 1959). This anomaly (Fig. 20.5), which appears to be produced by an east-west concentration of current, results in a change in the sign of Z variations across the area. The application of Equation 20.4.7 is obviously most appropriate in those cases where a dense array of observations is available.

If simultaneous measurements at an array are not available, recourse must be made to statistical relations between the field components. As suggested above, variations in the vertical component Z across an area are most diagnostic of lateral variations in conductivity. A simple method of analysis is to compute the ratio Z/H, for disturbances of given period, for each station.

The classic area of anomalous variation of Z/H is Japan (Fig. 20.6), which appears to be underlain by a major region of enhanced electrical conductivity (Rikitake, 1966). The situation is more complicated than the simplified outline above would indicate, because, as Rikitake has shown, the magnitude of the Z/H variation requires a distribution of both highly conducting and highly insulting rocks in order to concentrate the induced current. In the western Canadian arctic, Whitham (1964) has described a broad area in which Z-variations are suppressed (the magnetogram shown in Figure 18.2 is from this area), corresponding to the second effect described above. Hermance and Garland (1967) have employed Z/H ratios to infer that a region of high conductivity lies at shallow depths beneath Iceland.

A more quantitative approach, due to Parkinson (1959) and Schmucker

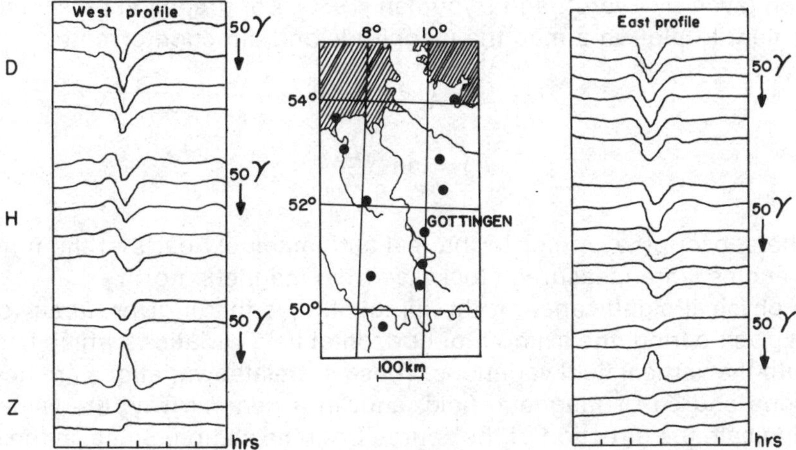

Figure 20.5 Variation in the magnetic elements on two profiles across northern Germany. Note the change in sign of the variation in Z between stations. (After Schmucker)

(1970), is to express the vertical field Z as a linear function of horizontal components, say H and D:

$$Z = AH + BD$$

Here Z, H, and D are to be understood as pertaining to a given period; normally they will be obtained as Fourier transforms of a length of record over some time interval. The parameters A and B are complex numbers, because of phase differences between the components, and are known as *transfer functions.* Methods of computing them by a least squares fitting of the transform

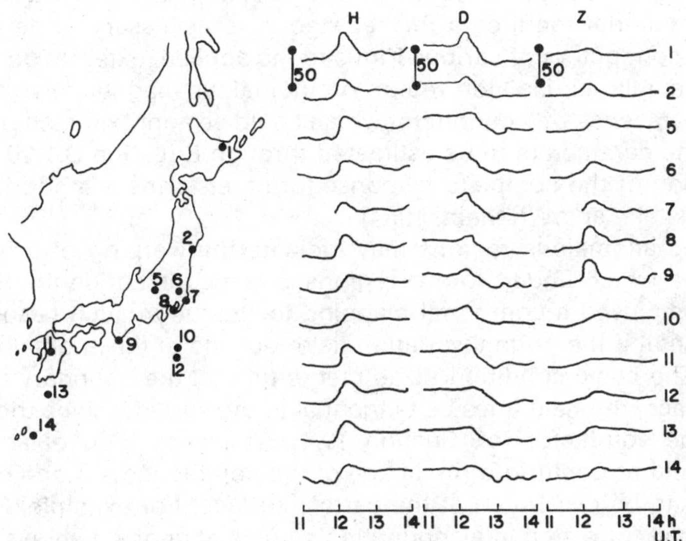

Figure 20.6 Variation in the magnetic elements at 12 stations in Japan. Note the similarity in H and D, contrasted with the considerable difference in response for Z. (After Rikitake)

have been given by Everett and Hyndman (1967). For graphical presentation, it is convenient to plot on a map the magnitude and direction given by

$$M_{r,i} = (A_{r,i}^2 + B_{r,i}^2)^{1/2}$$

$$\theta_{r,i} = \tan^{-1} \frac{B_{r,i}}{A_{r,i}}$$

where the subscripts r, i refer to the real and imaginery parts of the functions and the angles θ are measured clockwise from magnetic north.

The physical significance of the direction θ_r is that it gives, at any station and at a given period, the azimuth of horizontal field variations which best correlate with the vertical field variations. These correlated variations are normally due to one source of magnetic field, and, in a general way, the value of θ_r should indicate the direction of the source from the station. Since, in the northern hemisphere, Z is taken positive downward, consideration of the flux lines around an induced current (Fig. 20.3) shows that an arrow plotted at a station in the direction θ_r will point away from the current. For this reason, it is usual to reverse the sense of the arrows when they are plotted. The magnitude M_r is a measure of the vertical-to-horizontal field ratio. In the case of a single line conductor, it can be expected to be zero over the conductor and to reach maximum values on either side, at distances equal to the depth of the line source. In an analogous way, θ_i and M_i refer to horizontal fields which are in quadrature to the vertical field; they can indicate the presence of currents in systems whose circuit parameters (e.g., resistance) produce this shift in phase.

An example of a display of "induction arrows," as they are called, is shown in Figure 20.7. The explanation given (Bailey et al., 1974) for the change in direction with decreasing period is that at longer periods currents induced in the ocean are dominant, while at short periods the arrows are indicating the presence of local currents, possibly flowing along faults associated with the St. Lawrence River.

Separation of the fields into external and internal parts may show that an anomalous region is present, but the interpretation in terms of a distribution of conducting material requires a further step. It is necessary to determine the response of conductors of various forms to the applied, external portion of the field. For example, an isolated region of anomalous magnetic variation might suggest the presence of a conductor which could be approximated by a sphere, permitting the parameters to be estimated through Equation D.1.20 (Appendix D). Calculation of the complete response for other forms is a good deal more complex (Rikitake and Whitham, 1964).

Extended anomalous regions may indicate the warping of a highly conducting layer, which under normal regions is at a constant depth. Schmucker (1970) has employed a conformal mapping technique in such cases. This approach is useful if the normal variation field, outside of the anomalous area, is horizontal. The basic condition to be met is that, at the boundary of the conducting surface, the field lines be tangential to the surface, since the field cannot enter the conductor (cf. Section 19.4). Transformation of an originally horizontal field at depth into the shape of the conducting surface permits the components to be calculated at the earth's surface. For example, Figure 20.8 gives the result for a step in a conducting surface at depth. It shows computed values of the vertical and horizontal fields compared to those observed by

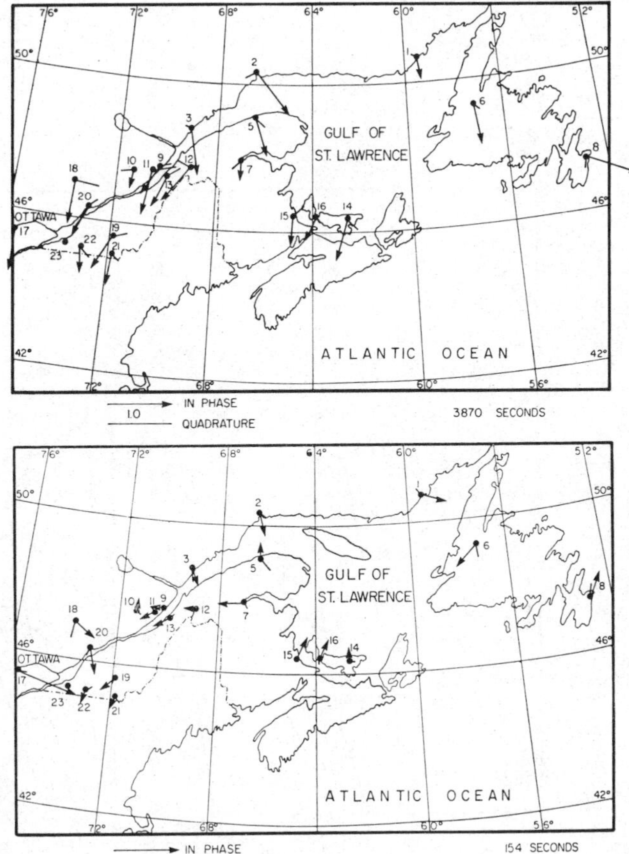

Figure 20.7 Induction arrows measured over the Gulf of St. Lawrence area for 2 periods. Notice how the in-phase arrows at the longest periods are influenced chiefly by the ocean; with decreasing period they rotate, apparently responding to the effect of currents flowing along more local structures. (After Bailey et al, 1974)

Schmucker (1970) on a profile from southern California to Texas. The step-like structure in the conducting sheet is believed to represent a change in conductivity distribution between the Cordillera and the Texas foreland. On the profile shown, the change takes place in the vicinity of the Rio Grande. By means of a relaxation technique Schmucker was able to derive, from the step, a curved form which gave a still closer fit to the observations.

While conformal mapping appears to be a very useful tool in this type of interpretation, its limitations must be recognized. These include the assumptions regarding the normal field (horizontal) and the fact that the anomalous fields must be in phase with it. Any observations on frequency response or phase shift are not used in the interpretation.

The principle involved in the calculation of the response of isolated bodies, bounded by plane surfaces, in a uniform inducing field is that Maxwell's equations may often be applied to the electric and magnetic fields separately. The great majority of the calculations have been made for two-dimensional forms, in which no properties vary along the strike direction and in which the uniform applied field is either normal or parallel to strike. Let us take the $x-y$ plane to represent the surface of the earth, with the x-axis parallel to strike and the z-axis vertically downward. As is done for Equation D.2.1, Maxwell's equa-

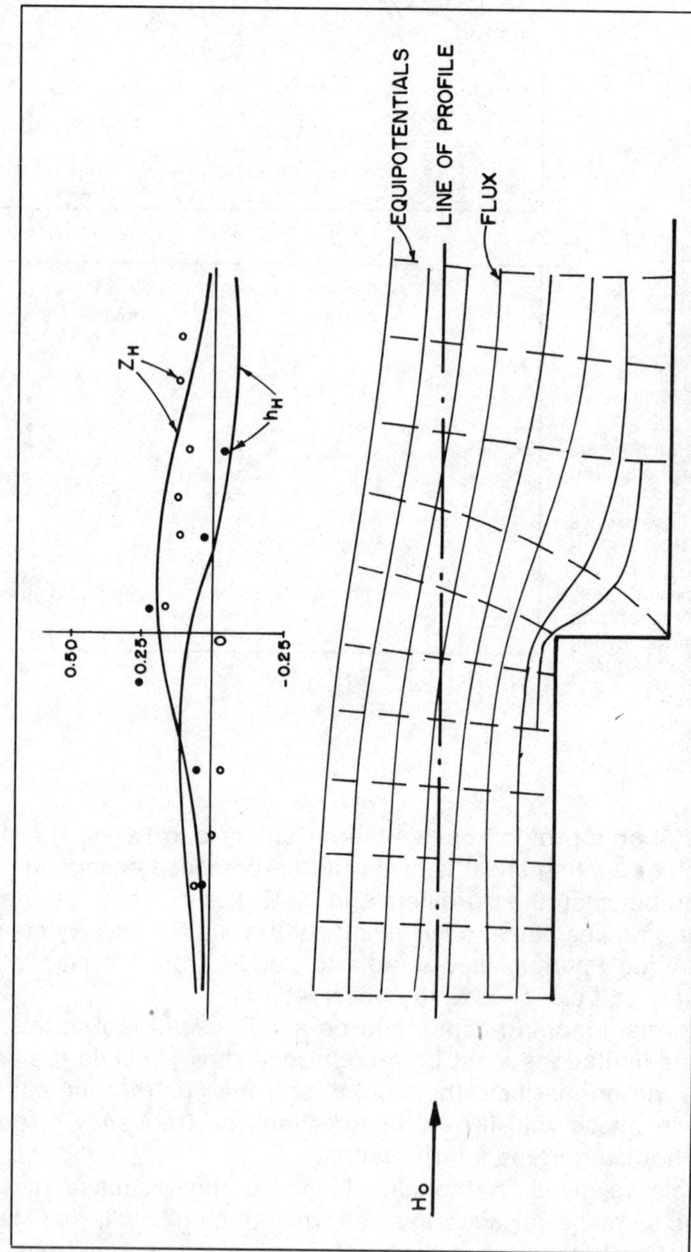

Figure 20.8 Distortion of the flux lines and equipotentials above a step in a highly-conducting surface. At top, computed values of the operators z_H and h_H (i.e., vertical and horizontal disturbance fields for a unit inducing field H_o) compared to observations across the Rio Grande anomaly. (After Schmucker)

tions may be separated into an equation for each of the magnetic and electric fields:

$$\left.\begin{aligned} \nabla \times \nabla \times \underline{E} &= -4\pi\sigma\, \dot{\underline{E}} \\ \nabla \times \nabla \times \underline{H} &= -4\pi\sigma\, \dot{\underline{H}} \end{aligned}\right\} \qquad 20.4.8$$

Allowing all field quantities to vary as $e^{i\omega t}$, and noting that $\partial/\partial x = 0$, these reduce to

$$\left.\begin{aligned} \frac{\partial^2 E_x}{\partial y^2} + \frac{\partial^2 E_x}{\partial z^2} &= 4\pi i\sigma\omega\, E_x \\ \frac{\partial^2 H_x}{\partial y^2} + \frac{\partial^2 H_x}{\partial z^2} &= 4\pi i\sigma\omega\, H_x \end{aligned}\right\} \qquad 20.4.9$$

where σ is the value of conductivity appropriate to each region. At each of the plane boundaries between regions, tangential components of $\underline{E}$, and all components of $\underline{H}$, must be continuous. However, the boundary conditions at large distances from the anomalous structure depend upon the orientation of the applied field relative to the strike direction. The case in which the applied magnetic field is perpendicular to the strike is known as "*E*-polarization," and that in which it is parallel to strike is called "*H*-polarization." Jones and Pascoe (1971) have discussed the appropriate conditions at large distances for both polarizations. Numerical solutions for the field at the earth's surface have been obtained (Jones, 1973) by integration of Equations 20.4.8 and 20.4.9 over a grid of points describing the inhomogeneities. The computation, straightforward in principle, represents a formidable computing problem, even for simple structures. The great limitation of the two-dimensional solutions is that they do not permit the channelling of current into a region of high conductivity from the surroundings, since any conductor is assumed to extend to infinity in both directions along strike. That channelling of current induced over a large volume of the background material may be more important than local induction, for at least a part of the frequency spectrum, has been shown by Lajoie (1973) and Lajoie and West (1976). As shown in Figure 20.9, these authors have computed the vector current density for points in a conducting plate of finite dimensions, in the field of a vertical oscillating magnetic dipole. Currents in phase with, and in quadrature with, the source are shown separately. Note, from the figure, that when the plate conductivity is ten times that of the background, the pattern of current density shows that the plate is essentially channelling the quadrature current induced in the host medium. With increasing conductivity, current vortices—the result of local induction—begin to form, first with quadrature current and finally with in-phase current. The shift in phase results from the fact that with very high conductivity, the anomalous body acts primarily as an inductor, producing a second phase shift of $\pi/2$ from the quadrature current induced in the host medium. To return to the case of structures in uniform applied fields, Jones and Pascoe (1972) have developed a program for the case of general three-dimensional structures. Current channelling is, of course, automatically included in the solutions in this case, but the computation is considerably more time-consuming than in the case of two-dimensional bodies.

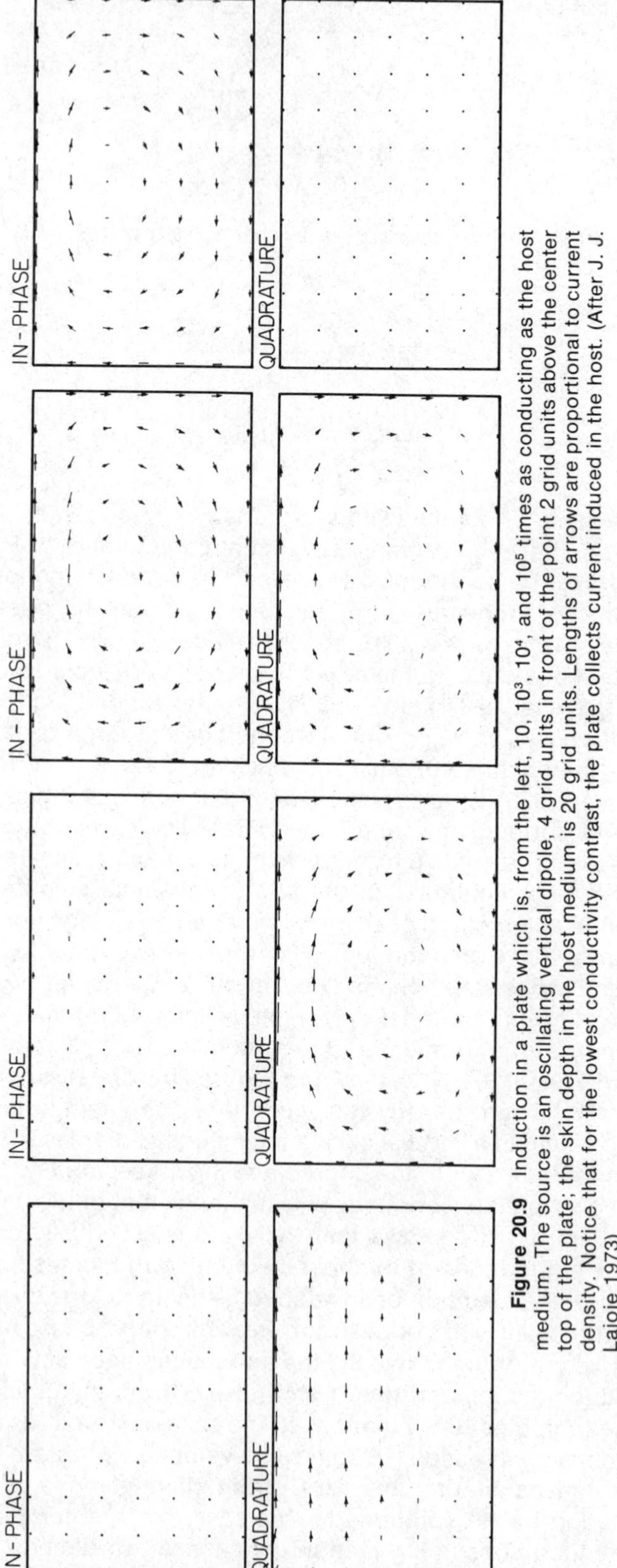

Figure 20.9 Induction in a plate which is, from the left, 10, 10^3, 10^4, and 10^5 times as conducting as the host medium. The source is an oscillating vertical dipole, 4 grid units in front of the point 2 grid units above the center top of the plate; the skin depth in the host medium is 20 grid units. Lengths of arrows are proportional to current density. Notice that for the lowest conductivity contrast, the plate collects current induced in the host. (After J. J. Lajoie 1973)

Because of the cost of computing the response of three-dimensional structures, even in a uniform applied field, there remains a useful place for laboratory scale models. As a dimensional examination of Equation 20.4.9 suggests, a scale model will produce the same ratio of secondary to primary field at all points if

$$(\sigma\omega L^2)_{\text{model}} = (\sigma\omega L^2)_{\text{earth}}$$

where L is the same characteristic length in both cases. Thus, if a laboratory scale model is made in which all linear dimensions are reduced by a factor of 10^4 (i.e., 1 m in the model representing 10 km in the earth), the product $\sigma\omega$ must be increased by 10^8, through the use of more conductive materials and higher frequencies in the laboratory. A review of the use of analogue models has been given by Dosso (1973).

Methods based upon the simultaneous measurement of H and the electric field E at the same station were derived by Cagniard (1953), and were proposed initially for geophysical prospecting. In practice, recordings can be made with any type of recording magnetometer which responds to a component of the magnetic field, together with an electric field recording system consisting of two electrodes, separated by perhaps 0.5 km, connected by cable to a recording potentiometer. We imagine first a plane electromagnetic wave incident vertically on the surface of a uniform earth, take the wave to be polarized with E in the direction of x and H in the direction of y, and assume a time factor $e^{-i\omega t}$. Maxwell's equations reduce to

$$\left.\begin{aligned} \frac{\partial^2 E_x}{\partial x^2} + 4\pi\sigma\omega i E_x = 0 \\ H_y = -\frac{i}{\omega}\frac{\partial E_x}{\partial z} \end{aligned}\right\} \quad 20.4.10$$

of which a solution (no longer a traveling wave) is

$$E_x = A \exp\,(a\sqrt{\sigma} z) + B \exp\,(-a\sqrt{\sigma} z) \quad 20.4.11$$

where

$$a = 2\pi / T^{1/2}\sqrt{1-i} \text{ and } T = 2\pi/\omega$$

If the earth is taken to be the half-space $z > 0$, we must have $A = 0$ for the fields to remain finite, and at any value of z,

$$H_y = (2\sigma T)^{1/2} e^{i\pi/4} E_x \quad 20.4.12$$

In particular, H_y and E_x could be the fields measured at the earth's surface. The true conductivity of the half-space would be determined from

$$\sigma = \frac{1}{2T}\left(\frac{H_y}{E_x}\right)^2 \quad 20.4.13$$

In the case of the actual earth, the value of σ calculated by inserting measured values of the field changes into Equation 20.4.13 is called the apparent conductivity. It represents some mean value of the conductivity seen by the fields, and its variation with period T is therefore diagnostic of the variation of conductivity with depth in the earth. Cagniard presented curves showing the variation of apparent conductivity with period for an earth consisting of a layer over a half-space; other workers (e.g., Srivastava, 1967) have computed model curves for more layers. In application of the method, records of the variations of H and E are analyzed to give their spectral components, and Equation 20.4.13 is evaluated for a series of values of T. The curve of σ thus obtained is compared to computed model curves.

The important ratio H_y/E_x which appears in Cagniard's solution may be obtained in terms of magnetic field components only, eliminating the need for electric field measurements. This may be of convenience in special cases, such as measurements made at sea. Suppose that recording magnetometers are operated at the sea surface and vertically beneath on the sea floor. Returning again to Maxwell's equations, we can write for the field components in the ocean,

$$\frac{\partial H_z}{\partial y} - \frac{\partial H_y}{\partial z} = \sigma_0 E_x \qquad 20.4.14$$

where σ_0 is the conductivity of sea water. In the case of the plane-wave source and horizontally layered structure, which has been assumed, H_z is small and its horizontal gradient even more so. Thus,

$$E_x = -\frac{1}{\sigma_0}\frac{\partial H_y}{\partial z} \qquad 20.4.15$$

The vertical gradient $\partial H_y/\partial z$ for a given period can be obtained from comparison of the sea-surface and sea-floor magnetometers, and Equation 20.4.15 then gives the value of E_x that would have been measured on the sea floor. When this is combined with the measurements of H_y itself from the sea-floor magnetometer, a Cagniard-type sounding of the oceanic crust becomes possible. There are two important additional assumptions implicit in this approach: that no extraneous currents are flowing in the ocean between the magnetometers, and that the sea water is at rest in the earth's main field.

The Cagniard approach is beautifully simple in principle, but this very simplicity means that there are severe limitations upon its applicability. We note that a plane-wave source is assumed. If the source of the disturbance is at a finite distance, the incoming wave will not be plane, and Price (1962) has shown the corrections necessary for sources of finite dimensions. The second limitation is the assumption of horizontal layering within the earth. Vertical boundaries, or anisotropy of conductivity in horizontal layers, produces departures from the ideal curves. On the other hand, the magneto-telluric method, as it is called, does provide a means of bridging the gap between the surface of the earth, where the conductivity can be measured, and the region of the mantle in which the global average curve becomes valid. An example of a curve of apparent conductivity for a continental area together with the suggested interpretation is shown in Figure 20.10. We note that, at the longest periods, the increase in conductivity of the mantle is becoming apparent.

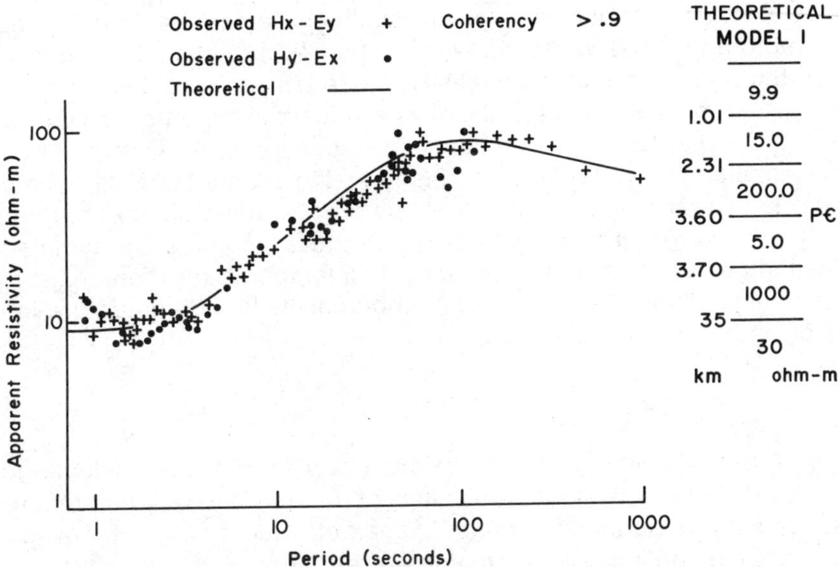

Figure 20.10 Magneto-telluric sounding in western Canada. The solid curve is computed for the model shown at the right, which is based on known thicknesses and resistivities above the Precambrian surface. (Courtesy of K. Vozoff and R. M. Ellis)

20.5 THE ELECTRICAL CONDUCTIVITY OF ROCKS

The interpretation of areas of anomalous conductivity in either the crust or the mantle requires a knowledge of the range of conductivity among common earth materials. At low temperatures, virtually all rock-forming minerals except a few metallic ores are very poor conductors. Electrical conduction at shallow depths is therefore chiefly by ionic conduction in the saline solutions which fill the pore spaces or fractures of the rocks. On this basis, igneous rocks are less conductive than sedimentary rocks, and the decrease in conductivity shown by Figure 20.10 indicates the transition from a sedimentary basin to the crystalline crust.

Notice that the range of values of the resistivity, by a factor of more than 10^2 within the section shown on the figure, is much greater than that for other physical properties such as density or seismic wave velocity. It is not unusual, in the laboratory, to measure a resistivity as high as 10^4 ohm m for a dry igneous rock and as low as 10 ohm m for a saturated sedimentary rock. This great range of properties makes the electrical property of the crust and mantle a sensitive indicator of composition and state. The laboratory measurement of resistivity is normally made by determining the resistance of a sample of regular geometry and computing the resistivity ρ from the relation

$$R = \rho L / A$$

where R is resistance in ohms, L the length of the sample, and A the cross-sectional area. This relationship shows immediately why experimental workers usually express ρ in ohm cm or ohm m, whereas in the theoretical relations we have introduced it must be in emu.

As temperature increases, silicate materials such as olivine show an increase in electrical conductivity, which has been demonstrated in the laboratory (Akimoto and Fujisawa, 1965). We note (Fig. 20.11), in addition to the main trend, a sudden increase in conductivity at one temperature. This is related to the change in atomic structure of the olivine, and we shall return to a consideration of it in Chapter 26. The general increase appears to result from three independent effects (Tozer, 1959). First, many of the silicates are intrinsic semiconductors, in that electrons in the atoms may be thermally excited into the conducting band. At sufficiently high temperatures, conduction by ions freed from the lattice becomes significant, and at all temperatures there may be conduction by impurities. The variation in conductivity with temperature in each case is given by an equation of the form

$$\sigma = \sigma_0 e^{-E/kT}$$

where E is the energy necessary to excite a carrier into the conducting state and k is Boltzmann's constant. The values of σ_0 and E are very different for the different processes; ionic conduction is characterized by a greater σ_0, but becomes dominant only at higher temperatures. Tozer has suggested that intrinsic semiconduction is the most important effect throughout most of the mantle. The effect of increasing pressure is different for the different mechanisms (Section 25.4).

Within the crust, conduction may be by a series-parallel complex of minerals and saturated pores. Brace (1971), on the basis of laboratory measurements, has estimated the ranges of effective resistivity in typical crustal rock for both mechanisms, for different geographical areas. An example of his curves is shown in Figure 20.12. Pore conductivity depends upon porosity and the salinity of the pore fluids; it decreases with increasing pressure, rapidly at first, as pores and cracks are closed. Mineral conduction by the semiconduction mech-

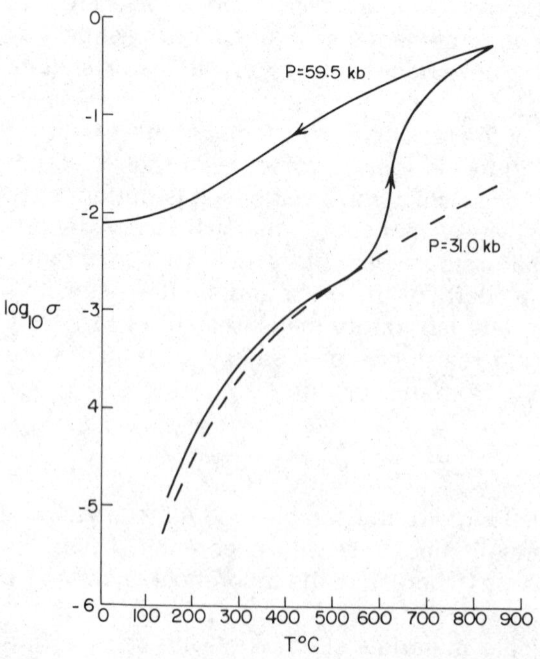

Figure 20.11 Variation with temperature of the electrical conductivity of olivine, as measured in the laboratory under pressures of 31.0 and 59.5 kb. The sudden increase in conductivity in the latter case coincides with a phase change in the olivine. (After Akimoto and Fujisawa)

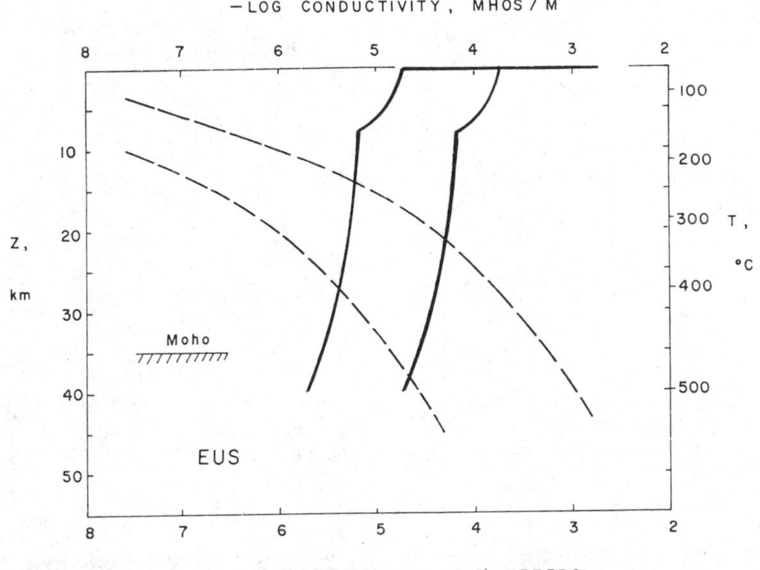

Figure 20.12 Estimate of conductivity in the crust of the eastern United States for pore conduction (solid lines) and electronic conduction (broken lines). (From W. F. Brace, 1971; courtesy of the author and the American Geophysical Union.)

anism depends upon the mineralogy and the temperature gradient in the region. Brace suggests that above the region of intersection of the curves, conduction is predominantly through pore fluids, while below, it is predominantly by minerals. Resistivity in the crust should reach a maximum of approximately 10^5 ohm m at depths of between 10 and 20 km.

Field measurements, by the methods of the previous section, very often give a maximum resistivity for the crust which is much less than 10^5 ohm m. Figure 20.10, for example, shows a maximum resistivity in the crystalline crust of 1000 ohm m. A possible explanation is the recent discovery (Stesky and Brace, 1973) of serpentinized rocks, from the floor of the Indian Ocean, with very low resistivity (100–1000 ohm m at low pressure and temperature). Stesky and Brace suggest that the presence of metallic oxides, such as magnetite, may provide the enhanced conductivity. It may be that the effect of such minerals in the lower crust, particularly in regions that represent ancient ocean floor, has been underestimated. By contrast, Hyndman (1967) has suggested that old, metamorphosed portions of continental crust may come to have a relatively high resistivity, through loss of water from the crystal lattice during metamorphism.

20.6 CONDUCTIVITY ANOMALIES AND TECTONICS

The global pattern of lateral variations in electrical conductivity is beginning to display a correlation with tectonics, as a number of reviews have pointed out (Gough, 1973; Porath and Dziewonski, 1971; Garland, 1975). In seeking a relationship, it is necessary to attempt to separate those effects which are obviously crustal in origin, and probably not temperature-controlled, from conductivity variations in the mantle which may well indicate lateral tempera-

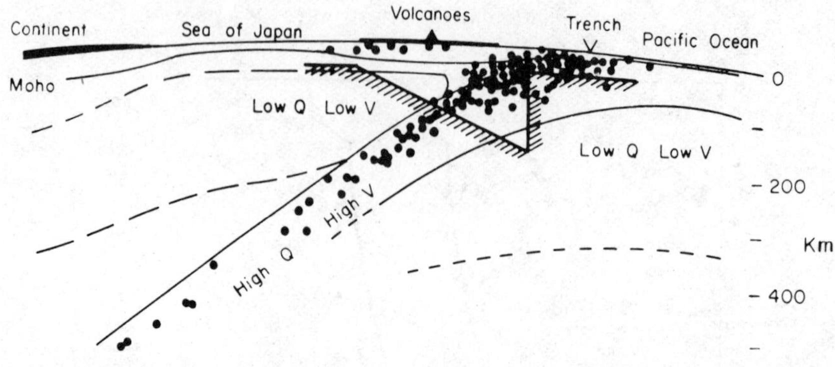

Figure 20.13 Conductivity structure beneath Japan in relation to zones of high and low seismic velocity and seismic quality factor Q (cf. Section 27.5). The hatched line is the inferred surface of the highly-conducting mantle. Seismic information based on Utsu (1971); conductivity by Rikitake (1969).

ture gradients. Unfortunately, no measurements are yet available for suboceanic ridges, where the highest temperatures would be expected, and only limited measurements from areas where ridges come ashore, as in Iceland, mentioned above. Measurements over subduction zones (Rikitake, 1966; Schmucker et al., 1964) suggest a juxtaposition of both high and low conductivity. In the Japanese case in particular, the region of low conductivity appears to be coincidental with the upper part of the descending cold slab (Fig. 20.13). Under the North American continent, a region of enhanced conductivity in the upper mantle appears to correlate rather well (Fig. 20.14) with the thickness

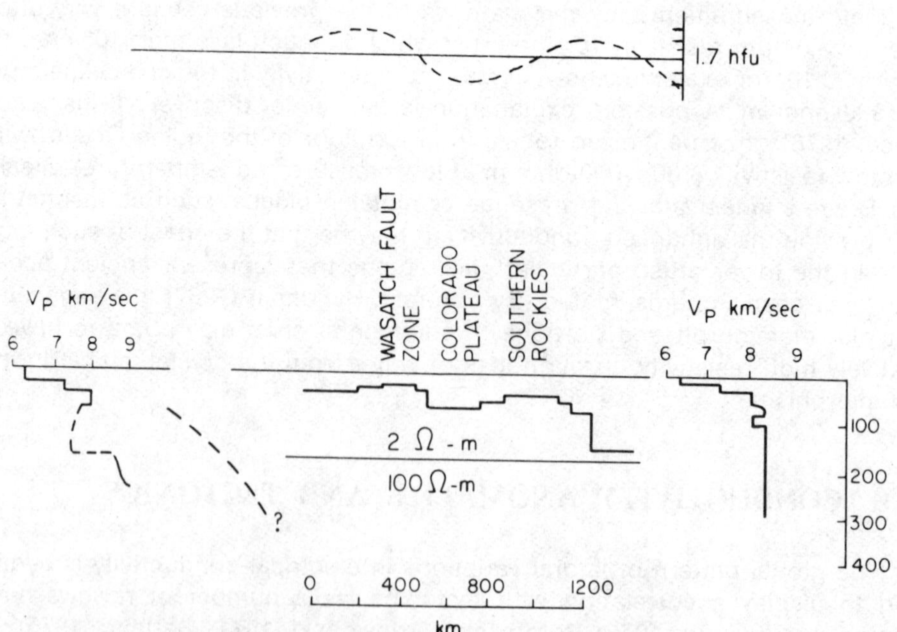

Figure 20.14 An inferred area of high conductivity beneath western North America (after Gough 1973). The seismic velocity sections are those of Johnston (to the west) and Massé (to the east). Heat flow is shown at the top. The broken line suggests the possible location of a former subduction zone.

and severity of the seismic low velocity layer (Porath, 1971). It may be the case that the curve of average global conductivity as a function depth should show a local maximum at a depth of about 100 to 200 km, corresponding with the low velocity layer. However, global studies, as summarized in Figure 20.1, so far have not had the resolution to detect this. A major conductivity anomaly in southern Africa (Gough et al., 1973) probably also has its source in the mantle.

There have been indications of high conductivity beneath continental rifts in the case of the Rhine Graben (Winter, 1973) and the East African rifts (Banks and Ottey, 1974). In both of these cases, the resistivity drops to values of the order of 10 ohm m at depths which appear to transcend the base of the crust. There is a suggestion, therefore, of elevated temperatures, possibly associated with the intrusion of new volcanic rock into the crust.

Conductivity anomalies which probably have their origin completely in the crust include those related to sedimentary basins in the Canadian Arctic (Praus et al., 1971) and northern Germany (Vozoff and Swift, 1968). The electrical conductivity of the continental crust may also be enhanced by major fractures or shear zones and perhaps also by the entrapment of ancient ocean floor. As noted in the previous section, rocks with anomalously high conductivity have been obtained from the modern ocean floor. Gough and Camfield (1972) have proposed that an extensive, linear anomaly in conductivity in central North America is related to a metamorphic belt in the crust. On a smaller scale, shear zones are known to be accompanied by the development of the mineral graphite, which has a high conductivity. A portion of this anomaly can be seen crossing the array shown in Figure 20.4. A similar case is a linear increase in conductivity in southern Australia, described by Tammemagi and Lilley (1973). There is a remarkable correlation in position of the conductor with seismicity (Figure 20.15), and the authors suggest that it may represent an intra-plate boundary, with entrapped conductive geosynclinal rocks or ancient ocean floor.

The overall conclusion is that lateral variations in electrical conductivity can be most useful in the outlining of major structures, whether the anomalies

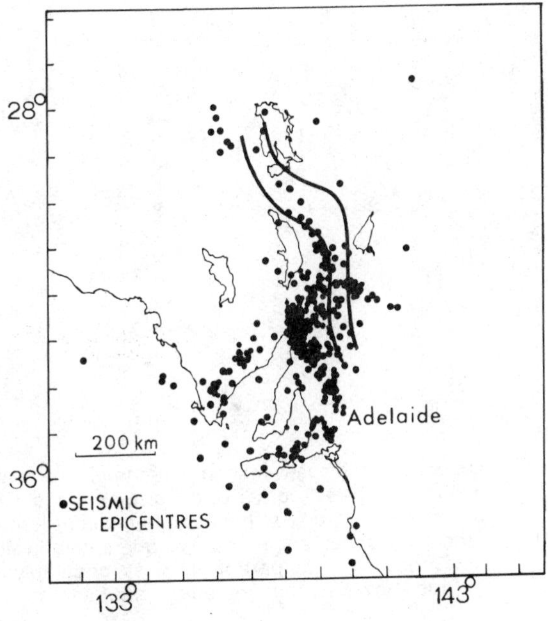

Figure 20.15 Location of a conductivity anomaly in Australia (between the heavy lines) compared with the distribution of earthquake epicentres. (After Gough, McElhinny and Lilley, 1974)

have their source in the mantle or in the crust. There remain difficulties in the study of structures near the sea coast, because of the large effect of the ocean-land boundary itself. The coastline effect for particular geometries has been evaluated theoretically by Rikitake (1967) and by means of models (Roden, 1964), but it remains a difficulty, particularly since real differences in mantle conductivity may be associated with the borders of continents. For the study of the electrical structure of the oceanic crust and sub-oceanic mantle, the best hope lies in sea-bottom recording magnetometers (Poehls and Von Herzen, 1976) which could be operated away from any coastline. While short-period variations are screened by the ocean itself, longer-period changes are detected and can be used to infer the conductivity beneath the ocean floor. The chief difficulties are the expense of the instrumentation and of deploying and retrieving an ocean-floor array of instruments, and the correction for the magnetic effects of electric currents induced in the ocean above the instruments by motions of the sea water through the earth's main magnetic field. Poehls and Von Herzen, from measurements made in the Atlantic Ocean south of Bermuda and away from any ridge, infer a model in which a layer of relatively highly conducting material (resistivity 10 ohm m) extends in depth from the sea floor to 100 km, and overlies material of resistivity 20 ohm m. They suggest that the upper region may consist of two unresolved, thinner conductive regions: the sea floor itself, possibly saturated with sea water, and a layer of partial melting just above the depth of 100 km.

Problems

1. An array of recording magnetometers was operated at points of a rectangular grid, with a grid spacing of 80 km. During an isolated magnetic disturbance, with a period of about 1 hour, the following variations in vertical component, northward

		Z (downward)	Y (eastward)	X (northward)
$x = 0$	$y = 0$	28	−10	18
	1	10	−28	36
	2	−8	−9	19
	3	−2	−2	10
	4	2	0	8
$x = 1$	$y = 0$	20	−2	15
	1	26	−12	22
	2	8	−30	40
	3	−10	−9	22
	4	−4	−3	16
$x = 2$	$y = 0$	8	0	20
	1	12	−2	21
	2	19	−9	30
	3	0	−26	48
	4	−18	−12	28
$x = 3$	$y = 0$	−8	0	12
	1	0	0	12
	2	4	−2	14
	3	10	−11	23
	4	−8	−27	40

component, and eastward component were observed. (Values are in gammas; the origin is at the southwest corner of the array with the axis of x extending northward and the axis of y eastward.) Try to separate the contribution of external and internal sources during this disturbance. You may attempt an evaluation of the operators of Equation 20.4.5 for the central part of the array, and apply these by numerical integration over the grid. However, this procedure is difficult if the anomalous disturbance does not decrease to zero at the outer boundaries. A qualitative picture of the conductors is often obtained by plotting disturbance vectors in the horizontal plane (cf. Parkinson, 1959); these vectors tend to be normal to conductors. Can you then consider the sense of the vertical variations, to determine whether the sources are above or below the earth's surface?

2. Taking the mean values of the parameters for intrinsic semi-conduction and ionic conduction in olivine (Sections 20.5 and 25.4), estimate the temperature at which, in the absence of pressure effects, the conductivity for the two mechanisms would be equal. Assuming that the pressure coefficient of E_1 is -0.5×10^{-3} electron volts per kilobar, and E_2, 5×10^{-3} electron volts per kilobar, and neglecting all other variations of the parameters, estimate how this temperature might vary through the mantle. Compare your results with the proposed actual temperature distribution in the earth (Fig. 25.2).

*3. By investigating the behavior of the response parameter in Equation D.1.20, or by using Wait's curves, plot the anomaly in vertical magnetic variation field, for a period of 30 minutes, above a spherical body of conductivity 1×10^{-11} emu and radius 10 km. The center of the sphere is at a depth of 20 km in a non-conducting crust, and the inducing field is horizontal, with an amplitude of 100 gammas. For what period of inducing field would the secondary field have the greatest departure in phase?

BIBLIOGRAPHY

Akimoto, Syun-iti, and Fujisawa, Hideyuki (1965) Demonstration of the electrical conductivity jump produced by the olivine-spinal transition. *J. Geophys. Res.*, **70**, 443.

Alabi, A. O., Camfield, P. A., and Gough, D. I. (1975) The North American Central Plains conductivity anomaly. *Geophys. Journ.* **43**, 815–833.

Alldredge, L. R. (1977) Deep mantle conductivity. *J. Geophys. Res.*, **82**, 5427–5431.

Bailey, R. C. (1970) Inversion of the geomagnetic induction problem. *Proc. Roy. Soc. Lond. A*, **315**, 185.

Bailey, R. C., Edwards, R. N., Garland, G. D., Kurtz, R., and Pitcher, D. (1974) Electrical conductivity studies over a tectonically active area in eastern Canada. *J. Geomag. Geoelectr.*, **26**, 125–146.

Banks, R. J. (1969) Geomagnetic variations and the electrical conductivity of the upper mantle. *Geophys. J.*, **17**, 457.

Banks, R. J., and Ottey, P. (1974) Geomagnetic deep sounding in and around the Kenya rift valley. *Geophys. J.*, **36**, 321–335.

Brace, W. F. (1971) Resistivity of saturated crustal rocks to 40 km based on laboratory measurements. In *The structure and physical properties of the earth's crust*, ed. John G. Heacock. Amer. Geophys. Un. Monograph 14.

Cagniard, Louis (1953) Basic theory of the magneto-telluric method of geophysical prospecting. *Geophysics*, **18**, 605.

Camfield, P. A., Gough, D. I., and Porath, H. (1971) Magnetometer array studies in the north-western United States and south-western Canada. *Geophys. J.*, **22**, 201–221.

Carslaw, H. S., and Jaeger, J. C. (1959) *Conduction of heat in solids.* Clarendon Press, Oxford.

Chapman, S. (1919) The solar and lunar diurnal variations of terrestrial magnetism. *Phil. Trans. Roy Soc. A*, **218**, 1.

Dosso, H. W. (1973) A review of analogue model studies of the coast effect. *Phys. Earth Plan. Int.*, **7**, 294–302.

Eckhardt, D., Larner, K., and Madden, T. (1963) Long-period magnetic fluctuations and mantle electrical conductivity estimates. *J. Geophys. Res.*, **68**, 6279.

Everett, J. E., and Hyndman, R. D. (1967) Geomagnetic variations and the electrical conductivity structure in south-western Australia. *Phys. Earth Plan. Int.*, **1**, 24–34.

Garland, G. D. (1975) Correlations between electrical conductivity and other geophysical parameters. *Phys. Earth Plan. Int.*, **10**, 220–230.

Gough, D. I. (1973) The geophysical significance of geomagnetic variation anomalies. *Phys. Earth Plan. Int.*, **7**, 379–388.

Gough, D. I., and Camfield, P. A. (1972) Convergent geophysical evidence of a metamorphic belt through the Black Hills of South Dakota. *J. Geophys. Res.*, **77**, 3168–3170.

Gough, D. I., De Beer, J. H., and Van Zijl, J. S. (1973) A magnetometer array study in Southern Africa. *Geophys. J.*, **34**, 421–433.

Gough, D. I., McElhinny, M. W., and Lilley, F. E. M. (1974) A magnetometer array study in Southern Australia. *Geophys. J.*, **36**, 345–362.

Gough, D. I., and Reitzel, J. S. (1967) A portable three-component magnetic variometer. *J. Geomag. Geoelectr.*, **19**, 203.

Hartman, Olaf (1963) Behandlung lokaler erdmagnetischer Felder als Randwertaufgabe der Potentialtheorie. *Ab. Akad. Wiss. Göttingen, Math-Phys. Kl.* Heft 9.

Hermance, J. F., and Garland, G. D. (1968) Deep electrical structure under Iceland. *J. Geophys. Res.*, **73**, 3797.

Hyndman, R. O., and Hyndman, D. W. (1968) Water saturation and high electrical conductivity in the lower continental crust. *Earth Plan. Sci. Letters*, **4**, 427.

Jones, F. W. (1973) Induction in laterally non-uniform conductors: theory and numerical models. *Phys. Earth Plan. Int.*, **7**, 282–293.

Jones, F. W., and Pascoe, L. J. (1971) A general computer program to determine the perturbation of alternating electric currents in a two-dimensional model of a region of uniform conductivity with an embedded inhomogeneity. *Geophys. J.*, **24**, 3–30.

Jones, F. W., and Pascoe, L. J. (1972) The perturbation of alternating geomagnetic fields by three-dimensional conductivity inhomogeneities. *Geophys. J.*, **27**, 479–485.

Lahiri, B. N., and Price, A. T. (1939) Electromagnetic induction in non-uniform conductors, and the determination of the conductivity of the earth from terrestrial magnetic variations. *Phil. Trans. Roy Soc. A*, **237**, 64.

Lajoie, Jules J. (1973) Ph.D. thesis, University of Toronto, Toronto.

Lajoie, J. J., and West, G. F. (1976) The electromagnetic response of a conductive inhomogeneity in a layered earth. *Geophysics*, **41**, 1133–1156.

Lamb, H. (1932) *Hydrodynamics* (6th ed.). Cambridge Univ. Press, Cambridge.

Lambert, A., and Caner, B. (1965) Geomagnetic depth-sounding and the coast effect in western Canada. *Can. J. Earth Sci.*, **2**, 485.

McDonald, K. L. (1957) Penetration of the geomagnetic secular variation through a mantle with variable conductivity. *J. Geophys. Res.*, **62**, 117.

Parkinson, W. D. (1959) Directions of rapid geomagnetic fluctuations. *Geophys. J.*, **2**, 1.

Poehls, K. A., and Von Herzen, R. P. (1976) Electrical resistivity structure beneath the north-west Atlantic Ocean. *Geophys. J.*, **47**, 331–346.

Porath, H. (1971) Magnetic variation anomalies and seismic low-velocity zone in the western United States. *J. Geophys. Res.*, **76**, 2643–2648.

Porath, H., and Dziewonski, A. (1971) Crustal resistivity anomalies from geomagnetic deep-sounding studies. *Rev. Geophys. Space Phys.*, **9**, 891–915.

Porath, H., Oldenburg, D. W., and Gough, D. I. (1970) Separation of magnetic variation fields and conductive studies in the western United States. *Geophys. J.*, **19**, 237–260.

Praus, O., De Laurier, J. M., and Law, L. K. (1971) The extension of the Alert geomagnetic anomaly through northern Ellesmere Island, Canada. *Can. J. Earth Sci.*, **8**, 50–64.

Price, A. T. (1962) Electromagnetic induction within the earth. In *Physics of geomagnetic phenomena*, ed. S. Matsushita and Wallace H. Campbell. Academic Press, New York.

Rikitake, T. (1966) *Electromagnetism and the earth's interior.* Elsevier, New York.

Rikitake, T., and Whitham, K. (1964) Interpretation of the Alert anomaly in geomagnetic variations. *Can. J. Earth Sci.*, **1**, 35.

Roden, R. B. (1964) The effect of an ocean on magnetic diurnal variations. *Geophys. J.*, **8**, 375.

Runcorn, S. K. (1955) The electrical conductivity of the earth's mantle. *Trans. Amer. Geophys. Un.*, **36**, 191.

Schmucker, I. (1959) Erdmagnetische Tiefensondierung in Deutschland 1957/59; Magnetogramme und erste Auswerkung. *Ab. Akad. Wiss. Göttingen, Math-Phys. Kl.* Heft 5.

Schmucker, U. (1964) Anomalies of geomagnetic variation in the southwestern United States. *J. Geomag. Geoelectr.*, **15**, 193.

Schmucker, U. (1970) Anomalies of geomagnetic variations in the southwestern United States. *Bull. Scripps Inst. of Oceanography*, Vol. 13. University of California Press, Berkeley and Los Angeles.

Schmucker, U., Hartmann, O., Giesecke, A. A. Jr., Casaverde, M. and Forbush, S. E. (1964) Electrical conductivity anomalies in the earth's crust in Peru. *Carnegie Inst. Wash. Yr. Book*, **63**, 354.

Schuebe, Hans-Georg (1958) Die lösungen der Dirichletschen und Neumannschen Randwertaufgen als Hilfsmittel zur behandlung von Problemen des Erdmagnetismus. *Ab. Akad. Wiss. in Göttingen, Math.-Phys. Kl*, Heft 4.

Schuster, A. (1908) The diurnal variation of terrestrial magnetism. *Phil. Trans. Roy. Soc. A*, **208**, 163.

Siebert, Manfred (1958) Die Zerlegung eines Lokalen erdmagnetisches Feldes in äusseren und inneren Anteil mit hilfe des Zweidimensionalen Fourier-Theorems. *Ab. Akad. Wiss. Göttingen, Math-Phys. Kl.* Heft 4.

Srivastava, S. P. (1967) Magnetotelluric two- and three-layer master curves. *Pub. Dom. Obs. Ottawa,* **35**, No. 7.

Stesky, R. M., and Brace, W. F. (1973) Electrical conductivity of serpentinized rocks to 6 kilobars. *J. Geophys. Res.,* **78**, 7614–7621.

Tammemagi, H. Y., and Lilley, F. E. M., (1973) A magnetotelluric traverse in southern Australia. *Geophys. J.,* **31**, 433–445.

Tozer, D. C. (1959) The electrical properties of the earth's interior. In *Physics and chemistry of the earth.* ed. L. H. Ahrens, Frank Press, Kalewo Rankama, and S. K. Runcorn. Pergamon Press, London.

Utsu, T. (1971) Seismological evidence for anomalous structure of island arcs (with special reference to the Japanese region). *Rev. Geophys. Spac. Phys.* **9**, 839–890.

Wait, J. R. (1951) A conducting sphere in a time varying magnetic field. *Geophysics,* **16**, 666.

Walker, G. B., and O'Dea, P. L. (1952) Geomagnetic secular-change impulses. *Trans. Amer. Geophys. Un.,* **33**, 797–800.

Whitham, K. (1964) Anomalies in geomagnetic variations in the Arctic archipelago of Canada. *J. Geomag. Geoelectr.,* **15**, 227.

Whitham, K., and Andersen, F. (1965) Magneto-telluric experiments in northern Ellesmere Island. *Geophys. J.,* **10**, 317.

Winter, R. (1973) *Der Oberrheingraben als Anomalie der Elektrischen Leitfähigkeit.* Dissertation, Universität Göttingen, 117 p.

21 THE MAGNETISM OF ROCKS

21.1 MAGNETIC PROPERTIES OF ROCKS

Within the outermost few kilometers of the earth there are distributed minerals with ferromagnetic properties. The study of rocks containing these minerals has two broad applications in geophysics: the use of the resulting local distortions of the magnetic field to trace structures in the crust, and the analysis of the past history of the earth's magnetic field. In the first application, the magnetic properties of rocks, including susceptibility and permanent magnetization, play a role analogous to density in the study of local gravity anomalies. The second application, which makes use of permanent magnetization alone, constitutes the field of paleomagnetism.

The physics of the magnetization of a system as complicated as an assemblage of crystalline minerals is by no means simple, and a great deal has been written (Nagata, 1961) on this subject alone. However, for our purpose only the outstanding characteristics are required, and these can be traced fairly briefly.

Although most minerals are paramagnetic, virtually all detectable magnetic effects in rocks owe their origin to a very few minerals with ferromagnetic properties. It will be recalled that ferromagnetism is characterized by the parallel coupling of atoms within small volumes known as magnetic domains. Iron is the type case for the effect, and it is not surprising that the iron oxides, particularly magnetite, Fe_3O_4, are among the most important magnetic minerals. Within the crystal lattice of magnetite, iron ions are held in two types of site, in which the directions of atomic magnetic moments are oppositely directed (Fig. 21.1). There is a net effect in one sense, and magnetite in fact belongs to the class of materials known as ferrimagnetic.

In addition to magnetite, other important magnetic rock-forming minerals are hematite, Fe_2O_3, ilmenite, $FeTiO_3$, and a range of compositions between them. Almost the only other ferromagnetic mineral to be encountered is pyrrhotite, $Fe_{1-y}S$, whose magnetic property is important in geophysical exploration for ores.

Rocks are assemblages of minerals, in which the magnetic minerals would normally constitute a small proportion of the whole. These minerals are in the form of grains, which may range in dimension from a few hundred Angstrom units to centimeters. The smallest grains can consist of a single magnetic domain, while larger grains will be composed of several domains. Néel (1955) has shown that many of the magnetic characteristics of rocks may be explained by considering the rock to contain only non-interacting single-domain grains.

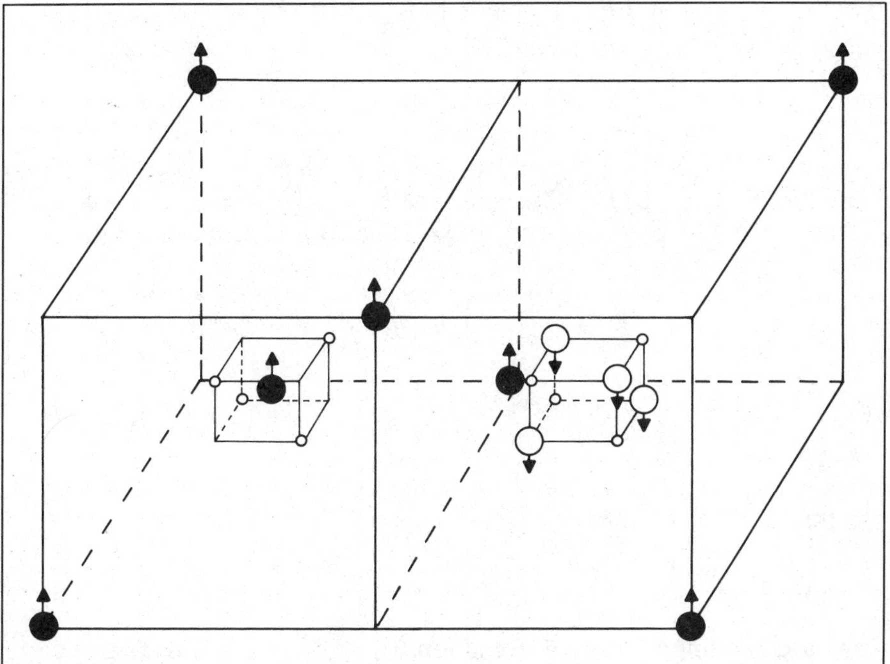

Figure 21.1 One fourth of the unit cell of a crystal of magnetite. Large black circles are iron atoms surrounded by four oxygen atoms; large open circles are iron atoms surrounded by six oxygen atoms. Arrows show direction of spin vectors. Not all oxygen atoms are shown. (After Nagata)

To appreciate Néel's theory, it is appropriate to consider the application of a magnetic field to single-domain particles in four different situations (Fig. 21.2). In the first, we take a distribution of grains, each with complete spherical symmetry and magnetic moment $\underline{\mu}$. These grains behave as paramagnetic atoms, in that the application of a magnetic field tends to rotate the moments, reversibly, from an original random state to the direction of the field. The potential energy of a grain in the field is given by $-\mu H \cos\theta$, where θ is the angle between the

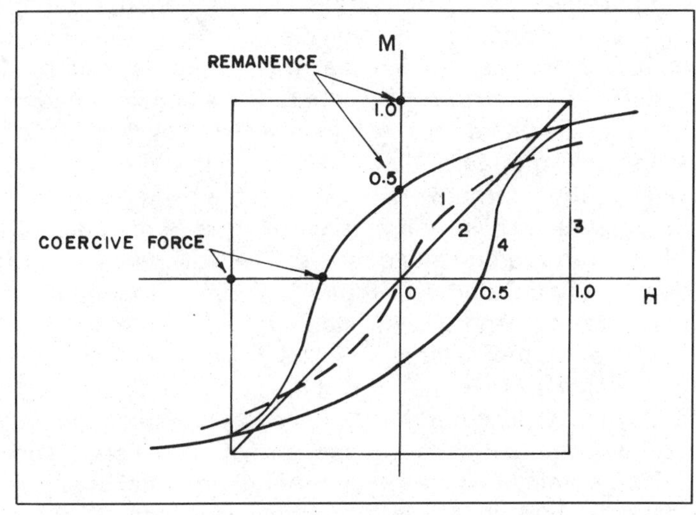

Figure 21.2 Variation of magnetization M (normalized) in the direction of applied field H (normalized) in the case of (1) the Langevin function; (2) a single grain, perpendicular to the easy direction; (3) a single grain, in the easy direction; and (4) a distribution of ellipsoidal particles, as measured by Stoner and Wohlfarth (1948).

moment and field. For an assemblage of N grains, the total magnetization will be

$$M = N\mu \overline{\cos \theta}$$

where $\overline{\cos \theta}$ is the average value over the assemblage. If the grains are in thermal equilibrium and obey a Boltzmann distribution law, the probability of finding a grain with its moment in a given solid angle $d\Omega$ is proportional to $e^{\mu H \cos \theta / kT}$, so that

$$\overline{\cos \theta} = \int_{4\pi} e^{\mu H \cos \theta / kT} \cos \theta \, d\Omega \Big/ \int_{4\pi} e^{\mu H \cos \theta / kT} \, d\Omega \qquad 21.1.1$$

$$= \text{ctnh } a - \frac{1}{a} \quad \text{or} \quad L(a)$$

where $a = \mu H / kT$ and $L(a)$ is known as the Langevin function. The magnetization is then

$$M = N\mu L(a) \qquad 21.1.2$$

For a given temperature T, Equation 21.1.2 gives the (reversible) variation of magnetization M with applied field H. The assemblage is said to be superparamagnetic, because it exhibits the properties of paramagnetism, but with larger individual moments.

Most grains, however, have an asymmetry which, in the simplest case, leads to one "easy" direction of magnetization. The asymmetry may arise from anisotropy of the crystal lattice, or from the shape of the grain itself. The shape effect is intimately connected with *demagnetization,* a property of all permanent magnets. It arises from the tendency of the magnetic poles of the magnet to reduce the intensity of magnetization within itself. For example, in the case of a flat sheet, magnetized transversely to intensity J, the pole strength per unit area of each face is J, and the demagnetizing field within, by analogy to the gravitational case for the attraction of a flat sheet, but with the difference that two sheets are contributing to the field, is $4 \pi J$. For the sphere, we saw, in Section 17.4, that the internal demagnetizing field was $\frac{4}{3} \pi J$. Only in the case of a very long, needle-shaped body does the demagnetization approach 0, for the poles may be considered infinitely distant. Apart from the cases mentioned, only ellipsoids possess an internal field which is uniform through the body and which may be written as NJ, where N is a constant. Suppose that we have such a grain, in the process of being magnetized by the rotation of elementary dipoles (*domains*) from an initially random state. When the intensity has reached the value J, the internal field is NJ, and the average work required to increase the magnetization of a unit volume by dj is $NJdj$, since dj is the moment rotated through an average angle of $\pi/2$. The work required to increase the magnetization to a final value J_f is thus $\frac{1}{2} NJ_f^2$ per unit volume, and this is known as the magnetostatic energy. For stability, the total energy is a minimum, and in the absence of other factors, this is achieved when N is a minimum. Hence, grains of the shape of elongated ellipsoids or needles have the greatest ability to be stably magnetized. In the absence of an applied field, the magnetization direction will coincide with the easy direction. When a field is applied, equilibrium will be obtained when the sum of the potential energy ($-\mu H \cos \theta$) and the demagnetization energy is a minimum; in general, this requires the magnetization direction to be inclined to the easy direction.

For our second situation, consider a single grain with the external field H applied perpendicular to its axis of easy magnetization. As H is increased, the magnetization is rotated, reversibly, toward the direction of H, so that the component of M in the H direction increases linearly with H. If the applied field is along the easy direction of magnetization, a very different situation arises. Even if the field is applied in the opposite sense to the magnetization, if the alignment is perfect, no torques act on the magnetization. However, small perturbations such as thermal agitation cause the magnetization to vary slightly in direction. Then, at a critical value of the applied field, it is rotated through 180°. When magnetization in the field direction is plotted against field, a rectangular hysteresis loop is obtained.

The fourth, and most realistic, situation is that of an assemblage of asymmetric grains, with their easy directions randomly oriented. Application of a magnetic field produces a rotation of the moments, which is partly reversible and partly irreversible. When the field is reduced to zero, the magnetization vectors revert to the easy direction in each grain, but the sense of these vectors is no longer random, being directed toward the sense of H. A hysteresis loop results, with the remanence and coercive force of the assemblage taking values very nearly one-half those found for a single grain parallel to the field.

The effect of grain anisotropy depends upon the volume of the grains. If the volume is small, or the temperature high, the energy associated with anisotropy may be so small compared with that of thermal agitation that the grains behave as superparamagnetic. If an assemblage is given a magnetization at time t_0, the magnetization will be found to decay as e^{-t/τ_0}, where τ_0, the relaxation time, varies directly as grain volume and inversely as the temperature. For a given grain the relaxation time is shortest just below the Curie temperature, the temperature at which ferromagnetic properties are lost. A grain for which τ_0 is long compared to the time of an experiment is said to be blocked or frozen; one for which τ_0 is very short is in equilibrium with the applied field. When an assemblage of grains is cooled through the Curie temperature, it acquires a magnetization in the direction of the field, which on further cooling becomes frozen in. This is known as thermoremanent magnetization (TRM) and is one of the bases of rock magnetism. The process in fact consists of the acquisition of a series of partial thermoremanent magnetizations, each acquired in a given range of temperature, say $T_2 > T_1$, and each independently preserving a record of magnetic field conditions during that interval of temperature (Thellier, 1946). Reheating the sample to temperature T_1 does not affect the particular partial TRM, while heating to T_2 destroys it. For most rocks containing a single magnetic constituent, however, the partial TRM associated with the 50° below the Curie point is dominant.

Néel's theory of single domain particles explains many of the observed properties of TRM in rocks. The greatest limitation appeared to be in the very assumption of single domains, which for magnetite would require the grains to have diameters as small as 0.03 micron. Stacey (1962, 1963), emphasizing this limitation, extended the theory to include the effect of multidomain grains and the transition from a single to a multidomain state. Recently, however, Dunlop (1968), working with artificial assemblages of known grain distribution as well as with natural rocks, has shown that the predictions of the single-domain theory agree very well with experiment, provided the effects of interactions between the grains are included.

The mechanism by which a rock becomes magnetized depends, obviously, on the history of the particular rock. Most igneous rocks have cooled in place through Curie temperatures of their constituent magnetic minerals and any permanent or remanent magnetization observed is thermoremanence. In the case of sedimentary rocks, magnetic grains, whose magnetization is often the legacy of a previous thermoremanence, may be deposited along with other minerals. The assumption made in paleomagnetism is that these grains are mechanically rotated, during deposition, so that their axes are parallel to the ambient magnetic field. It is recognized that subsequent rotation and compaction may produce a departure in the direction of magnetization of a few degrees from that of the field, although this has been observed only in the case of artificially deposited sedimentary rocks (King, 1955).

While the above two processes are the most important for the initial magnetization of rocks, during geological time other causes of magnetization are possible. A non-magnetic mineral may be altered to a magnetic one, resulting in chemical magnetization; and a magnetization, known as viscous magnetization, may be acquired at low temperatures, even in a relatively short time. The latter effect, like demagnetization, results from thermal agitations, and the magnitude of intensity acquired in a given time depends upon the relaxation time τ_0 for the grains involved. The fact that TRM results in the alignment of grains whose time constant at normal temperatures would be very long means that it is a less destructible (i.e., "harder") magnetization than viscous magnetization. It is well to note that while the time constant for single-domain grains at a given temperature increases with grain size, the single-domain theory applies only to grains which are very small, compared to all mineral grains normally found in even a fine-grained rock. Larger grains of magnetic minerals are invariably formed of a number of magnetic domains, and these do not possess the stability of single-domain particles. In general, it may be assumed that if the grain length is less than 0.03 micron, the grain is superparamagnetic; if the grain length is greater than 1 micron, the grain is almost certain to be multi-domain, but the precise transition size depends upon the grain shape.

We have mentioned the important magnetic mineral parameters of saturation, intensity of magnetization, Curie temperature, and coercive force. Unless a mineral is magnetized to saturation, it also acquires an induced magnetization (J_i) in a magnetic field. For small fields, J_i is related to the inducing field F by the linear relation

$$J_i = \kappa F$$

where κ is the (dimensionless) susceptibility of the mineral. The susceptibility of a rock containing a magnetic mineral is given approximately by the susceptibility of the mineral multiplied by the percentage by volume of that mineral in the rock (for a magnetite-bearing rock, κ approaches a value of one in the c.g.s. system as the percentage of magnetite approaches 100), although here again the size and shape of the mineral grains enter into consideration, since demagnetization has an effect on the induced magnetization. It is sometimes convenient to express the remanent magnetization of a rock in terms of the induced magnetization by means of the Koenigsberger ratio Q, defined by

$$Q = \frac{J_r}{\kappa F}$$

where J_r is the intensity of remanent magnetization and F is the total earth's magnetic field where the rock is located.

Because of the importance of the system Fe_2O_3–Fe_3O_4–$FeTiO_3$, and the continuous range of solid solutions between them, the most satisfactory way of displaying the magnetic properties is by contours on ternary diagrams (Fig. 21.3), a method extensively employed by Nagata.

It must be apparent that in dealing with rock magnetism we are faced with a double complexity. Any particular mineral, such as magnetite, may be ferrimagnetic, with oppositely polarized sublattices, while any given rock may contain magnetic minerals, representing different points in the system noted above, or different systems. The fact that a range of Curie temperatures and saturation magnetizations may be represented in one rock leads to a number of possible consequences, of which the most important is reversed thermoremanence (Néel, 1955). Although Néel proposed a number of mechanisms, it is perhaps simplest to consider the case of an igneous rock containing a close intergrowth of elongated grains of two magnetic constituents. When the rock was cooled from a high temperature, grains with the higher Curie temperature were magnetized first. When the temperature passed the Curie point of the second constituent, grains of it would find themselves in the external field of neighboring magnetized grains, and this field would be directed oppositely to that of the earth at the site. If, at low temperatures, the magnetization of the second constituent is the greater, the rock as a whole will display a thermoremanent

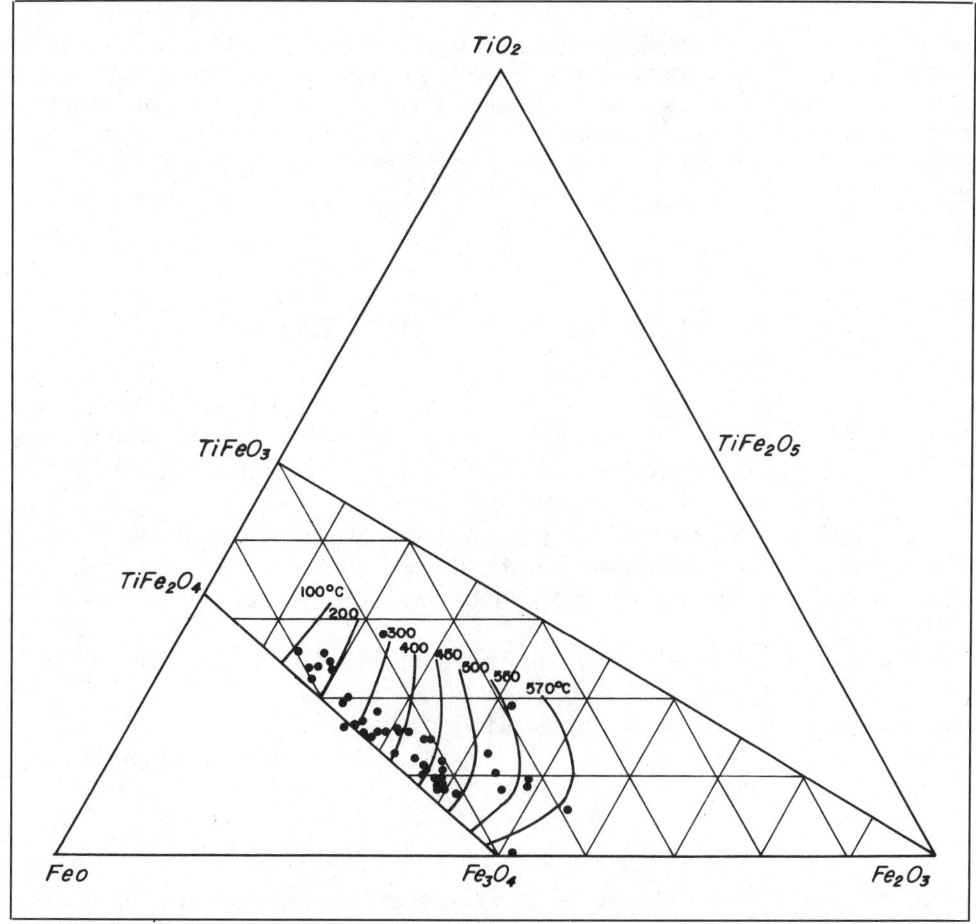

Figure 21.3 Contours of the Curie temperature as a function of composition for the system TiO_2, FeO, Fe_2O_3. (After Nagata)

magnetization antiparallel to the ambient field at the time of cooling. A rock which has been shown in the laboratory to possess this characteristic is the Haruna dacite of Japan (Nagata, 1952). Laboratory tests on other rocks which may be self-reversed have given negative results, possibly because of a change in mineral components upon heating to the Curie temperature. Obviously the possibility of self-reversals is of the greatest importance in considering the reality of apparent reversals in the earth's magnetic field, and we shall consider this in a later section.

The fact that elastic stress can change the susceptibility and remanence of rocks is the reason for an apparent relation between magnetic events and earthquakes. It is known from laboratory investigations (Nagata, 1970a, 1970b) that a uniaxial compression applied in the direction of magnetization reduces both the susceptibility and remanence, while the application of a perpendicular compression increases them. The effects appear to be related to the rotation of magnetic domains, but they are complicated by containing both reversible and irreversible components. Using earlier values of the parameters connecting magnetization and stress, Stacey (1964) calculated the change in magnetic field intensity to be expected over various models of faults. Assuming an average intensity of magnetization in the local crust, he predicted peak-to-peak changes in the magnetic field of the order of 10 gammas. Small changes in the magnetic field associated with, or preceding, seismic activity have been reported for the Matsushiro earthquake swarm of Japan (Rikitake, 1968), the Alaska earthquake of 1964 (Breiner, 1964) and the San Andreas fault (Breiner and Kovach, 1970). There is obviously great promise for the phenomenon as a predictor of earthquakes in active areas, but the isolation of small magnetic signatures from the background of magnetic variations is not simple. It is also not completely clear whether the principal effect will be found to be related to the increase in stress before a shock, or the release of stress at the time of faulting.

21.2 INSTRUMENTS FOR MEASURING ROCK MAGNETISM

The measurement of the remanent magnetization of a rock requires the determination of the direction and magnitude of the magnetic moment of a rock sample of some convenient volume, perhaps a few cubic centimeters. In many cases, this will require the measurement of a moment of the order of 1×10^{-6} emu, for which the most satisfactory techniques that have been employed are the astatic deflection magnetometer and the spinner or rock generator.

If a magnetic needle of moment M is suspended on a fiber with a torsion constant σ in the horizontal component H of the earth's field, it will oscillate about the meridian, when deflected, with a period

$$T = 2\pi\sqrt{\frac{I}{MH+\sigma}} \qquad 21.2.1$$

where I is the moment of inertia of the suspended system. To increase the sensitivity of deflection by a sample brought near it, we should make M large,

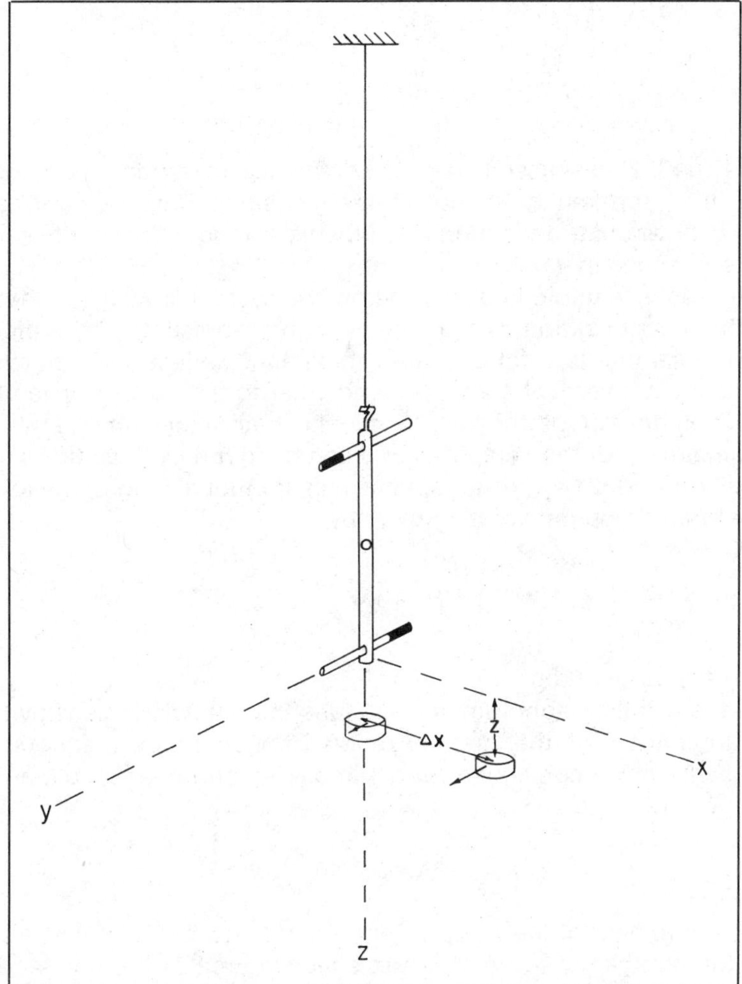

Figure 21.4 The astatic magnetometer, with samples in the central and displaced positions.

but this increases the restoring couple also, unless, as in Figure 21.4, two magnets of equal and opposite moment are suspended. If the value of H is the same at both magnets, the restoring torque is controlled only by σ, and the period becomes

$$T = 2\pi\sqrt{\frac{I}{\sigma}} \tag{21.2.2}$$

In this condition, the system is said to be astatic. If a sample is brought near the lower magnet, and if the dimensions are such that the sample has no effect on the upper magnet, the system will be deflected through an angle θ where

$$\sigma\theta = MF, \tag{21.2.3}$$

F being the horizontal field at the lower magnet produced by the sample. The sensitivity of the system is

$$\frac{\theta}{F} = \frac{T^2 M}{4\pi^2 I} \quad 21.2.4$$

Blackett (1952) discussed in great detail the methods of optimizing the sensitivity. It is apparent that the highest sensitivity involves a long natural period. As in all angular deflection instruments, a long optical path is employed to measure changes in θ.

In use, a small sample is placed below the system, initially on the axis. We suppose the magnetization of the sample to be represented by a dipole at its center. If the sample is rotated until a maximum deflection is observed, this dipole must lie in a vertical plane perpendicular to the needle system, and the horizontal component of the dipole can be determined. It follows that the vertical component of the sample dipole can be found by measuring in the off-center position (Fig. 21.4), for in that case an additional field at the lower magnet, normal to it, is operative; it is given by

$$H = 3p_z \frac{\Delta x}{z^4} \quad 21.2.5$$

where p_z is the vertical dipole moment.

While the astatic magnetometer provides the ultimate in sensitivity, for the rapid determination of the magnetization of moderately magnetic rocks a *spinner* may be more convenient. If a sample is spun near a coil, an emf E is generated, where

$$E = FNA\omega \times 10^{-8} \text{ volts}$$

in which F is the field of the sample perpendicular to the axis of rotation at the coil, N is the number of turns, A is the effective area of the coil, and ω is the angular velocity of rotation. A single coil with sufficient amplification would be sensitive to transient magnetic variations; but suppose two identical coils are arranged, so that the sample may be spun near one, while a reference magnet of known moment is rotated near the other. The differential output of the coils is insensitive to transients, and gives the ratio of magnetic moments perpendicular to the axis. Furthermore, if the difference in phase of the individual outputs is measured, the orientation of this component of the sample moment, relative to the reference magnet, is determined. Rotations about two perpendicular axes in the sample are therefore sufficient to completely determine the vector magnetization.

For the highest sensitivity in the measurement of remanence, a superconducting magnetometer (Collinson, 1975) is now available. In this instrument, the sample is introduced into a coil maintained at liquid helium temperature. The component of magnetization of the sample induces a persistent current in the coil, with the magnitude of the current independent of the rate at which the sample is introduced. This current is measured through the magnetic field it produces in a second coil and the further use of superconductor technology. Finally, superconducting shields are maintained around the detecting head, to reduce external fields to the order of 10^{-6} gamma. It is argued that intensities as low as 10^{-10} emu can be measured on a sample of 1 cm^3 volume.

For the determination of the susceptibility of a sample, some form of inductance bridge is normally used. If a sample completely fills a coil which was initially air-cored, the coil's self-inductance is increased by a factor equal to the permeability (μ) of the sample. The susceptibility κ is then determined from the standard relation

$$\mu = 1 + 4\pi\kappa \qquad 21.2.6$$

The important factor to consider in susceptibility determinations is the field to which the sample is actually exposed, since susceptibility is field-dependent. For most purposes, of course, it is susceptibility in the earth's field which is desired.

21.3 THE TECHNIQUES AND RESULTS OF PALEOMAGNETISM

We have seen that both the thermoremanent magnetization of igneous rocks and the detrital magnetization of sedimentary rocks preserve, in most cases, a record of the direction of the magnetizing field at the time the rock was formed. The first requirement for the establishment of the earth's field at some point in geological time is the collection of a set of samples, oriented in space, from the formation to be studied. Usually, these will be small cubes cut out of the rock in place, with the horizontal plane and geographic north marked on each cube. If it is obvious that the rock has been deformed since formation, as in the case of a dipping sedimentary layer, the original horizontal plane, as indicated by the bedding, is also marked. It is also usual, in the case of a layered rock, to sample over a vertical interval which would represent a few thousand years in time, so that when the results of the samples are averaged, the effect of secular change is removed. Core drilling is an effective way of sampling over a vertical interval, but the cores are rarely oriented in azimuth, and only the dip of the magnetization vector can be determined.

Measurement of the vector magnetization of the samples by the methods of the previous section yields a number of directions in space. These may be conveniently represented by a stereographic projection of the end points of a unit vector aligned along the directions of magnetization (Fig. 21.5). The grouping of points on the diagram then expresses the degree of consistency in the determinations.

Because a rock may acquire a "soft" component of magnetization at any time after its formation, it is normal for the initial results to show a good deal of scatter. Fortunately, it is possible to remove the soft components, without destroying the harder detrital or thermoremanent magnetization, by subjecting the sample to an alternating magnetic field which is gradually decreased to zero while changing its orientation in a random way. This operation, known as cleaning, is of fundamental importance to the obtaining of useful paleomagnetic results, and the acceptance of paleomagnetism as a scientific technique may be said to date from the recognition of this importance. An example of the effect of cleaning is shown in Figure 21.5.

If one obtains a consistent set of directions, different from the present direction of the earth's field at the site, this is taken to indicate stability of the remanence. Actually, there are a number of more elaborate tests for stability (Irving, 1964) which may be applied in specific cases. Perhaps the most convincing of these is the folded sedimentary layer, in which samples taken from differ-

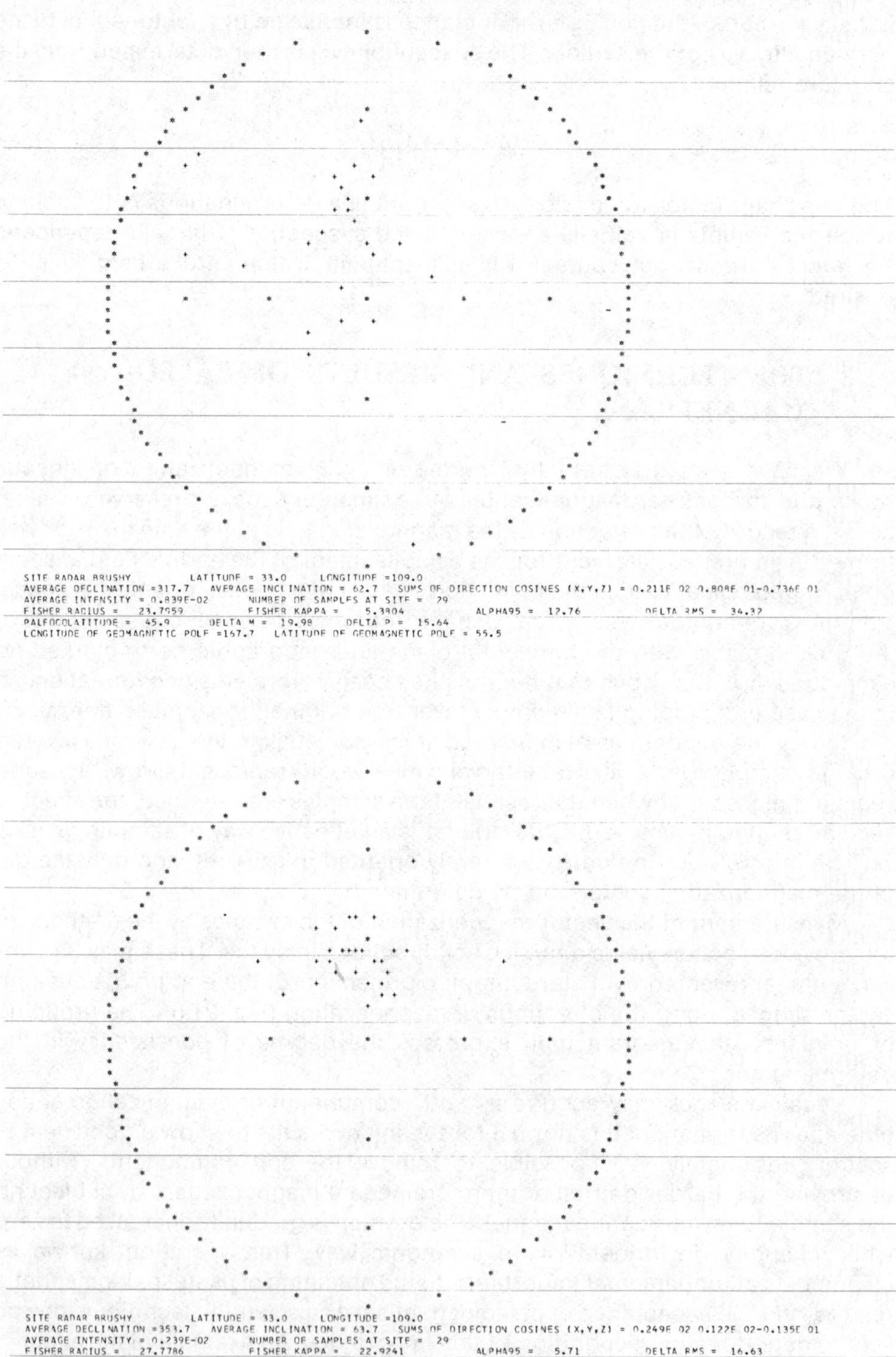

Figure 21.5 Computer plot of the projection of magnetization vectors, before (*Top*) and after cleaning. The computer program also gives the equivalent pole position. (Courtesy of D. W. Strangway)

ent limbs of a fold show random directions. When the samples are rotated by the amounts needed to "unfold" the stratum, the directions become parallel.

The mean direction of magnetization from a consistent group is used to infer the equivalent pole positions, on the assumption of a dipole field. It follows from the potential of the field on a sphere (Equation 17.4.2) that the inclination I and the colatitude θ of the site are related by

$$\tan I = 2 \cot \theta \qquad 21.3.1$$

This equation may be said to be the basic one of paleomagnetism. It shows that, in principle, an *ancient pole position* may be determined from a single paleomagnetic measurement, by measuring, from the sampling site, an arc equal to the colatitude θ along the azimuth of the intensity vector (Fig. 21.6). If, as in the example shown, this location does not fall in the region of the present poles, there is evidence of polar wandering, or motion (subsequent to magnetization) of the land mass from which the sample was taken, or both. We shall return below to the question of separating true polar wandering from plate motions. For the moment, let us see how much information can be obtained on the ancient orientation and location of the sampled plate. It was mentioned, in Section 1.4, that all motions on a sphere are equivalent to rotations; there must be some rotation of the plate which would restore the apparent pole position to the north pole. The necessary rotation is indicated in Figure 21.6, and the center for this rotation is a point on the equator, 90° in longitude west of the ancient pole position (McElhinny, 1973). This rotation, applied to the plate, puts the land mass in its correct orientation relative to the north pole. The plate could, however, be rotated about the north pole, resulting in a change of longitude, with no change of pole position occurring. The maximum information that is obtainable from a single determination of ancient pole position is therefore the latitude and orientation, but not the longitude, of the sampled plate. In practice, the magnetization directions on the original stereogram define an area rather than a point, and the ancient pole is determined as an area on the globe; statistical methods are available to determine the size and shape of this area for adopted confidence limits (Irving, 1964).

Although it had been known for many years that the directions of magnetization of many rocks did not correspond to the present directions of the geomagnetic field at their sites, it was not until the mid-1950s that sufficient paleomagnetic data became available to suggest that the poles had moved in a systematic way over the surface of the globe. The improvement in the data avail-

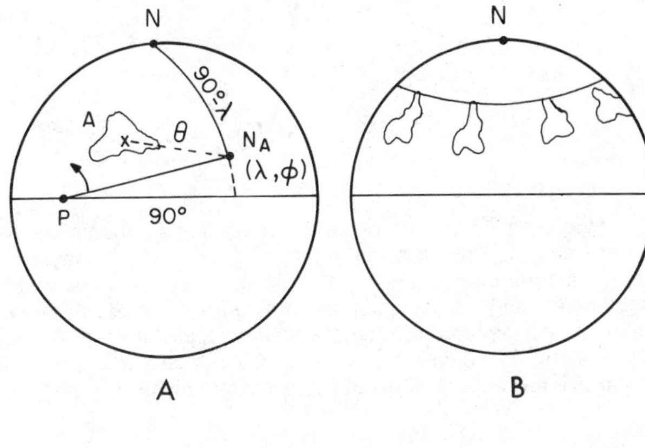

Figure 21.6 (a) The rotation necessary to bring a pole position N_A, determined from a sample site on continent A, to the north geographic pole. (b) Possible positions and orientations of A. (After McElhinny 1973).

able came as a result of a number of factors; for example, the realization of the fact that the most stable rocks are often those with fairly low values of magnetization, coupled with the availability of more sensitive magnetometers to permit the measurements. In any case, a growing number of determinations of pole positions for samples of different geological ages permitted the constructions of *apparent polar wandering curves* (Fig. 21.7), suggesting that the poles had moved relative to the earth's surface. Further determinations, in which results from different continents were separated, led to different apparent polar wandering curves for different continents, also shown on Figure 21.7. It is characteristic of these curves that they are not well-determined before the beginning of the Paleozoic, because of the relative scarcity of suitable Precambrian samples. The path of one of the magnetic poles over the earth is a global phenomenon, and should be independent of the site of the observer. The fact that polar wandering curves for different continents do not agree is the essential paleomagnetic evidence for continental drift. It is possible to infer the relative movement of different continental blocks, in different intervals of geologic time, by forcing the polar wandering curves to come together.

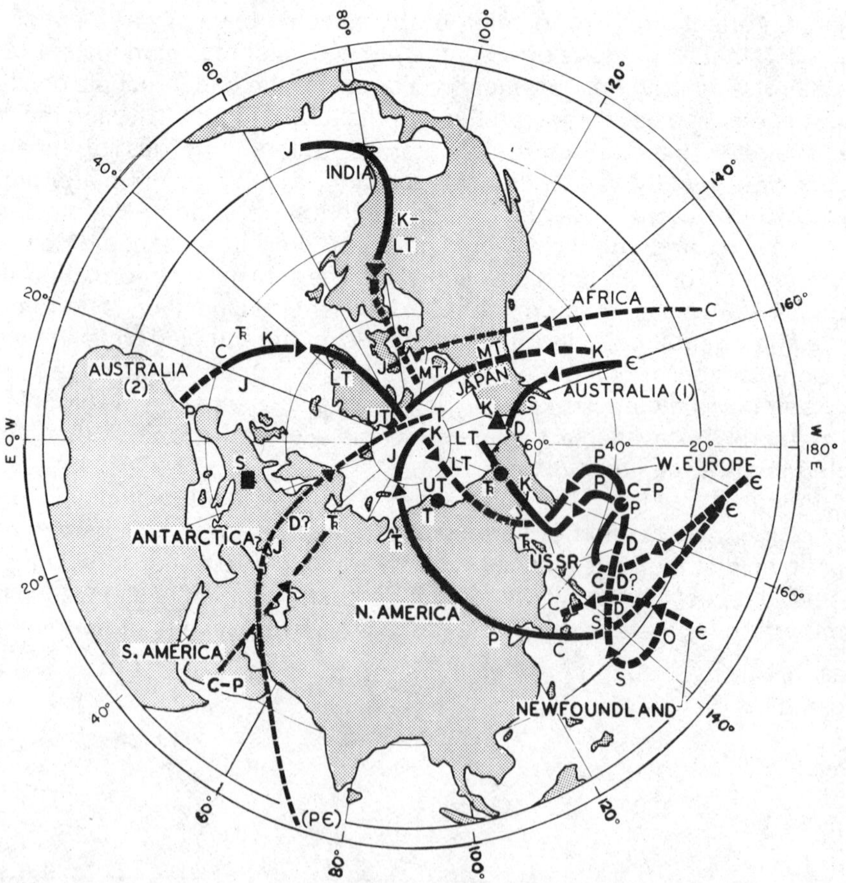

Figure 21.7 Curves of polar wandering for samples from different continents. Polar movement since the Precambrian, relative to various land masses. Solid curves have been traced where the palaeomagnetic data from three or more levels in geological time follow a fairly consistent sequence. Single pole positions are relative to: ■ China; ● Greenland; ▲ Madagascar. Letters refer to: (PЄ), Precambrian; Є, Cambrian; O, Ordovician; S, Silurian; D, Devonian; C, Carboniferous; P, Permian; Tr, Triassic; J, Jurassic; K, Cretaceous; LT, MT, UT, Lower, Middle, and Upper Tertiary. Polar azimuthal projection of the present Northern Hemisphere. (Courtesy of E. R. Deutsch)

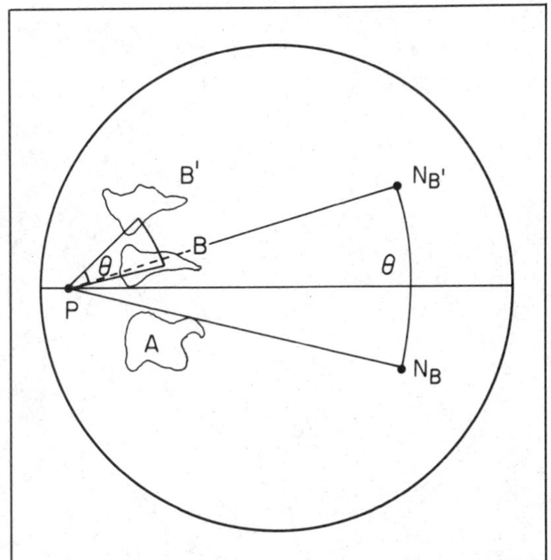

Figure 21.8 Apparent polar wander versus continental displacement. Because continent B is close to the pole of relative rotation between A and B, its linear displacement is much less than that of the pole N_B. (After McElhinney 1973)

In assessing relative plate motions from apparent polar wandering curves, it is again necessary to keep in mind that these motions are relative rotations, and that the location of the rotation pole for plate motion has a great effect in determining the differences between two different apparent polar wandering curves (McElhinny, 1973). For example, Figure 21.8 shows two plates, *A* and *B*, in relative rotation about a point on the equator close to *B*. If plate *A* remains fixed relative to the north pole while *B* rotates, the pole positions determined from samples on *B* change through large arcs. Greatly different apparent polar wandering paths would result, whereas the linear displacements of *B* relative to *A* are rather small. The converse is the situation already mentioned in connection with Figure 21.6: plate rotations about a pole at the magnetic pole produce no change in the apparent polar wandering path.

Apparent polar wandering paths from different plates may be identical up to some time in geological history and then diverge (Fig. 21.9), giving evidence of the break-up of a plate or supercontinent. Conversely, two paths originally distinct in character may come together, pointing to the coalescence of separate plates.

For the unambiguous interpretation of apparent polar wandering paths, it is imperative that the samples be adequately dated and that the dates refer to the same point of time in the sample's history as the magnetization. When the dating is less certain, segments of the paths may often be connected in different ways. An outstanding example is the track of the north pole determined from late Precambrian samples taken from North America (Fig. 21.10). Most samples give pole positions falling on a single track involving a number of "loops" or hairpin turns. However, samples taken from the structurally distinctive Grenville province of eastern North America yield pole positions which appear to form a distinct segment. The age of the Grenville magnetization is not well established, but it is believed to be between 900 and 1100×10^6 years. However, until further information is available, there are at least three obvious ways to connect the segments shown in Figure 21.10 (Stewart and Irving, 1974; Irving, Emslie and Ueno, 1974; Buchan and Dunlop, 1976). The Grenville segment could be incorporated into a single path if the magnetization dated from be-

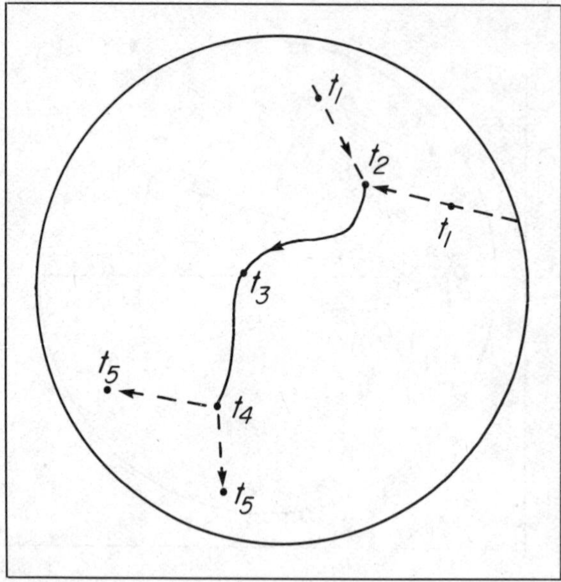

Figure 21.9 Apparent polar wander paths for successive times $t_1 \ldots . t_5$. At time t_2, two plates merged; at time t_4 the new plate divided.

tween 1200 and 1400 $\times$ 10^6 years (Case 1) or from between 1000 and 800 $\times$ 10^6 years (Case 2). Both of these connections involve considerable excursions in the path and would require rapid drifting of the inferred single North American Precambrian plate. The third possibility is that the Grenville poles really do re-

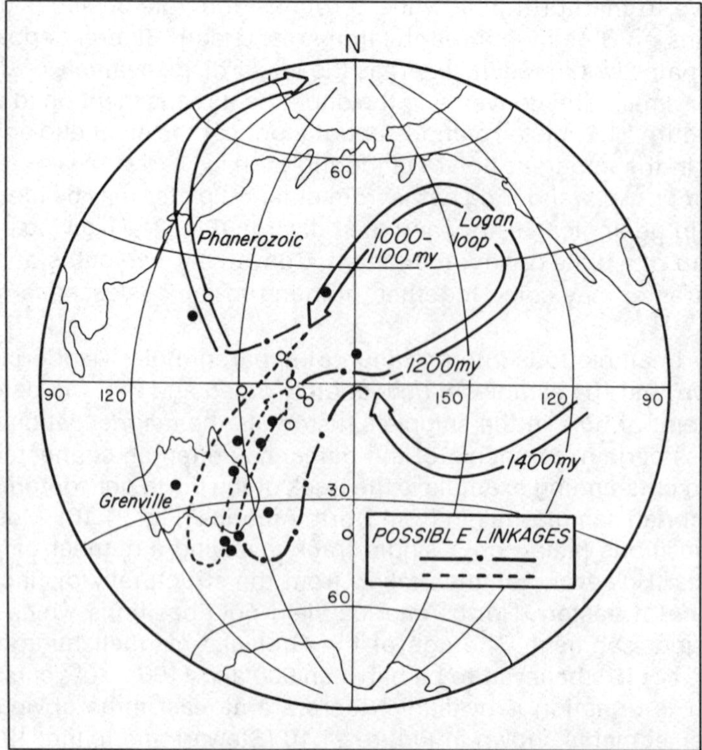

Figure 21.10 Apparent polar wander paths determined from North American samples for the late Precambrian. Samples from the Grenville structural province define the broken arrow. Whether or not the Grenville is assumed to represent a distinct plate depends upon the linkage adopted.

flect the motion of a separate plate and that the junction of the paths reflects the coalescence of the Grenville province against the rest of the North American Precambrian. On this hypothesis, poles from near the junction could represent magnetization acquired by metamorphism accompanying the collision. The structural implication of this hypothesis (Irving, Emslie and Ueno 1974) is that the Grenville was formed as much as 100° in arc from the rest of North America and that it moved through this distance to a collision. Choice between these (and other) hypotheses awaits more definitive dating of the age of magnetization. It is difficult to overstress this point, since both the apparent age of a rock and its magnetization can be affected by metamorphic and thermal events during its history.

We should emphasize also that apparent polar wandering paths are determined only at discrete points in time. It is erroneous to assign a uniform apparent velocity to them, as the apparent polar motion (i.e., plate motion) might better be represented by long fixed periods, separated by intervals of rapid motion (Irving, 1968).

The most important early results from paleomagnetism were an apparent westward movement of North America by perhaps 4000 km since Carboniferous time, a complicated movement by Australia around the south geographic pole, and a northward movement of India through 50° of latitude since Cretaceous time. These indications appeared to be established beyond any uncertainty in the positions of the individual polar wandering curves. On a global basis, a major contribution of paleomagnetism has been in providing control on the reassembling of continents as they were before the current episode of spreading, and we shall return to this point in Chapter 27. It should be kept in mind, however, that all samples older than 200,000 years are indeed from present *continents* (although they may include portions of *ancient* ocean floor) so that they may not provide the optimum distribution of points for reassembling ancient *plates.* There are, in addition, a great number of problems of major tectonic interest, but of less than global significance, which lend themselves to paleomagnetic investigations. These include, for example, studies on the possible rotation of a subcontinental land mass such as Spain or Newfoundland. It is impossible to comment here upon all of these investigations, and the detailed interpretations will, in all probability, be subject to future change as more results are obtained.

We return now to the question of "true polar wander." The early paleomagnetic measurements were made before the full implications of plate tectonics were recognized, and magnetic pole positions relative to sample sites were thought to define a movement of the poles in relation to the entire earth's surface. Acceptance of the fact that portions of the lithosphere have had very large displacements with respect to each other led to emphasis on the point that a polar wandering path based on samples from one plate is only "apparent," and gives essentially the history of displacement of that plate relative to the magnetic pole. The question that may still be asked is whether the magnetic poles have moved relative to the earth as a whole. It is important to recognize the significance of the question. First, the poles of the dipole magnetic field are believed to closely coincide with the geographic poles, when the effects of secular change are removed by averaging over several centuries. This is based partly on paleomagnetic observations on very young geological and archeological material and partly on the probable role of the earth's rotation in determining the nature of the core dynamo. Secondly, apart from the known effects of precession (Section 1.3), the earth's spin axis is taken to be along a direction fixed in space. True polar wandering is therefore equivalent to a "slipping" or rotation of the body of the earth relative to this fixed direction. However, there is a further restriction, in that there is no completely accepted frame of reference for the body of the earth beneath the lithosphere (this is equivalent to the statement that absolute plate motions cannot be determined: Chapter 27), and all of our observations relate to points on the lithosphere. Finally, therefore, true polar wander must be defined as the rotation of either the body of the earth or the lithosphere as a whole, relative to the spin axis.

McKenzie (1972) pointed out that if there is no true polar wander, the vector sum of all plate velocities relative to the pole will be zero. Consequently, the vector sum of all plate displacements relative to the pole, for a given interval of geological time, will be zero. He suggested that polar wandering is a useful concept only if, when the mean of plate velocities relative to the pole is subtracted from each, relative plate motions are greatly reduced. This analysis showed that polar wandering over a given interval would be determined, provided the positions of all plates relative to the pole were known at the beginning and end of the interval. In forming vector sums and means, McKenzie proposed that velocities be weighted by the area of their respective plates, so that a small, fast-moving plate would not have an undue influence. Jurdy and Van der Voo (1974, 1975) have extended the principle to provide a method that requires only one reliable pole position, relative to one plate, at the beginning and end of the time interval, provided the relative displacements of the plates from each other are known. For the past 100 million years, the latter are given by the ocean-floor magnetic spreading pattern. The method of Jurdy and Van der Voo involves reassembling the lithospheric plates, on the basis of the spreading information, for selected times and determining a vector displacement $\underline{F_i}$ relative to the mean magnetic pole, for each of N plates of area Δs for the intervals between these times. They then sought a vector displacement $\underline{G}$, resulting from a rigid-body rotation of the entire lithosphere to the spin axis, such that the sum

$$\sum_{N} (\underline{F_i} - \underline{G})^2 \, \Delta s$$

over the entire earth is minimized. From the values of $\underline{G}$ so found, the true displacement of the pole could be expressed in terms of arc length and direction. Their analysis indicated surprisingly small components of true polar wandering: 2.0° for the past 55 million years and 6.3° for the past 115 million years. The extension of their method to earlier times requires a knowledge of the sea-floor spreading history for those times, and this is not yet available in sufficient detail.

The small amount of true polar wandering is remarkable when one considers the very large relative displacement of plates for the same period. We have already noted that the spreading velocities are of the order of centimeters per year, so that plate displacement of thousands of kilometers must be expected in a period of 115 million years, and this will be illustrated in the reconstructions of Chapter 27. A relation between plate motion and true polar wandering might well be expected, if the earth's spin axis is the axis of greatest moment of inertia (Section 1.3) and if plate motion produces a redistribution of mass and an alteration of the principal moments of inertia. We can only conclude that, perhaps because of the operation of isostasy, large scale plate motions can take place without significant alteration of these moments.

The great interest stimulated by paleomagnetic measurements on rocks should not be allowed to obscure the important results of research conducted on other materials. For the study of the time-changes in the field through the centuries immediately preceding the availability of direct observations, measurements on archeological materials (Aitken, 1974), lake sediments (Creer et al, 1976) and varve clays (Johnson et al, 1948) have all proven useful. For the first, the most useful are the rocks of a kiln, still in place, which have acquired a TRM during the operation of the kiln. Materials fired in the kiln, such as pots, may also be found to possess thermo-remanent magnetization, from which the ancient field inclination, but not azimuth, may be determined.

Measurements on lake cores giving inclination, relative declination and intensity have been important in establishing a time variation in the field, with periods longer than that of the conventional secular change. For example, from 9 meters of core from Lake Michigan, spanning approximately 11,000 years, Creer et al (1976) report swings in the relative declination with an average period of approximately 2000 years (based on C^{14} dating). If the non-dipole portion of the geomagnetic field is represented by a distribution of vertical dipoles, near the core boundary (Section 18.4), these swings could result from an oscillation in the strength of these dipoles, but, in contrast to the conventional secular change, of shorter period, a westward drift (Section 18.5) of the sources of disturbance is not suggested by the observations.

Varve clays are glacial meltwater deposits, marked by annual layers, which may be found to have a depositional remanent magnetization. They offer the unique advantage that the sampling of successive bands permits the year-by-year record of changes in the field direction to be traced. The pioneer work referred to above, based on varve clays from New England, dated at 9,000–15,000 years BC, suggested that the mean geomagnetic pole position, averaged over several centuries, is close to the geographic pole. In other words, because of the westward drift of the secular change and its integrated effect at the rate of about 0.2 degrees of longitude per year, the field averaged over periods in excess of 1800 years is close to that of an axial dipole. But a completely satisfactory reconciliation between this result and the characteristics of the longer-period variations found in lake sediments, has not yet been accomplished.

For the complete description of the ancient field, an estimate of intensity as well as of pole position is required. Koenigsberger (1938) proposed that, in the case of TRM, the ancient intensity T_E should be obtainable by heating a sample to above the Curie temperature in the laboratory, and cooling to room temperature in a known field T_L. Then, if J_N is the original TRM intensity, and J_L is that acquired in the laboratory, the following proportionality should apply:

$$J_L / J_N = T_L / T_E$$

Unfortunately, very few natural materials have a sufficiently simple magnetization history for this proportionality to be valid, and to yield accurate estimates of the ancient field strength. Thellier and Thellier (1959) introduced a modification to the technique, which has the great advantage that the observations themselves may be used to confirm the reliability of the result. The Thellier-Thellier method is based on the concept of independent partial TRMs (Section 21.1). For any interval of temperature, T_1 to T_2, the partial TRM acquired in the laboratory, J_e, should bear a constant ratio K to the partial TRM J_n that was acquired when the rock originally cooled through the same interval. The partial TRM J_n can be determined in the laboratory by noting the magnetization *lost* when the sample is heated through the interval (or heated and then cooled in zero field), while J_e is obtained by cooling through the interval in the laboratory field T_L. Coe (1967a, 1967b) has discussed the procedure and its limitations in detail. If the reduced magnetization found on heating through successively greater intervals of temperature plots linearly against J_e for the same intervals, the slope of the line is $(-K)$, and the ancient field is given by

$$T_E = K\,T_L$$

Samples which yield a straight line over the entire temperature range from room temperature to the Curie temperature are considered ideal for paleointensity estimates. Those which show departures from linearity at the high- or low-temperature end probably have a complex TRM or contain secondary magnetizations, but estimates of the ancient field strength may be made from the linear portion of the plot.

The principal operational disadvantage of the Thellier-Thellier method is that it is very time-consuming, and relatively few rock samples have been subjected to it. Evidence of an ancient field, reduced to the contemporaneous equator, significantly less than the present value, could be related to the decrease in field during a reversal (Section 21.4). There have been very few, if any, suggestions of an ancient field greater by as much as a factor of two from the present value. It appears that, at least since late Precambrian time, the geomagnetic dipole moment, when the field is not reversing, has been a relatively constant parameter of the earth.

21.4 REVERSALS OF THE MAGNETIC FIELD

An example of a mechanism by which a rock could acquire a remanence in the opposite sense to that of the ambient field was given in Section 21.1. When paleomagnetic directions are analyzed, it is found that there are many examples which give an intensity vector of opposite sense to that expected. Many of these rocks do not appear to have the particular mineralogical characteristics necessary for self-reversal, and this suggested that the field itself had reversed at intervals through geological time. However, the mere fact that self-reversal is possible means that isolated observations of reversed polarity are not sufficient to establish an alternating geomagnetic field. The proof of an alternating field lies in the correlation in age of reversely magnetized samples from different areas, and the establishment of a time scale for reversals.

For the establishment of such a scale, the ideal situation would be to date each reversal by the methods of geochronology (Chapter 23). Cox and Dalrymple (1967) analyzed the potassium-argon ages and polarities of 88 samples of volcanic rocks, ranging in age up to 3.6×10^6 years. In so doing, they made a careful estimate of the standard deviation of the best potassium-argon age determination, and this turned out to be approximately 2 per cent of the measured age. There is thus a fundamental obstacle to extending the potassium-argon dating to very much older samples: the absolute uncertainty, which is thus 72,000 years at an age of 3.6×10^6 years, becomes greater than some of the polarity epochs which are being studied. The most detailed available scale based on radioactive dating results from an extension to 4.5×10^6 years, published by Cox (1969) and shown in Figure 21.11.

For the establishment of a reversal scale for earlier times, other methods of dating must be used. Some extension is possible through the use of deep-sea

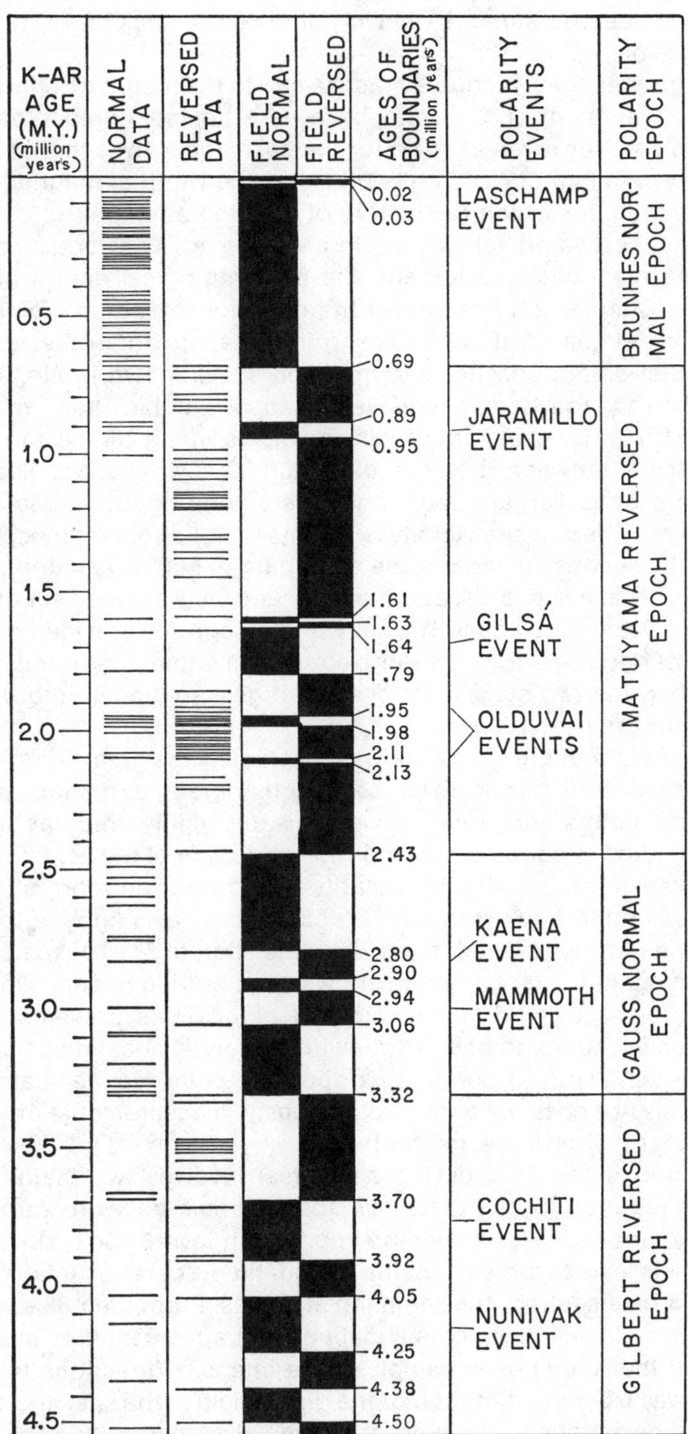

Figure 21.11 Geomagnetic polarity time scale for the past 4.5 million years, based on potassium-argon dating of volcanic rocks. (From Cox, *Science* **163**; copyright 1969, by the American Association for the Advancement of Science)

sediment cores (Foster and Opdyke, 1970). Sedimentation in the oceans proceeds at rates from less than 5 to about 100×10^{-6} m/yr, and cores up to 28 m in length have been obtained. Extension of the scale back to 9×10^6 years was therefore possible.

For still greater ages, recourse must be had to the magnetic character of the ocean floor, with control provided by a very limited number of terrestrial samples. The pioneer reversal scale for the past 80 m.y. was that of Heirtzler et al (1968), shown in Fig. 21.12a. It was based on the width of magnetic stripes in the South Atlantic, for which the history of opening appeared to be simple, so that a constant spreading velocity could be assumed. The acquisition of many more oceanic magnetic profiles and the paleontological dating of deep sea drill samples (Chapter 27) has shown the need for revision of some details of this scale. LaBrecque et al (1977), using information from other oceans, and some terrestrial data, published the modified scale shown in Fig. 21.12b. The terrestrial data is from a sequence exposed at Gubbio, Italy, measured by Lowrie et al (1976). This sequence yields reversals which can be correlated with the marine scale, between the ages of 65 and 80 m.y., and it is also known to span the Cretaceous-Tertiary boundary. A very good tie to the geological time scale is thus provided, in the vicinity of anomaly 29. For older times, Larson and Hilde (1975) have constructed a scale extending to 160 m.y. in age, and shown in Fig. 21.12c. This scale is based on a clear and well-mapped set of lineations in the Pacific Ocean floor near Hawaii, with age control provided by paleontological ages of deep sea drill cores into equivalent anomalies found near Japan. Taken together, the two scales just described give an almost unbroken record of reversals for the past 160 m.y.

An examination of Fig. 21.12 shows immediately the great complexity of the reversal process. Normal and reversed periods of long and short duration are intermixed in an apparently random way. It is this highly irregular behaviour in time which studies of coupled dynamos for the source of the field (Section 17.5) have attempted to model. The irregularity would be even more striking if the scale were extended back through Paleozoic time. It is known, for example, that within the Permian period, there is a time span of 5×10^7 years, known as the Kiaman magnetic interval, which shows no alteration in field sense. But the construction of a scale for Paleozoic and older times presents many difficulties. Not only is there no ocean floor with preserved lineations, but for earlier Precambrian there are no continuous apparent polar wander paths. Without these paths it is not possible to tell from intensity measurements on an isolated sample whether it shows reversal or not.

The nature of the field during a reversal involves two major questions: What time is required for the reversal? Does the field decay to zero to grow in the opposite direction, or does the dipole flip over? Cox and Dalrymple (1967), in their establishment of the dated time scale, estimated the time required for a reversal from the sampling statistics. From their 88 samples, they searched for those samples actually dating from a reversal. For this selection, they required that the chosen samples have intensity directions two standard deviations away from the direction of the dipole field at the site, and ages within two standard deviations of the date of a reversal on their time scale. Only one sample of the 88 met both criteria, showing that the total time during which the field is transitional is very short. In fact, if T is the total time interval involved and ΔT the time during which the field is transitional, it may be shown that

$$\frac{\Delta T}{T} = \frac{R}{N}$$

where R is the number of samples dating from a reversal and N is the total number of samples. Applying the method to the additional data of Cox (1969) gives the remarkably short time of 2000 years within which a field reversal is accomplished.

Of importance to the second question are measurements of the intensity of magnetization of rocks from a sequence which includes a reversal. Measurements on lavas from Africa (Van Zijl et al., 1962), Japan (Momose, 1963), and elsewhere have shown that the intensity does not drop to zero, but to 20 to 40 per cent of normal. This has been interpreted as evidence that the non-dipole portion of the field does not vanish during a reversal, and that the rocks become magnetized in the local non-dipole field. On the other hand, there is evidence that the dipole portion decayed to zero before growing in the opposite sense, and did not flip over without change in magnitude. In the latter case, the maximum decrease would be 50 per cent (the difference in dipole field between equator and poles), which would not be compatible with the observations on intensity (Smith, 1967). The relation of the rapidity of reversal to the rate of polar wandering is not well known. Cox (1968), by considering the probability of a reversal to be greatest when the non-dipole field is large relative to the dipole field, has proposed a model in which the probability of reversals is variable through geological time. However, there is as yet no independent evidence on the variability of the ratio of non-dipole to dipole field.

Statistical examination of the reversal sequence to locate hidden periodicities is an obvious approach, but Phillips and Cox (1976), discussing previous studies and generalizing the method of analysis, conclude that there is no evidence for the periodic occurrence of reversals. Lengths of intervals are randomly distributed; the probability of any given length occurring is extremely small, but the relatively great expanse of geological time leads to some occurrence of that length. On the other hand, if the average length of interval is computed for time spans of 10^6 years, the averages are found to change through geological time (Cox 1975). For example, approximately 45 m.y. ago the mean frequency of reversal changed from approximately one to three reversals per million years.

The lack of any regular periodicity in the reversal time scale has led to some difficulty in nomenclature. When reversals were first recognized, it was believed that a rather regular succession of periods of about 10^6 years was present, with polarity predominantly in one sense during each period. These intervals were termed epochs and the first four, counting back from the present "normal" polarity direction, were named: Brunhes (Normal), Matuyama (Reversed), Gauss (N), Gilbert (R). The term *event* was applied to shorter period reversals within the epochs. It is now recognized, as Figures 21.11 and 21.12 show, that it is impractical to distinguish between epochs and events.

Finally, it is worth noting that the magnetization vector of the dipole field at present, from geographic north to geographic south, is opposite to the angular momentum vector and for this reason the present "normal" polarity has been called "negative."

21.5 MAGNETIC ANOMALIES

In this section we consider the local effect on the geomagnetic field of a body of rock in the outer part of the earth, when the magnetization of that body

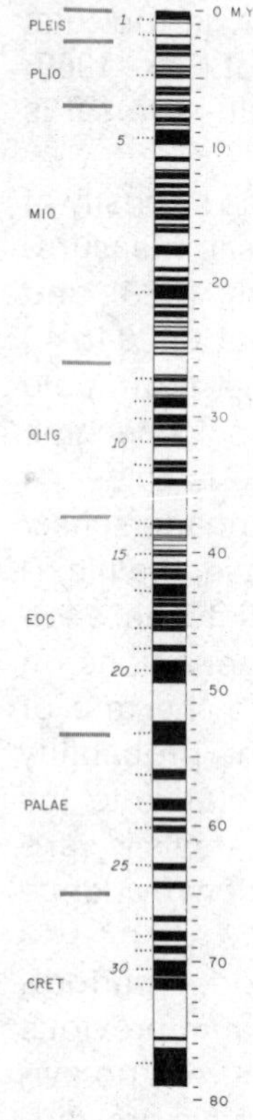

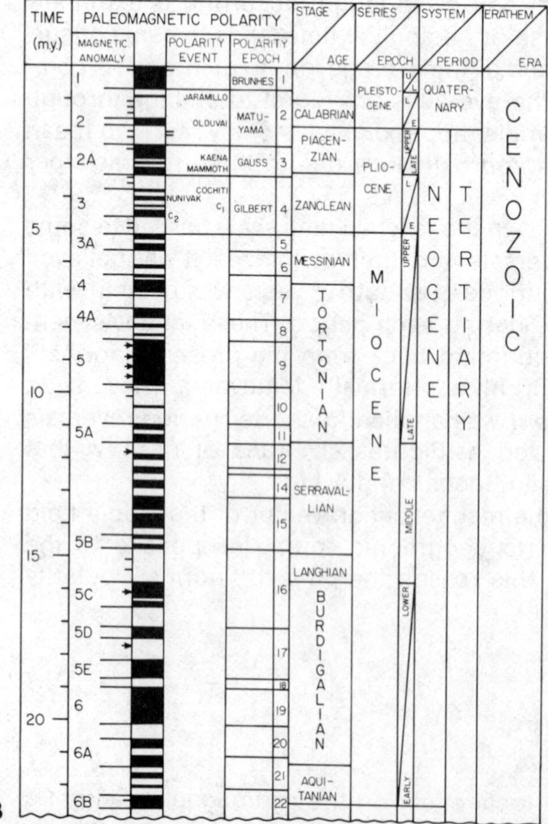

Figure 21.12 Geomagnetic polarity time scales. (*A*) for the past 80 million years, based on ocean-floor magnetic anomalies (From Heirtzler et al 1968) (*B*) for the time since late Cretaceous (From La Brecque et al 1977) (*C*) for the late Jurassic and early Cretaceous (From Larson and Hilde 1975). Courtesy of the authors, and for *A*, the American Geophysical Union.

Illustration continued on opposite page

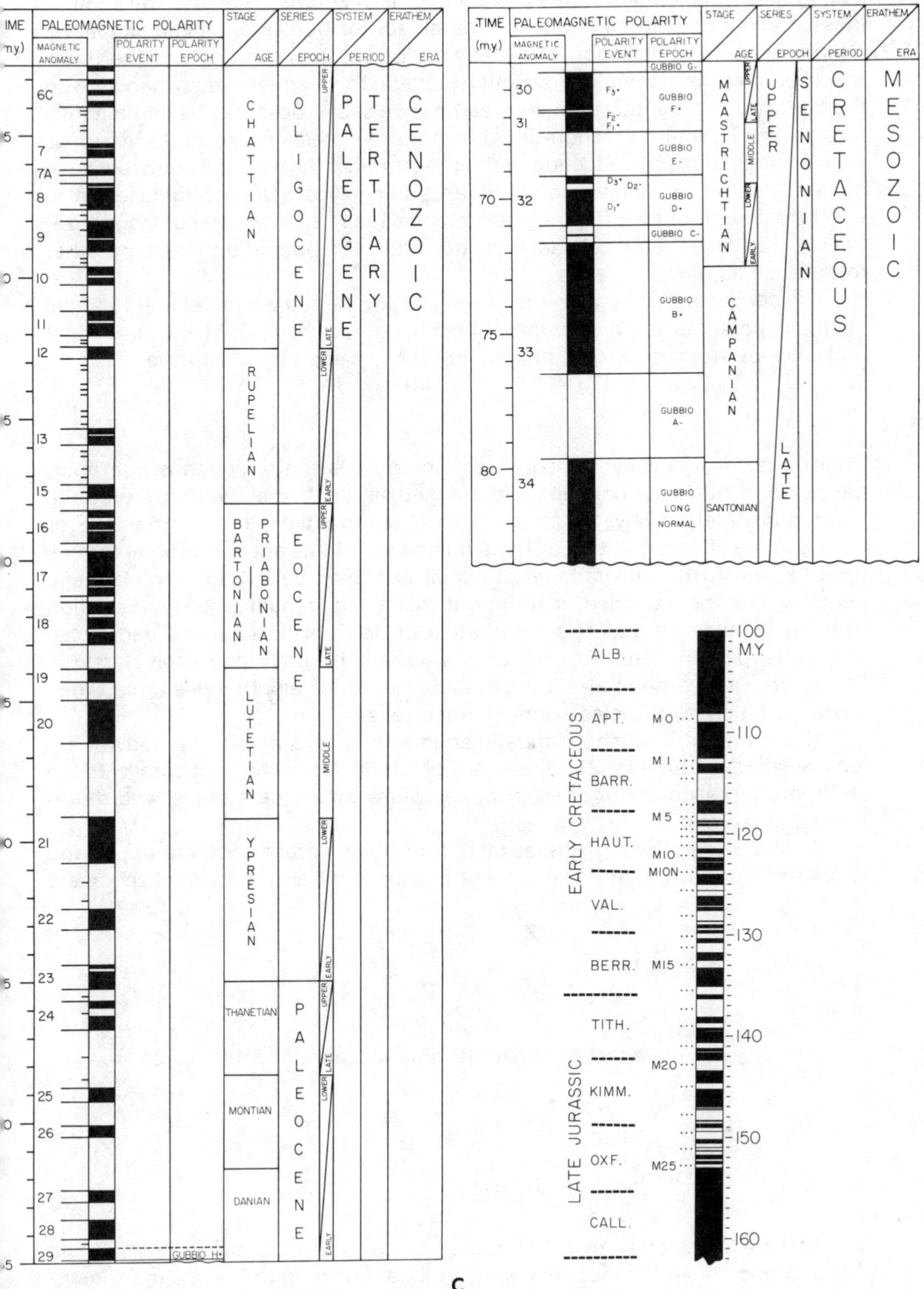

C

Figure 21.12 *Continued.*

differs from that of its surroundings. The effect is estimated in a manner which is analogous to that employed in computing gravity anomalies, in that the anomalous potential or field component is obtained, for bodies of simple shape, by integration of the attraction of a volume element over the limits of the body. However, there is a complication in the fact that the magnetic effect depends on the direction of magnetization, and, as we have seen, rocks may be permanently magnetized in some direction different from the present field of the earth, as well as magnetized by induction in the present field. Strictly speaking, because of demagnetization effects, only spheres and ellipsoids lend themselves to exact treatment, but for the magnetizations usually encountered demagnetization is neglected, and calculations are made for bodies bounded by other forms such as plane surfaces.

It is convenient to begin with a result already mentioned in Section 15.2: the magnetic potential V of a uniformly magnetized body is given by the directional derivative of the gravitational potential U of the same body. We have

$$V = -J\frac{\partial U}{\partial i} \tag{21.5.1}$$

where J is the intensity of magnetization and i is the direction of the magnetization. If the body contains two magnetizations, J may be considered the vector sum; alternatively, the calculation of the potential may be carried out for each and the effects combined. The significance of this equation is that for any form for which the gravitational potential has been calculated, the magnetic potential may be obtained. It turns out that the potential given by Equation 21.5.1 is identical to that of a distribution of poles, with a surface density $\pm J$ unit poles per unit area, over each surface of the body perpendicular to J. Planes on which south poles are distributed produce an attractive force, while those on which north poles are distributed repel.

The expression for the magnetic anomalies due to a dike-like body of extensive strike length (Fig. 21.13) is easily obtained. Magnetization parallel to the strike produces no distribution of poles over a bounding surface, and therefore no attraction.

Simple integrations of the attraction of poles spread over the upper and lower surfaces show that the anomalies produced by vertical magnetization are

$$\left.\begin{aligned} \Delta Z_z &= 2J_z(\theta_2 - \theta_1) \\ \Delta X_z &= 2J_z \ \ln\frac{r_2 r_3}{r_1 r_4} \end{aligned}\right\} \tag{21.5.2}$$

Similarly, for the effect of horizontal magnetization the anomalies are

$$\left.\begin{aligned} \Delta Z_x &= 2J_x \ \ln\frac{r_2 r_3}{r_1 r_4} \\ \Delta X_x &= 2J_x(\phi_1 - \phi_2) \end{aligned}\right\} \tag{21.5.3}$$

These expressions can be combined to give the total attractions, ΔZ and ΔX, of a body in an inclined magnetizing field. Furthermore, since many magnetic surveys, particularly over the oceans, are made with instruments which measure the scalar change in the total intensity, it is necessary to express the

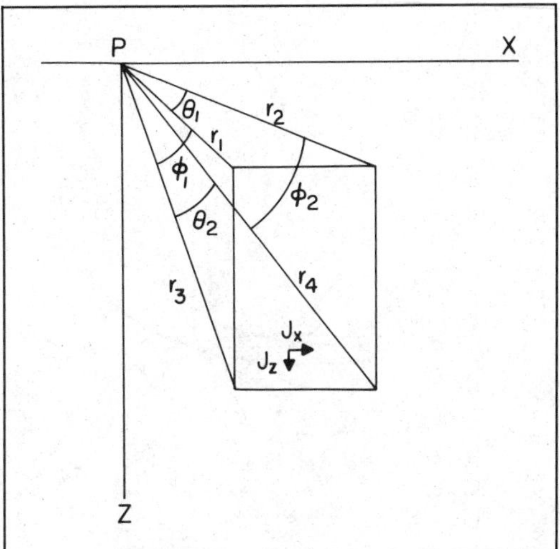

Figure 21.13 Quantities involved in the calculation of the magnetic anomaly over a tabular body, of infinite extent normal to the page.

latter quantity in terms of ΔZ and ΔX. It is

$$\begin{aligned}\Delta T &= \Delta H \cos i + \Delta Z \sin i \\ &= \Delta X \cos i \cos A + \Delta Z \sin i\end{aligned} \qquad 21.5.4$$

where i is the inclination of the field, and A is the angle between the x-axis (perpendicular to the strike) and the magnetic meridian. Equations 21.5.2, 21.5.3, and 21.5.4 are the basic ones applied for the computation of the anomalies to be expected over any section of the crust characterized by tabular bodies with steeply dipping boundaries, and have been widely employed in the interpretation of ocean surveys. The chief feature to note is that a given geometry can produce very different anomalies, depending on the relative importance of J_x and J_z; furthermore, the response of the total field magnetometer depends upon the local value of i, and the orientation of the structure. In general, the angular terms of Equations 21.5.2 and 21.5.3 are symmetric, while the logarithmic terms are antisymmetric. Vertical component anomalies produced by steeply-dipping bodies are symmetric when the bodies are magnetized by induction in high latitudes, or when the strike is parallel to the magnetic meridian, for in these cases there is no horizontal magnetization. Furthermore, Equation 21.5.4 indicates that in high latitudes ($i \geqslant 60°$, say), the total field anomaly resembles the vertical component anomaly, while near the magnetic equator ΔT resembles the horizontal component anomaly. Also, in the case of a body magnetized by induction near the magnetic equator, it is the contribution ΔX_x which is dominant. Typical forms of anomaly curves in these cases are shown in Figure 21.14. The caution to be applied is to recognize that intensity J resulting from remanence may be in a very different direction from the present earth's field. This is particularly applicable to the ocean floors, whose magnetization appears to be dominantly from remanence.

Expressions for more general forms, suitable for machine calculation, have been given in works on geophysical prospecting (Bahattacharyya, 1964).

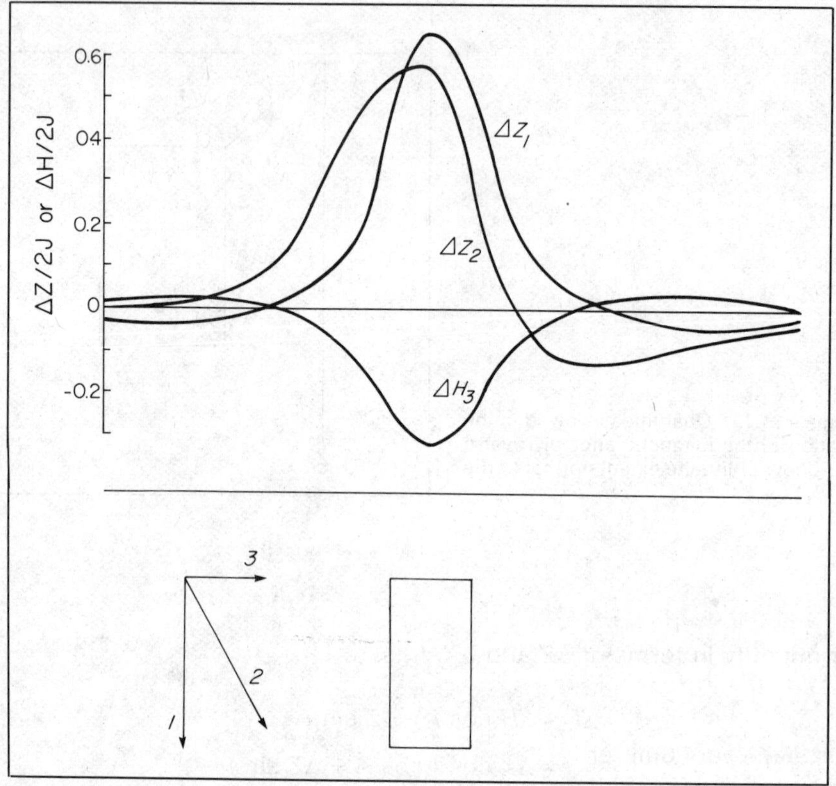

Figure 21.14 Typical magnetic anomaly component profiles across an east-west striking block, for the inclinations shown. Total-field magnetic anomaly profiles are identical to cases 1 and 3, and resemble 2.

21.6 THE MAGNETIC CHARACTER OF CONTINENTS AND OCEANS

It follows from Equations 21.5.2 and 21.5.3 that for a rock mass to produce a magnetic anomaly of, say, 500 gammas (1 per cent of the earth's main field), the intensity of magnetization must be of the order of 5000×10^{-6} emu. Virtually no sedimentary rocks possess a remanence of this magnitude, nor do they have sufficient susceptibility to acquire the magnetization by induction. Igneous and metamorphic rocks may possess both types of magnetization.

Within the continents, the largest areas of igneous and metamorphic rocks, either exposed or concealed beneath sedimentary deposits, are of Precambrian age. These areas form the shields of many continents. It is found that the Koenigsberger ratio of these old rocks is relatively low, and that, in general, remanent magnetization is small compared to induced magnetization. Even in rock units containing significant remanence, it appears to vary in direction to such an extent that its contribution to the anomaly produced by a large mass of rock is small. There are exceptions, of course, for certain inversely magnetized Precambrian dikes are known to produce extensive negative anomalies. But in general, the magnetic character of the continents appears to be governed by variations in susceptibility.

It was mentioned in Section 21.1 that the susceptibility of many rocks could be obtained in terms of the magnetite content. An empirical relation which may be useful is

$$\kappa = 3000 \times 10^{-6} \times (\text{volume percentage of magnetite})$$

Tabulated values of susceptibilities (Clark, 1966) for igneous and metamorphic rocks range from 0 to 30,000 $\times$ 10^{-6}, with a general increase toward more basic rocks. However, it must not be thought that representative values can be given to the particular rock types. The evidence for this is in the continental magnetic maps themselves (Fig. 21.15), which show an extremely irregular pattern, even when reduced to a small scale. If magnetic properties were uniform throughout various rock types, the maps would divide into relatively few regions of different magnetic field strength. In spite of this, magnetic maps of continental areas have the ability to depict major structures, as Figure 21.16 indicates, and can be extremely useful in tracing these structures, even to the extent of suggesting the former fit of separated continents (Strangway and Vogt, 1969). It is probable that more use of them will be made in the future, when the techniques developed for geophysical prospecting are applied to broader-scale studies. The very complexity of the field, on all scales, demands that statistical methods of filtering and trend detection be applied.

An example of a possible application of magnetic surveys is the mapping of the Curie point isotherm. It has been shown (Alldredge and Van Voorhis, 1961) by a study of anomaly wavelengths that there are probably no sources of magnetic anomaly between a depth of a few tens of kilometers and the core boundary. The lower limit of crustal anomalies should correspond to the isothermal

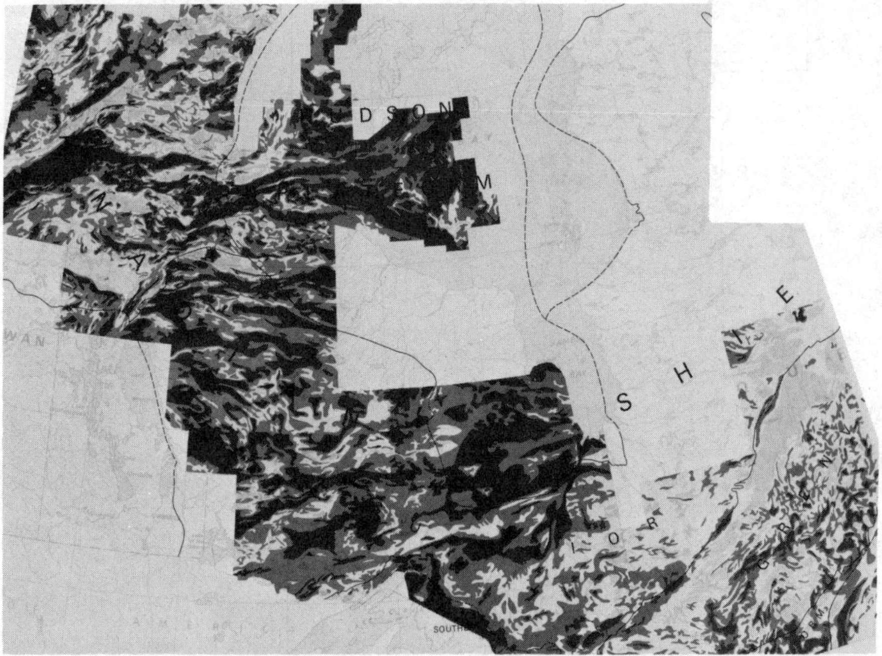

Figure 21.15 Composite aeromagnetic map of a portion of the Canadian Precambrian shield. The map is produced from large-scale sheets, which were based upon surveys with flight lines at a spacing of about one mile. Intervals of 200 gammas between shades; darkest represents greatest total field. (Courtesy of the Geological Survey, Dept. of Energy, Mines and Resources, Canada)

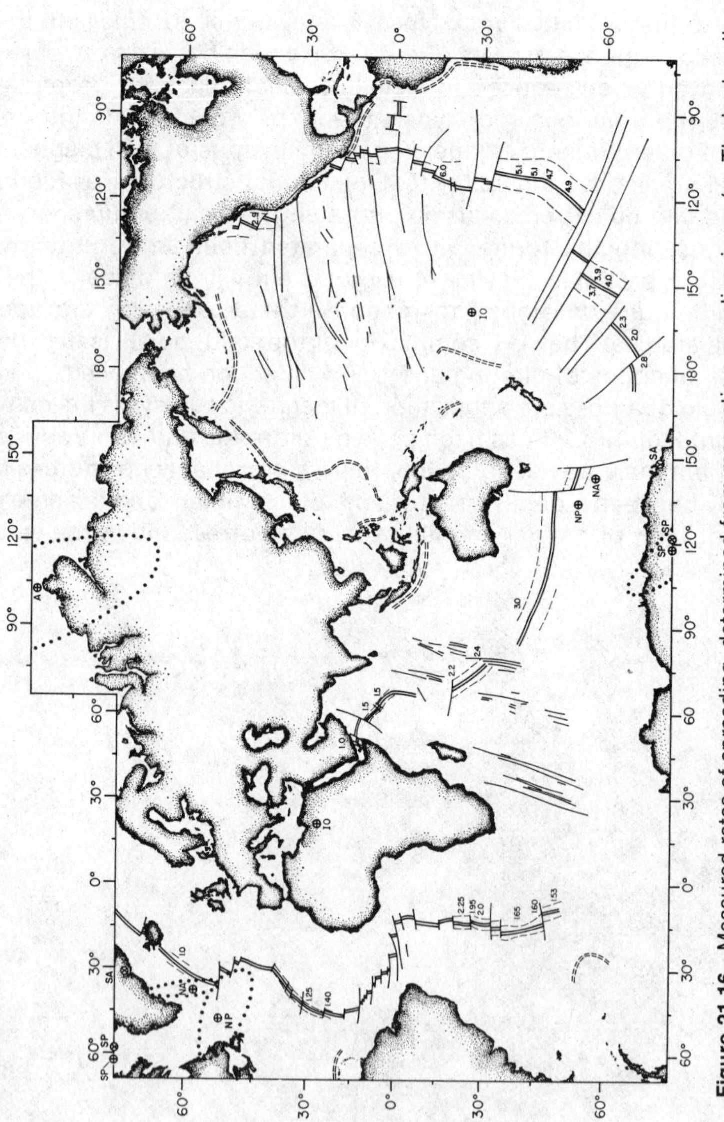

Figure 21.16 Measured rates of spreading, determined from magnetic anomalies, in cm/yr. The broken line on each side of ridges gives the position of "anomaly 5," 10 million years old. Single lines are fracture zones. (Courtesy of X. LePichon and the American Geophysical Union)

surface for the Curie point of magnetite, and this surface could be expected to be warped, depending on geothermal conditions. An upward warping would be equivalent to a slab of material negatively polarized relative to the crust above, and its effect could be estimated by Equation 21.5.4. Published examples of the possibility of Curie point effects, such as on the anomalies measured during a transcontinental survey in the United States (Pakiser and Zietz, 1965), are suggestive but not yet conclusive. Pakiser and Zietz observed that crustal magnetic anomalies were stronger east of the Cordillera, and they suggested that from the mountain front westward, the Curie point isotherm could be warped into the lower crust.

While the continents display a pattern of magnetic anomalies that is irregular, the ocean floors produce quite the opposite effect. When magnetic surveys of the ocean first became available, a linear trend of magnetic anomalies was recognized (Vacquier, 1959). Vine and Matthews (1963) explained these in terms of the permanent magnetization of ocean floor basalt, in contrast to the induced magnetization of continental rocks. The fact that the oceanic anomaly trends were parallel to mid-ocean ridges, and symmetrical on either side of these ridges, suggested to Vine and Matthews that oceanic basalt, injected along the central rifts and magnetized in the direction of the earth's field at the time of cooling, is divided into strips moving in either direction away from the rift. If this is the case, the width of successive strips could be expected to be proportional to the length in time of successive magnetic polarity epochs. Conversely, the age of a particular reversal, taken from the potassium-argon dating of young rocks, together with the distance from the ridge axis to the corresponding strip, yields a velocity of spreading of the sea floor. Velocities of 1 to 8 cm per year for different ridge systems have been found. Spreading rates estimated magnetically for the principal ridges are shown on Figure 21.16. In addition to those shown, a number are now available for more minor spreading centers. A corollary of the principle of ocean-floor spreading is that the age of the oceanic crust is given by the correlation of magnetic stripes. (The 10 million year old line is shown on Figure 21.16.) Pitman et al. (1974) have published a "geological map of the oceans" on which the age of oceanic crust is indicated. The oldest ocean floor is probably located in the northwest Pacific.

The original predictions of the magnetic character of the ocean floor have been amply verified by subsequent surveys (Heirtzler et al., 1968). In all cases, a most striking symmetry of anomalies has been found, together with a correlation of wide and narrow stripes with the measured lengths of epochs. Indeed, in a subject where order-of-magnitude agreement has often been accepted as a confirmation of an argument, the beautifully simple behavior of the ocean floors is remarkable (Fig. 21.17).

Most workers on ocean-floor magnetic anomalies have calculated model cross sections whose magnetic effects, computed according to Equation 21.5.4, agree with the observed profiles (Fig. 21.18). Typical parameters which must be inserted to obtain a fit (Loncarevic, Mason, and Matthews, 1966) are 5×10^{-3} emu for the polarization, and 2 to 3 km for the depth to the bottom of the magnetized blocks. Over the axis of all ridges a positive anomaly, corresponding to the youngest material, magnetized during the present epoch, is observed. Away from the axis, the amplitude decreases, as might be expected, since the older material has been subject to the demagnetizing effect of field reversals. Before drill core samples from the ocean floor were available, the magnetization of dredged samples was studied (Opdyke and Hekinian, 1967; Irving, Robertson,

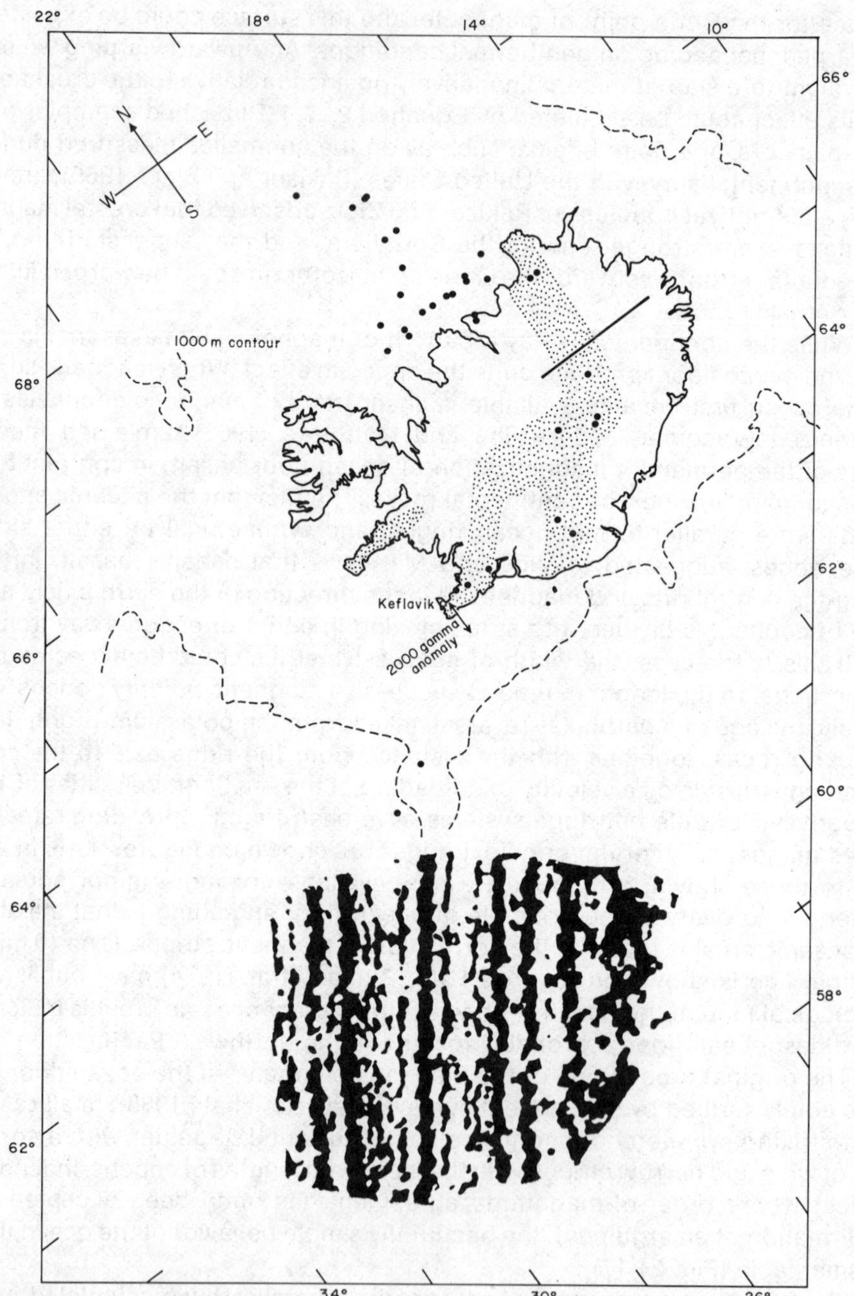

Figure 21.17 Pattern of magnetic anomalies observed over the Reykjanes ridge, south of Iceland. (Courtesy of J. R. Heirtzler)

and Aumento, 1970). The measurements of Irving et al. indicated that the mean intensity of magnetization of basalts from the Mid-Atlantic Ridge was 92×10^{-4} emu; in other words, ample to explain the observed pattern of magnetic anomalies. Samples from the median valley, within 6 km of the ridge axis, had a mean intensity of 574×10^{-4} emu, in accord with the greater amplitude of the central anomaly. It was suggested that viscous demagnetization, probably accelerated by the increased temperature of the axial zone, decreased the magnetization rather rapidly in this zone.

There is, however, an objection to the hypothesis of viscous demagnetization as the cause of the decrease in intensity away from the axial zone, pointed out by Marshall and Cox (1972), who offered an alternative explanation. These authors noted that most of the decrease in intensity occurs within 10 km of the ridge axis, and that within this limit the oceanic crust has throughout its lifetime been exposed to a field in the present direction. For viscous demagnetization to take place, a magnetized body must stand in zero field or in a field opposed to the magnetization. As an alternative explanation, they propose the

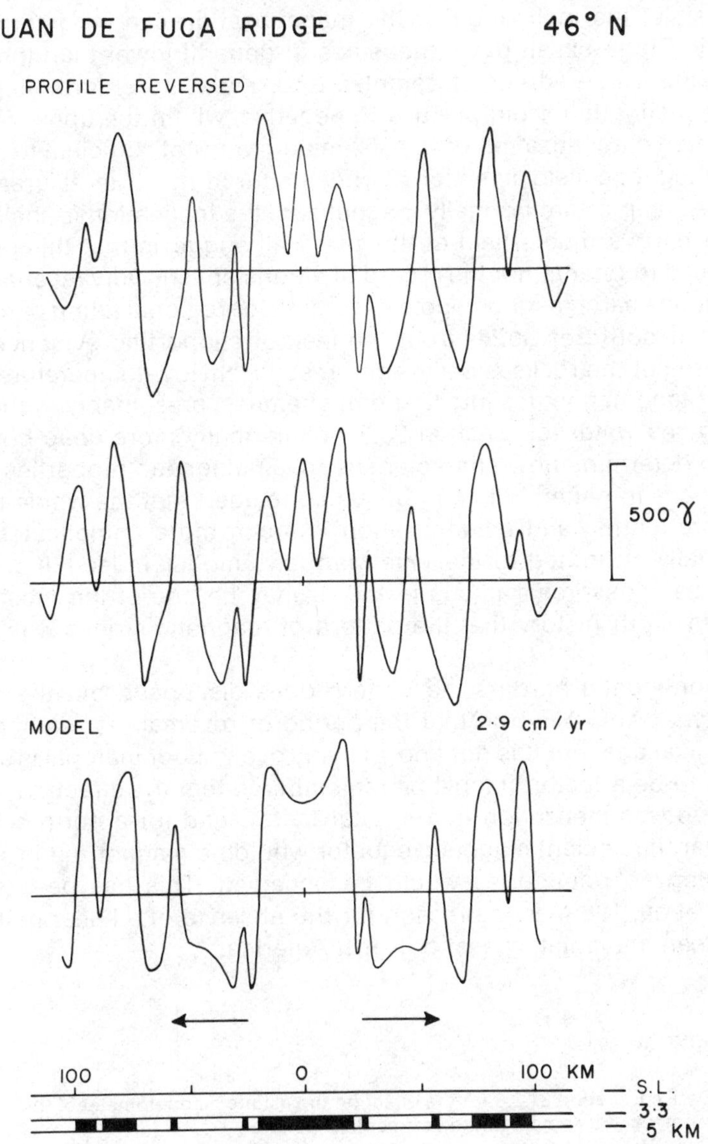

Figure 21.18 Total field magnetic profile across the Juan de Fuca ridge (*Middle*), with its mirror image (*Top*), to indicate the high degree of symmetry. The lower profile is computed for the model shown at bottom; a spreading rate of 2.9 cm/year is indicated by comparison with the time-scale of geomagnetic reversals. (Courtesy of F. J. Vine)

decrease in intensity through oxidation of titanomagnetite to titanomaghemite during submarine weathering. Measurements on 12 samples selected over a total range of about 8 cm on a fragment of partially weathered pillow basalt showed a decrease in remanent magnetization from about 320×10^{-4} emu for an unweathered sample to 180×10^{-4} emu in the weathered rim. Submarine weathering must therefore be accepted as a major factor in determining the magnetic properties of the ocean floor.

The first samples of oceanic basalt provided by the Deep Sea Drilling Project confirmed the importance of weathering, but also pointed out the hitherto unsuspected complexity of the magnetized ocean floor (Hall, 1976). In the vicinity of the Mid-Atlantic Ridge, samples were obtained at depths up to 582 meters below the top of the basalt. While complete orientation of drill samples is not possible, the inclination of the magnetization can be determined. The Mid-Atlantic Ridge basalt gave intensities in general lower than those quoted above for the earlier dredged samples and, even more surprising, showed changes in inclination, from positive to negative, within the upper 500 meters. The integrated magnetization of the 500 meters was not sufficient to explain the observed magnetic field lineations, whose source must lie at greater depth. Even where samples are normally magnetized, the inclination is shallower than that of the earth's dipole field at the site. Hall suggests that this observation could be used to infer either the record of a transient and very recent (within the past 10 million years) magnetic pole excursion, or regional tilting of the oceanic crust. Both hypotheses suffer from the lack of supporting evidence. The low magnetization of the rocks is believed to result from low-temperature oxidation of original titanomagnetites into titanomaghemites, presumably by the action of percolating sea water (cf. Section 25.5). While many more deep holes will be required to determine how characteristic these magnetic properties are of the world's oceans in general, it can now be concluded that the whole process of sea-floor production and magnetization is much more complicated than the picture usually attributed to the Vine-Matthews model. Indeed, if chemical alteration is as widespread as suggested, it may be one of the most fortunate accidents in earth history that the pattern of magnetic stripes was preserved at all.

Near continental borders the pattern does disappear, but the reason for this is not yet clear. It is true that the period of reversals appears to increase with geological age, but it is not known if any ocean floor material is sufficiently old for this to be a factor. It must be remembered that the direction of the field at the time of magnetization is a critical factor, and for a north-south ridge, material near the ancient magnetic equator would be magnetized in such a way that no polarized boundaries would be observed. This has been suggested (Schneider et al., 1969) as the reason for the absence of a linear pattern in the Atlantic Ocean immediately east of North America.

Problems

1. Derive Equation 21.2.5 for the effect on the astatic magnetometer of the vertical dipole moment of a sample in the "off-center" position.

2. Establish the "working equation" of paleomagnetism (Section 21.3):

$$\tan I = 2 \cot \theta$$

3. It is determined that the inclination of the intensity vector in a sample is 60°; measurements of the inclination and azimuth of the vector each have an uncertainty (according to a chosen criterion) of 1°. Assuming there has been no rotation of the land mass from which the sample was taken, what is the size and shape of the region in which the ancient magnetic pole is indicated?

4. The direction and magnitude of the vector of remanent magnetization of a sample of volcanic rock were determined with a spinner magnetometer, both in the original condition and after "magnetic cleaning" in alternating fields of different initial strength. The observations were:

A.C. field (oersted)	Azimuth	Dip	Magnitude (10^{-6} c.g.s.)
0	10°E.	69°	120
100	16°E.	68°	68
200	21°E.	66°	49
300	27°E.	65°	40
500	30°E.	63°	29
700	45°E.	61°	20

Plot on a stereographic projection the points of intersection of the intensity vector with a sphere of radius 5.0 cm. The present magnetic field at the collection site dips 70° with zero declination.

How do you explain the observed pattern? Can you show quantitatively the effect of the cleaning, by determining the magnitude of the "soft" component of magnetization after each step?

5. If the Americas have moved westward relative to Europe by 4000 km since Carboniferous time, how does the average spreading rate compare with that deduced for recent times, from magnetic lineations (Fig. 21.16)?

6. The average intensity of magnetization of the oceanic crust is 5×10^{-3} c.g.s. units. Assume that the uppermost mantle has similar magnetic properties. If the Curie point isotherm were upwarped by 5 km from its normal depth, across a strip 100 km wide, what magnitude and sign of magnetic anomaly would be observed over the strip? Are there reasons for believing the mantle to be more or less magnetic than the oceanic crust?

7. A section of oceanic crust is composed of strips, averaging 20 km in width, of alternate polarization. Taking the intensity of magnetization to be 5×10^{-3} c.g.s. units, compare the amplitude of the resulting total-field magnetic anomalies, observed 5 km above the sea floor, if the section of crust were formed (a) at a place where the magnetic dip was 70° with the ridge oriented east-west; (b) at the same place with the ridge oriented north-south; (c) at the magnetic equator with the ridge east-west; (d) at the magnetic equator with the ridge north-south.

8. For some rock sequences, such as the late Precambrian Keweenawan of the Lake Superior region (Palmer, 1970), the magnetic reversal is asymmetric, in that normal and reversed intensity vectors are not antiparallel. In the case mentioned, the normal intensity vector is inclined downward at 39° to the horizontal, the reversed upward at 69°, in the same vertical plane. Discuss the following as possible causes of this behaviour: (a) the addition of a secondary component, to normal and reversed rocks equally, following an original symmetric reversal; (b) plate motion between the times of formation of normal and reversed rocks; (c) a non-geocentric dipole as the source of the earth's field. If you had available both paleolatitudes and paleointensities for normal and reversed samples, to what extent could you distinguish between these causes?

9. Detailed measurements of inclination and declination are made on a section of very young lake sediments, at a site where the mean inclination is 70°, declination is 0°, and the total intensity is 0.50 Gauss. These values may be assumed for the time span of deposition. If during deposition, a focus of maximum positive non-dipole field, of central magnitude 0.10 Gauss, had drifted west at 0.20° of longitude per year, along a latitude 30° south of the site, what precisely would be the record of inclination and declination changes in the section? Assume the focus of non-dipole field results from a vertical dipole at the core boundary (Section 18.4).

BIBLIOGRAPHY

Aitken, M. J. (1974) *Physics in Archaeology.* 2nd Ed., Clarendon Press, Oxford.

Alldredge, Leroy R., and Van Voorhis, Gerald D. (1961) Depth to sources of magnetic anomalies. *J. Geophys. Res.*, **66**, 3793.

Bhattacharyya, B. K. (1964) Magnetic anomalies due to prism-shaped bodies with arbitrary polarization. *Geophysics,* **29**, 517.

Blackett, P. M. S. (1952) A negative experiment relating to magnetism and the earth's rotation. *Phil. Trans. Roy. Soc. A,* **245**, 309.

Breiner, S. (1964) Piezomagnetic effect at the time of a local earthquake. *Nature,* **202**, 790.

Breiner, S., and Kovach, R. L. (1967) Local geomagnetic events associated with displacements on the San Andreas fault. *Science,* **158**, 116.

Buchan, Kenneth L., and Dunlop, David J. (1976) Palaeomagnetism of the Haliburton intrusions: superimposed magnetizations, metamorphism and tectonics in the Late Precambrian. *J. Geophys. Res.*, **81**, 2951–2967.

Clark, S. P. (1966) Handbook of physical constants. *Geol. Soc. Amer., Memoir 97.*

Coe, Robert S. (1967a) Paleo-intensities of the earth's magnetic field determined from Tertiary and Quaternary beds. *J. Geophys. Res.,* **72**, 3247–3262.

Coe, Robert S. (1967b) The determination of palaeintensities of the earth's magnetic field with emphasis on mechanisms which could cause non-ideal behaviour in Thellier's method. *J. Geomag. Geoelect.,* **19**, 157–179.

Collinson, D. W. (1975) Instrument and techniques in palaeomagnetism and rock magnetism. *Rev. Geophys. Space Phys.,* **13**, 659–686.

Cox, Allan (1968) Lengths of geomagnetic polarity intervals. *J. Geophys. Res.,* **73**, 3247.

Cox, A. (1969) Geomagnetic reversals. *Science,* **163**, 237–245.

Cox, Allan (1975) The frequency of geomagnetic reversals and the symmetry of the nondipole field. *Reviews of Geophys. and Space Phys.,* **13**, 35–51.

Cox, A., and Doell, R. R. (1960) Review of palaeomagnetism. *Bull. Geol. Soc. Amer.,* **71**, 645.

Cox, A., and Dalrymple, G. B. (1967) Statistical analysis of geomagnetic reversals data and the precision of potassium-argon dating. *J. Geophys. Res.,* **72**, 2603–2614.

Creer, K. M., Gross, D. L., and Lineback, J. A. (1976). Origin of regional geomagnetic variations recorded by Wisconsinan and Holocene sediments from Lake Michigan, USA and Lake Windermere, England. *Bull. Geol. Soc. Amer.,* **87**, 531–540.

Deutsch, Ernst (1966) The rock magnetic evidence for continental drift. In *Continental drift,* ed. G. D. Garland. Royal Society of Canada, Spec. Pub. 9, Univ. of Toronto Press, Toronto.

Dunlop, D. J. (1968) Monodomain theory: experimental verification. *Science,* **162**, 256.

Foster, J. H., and Opdyke, N. D. (1970) Upper Miocene to Recent magnetic stratigraphy in deep-sea sediments. *J. Geophys. Res.,* **75**, 4465–4473.

Hall, J. M. (1976) Major problems regarding the magnetization of oceanic crustal layer 2. *J. Geophys. Res.,* **81**, 4223–4230.

Heirtzler, J. R., Dickson, G. O., Herron, E. M., Pitman, W. C. III, and Le Pichon, X. (1968) Marine magnetic anomalies, geomagnetic field reversals and motions of the ocean floor and continents. *J. Geophys. Res.,* **73**, 2119.

Irving, E. (1964) *Palaeomagnetism.* John Wiley and Sons, New York.

Irving, E. (1968) The distribution of continental crust, and its relation to ice ages. In *The history of the earth's crust,* ed. Robert A. Phinney. Princeton Univ. Press, Princeton.

Irving, E., Robertson, W. A., and Aumento, F. (1970) The Mid-Atlantic Ridge near 45°N. VI. Remanent intensity, susceptibility, and iron content of dredged samples. *Can. J. Earth Sci.,* **7**, 226.

Irving, E., Emslie, R. F., and Ueno, H. (1974) Upper Proterozoic palaeomagnetic poles from Laurentia and the history of the Grenville structural province. *J. Geophys. Res.,* **79**, 5491–5502.

Johnson, E. A., Murphy, T., and Torreson, O. W. (1948) Pre-history of the Earth's magnetic field. *Terr. Magn. and Atmos. Elect.,* **53**, 349–372.

Jurdy, Donna M., and Van der Voo, Rob (1974) A method for the separation of true polar wander and continental drift, including results for the last 55 m.y. *J. Geophys. Res.,* **79**, 2945–2952.

Jurdy, D. M., and Van der Voo, R. (1975) True polar wander since the early Cretaceous. *Science,* **187**, 1193–1196.

King, R. F. (1955) The remanent magnetism of artificially deposited sediments. *Month. Not. R. A.S., Geophys. Suppl. 7,* 115.

Koenigsberger, J. G. (1938) Natural residual magnetism of eruptive rocks, 1 and 2. *Terrest. Magnetism and Atmospheric Elect.,* **43**, 119–127 and 299–320.

La Brecque, John L., Kent, Denis V., and Cande, Steven C. (1977) Revised magnetic polarity time scale for Late Cretaceous and Cenozoic time. *Geology,* **5**, 330–335.

Larson, Roger L., and Hilde, Thomas C. (1975) A revised time scale of magnetic reversals for the Early Cretaceous and Late Jurassic. *J. Geophys. Res.,* **80**, 2586–2594.

Loncarevic, B. D., Mason, C. S., and Matthews, D. H. (1966) Mid-Atlantic Ridge near 45°N. The median valley. *Can. J. Earth Sci.,* **3**, 327.

Lowrie, W., Alvarez, W., and Premoli-Silva, I. (1976) Late Cretaceous geomagnetic reversal sequence. *Trans. Amer. Geoph. Un.,* **57**, 238.

Marshall, Monte, and Cox, Allan (1972) Magnetic changes in pillow basalt due to sea floor weathering. *J. Geophys. Res.,* **77**, 6459–6469.

McElhinny, M. W. (1973) *Palaeomagnetism and plate tectonics.* Cambridge Univ. Press, Cambridge.

McKenzie, D. P. (1972) Plate tectonics. In *The nature of the solid earth,* ed. Eugene C. Robertson. McGraw-Hill, New York.

Momose, K. (1963) Studies in the variation of the earth's magnetic field during Pliocene time. *Bull. Earthquake Res. Inst.,* **41**, 487.

Munk, W. H., and MacDonald, G. J. F. (1960) *The rotation of the earth.* Cambridge Univ. Press, Cambridge.

Nagata, Takesi (1961) *Rock magnetism.* Marozen, Tokyo.

Nagata, T. (1970a) Anisotropic magnetic susceptibility of rocks under mechanical stresses. *Pure and Appl. Geophys.,* **78**, 110.

Nagata, T. (1970b) Effects of a uniaxial compression on remanent magnetization of igneous rocks. *Pure and Appl. Geophys.,* **78**, 100.

Néel, L. (1955) Some theoretical aspects of rock magnetism. *Adv. Phys.,* **4**, 191.

Opdyke, N. D., and Hekinian, R. (1967) Magnetic properties of some igneous rocks from the Mid-Atlantic Ridge. *J. Geophys. Res.,* **72**, 2257.

Pakiser, L. C., and Zietz, Isidore (1965) Transcontinental crustal and upper mantle structure. *Rev. Geophys.,* **3**, 505.

Palmer, H. C. (1970) Palaeomagnetism and correlation of some Middle Keweenawan rocks, Lake Superior. *Can. J. Earth Sciences,* **7**, 1410–1436.

Phillips, Jeffrey D., and Cox, Allan (1976) Spectral analysis of geomagnetic reversal time scales. *Geophys. Journ.,* **45**, 19–33.

Pitman, Walter C. III, Larson, Roger L., and Herron, Ellen M. (1974) *The age of the ocean basins* (map). Geol. Soc. Amer.

Rikitake, T. (1968) Geomagnetism and earthquake prediction. *Tectonophysics,* **6**, 59.

Schneider, E. D., Vogt, P. R., and Lowrie, A. (1969) Diapiric structures and magnetic anomalies of the pre-Cenozoic Atlantic Ocean (abst.) *Trans. Amer. Geophys. Un.,* **50**, 212.

Smith, P. J. (1967) The intensity of the ancient geomagnetic field: a review and analysis. *Geophys. J.,* **12**, 321.

Stacey, F. D. (1962) A generalized theory of remanence, covering the transition from single domain to multi-domain magnetic grains. *Phil. Mag.,* **7**, 1887.

Stacey, F. D. (1963) The physical theory of rock magnetism. *Adv. Phys.,* **12**, 46.

Stacey, F. D. (1964) The seismomagnetic effect. *Pure and Appl. Geophys.,* **58**, 5.

Stewart, A. D., and Irving, E. (1974) Palaeomagnetism of Precambrian sedimentary rocks from N. W. Scotland and apparent polar wandering path of Laurentia. *Geophys. J.,* **37**, 51–72.

Stoner, E. C., and Wohlfarth, E. P. (1948) A mechanism of magnetic hysteresis in heterogeneous alloys. *Phil. Trans. Roy. Soc. Lond. A,* **240**, 599.

Strangway, D. W., and Vogt, P. R. (1969) Aeromagnetic tests for continental drift in Africa and South America (abst.) *Trans. Amer. Geophys. Un.,* **50**, 136.

Thellier, E. (1946) Sur la thermorémanence et la théorie du métamagnétisme. *C. R. Acad. Sci. Paris,* **223**, 319.

Thellier, E., and Thellier, O. (1959) Sur l'intensité du champ magnétique terrestre dans le passé historique et geologique. *Ann. Geophys.,* **15**, 285–376.

Vacquier, V., Raff, A. D. and Warren, R. E. (1961) Horizontal displacements in the floor of the Pacific Ocean. *Bull. Geol. Soc. Amer.,* **72**, 1251.

Van Zijl, J. S. V., Graham, K. W. T., and Hales, A. L. (1962) The palaeomagnetism of the Stromberg lavas II: The behavior of the magnetic field during a reversal. *Geophys. J.,* **1**, 169.

Vine, F. J., and Matthews, D. H. (1963) Magnetic anomalies over oceanic ridges. *Nature,* **199**, 947.

22 RADIOACTIVITY

22.1 FUNDAMENTAL PRINCIPLES

The fact that a number of naturally occurring elements in the earth are radioactive has important consequences in geophysics, in at least three ways. First, the heat produced by radioactive disintegrations probably represents the most important factor in the establishment of thermal conditions within the earth. Secondly, the rate of radioactive decay of certain elements provides the only physical means of establishing a time scale for events in the past history of the earth. Finally, the varying distribution of the products of radioactive disintegration provides a means of tracing the history of minerals.

We begin by recalling a few principles of nuclear physics. An element X has chemical properties determined by the number Z of its electrons, which is equal to the number of protons in the nucleus, and is known as the atomic number of the element. The nuclei of all elements heavier than hydrogen also contain neutrons, and we denote the total number of protons and neutrons by A, the mass number of the element. Most elements can exist with different values of A, and the corresponding atoms are known as *isotopes* of the substance. A particular isotope is designated $_ZX^A$.

Radioactivity is the spontaneous emission of particles or electromagnetic radiation by the atoms of certain elements. All of the isotopes of elements with $Z > 83$ are radioactive, as are some isotopes of lighter elements. The largest particle emitted by the nuclei of radioactive elements is one of mass number 4 and positive charge 2; it is in fact a helium nucleus, but is designated an α particle. The release of an α particle from a nucleus leaves the original atom with Z decreased by 2 and A by 4. A new chemical element therefore results, and its nucleus may initially be in an excited state. The emission of electromagnetic radiation, that is, γ rays, brings it to the ground state. The α particle which is emitted, although it carries appreciable kinetic energy, is large in comparison with other particles, and is normally brought to rest within less than a millimeter of rock. When it captures two electrons it becomes a neutral atom of helium.

The nuclei of other radioactive elements emit electrons, known as β-particles. These can be shown to be formed immediately before they are emitted—since the nucleus cannot contain free electrons—and their emission results in the conversion of one neutron to a proton. The atomic number of the resulting atom is therefore greater by one than that of the parent, while the mass number remains the same. Beta emission is accompanied by the emission of a neutrino, a particle with vanishingly small rest mass and no charge. Neutrinos carry angular momentum and energy; they have the distinctive property of passing through a mass of terrestrial dimensions without yielding this energy to the surrounding matter. The β particle travels farther through matter than an α particle, but is normally brought to rest within a few millimeters of rock.

Some nuclei have the ability to capture one of their own electrons from the innermost or k shell. This phenomenon, which is somewhat the reverse of β emission, results in a new element with Z decreased by one and the mass number A unchanged. The process is usually accompanied by the emission of a photon of X-radiation.

22.2 DECAY CONSTANTS AND HALF-LIVES

According to the principles of quantum mechanics, both α and β emissions are determined by a probability which is characteristic of a given radioactive nucleus. The probability is unaffected by physical conditions, and it is independent of the time since the nucleus was created. For an assemblage of atoms of some radioactive substance, the probability that a disintegration will take place in a certain time interval is therefore proportional to the number of atoms present.

If we let n be this number at time t, we have

$$\frac{dn}{dt} = -\lambda n \qquad 22.2.1$$

of which the solution is

$$n = n_0 e^{-\lambda t} \qquad 22.2.2$$

where n_0 is the number present at $t = 0$.

The constant λ, which is characteristic of a given decay scheme, is known as the *decay constant,* and, for long-lived atoms, is usually quoted in years^{-1}. Another quantity, less convenient for calculations but often quoted, is the *half-life,* or the time required for one-half of the initial number of atoms to disintegrate. The half-life T is given by

$$\frac{n_0}{2} = n_0 e^{-\lambda T}$$

or

$$T = \frac{\ln 2}{\lambda} \qquad 22.2.3$$

For a simple decay scheme, in which the end or daughter product is stable, the number of atoms of the daughter product formed in time t is equal to the number of parent atoms that have disintegrated, that is, $(n_0 - n)$. If λ is known, measurement of the abundance of a parent and daughter permits the determination of t. This is the basis for determinations of the ages of mineral samples.

In some cases of geophysical importance, the initial daughter product is radioactive, and it decays, perhaps through a series of radioactive elements, until the stable end product is reached. Each intermediate radioactive element has a characteristic decay constant, but after a certain time interval series of this type reach an equilibrium in which the amounts of the intermediate members remain constant. For each disintegration of the first member of the series, an atom of the end product is formed. However, since the members of the series may be different chemical elements, chemical reactions may destroy the equilibrium.

22.3 HEATING BY RADIOACTIVITY

When α or β particles emitted by a radioactive atom are brought to rest by the surrounding medium, their kinetic energies are converted to heat in that medium. For different disintegrations, the energies carried by the particles fall over a wide spectrum, but the kinetic energy of α particles is considerably greater than that of β particles. An empirical relation for the energy of emitted α particles is

$$\ln T = AK^{-1/2} - B \tag{22.3.1}$$

where T is the half-life of the emitter, K is the kinetic energy of the α particle and A and B are constants.

As far as significant contributions to the heat of the earth are concerned, the number of radioactive elements to be considered are relatively few (Birch, 1954). Two isotopes of uranium, U^{238} and U^{235}, and thorium, Th^{232}, have long half-lives, and are widely distributed through the rocks of the earth. One isotope of potassium, K^{40}, is important, and was even more so in past geological

TABLE 22.1 DECAY SCHEMES. (FROM DODD, et al. IN *MINING AND GROUNDWATER GEOPHYSICS*/1967, ED. L. W. MORLEY. BY PERMISSION OF GEOLOGICAL SURVEY OF CANADA.)

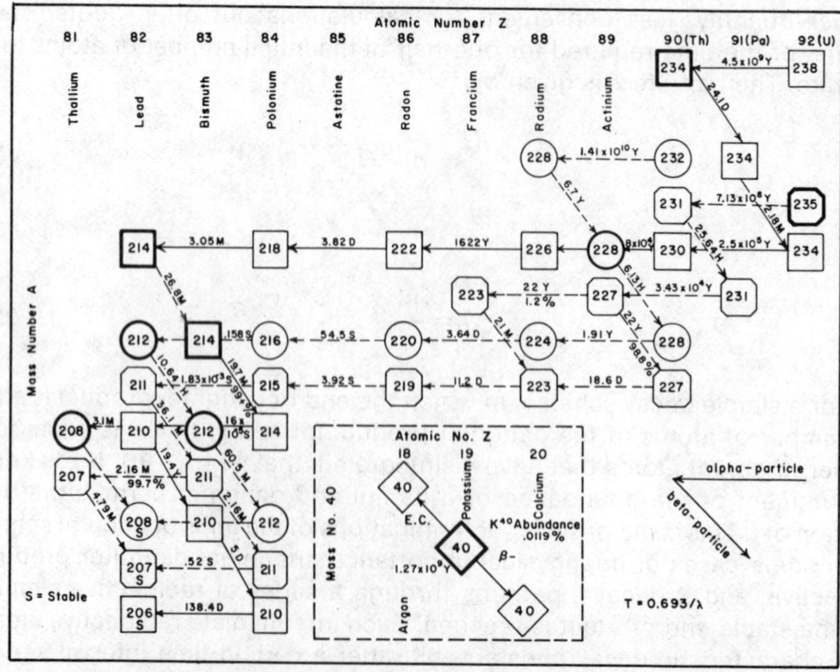

TABLE 22.2 HEAT PRODUCTIVITIES OF RADIOACTIVE MINERALS (AFTER BIRCH) AND TYPICAL ROCK TYPES (AFTER CLARK AND RINGWOOD)

Isotope or Mineral	Productivity cal/gm/yr
U^{238}	0.71
U^{235}	4.3
U (ordinary)	0.73
Th^{232}	0.20
K^{40}	0.22
K (ordinary)	27×10^{-6}
Rb^{87}	130×10^{-6}
Rb (ordinary)	36×10^{-6}

Region	U ppm	Th ppm	K per cent	Heat Production 10^{-13} cal/cm³ sec
Oceanic crust	0.42	1.68	0.69	0.714
Continental shield	1.00	4.00	1.63	1.67
Younger continental upper crust	1.32	5.28	2.15	2.2

time when it was more abundant. Other radioactive elements, such as certain isotopes of the rare earths, and the isotope of rubidium, Rb^{87}, which is useful in geochronology, appear to be distributed in too small amounts to contribute significantly in this respect.

The important uranium and thorium decay schemes, which end with the different isotopes of lead, are shown in Table 22.1. It is possible in principle to determine the amount of energy converted to heat by simply taking the mass difference between the parent and the end product plus the mass of the emitted α particles, and applying Einstein's relation $E = mc^2$. It is experimentally easier to sum the energies, as measured in the laboratory, for the individual emissions. In any case, it is necessary to separate the contribution of α's and β's in order to deduct from the latter the energy carried by the neutrino, which is not absorbed by the earth. Fortunately, in the case of the uranium-thorium series, 90 per cent of the energy is provided by the α particles; Birch assumed that two-thirds of the β decay energy was carried by neutrinos. In the case of the decay of K^{40}, energy is carried both by β particles and by a γ emission which is associated with a *k*-capture process. The relative rates of the two processes was a matter of some uncertainty, and led to conflicting values of heat production as well as to difficulties in geochronology, although this appears to be resolved (Section 23.2). Birch's values for the heat production, in calories per gram per year, are shown in Table 22.2. In this table, the heat productions for ordinary uranium and ordinary potassium are based on the known abundance ratios of the isotopes of these elements as they are at present. The heat production of a given rock type depends upon the concentration of the radioactive element in the rock, and, as the table indicates, is of the order of 10^{-13} cal/cm³ sec (the more convenient unit for computations). Uranium, thorium, and potassium are all more abundant in the intermediate rocks of the continental crust than in the basaltic oceanic crust.

BIBLIOGRAPHY

Birch, F. (1954) Heat from radioactivity. In *Nuclear geology,* ed. H. Faul. John Wiley and Sons, New York.

Clark, Sydney P. Jr., and Ringwood, A. E. (1964) Density distribution and constitution of the mantle. *Rev. Geophys.*, **2**, 35.

23 GEOCHRONOLOGY

23.1 THE RUBIDIUM-STRONTIUM METHOD

Age determinations based on radioactive decay depend upon the measurement of the abundances of a parent isotope and its stable end product, and upon a knowledge of the decay constant. Although the method based upon the radioactivity of Rb^{87} was not the first to be applied, it is convenient to consider it first, because the decay scheme is relatively simple.

The radioactivity of rubidium was discovered in 1905; in 1937 it was attributed to the isotope Rb^{87}, which forms 27.8 per cent of naturally occurring rubidium. The other isotope, Rb^{85}, is not radioactive. Radioactive decay of Rb^{87} proceeds by a highly forbidden (i.e., low probability) β emission to produce Sr^{87}. While rubidium is not an abundant element of the earth's crust, it is more widely distributed among minerals such as the micas and feldspars than was once supposed. Originally, it was detected in the pink mica, lepidolite, and the suggestion to employ the decay scheme for geochronology came from Goldschmidt (1937) and Hahn and Walling (1938).

Equation 22.2.2 is directly applicable to the single decay scheme. Suppose that at time $t = 0$ the daughter product strontium begins to be trapped in the rubidium-containing rock. This time is normally taken to be the time of solidification of the igneous rock. At time t after solidification, the number of atoms of strontium present is given by

$$[Sr^{87}]_t = [Rb^{87}]_0 - [Rb^{87}]_t = [Rb^{87}]_0(1 - e^{-\lambda t}) \qquad 23.1.1$$

where λ is the decay constant for Rb^{87}.

Putting all quantities in terms of their values at time t, we have

$$[Sr^{37}]_t = [Rb^{87}]_t(e^{\lambda t} - 1) \qquad 23.1.2$$

To determine t, it is necessary to measure the abundances of Sr^{87} and Rb^{87} and to know λ. The relative abundance of Rb^{87} varies with time, but is constant in space. This is typical of the isotope ratios of all but the light elements, so that, once determined, the ratio may be adopted without further measurement. For rubidium, as we have noted, Rb^{87} forms 27.8 per cent (atomic) of the element. Chemical analysis of a mineral for rubidium would therefore be sufficient to yield the Rb^{87} content. To determine the abundance of the daughter product Sr^{87}, the strontium in the sample must be isolated chemically, then subjected to mass spectrographic analysis to yield the abundance of Sr^{87} (Fig. 23.1). A complication immediately enters, in that common strontium contains the isotopes Sr^{86}, Sr^{87}, and Sr^{88} and a correction must be made for the contribution of Sr^{87} from common strontium. Until fairly recently, uncertainty in the isotope ratios to adopt for common strontium led to uncertainties in this correction.

Figure 23.1 Solid-source mass spectrometer in use for the isotopic analysis of strontium. The metal tube is visible at left, and a portion of the mass spectrum can be seen on the recorder near center.

Before we return to this question, it is appropriate to consider the decay constant. It has been known for some years that the half-life of Rb^{87} is approximately 5×10^{10} years. This large value implies that the rate of decay is very low, and direct counting of the β particles in the laboratory is difficult. For this reason, the half-life (or decay constant) was imperfectly known, and rubidium-strontium age determinations were similarly uncertain. Aldrich and others (1956) determined a value for the half-life of Rb^{87} by matching Rb–Sr ages on six rocks to the lead ages (by methods which we discuss below) of the same rocks. These ages ranged from 380 to 2640×10^6 years, and the best value of half-life was found to be $(5.0 \pm 0.2) \times 10^{10}$ years. Flynn and Glendenin (1959) measured the specific activity and energy spectrum of β emissions from Rb^{87} by using a liquid compound of Rb in a liquid scintillation counter. They pointed out that the spectrum contains an excess of low energy particles, which were probably not detected in earlier experiments. Their best value for the half-life was $(4.70 \pm 0.05) \times 10^{10}$ years.

These examples show the difficulty of determining the half-life (or decay constant) of a very long-lived radioactive isotope. It is obviously essential that ages be quoted with respect to consistent decay constants, in all physical dating methods. For this reason, the Subcommission on Geochronology of the International Union of Geological Science has adopted and recommended the set of decay constants and isotopic ratios shown in Table 23.1 (Steiger and Jäger, 1977); reasons for the adoptions, with primary references to measurements, are given in this reference.

The difficulty of correcting for common strontium was solved by Compston, Jeffrey and Riley (1960), who pointed out that isotope ratios could be deter-

TABLE 23.1 DECAY CONSTANTS AND ISOTOPE RATIOS RECOMMENDED BY THE INTERNATIONAL UNION OF GEOLOGICAL SCIENCES. FROM STEIGER AND JÄGER (1977).

Uranium	
$\lambda\ (^{238}U) = 1.55125 \times 10^{-10}/yr$	atomic ratio $^{238}U/^{235}U = 137.88$
$\lambda\ (^{235}U) = 9.8485 \times 10^{-10}/yr$	
Thorium	
$\lambda\ (^{232}Th) = 4.9475 \times 10^{-11}/yr$	
Rubidium	
$\lambda\ (^{87}Rb) = 1.42 \times 10^{-11}/yr$	atomic ratio $^{85}Rb/^{87}Rb = 2.59265$
Strontium	
atomic ratios $^{86}Sr/^{88}Sr = 0.1194$	
$^{84}Sr/^{86}Sr = 0.056584$	
Potassium	
$\lambda(^{40}K_\beta-) = 4.962 \times 10^{-10}/yr$	isotopic abundances $^{39}K = 93.2581$ atom %
$\lambda(^{40}K_e) + \lambda'(^{40}K_e) = 0.581 \times 10^{-10}/yr$	$^{40}K = 0.01167$ atom %
	$^{41}K = 6.7302$ atom %
Argon	
atomic ratio $^{40}Ar/^{36}Ar$ atmospheric = 295.5	

mined for several constituent minerals of the same rock. Expressing all concentrations in terms of the non-radiogenic Sr^{86}, it is possible to write

$$\left(\frac{Sr^{87}}{Sr^{86}}\right)_{total} = \left(\frac{Sr^{87}}{Sr^{86}}\right)_{initial} + \left(\frac{Rb^{87}}{Sr^{86}}\right)(e^{\lambda t} - 1) \qquad 23.1.3$$

for each mineral, with t the same for all. If the measured ratio Sr^{87}/Sr^{86} is plotted against Rb^{87}/Sr^{86}, a straight line results (Fig. 23.2), whose slope is $(e^{\lambda t} - 1)$ and whose intercept yields the *initial ratio* of Sr^{87}/Sr^{86} for the common strontium in the rock. Such a straight line is called an *isochron,* and the age of the rock is determined from its slope. This method is known as whole-rock dating, and it can be said to be responsible for the great increase in the usefulness of the rubidium-strontium method.

The method of expressing the isotope ratios used in Figure 23.2 emphasizes the importance of the ratio Sr^{87}/Sr^{86} in common strontium. This ratio commonly lies between 0.700 and 0.725, but because Rb^{87} is highly concentrated in the crust, it is higher for crustal than for mantle material. The full significance of the ratio is discussed in Section 23.4 below.

The operational difficulty of the method is in the determination of the slope $(e^{\lambda t} - 1)$. For most rocks, $e^{\lambda t}$ is very close to unity, and unless the slope is determined with great care, the age t is subject to large error. Both coordinates of the plot, Sr^{87}/Sr^{86} and Rb^{87}/Sr^{86}, are subject to errors of measurement, and appropriate least-squares techniques must be employed. York (1967) designed a method for determining the slope which has since found wide application in other fields.

The great advantage of the Rb–Sr method is that it represents a solid-solid system. In contrast to other methods, in which a gaseous phase is involved, there is relatively little chance of loss of parent or daughter product. The disadvantages are that rubidium is not an abundant element of the crust, and the long half-life makes it difficult to apply to young rocks. It is, however, probably the most suitable method for dating large areas of Precambrian rocks, and it has been used to date stony meteorites and even silicate inclusions in iron meteorites (Burnett and Wasserburg, 1967).

23.2 THE POTASSIUM-ARGON METHOD

Mention has already been made, in connection with radioactive heating, of the fact that the isotope K^{40} of potassium is radioactive. This isotope forms 0.0117 per cent (with an uncertainty of 1 in the last digit) of the atoms of natural potassium; it decays by two schemes. Emission of β particles from ${}_{19}K^{40}$ produces ${}_{20}Ca^{40}$, while capture of an electron from the k shell produces ${}_{18}A^{40}$. Two decay constants are therefore involved, and we may write:

$$\frac{d}{dt}K^{40} = \lambda_\beta K^{40} \text{ for decay to calcium}$$

$$\frac{d}{dt}K^{40} = \lambda_k K^{40} \text{ for decay to argon}$$

The total rate of change is:

$$\frac{d}{dt}K^{40} = (\lambda_\beta + \lambda_k)K^{40} = \lambda K^{40}$$

Assuming that at time $t = 0$ daughter products begin to accumulate, and that there has been no loss since, the present ratios are

$$\left.\begin{aligned} \frac{Ca^{40}}{K^{40}} &= \frac{\lambda_\beta}{\lambda}(e^{\lambda t} - 1) \\ \frac{A^{40}}{K^{40}} &= \frac{\lambda_k}{\lambda}(e^{\lambda t} - 1) \end{aligned}\right\} \qquad 23.2.1$$

We note that, in either case, a knowledge of two decay constants is required, although these are often quoted in terms of one constant and the *"branching ratio,"* λ_k/λ_β. The decay constant for β decay can be determined in the laboratory by counting, while that for electron capture must be measured indirectly, usually by measurement of X-rays whose emission is associated with the capture. Damon (1968), in reviewing the determinations, suggested values of 0.588×10^{-10} yr^{-1} for λ_k and 4.78×10^{-10} yr^{-1} for λ_β. The corresponding total decay constant yielded a half-life of 1.3×10^9 years. Recommended values are shown in Table 23.1.

The isotope Ca^{40} forms 97 per cent of common calcium, so that it would be impossible, in almost any rock, to determine the concentration of radiogenic calcium. Age determinations are based upon the measurement of the concentration of potassium and of radiogenic argon. Usually, chemical analysis for total potassium is employed, and the percentage of K^{40}, as quoted above, is assumed. The rock sample is melted, the gases are extracted, and argon is separated for mass spectrographic analysis. A promising variation of the original technique for potassium is the use of a neutron flux from a nuclear pile to convert K^{39} to A^{39} to eliminate the need for chemical analysis (Merrihue and Turner, 1966). It was found by York and Berger (1970) that an exposure for 24 hours to a flux of 10^{13} neutrons/cm^{2}sec was sufficient to produce this conversion, and apparently reliable ages were obtained by means of mass spectrographic analysis alone.

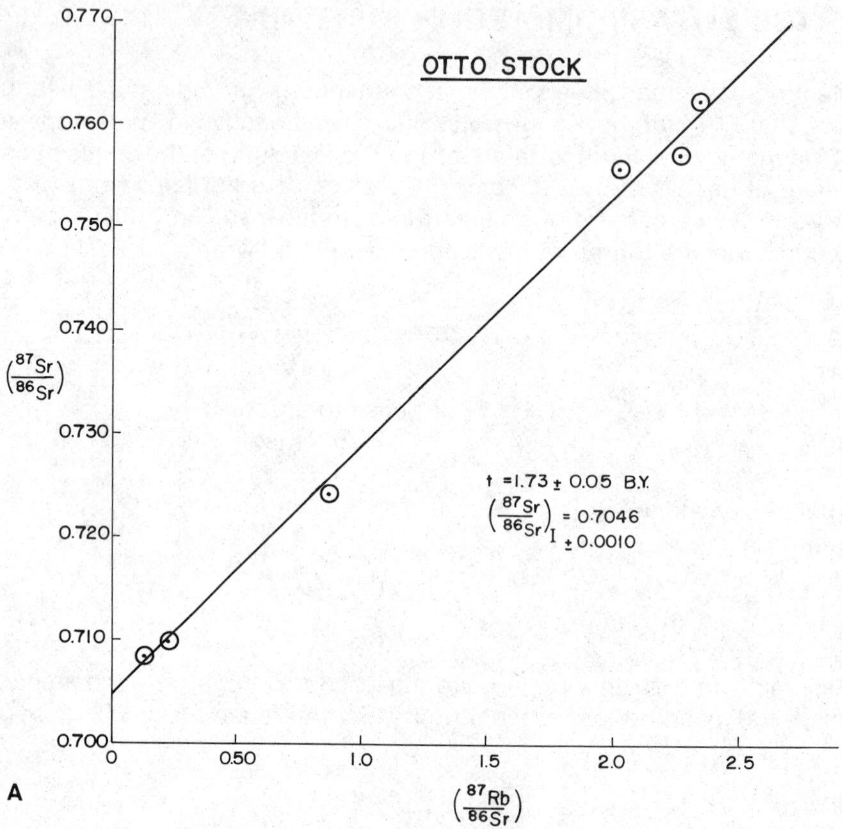

Figure 23.2 Isochrons for two rocks from adjacent igneous intrusions in the Canadian Precambrian shield. In spite of the apparent similarity in slopes, the ages given by the two straight lines are very different. (Courtesy of D. York)

Illustration continued on opposite page

A complication arises because atmospheric argon, with which the sample may be contaminated, contains A^{40} as well as A^{36} and A^{38}. If it were certain that the contamination was from the present atmosphere, a correction could be applied by measuring the abundance of A^{36} and applying the present atmospheric ratio, $A^{40}/A^{36} = 295.5$. However, it has been shown, from the dating of very young basalts from the ocean floors (Dalrymple and Moore, 1968), that these rocks may contain excess argon whose isotopic ratio is quite different from that of the present atmosphere. The rocks studied by Dalrymple and Moore were known to be less than 1 million years old, yet the excess A^{40}, after the normal correction for atmospheric argon had been applied, gave apparent ages up to 40×10^6 years. The excess appeared to correlate with depth of water at the sample sites, suggesting an effect of hydrostatic pressure on the trapping of magmatic gas at the time the rocks were formed. It appears that the uncertainty may be reduced by the use of an isochron plot (McDougall and Stipp, 1969; York, Baksi, and Aumento, 1969) similar to that discussed above for Rb/Sr dating. In this case the ratio A^{40}/A^{36} is plotted against K^{40}/A^{36} for a group of determinations from the same rock. As in the application for Rb/Sr dating, the slope of the resulting straight line gives the age, while the intercept gives the initial value of the ratio A^{40}/A^{36}. The assumption, of course, is that all samples analyzed are of the same age and were contaminated with gas of the same isotopic composition.

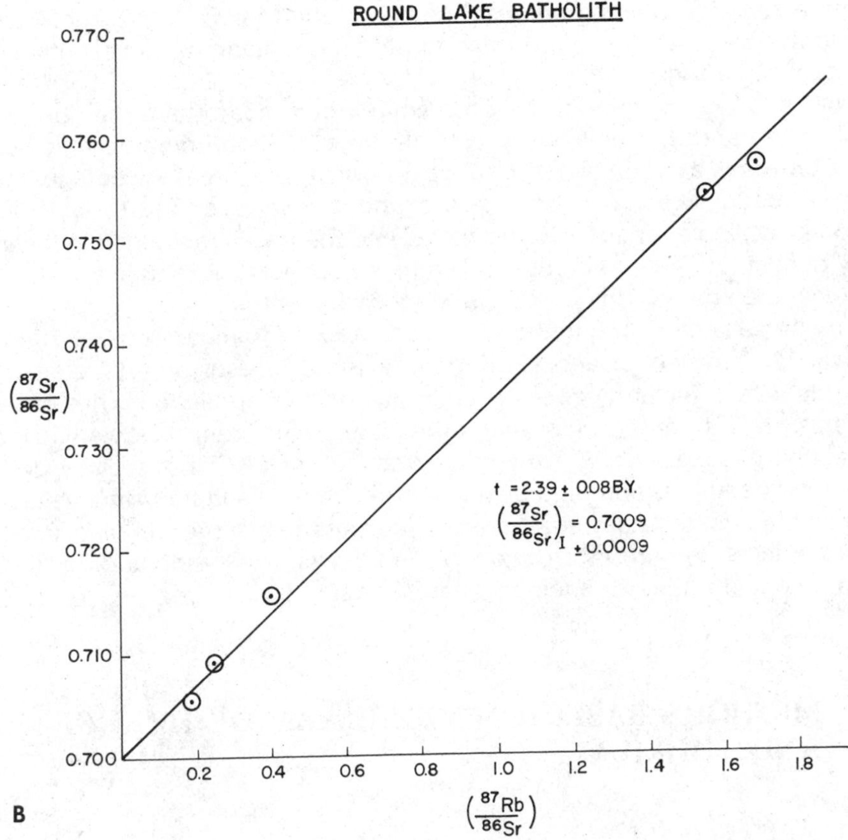

Figure 23.2 *Continued*

The success of potassium-argon dating depends critically upon the mineral chosen for analysis. Some minerals, such as beryl and tourmaline, are known to be characterized by large amounts of extraneous A^{40}. For other minerals, the difficulty is the loss of A^{40} by diffusion during the lifetime of the rock. Diffusion takes place at a rate which depends upon temperature, but is also very different for different minerals. Early work was done on feldspar minerals, for convenience, but these are now known to be less retentive of argon than the micas or hornblendes. Damon (1968) has ordered the retentivity at two temperatures in this way:

Below 250°C	hornblende > muscovite > biotite > phlogopite > sanidine > microcline > glauconite
Above 600 to 700°C	hornblende > muscovite > sanidine > orthoclase > biotite > phlogopite > glauconite

Even with the use of the most retentive mineral, it is recognized that if a rock is subjected for long periods to temperatures in excess of 700°C, much of the argon will be lost. Such heating could come from nearby igneous activity, but elevated temperatures could also be the result of burial in the earth's crust. Potassium-argon ages, particularly of older rock units, are therefore accepted with care as far as yielding an absolute age of the rock is concerned. By contrast, they may be very useful in dating particular metamorphic events during the lifetime of a rock.

The great advantages of the potassium-argon method are the abundance of potassium and the relatively short half-life of K^{40}. The method may be employed on rocks as young as 1×10^5 years, which is a great advantage, for example, in dating the recent reversals of the earth's magnetic field. No other method is available for such young rocks, and there is in fact a gap in the availability of absolute age methods between the youngest rocks dated by K–A and the oldest material which can be dated by carbon-14.

The potassium-argon method has been applied to meteorites (Gerling and Pavlova, 1951; Stoenner and Zähringer, 1958), to measure the time since the meteorite was sufficiently cool to retain gas. In this application, however, the problems are very great. Most meteorites, stony and iron, have very low concentrations of potassium. Furthermore, the isotopes K^{40} and A^{40} may both be formed in anomalous amounts by nuclear reactions resulting from cosmic rays to which the meteorite was exposed. Measurements on samples from the centers of large meteorites are probably most reliable. Gerling and Pavlova obtained a concentration of ages between 4.0 and 4.5×10^9 years.

23.3 METHODS BASED ON THE DECAY OF URANIUM AND THORIUM

The radioactive decay of the isotopes U^{238} and U^{235}, and of Th^{232}, has been discussed above in connection with heating by radioactivity. These decay schemes (Table 22.1) have always played a most important role in geochronology and related studies (Russell and Farquhar, 1960). Because two radioactive isotopes, with different decay constants, are available in any uranium mineral, a check on the consistency of deduced ages is available. This is not the case in the rubidium-strontium or potassium–argon methods. By consideration of the isotope ratios of lead itself, it is possible to deduce an age for the earth, and also to trace the history of a particular sample of lead.

It is necessary to add a word concerning the *equilibrium* of the decay systems. As Table 22.1 indicates, many unstable intermediate products, with various decay constants, are involved. However, it is apparent that in each case the decay constant of the parent is much smaller than those of the intermediate daughter products. In this situation, the amounts of the intermediate products become constant after a time which is long compared to their half-lives, but short compared to the half-life of the parent, and the end product accumulates at a rate which depends only upon the decay constant of the parent. To indicate the proof of this, suppose A is the number of atoms of the parent present at time t, and B is the number of atoms of the first daughter product, while λ_1 and λ_2 are the respective decay constants.

Then

$$\frac{dB}{dt} = \lambda_1 A - \lambda_2 B \qquad 23.3.1$$

It may be verified by substitution that the solution is

$$B = \frac{A\lambda_1}{\lambda_2 - \lambda_1}\left[1 - e^{(\lambda_1 - \lambda_2)t}\right] \qquad 23.3.2$$

If $\lambda_2 \gg \lambda_1$, and t is several half-lives of B, this becomes

$$B = \frac{\lambda_1 A}{\lambda_2} \qquad 23.3.3$$

Similar expressions result from all other intermediate products. The number of atoms of the stable end product P at time t is given by

$$\begin{aligned} P &= A_0 - B - C \cdots \\ &= Ae^{\lambda_1 t} - A\left(\frac{\lambda_1}{\lambda_2} + \frac{\lambda_1}{\lambda_3} + \cdots\right) \end{aligned} \qquad 23.3.4$$

The sum involved in the bracketed term of Equation 23.3.4 is small compared to $e^{\lambda_1 t}$ after a few million years, and the amount of P that has accumulated is given by the first term, which depends only upon λ_1.

Different quantities can be measured to determine the age of a mineral of uranium or thorium, and over the years, attention has been directed toward various procedures.

CHEMICAL ANALYSIS FOR URANIUM OR THORIUM AND HELIUM. We have referred to the isotopes of lead as the stable end products, but the α particles emitted at several stages of the decay scheme are helium nuclei, and helium gas also accumulates in the mineral. Some early age determinations were based on these measurements, but it will be apparent, following the discussion above of argon loss, that helium loss must be a serious factor. This was shown by Keevil (1941) and the method is no longer accepted.

CHEMICAL ANALYSIS FOR LEAD AND URANIUM OR THORIUM. The complication which we have not yet mentioned is the presence of common lead, which contains the isotopes Pb^{204}, Pb^{206}, Pb^{207}, and Pb^{208}. Although some early work was done by determining the atomic weight of lead in minerals, it is now accepted that mass spectrometry is essential.

DETERMINATION OF URANIUM AND MASS SPECTROGRAPHIC ANALYSIS OF LEAD. Ages are determined from the ratios Pb^{206}/U, Pb^{207}/U, and if thorium is present, from Pb^{208}/Th. This may be considered as the standard method of dating, although it is to be noted that an age can be determined from the lead isotope ratios alone. The present ratio of Pb^{207} and Pb^{206} is given by

$$\frac{Pb^{207}}{Pb^{206}} = \cdot\left(\frac{U^{235}}{U^{238}}\right)_{\text{present}} \cdot \frac{e^{\lambda_{235}t} - 1}{e^{\lambda_{238}t} - 1} \qquad 23.3.5$$

and the present ratio of the uranium isotopes is a well determined quantity.

The difficulty is that the possibility of the decay schemes becoming disturbed—for example, through the loss of uranium or an intermediate product such as the gas radon, or the gain of lead from another source—is sufficiently great that it is essential to consider individually the two values of t provided by U^{235} and U^{238}. Wetherill (1956) showed that if the ratios are plotted as in Fig-

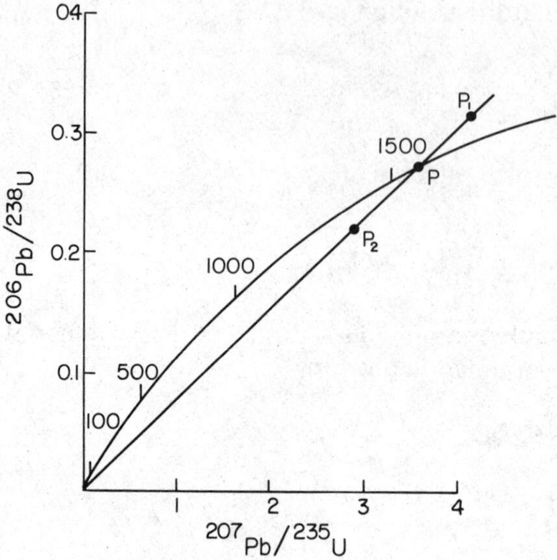

Figure 23.3 The concordia plot for lead isotopes, with the age of an undisturbed sample in millions of years. Recent loss of uranium or lead causes the ratios to plot at P_1 and P_2 respectively, instead of at P.

ure 23.3, an undisturbed system in which all the lead results from uranium decay in the mineral will yield a point on the curved line known as the *concordia,* and the resulting age is called a concordant lead age.

If the measured ratios on a mineral do plot on the concordia, the age of that mineral is determined with a high degree of certainty. The difficulty is that many minerals have a history of lead or uranium loss, giving ratios which are known as discordant. If a system very recently lost lead, the result would be equivalent to mixing with fresh uranium, and the Pb/U ratios would fall at points (such as P_2 in Figure 23.3) on the chord joining the undisturbed position to the origin. Recent loss of uranium would cause the ratios to plot at a point such as P_1. For loss of lead at a general time t_1 years ago, from a sample of age t_2, the point lies on a chord joining the points t_1 and t_2 of the concordia. It is possible that loss of lead does not occur at discrete episodes in the history of the mineral, but rather as a continuous process, through diffusion. Tilton (1960) showed that in this case, for a mineral of age t_2, the ratios would fall on a smooth curve joining the origin to the point t_2 of the concordia.

The study of concordant and discordant ratios can obviously provide a very detailed history of a mineral. From the point of view of geochronology, the aim is to determine the minerals which are most often concordant. It has been found (Catanzaro, 1968) that zircon ($ZrSiO_4$) is particularly useful in this regard. Zircon occurs as an accessory mineral in many rocks, and appears to represent a very nearly closed system for uranium and lead. The uranium is present as a small fraction of the atoms, substituting for zirconium. Lead resulting from the uranium decay is trapped in the crystal or grain, and can be preserved through metamorphic events.

We turn now to a consideration of the istopes of lead itself, including the non-radiogenic Pb^{204} (Kanasewich, 1968). The aim is to determine the original or primeval isotope ratios of lead, and thereby an age for the earth as a whole, by the study of leads which were separated into lead minerals at different times in the earth's history. For this study, it is convenient to introduce symbols for the various ratios, as shown in Table 23.2; the numerical values listed in column 2 are not known initially. We let t_0 be the time of formation of the earth, and t_1 the time of separation of a particular lead, from a closed system containing uranium and thorium, into an isolated lead mineral; both times are considered relative to $t = 0$ at present, with time increasing toward the past. The time t_0 gives the date at which the mantle was first able to support regional differences in its uranium-lead ratios. With this convention, it is easy to show that

$$x = a_0 + \mu(e^{\lambda t_0} - e^{\lambda t_1})$$

$$y = b_0 + \frac{\mu}{137.8}(e^{\lambda' t_0} - e^{\lambda' t_1}) \qquad 23.3.6$$

$$z = c_0 + W(e^{\lambda'' t_0} - e^{\lambda'' 1_1})$$

In these equations, x, y and z are quantities that can be measured on samples of lead of different ages t_1 (the age of the deposit being established independently). The quantities a_0, b_0, and c_0 for the isotope ratios of primeval lead, and the time t_0, are initially unknown. Houtermans (1946) suggested the use of these equations to determine the age of the earth. In practice, the great difficulty has been the selection of leads with simple histories, and it was only with the inclusion of leads from meteorites, which are characterized by simple histories, that meaningful solutions were obtained (Murthy and Patterson, 1962). The best solution of the equation yields a value of t_0 of 4.55×10^9 years (Kanasewich, 1968) and this is the time corresponding to the primeval ratios shown in Table 23.2.

For any closed uranium-lead system, the ratios Pb^{207}/Pb^{204} and Pb^{206}/Pb^{204} will change with time along one of a family of curves known as growth curves (Fig. 23.4). These curves start at a point which corresponds to the isotope ratios or primeval lead; their curvature depends upon the initial ratio of uranium to Pb^{204} (μ) in the system. Lead separated from the system at time t_1 will have isotope ratios which plot on a straight line, known as an isochron, intersecting the growth curves and passing through their common initial point. This may be seen from Equations 23.3.6, since for constant t_1

$$\frac{y - b_0}{x - a_0} = \frac{e^{\lambda' t_0} - e^{\lambda' t_1}}{137.8(e^{\lambda t_0} - e^{\lambda t_1})} = \text{constant} \qquad 23.3.7$$

The values of x and y from a lead deposit which has had a simple history will

TABLE 23.2 (AFTER KANASEWICH, 1968).

	Value or Symbol	
Isotope ratio	4.55×10^9 *years ago*	*At present*
Pb^{206}/Pb^{204}	$a_0 = 9.56$	x
Pb^{207}/Pb^{206}	$b_0 = 10.42$	y
Pb^{208}/Pb^{204}	$c_0 = 30.0$	z
U^{238}/Pb^{204}	2.0124μ	μ
U^{238}/U^{235}	3.33	137.8
Th^{232}/Pb^{204}	$1.2548W$	W

Decay Constants (10^{-9} yr^{-1})	
U^{235}	$\lambda' = 0.9722$
U^{238}	$\lambda = 0.1537$
Th^{232}	$\lambda'' = 0.0499$

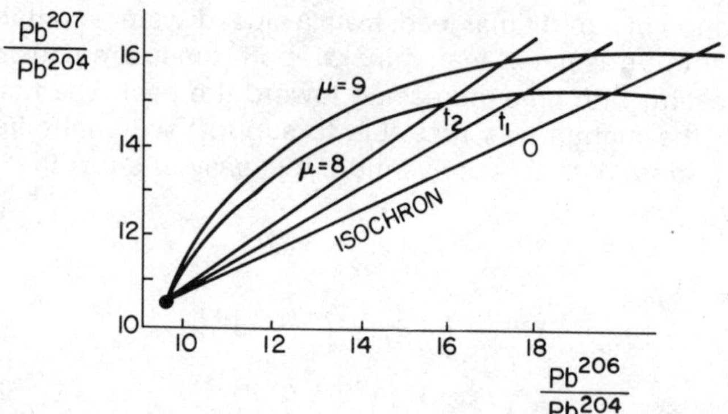

Figure 23.4 Growth curves for lead for two values of the initial ratio of uranium to lead (μ). Samples of lead separated from the source at time t (measured from 0 at the present) fall on straight lines known as isochrons.

plot as a point on a certain growth curve, provided the ratios are determined with sufficient accuracy. This has not always been the case, for the resolution of the mass abundances of relatively heavy isotopes such as those of lead requires precision mass spectrometry. Kanasewich (1968) has shown that many of the earlier interpretations were in error because of inaccurate determinations, particularly of the small Pb^{204} content. If the lead in a given deposit is anomalous, in that it is a mixture of contributions from the same region of the mantle or crust made at two different times, the values of x and y will plot on a straight line, joining the points of intersection of two isochrons with the growth curve. It can be seen that considerably more complicated histories are possible, which will lead to different distributions on the x-y plot. The present evidence suggests that single-stage isochrons are very rare, if present at all. While the mantle apparently does have the local inhomogeneities corresponding to different values of μ required by the simple theory, the isotope ratios from virtually all lead deposits plot on curves other than the simple isochrons. This suggests that the mantle material, from which the deposits were derived, was frequently mixed during the earth's history.

23.4 THE GEOLOGICAL TIME SCALE

Although the availability of physical methods for dating rocks has been of the greatest importance in the study of specific areas, it is difficult, on a global scale, to take account of more than a few critical dates: the age of the earth, the oldest rock age measured, the main points of the geological time scale, and perhaps the dates of the last few reversals of the geomagnetic field. It is particularly important to recognize that radiometric ages, apart from lead-zircon ages, almost always give the date of the last orogeny, rather than the age of the rock material.

From the time absolute ages first became available, an attempt has been made to relate these to the stratigraphic time scale. Kulp (1961) reviewed the reliable ages that were available for rocks whose stratigraphic position is known, and prepared a time scale. In general, the dates for the beginnings of periods were not greatly different from a scale deduced much earlier by Holmes (1947), and widely used. Recent minor revisions of Kulp's scales, by a Commission of the International Union of Geological Sciences (IUGS, 1967), are incorporated in Figure 23.5.

To the stratigraphic dates in Figure 23.5 have been added the ages of the earth, and of the oldest known rock. The very long time interval between these two should be noted. Apparently, in the early history of the earth, conditions such as the temperature were such that no relic of an original crust has been preserved. Very old rock ages have been reported from Africa and Greenland. In the case of the latter, Moorbath et al. (1972) give a Rb-Sr age of 3.75×10^9 years for rocks near Godthaab on the southwest coast.

We may note also the great length of the Precambrian, from the oldest rock to the beginning of the Cambrian. Although it is not convenient to show intermediate dates on a time scale (since divisions of the Precambrian do not

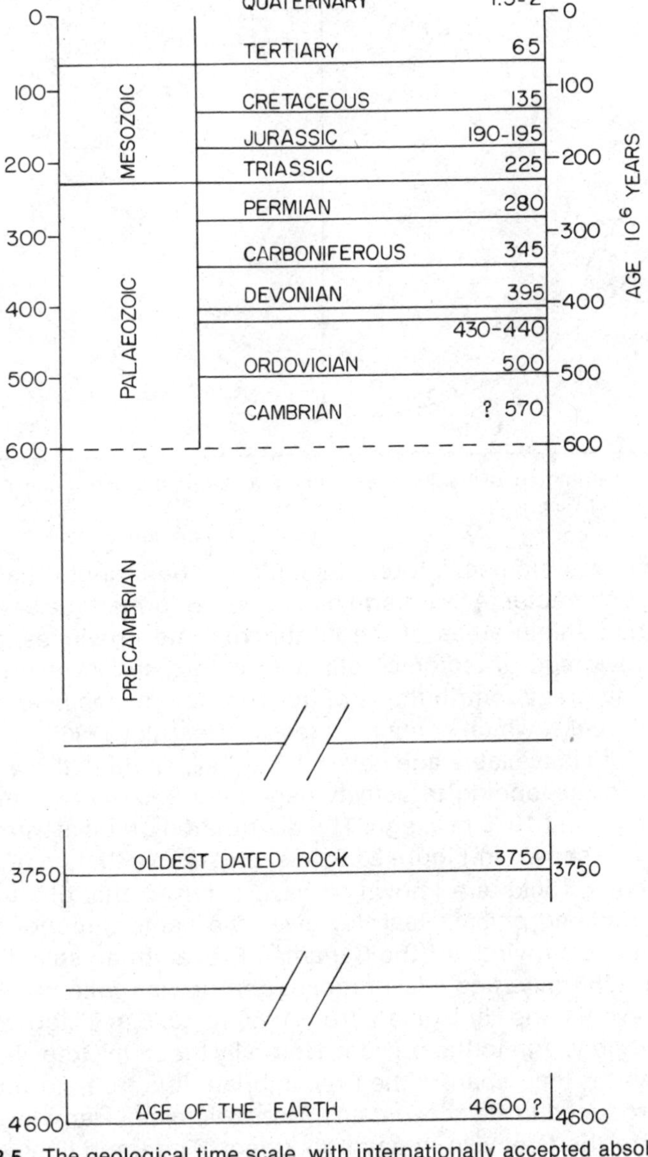

Figure 23.5 The geological time scale, with internationally accepted absolute dates.

Figure 23.6 Distribution of earlier Precambrian rocks of the earth—the Superior regime, with trend lines. (After Dearnley)

Illustration continued on opposite page

yet have a universal nomenclature), it should not be thought that this span of time is without character. Age determinations have formed the basis of dividing the Precambrian shield areas of the continents into provinces, each characterized by its own age of tectonic activity (Bickford and Wetherill, 1965). Certainly one of the great contributions of geochronology has been the dating of events in these rocks, which contain no fossils. Dearnley (1966) has shown that if a histogram of all available age determinations is plotted, three Precambrian peaks occur, corresponding to activity beginning 2750 m.y. (1 m.y. = 1×10^6 years), 1950 m.y., and 1075 m.y. ago. The distribution of rocks with ages greater than 1950 m.y. is shown in Figure 23.6. It is remarkable that, in every continent except Antarctica, rocks are known to have survived this great span of time without later metamorphism. Dearnley gives the name Superior regime (after the corresponding province of the Canadian Precambrian shield) to the event which began 2750 m.y. ago. He proposes that it was followed by two other major global events: the Hudsonian, from 1950 to 1075 m.y. ago, and the Grenville, from 1075 m.y. ago to the present. Normally the term Grenville is restricted to a much shorter time span in the Precambrian. It is true, however, that orogenic belts produced in Palaeozoic and Mesozoic times were developed along lines parallel to the Grenville in each continent. The trend lines (directions of

Figure 23.6 *Continued*

folding and faulting) of the two later regimes tend to swirl around the Superior nuclei, generally cutting across the older trends and forming a more complicated pattern. There is evidence, as Dearnley shows, that if the continents are reassembled according to the evidence of paleomagnetism (Section 21.3) and to theories of global tectonics (Chapter 27), both the relatively simple Superior and more complicated Grenville features form consistent global patterns. This might suggest that the continental masses remained as one block until the most recent period of separation, but other evidence (Section 27.1) makes this unlikely.

The question of whether continental crust is all a remnant of a primeval sialic crust, reworked at various times, or whether it has been added to from the mantle during episodes of orogeny, is one of the most fundamental problems of earth history. Much of the evidence is geochemical in nature and outside of the scope of this book, but there is one line of evidence which is the direct outcome of Rb-Sr age determinations, and it is convenient to consider it here.

In Section 23.1, the concept of the initial ratio of Sr^{87}/Sr^{86} was introduced, as the ratio characteristic of the source magma of a dated rock. We first investigate how the ratio Sr^{87}/Sr^{86} has varied, through geological time, in both mantle and crust. As we have noted, the half life of Rb^{87} is very long even compared to the age of the earth, and this permits a simplification of equation 23.1.3. Using the series expansion for $e^{\lambda t}$:

$$e^{\lambda t} = 1 + \lambda t + \frac{(\lambda t)^2}{2!} + \frac{(\lambda t)^3}{3!} + \cdots$$

we have, for λt small,

$$e^{\lambda t} \doteq 1 + \lambda t$$

$$\therefore \left(\frac{Sr^{87}}{Sr^{86}}\right) \doteq \left(\frac{Sr^{87}}{Sr^{86}}\right)_{\text{Initial}} + \left(\frac{Rb^{87}}{Sr^{86}}\right) \lambda t \qquad 23.4.1$$

A plot of Sr^{87}/Sr^{86} against time t, for a rock or mineral in a reasonably homogeneous environment, therefore yields a straight line whose slope is $(Rb^{87}/Sr^{86} \cdot \lambda)$, the isotope ratio being the present value in the local environment. With the introduction of a numerical factor determined by the atomic weights, the slope may also be expressed in terms of the present concentration (rather than isotopic) ratio, Rb/Sr. In the case of the earth's mantle, as a whole, the initial ratio of Sr^{87}/Sr^{86} would be the value at the date of formation of the earth, and the present value of Rb/Sr may be estimated as 0.025 (Faure and Powell, 1972). The ratio Sr^{87}/Sr^{86} must therefore have grown with time as suggested by the lower line in Figure 23.7. A magma extracted directly from the mantle at any time t should produce rocks whose initial ratio corresponds to the point on this line at time t. On the other hand, the present ratio Rb/Sr within the continental crust is known to be very much greater, perhaps 0.18 (Faure and Powell, 1972). Within the crust, the average growth rate of the ratio Sr^{87}/Sr^{86} is along a line whose slope is approximately seven times that of the mantle growth line. The upper line of Figure 23.7 suggests the growth for a volume of average crust extracted from the mantle 2700 years ago. A magma formed by melting of a portion of this crust at a subsequent time t' would form rocks whose average initial ratio corresponds to the point on the upper curve at time t'. There is thus the possibility of using initial ratios, as measured in geochronology, to determine the magma source region of a rock, an approach proposed by Hurley et al. (1962).

More recent work has demonstrated that a variety of igneous rocks of very different ages, from ancient shields to mid-ocean ridges (Figure 23.7), have initial ratios lying very close to the mantle line (Moorbath, 1975; O'nions and Pankhurst, 1978). The suggestion is clear that these rocks represent material newly added from the mantle, rather than from the reworking of an older or primeval crust. Whether the small scatter in these initial ratios from an ideal growth curve is a true measure of mantle chemical heterogeneity, or simply a manifestation of measurement error, is not yet resolved.

23.5 OTHER METHODS OF GEOCHRONOLOGY

The methods of age determination based upon the radioactive decay of the appropriate isotopes of uranium, potassium or rubidium have provided the key points on the time scale described above. However, certain other approaches have promise for the future, and are, in addition, of very considerable scientific interest. We shall mention three of these.

FISSION TRACK DATING. This method depends upon the spontaneous fission of an isotope such as U^{238}, which proceeds with the very long half-life of 8.0×10^{15} years (Kuroda, 1963). When the product particles of fission travel through a material, they leave records of their passing in the form of cylindrical regions in which atoms are displaced from their normal positions. Normally, this region is invisible, but Fleischer and colleagues (1965) showed that, for many crystalline materials as well as glass, etching would reveal the particle tracks. Furthermore, they showed that the density of tracks is a function of the age of the material, and can be used for dating. The requirement is that a natural material have a sufficient concentration of a fissionable element, usually uranium, for tracks to be produced. Counts of the density of etched tracks, per unit surface area of a thin sample, are made in the material in its natural state and again after it has been exposed to a flux of thermal neutrons. These induce fission of atoms of U^{235}, producing a greatly increased number of tracks, whose density is a measure of the concentration of uranium. The age, which is the time since the sample was cool enough to preserve tracks, can be obtained from the two densities.

Fission may also be induced in other heavy nuclei by a flux of cosmic ray particles. Counting of the tracks by the method above can be used to estimate an exposure age for meteorites (Fleischer, Price, and Walker, 1968).

EXTINCT RADIOACTIVITIES. It appears that certain radioactive elements formed in the early stages of the genesis of the elements of the galaxy have become extinct through radioactive decay. In some cases the measurement of the stable daughter products from these extinct species can be used for chronology, in particular, of meteorites.

Certain stony meteorites show an excess of the xenon isotope ${}_{54}Xe^{129}$. This isotope is known to be formed from the β decay of ${}_{53}I^{129}$ which has a half-life of 17×10^6 years (Reynolds 1963). The isotopes of iodine (I^{127}, stable, and I^{129}, radioactive) are believed to have been formed by fast neutron capture during the period of galactic synthesis. Wet let this time span be T. If trapping of Xe^{129} in the parent body of a meteorite began a time ΔT after synthesis was complete, the ratio of isotopes would be

$$\frac{Xe^{129}}{I^{127}} = \left(\frac{I^{129}}{I^{127}}\right)_0 e^{-\Delta t/\tau} = \left(\frac{\tau}{T}\right) e^{-\Delta t/\tau}$$

where the subscript 0 refers to the ratio at the end of nucleon synthesis, and τ is the mean life $\left(\frac{1}{\lambda} = 25 \times 10^6 \text{ years}\right)$ of I^{129}. Measurements of the abundance of radiogenic X^{129} thus lead to estimates of the relative magnitudes of T and ΔT. Reynolds suggests values of 2×10^{10} years for T, and in some cases the relatively short period of 50×10^6 years for ΔT. This would suggest that the bodies of the solar system evolved very quickly, compared to the time during which the elements were formed.

Meteorites also show variations in the other isotopes of xenon, including Xe^{131}, Xe^{132}, Xe^{134}, and Xe^{136}, which are known to be formed in the fission of the plutonium isotope Pu^{244}. It has been suggested that Pu^{244} was also formed by fast neutron capture during nucleogenesis, and has become extinct. The difficulty of using one of the above radiogenic isotopes for chronology is that there is no other isotope of plutonium available for a control, but Hohenberg, Munk, and Reynolds (1962) suggest that plutonium may always have been associated with uranium in a constant ratio. In that case, the use of a measurable

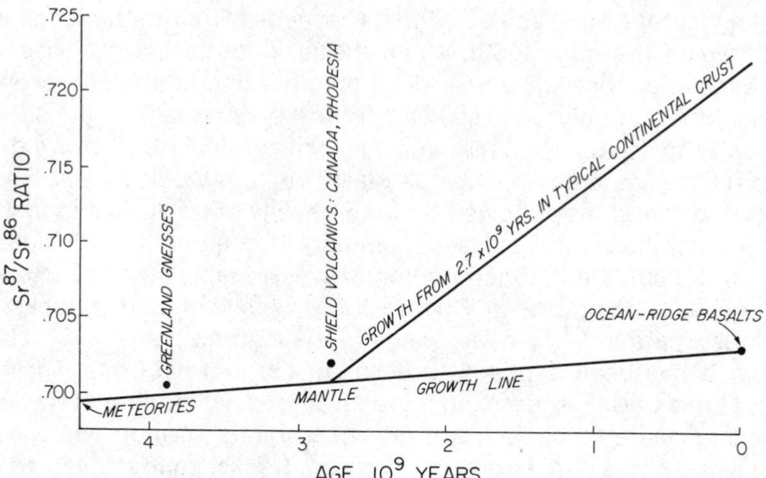

Figure 23.7 Probable variation with time of the Sr^{87}/Sr^{86} ratio in the mantle and in typical continental crust. Also shown are the measured initial ratios for meteorites, ancient gneisses, shield volcanics, and ocean-ridge basalts.

ratio such as Xe^{136}/U^{238} may be used to deduce the relative cooling histories of different meteorites. Measurements which have been made to date suggest that achondrites cooled sufficiently to retain gas as much as 4×10^8 years later than chondrites.

CARBON-14. The isotope C^{14} is formed in the atmosphere, at a height of about 1600 m, by cosmic ray collisions with N^{14}. It is radioactive, with a half-life of about 5730 years (Goodwin, 1962). The principle of the method of dating is that living organisms acquire a very small amount of C^{14} during their lifetimes, through the intake of CO_2; after their death, the C^{14} decays according to the usual rule, so that a measurement of the present abundance, in fossil material, provides the age. It is, of course, necessary to know the original concentration, and it is here that errors may be introduced. The rate of production of C^{14} depends upon the cosmic ray flux, which is influenced by the earth's magnetic field (Chapter 19). The rate of mixing in the atmosphere, and exchange of carbon with the oceans, both influence the distribution of C^{14} through the biosphere. Libby (1955) has discussed both the assumptions involved, and the counting techniques which are used in the laboratory; recent developments are given in Rafter and Grant-Taylor (1972). Since the method is applicable only to materials whose ages are a few half-lives of C^{14} (70,000 years at most), its application to geophysics is limited. Of particular importance is the dating of beach deposits, in order to determine the rate of vertical movement of the land surface (Chapter 27).

Problems

1. A rock of age 2.0×10^9 years is to be dated by the rubidium-strontium method. To what percent accuracy must the slope of the isochron be determined, if this age is to be accurate to 5 per cent?

2. Show that if a sample of lead is a mixture of two contributions, separated at times t_1 and t_2 from the same source region of the mantle, its ratios will plot on a chord joining the points of intersections of the corresponding isochrons with the growth curve (Fig. 23.4).

3. Replot the areas of Superior regime rocks, with their trend lines, on a map of the earth on which the continents are reassembled according to Figures 27.1 and 27.2. Is it possible to relate all of the trends to a single global system of orogeny?

4. In Section 23.5 the quantity "mean life" of a radioactive species was introduced, and defined as $\frac{1}{\lambda}$. Using Equation 22.2.2, show that this quantity does have the significance of a mean value, and determine its relation to half-life.

BIBLIOGRAPHY *(Chapter 23)*

Aldrich, L. T., Wetherill, G. W., Tilton, G. R., and Davis, G. L. (1956) Half-life of 87 Rb. *Phys. Rev.*, **103**, 1045.

Bickford, M. E., and Wetherill, G. W. (1965) Compilation of Precambrian geochronological data for North America. In *Geochronology of North America.* U.S. Nat. Acad. of Sciences, Pub. 1276, Washington, D.C.

Burnett, D. S., and Wasserburg, G. J. (1967) ^{87}Rb/^{87}Sr ages of silicate inclusions in iron meteorites. *Earth Plan. Sci. Letters,* **2**, 397.

Catanzaro, E. J. (1968) The interpretation of zircon ages. In *Radiometric dating for geologists,* ed. E. I. Hamilton and R. M. Farquhar. Interscience, London.

Compston, W., Jeffrey, P. M., and Riley, G. H. (1960) Age of emplacement of granites. *Nature,* **186**, 702.

Dalrymple, G. B., and Moore, J. G. (1968) Argon 40: Excess in submarine flow basalts from Kilauea volcano, Hawaii. *Science,* **155**, 1132.

Damon, Paul E. (1968) Potassium-argon dating of igneous and metamorphic rocks with applications to the Basin ranges of Arizona and Sonora. In *Radiometric dating for geologists,* ed. E. I. Hamilton and R. M. Farquhar. Interscience, London.

Dearnley, R. (1966) Orogenic fold belts and a hypothesis of earth evolution. In *Physics and chemistry of the earth,* 7, ed. L. H. Ahrens, Frank Press, S. K. Runcorn, and H. C. Urey. Pergamon Press, London.

Faure, G., and Powell, J. L. (1972) *Strontium Isotope Geology.* Springer-Verlag, Berlin.

Fleischer, R. L., Price, P. B., Symes, E. M., and Miller, D. S. (1964) Fission-track ages and track-annealing behavior of some micas. *Science,* **143**, 349.

Fleischer, R. L., Price, P. Buford, and Walker, R. M. (1968) Charged particle tracks: tools for geochronology and meteorite studies. In *Radiometric dating for geologists,* ed. E. I. Hamilton and R. M. Farquhar. Interscience, London.

Flynn, K. F., and Glendenin, L. E. (1959) Half-life and beta spectrum of 87 Rb. *Phys. Rev.,* **116**, 744.

Gerling, E. K., and Pavlova, T. G. (1951) Determination of the geological age of two stone meteorites by the argon method. *Doklady. Akad. Nauk. SSSR,* **77**, 85.

Goldschmidt, V. M. (1937) Geochemische Verteilungsgesetze der Elemente, IX. *Skrifter Norske Videnskaps. Akad. Oslo, Mat. Naturv. Kl.* **4**.

Goodwin, H. (1962) Dating conference, Cambridge University, England. *Nature,* **195**, 943.

Hahn, O., and Walling, E. (1938) Uber die Möglichkeit geologischer Alterbestimmungen Rubidiumhaliger Mineralen und Gesteine. *Z. anorg. u. allgem. Chem.,* **236**, 78.

Hohenberg, C. M., Munk, M. N., and Reynolds, J. H. (1967) Spallation and fissiogenic Xenon and Krypton from stepwise heating of the Pasamote achondrite; the case for extinct Plutonium 244 in meteorites; relative ages of chondrites and achondrites. *J. Geophys. Res.,* **72**, 3139.

Holmes, A. (1947) A revised estimate of the age of the earth. *Nature,* **163**, 453.

Houtermans, F. G. (1946) The isotope ratios in natural lead and the age of uranium. *Naturwissenschaften,* **33**, 185.

Hurley, P. M., Hughes, H., Faure, G., Fairbairn, H. W., and Pinson, W. H. (1962) Radiogenic strontium-87 model of continental formation. *J. Geophys. Res.,* **67**, 5315–5334.

IUGS (1968) International Union of Geological Sciences, Commission on Geochronilogy. A comparative table of recently published time-scales for the phanerozoic time—explanatory notice. *Geologie en Mijnbouw,* 406.

Kanasewich, E. R. (1968) The interpretation of lead isotopes and their geological significance. In *Radiometric dating for geologists,* ed. E. I. Hamilton and R. M. Farquhar. Interscience, London.

Keevil, N. B. (1941) Helium retentivities of minerals. *Amer. Mineralogist,* **26**, 405.

Kulp, J. L. (1961) Geological time scale. *Science,* **133**, 1105.

Kuroda, P. K. (1963) Dating methods based on the process of nuclear fission. In *Radioactive dating.* International Atomic Energy Agency, Vienna.

Libby, W. F. (1955) *Radiocarbon dating* (2nd Ed.) Univ. of Chicago Press, Chicago.

McDougall, I., and Stipp, J. J. (1969) Potassium-argon isochrons (abst.). *Trans. Amer. Geophys. Un.*, **50**, 330.

Merrihue, C. M., and Turner, G. (1966) Potassium-argon dating by activation with fast neutrons. *J. Geophys. Res.*, **71**, 2852.

Moorbath, S. (1975) Evolution of Precambrian crust from strontium isotope evidence. *Nature*, **254**, 395–398.

Moorbath, S., O'nions, R. K., Pankhurst, R. J., Sale, N. H., and McGregor, V. R. (1972) Further rubidium-strontium age determinations on the very early Precambrian rocks of the Godthaab District, West Greenland. *Nature*, (Physical Sciences), **240**, 78–82.

Murthy, V. R., and Patterson, C. C. (1962) Primary isochron of zero age for meteorites and the earth. *J. Geophys. Res.*, **67**, 1161.

O'nions, R. K., and Pankhurst, R. J. (1978) Early Archean rocks and geochemical evolution of the earth's crust. *Earth and Plan. Sci. Letters*, **38**, 211–236.

Rafter, T. A., and Grant-Taylor, T., eds. (1972) Proceedings of the 8th International Conference on Radiocarbon Dating. The Royal Society of New Zealand, Wellington.

Reynolds, J. H. (1963) Xenology. *J. Geophys. Res.*, **68**, 2939.

Russell, R. D., and Farquhar, R. M. (1960) *Lead isotopes in geology*. Interscience, New York.

Steiger, R. H. and Jäger, E. (1977) Subcommission on Geochronology: Convention on the use of decay constants in geo- and cosmochronology. *Earth. and Plan. Sci. Letters*, **36**, 359–362.

Stoenner, R. W., and Zähringer, J. (1958) Potassium-argon age of iron meteorites. *Geochimica et Cosmochimica Acta*, **15**, 40.

Tilton, G. R. (1960) Volume diffusion as a mechanism for discordant lead ages. *J. Geophys. Res.*, **65**, 2933.

Wetherill, G. W. (1956) Discordant uranium-lead ages I. *Trans. Amer. Geophys. Un.*, **37**, 320.

York, D. (1967) The best isochron. *Earth Plan. Sci. Letters*, **2**, 479.

York, D., Baksi, A. K., and Aumento, F. (1969) K-Ar dating of basalts dredged from the North Atlantic (abst.). *Trans. Amer. Geophys. Un.*, **50**, 353.

York, D., and Berger, G. W. (1970) $^{40}Ar/^{39}Ar$ age determinations in nepheline and basic whole rocks. *Earth Plan. Sci. Letters*, **7**, 333.

OUTFLOW OF HEAT FROM THE EARTH 24

24.1 FUNDAMENTAL RELATIONS

It has been known for many years that temperature, as measured in mines or boreholes, increases with depth. Before considering other sources of heat, we must investigate the possible effect of adiabatic compression.

The temperature gradient set up in a substance that has been compressed adiabatically is known as the adiabatic gradient, and is used very frequently as a standard of comparison. It is worth deriving it from fundamental principles of thermodynamics. When a quantity dQ of heat enters a unit mass of a substance, internal energy E and volume v change in such a way that

$$dQ = dE + p\,dv \qquad 24.1.1$$

where p is the pressure. We rewrite this as

$$\begin{aligned} dQ &= d(E + pv) - v\,dp \\ &= C_P\,dT + a\,dp \end{aligned} \qquad 24.1.2$$

where C_P is the specific heat at constant pressure, T is the temperature, and a is a quantity to be determined. It is left as an exercise to establish that

$$\frac{\partial}{\partial T}(a + v) = \frac{\partial C_P}{\partial p}$$

and

$$\frac{\partial}{\partial p}\left(\frac{C_P}{T}\right) = \frac{\partial}{\partial T}\left(\frac{a}{T}\right).$$

These yield

$$a = -T\frac{\partial v}{\partial T} \qquad 24.1.3$$

We may therefore write

$$dQ = C_P\, dT - T \frac{\partial v}{\partial T}\, dp$$

and for an adiabatic compression, $dQ = 0$, so that

$$\frac{dT}{dp} = \frac{T}{C_P} \frac{\partial v}{\partial T} = \frac{T\,\alpha}{C_P\,\rho} \qquad 24.1.4$$

where α is the volume coefficient of thermal expansion and ρ is the density. We obtain the gradient in terms of depth z through the relation

$$\frac{dp}{dz} = g\rho$$

so that

$$\frac{dT}{dz} = \frac{g\alpha T}{C_P} \qquad 24.1.5$$

The adiabatic gradient given by Equation 24.1.5 has special significance in the case of cooling fluids, and in establishing the possibility of convection (Section 25.6).

To estimate the numerical value of the adiabatic gradient at moderate depths in the earth, we may take $\alpha = (2\times 10^{-5})°C^{-1}$, $C_P = 8\times 10^6$ ergs/gm°C, and $T = 700°K$. The gradient is then only 0.15°/km, whereas the temperature gradient observed near the earth's surface is commonly 30°km. This in itself is evidence of a flux of heat from internal sources. The flux is maintained through the crust by thermal conduction, although convection may play a more important role in the mantle, and we turn next to a consideration of conduction.

Within a uniform, isotropic medium, the vector flux of heat q at any point is proportional to the gradient of the temperature, or

$$q = -K\nabla T \qquad 24.1.6$$

where K is a property of the material, known as the thermal conductivity, and q is measured in units of energy per unit area per unit time. If we now consider a small volume of the material, it is apparent that the change in temperature with time within it will depend upon three factors: the net flow of heat across its surface, the rate of heat generation within it, and the thermal capacity of the material. The statement of this is

$$\frac{1}{\kappa} \frac{\partial T}{\partial t} = \nabla^2 T + \frac{A}{K} \qquad 24.1.7$$

where $\kappa = K/\rho c$, ρ being the density and c the specific heat. The quantity κ is known as the diffusivity, and A is the heat generated per unit time per unit volume. In the case of the earth, A will normally represent the heat produced by radioactive decay.

The simplest solutions of Equation 24.1.7 are for steady-state conditions in a medium free of heat sources, for then

$$\nabla^2 T = 0; \qquad 24.1.8$$

in other words, T satisfies Laplace's equation, and the solutions are identical to results from other potential fields. For example, the effect on the temperature gradient of a sphere of uniform thermal conductivity imbedded in an infinite medium of different thermal conductivity is completely analogous to that of a sphere of anomalous magnetic permeability in an otherwise uniform magnetic field.

A simple but important case of time-variable temperatures concerns the penetration into the earth of diurnal, seasonal or long-period variations in the surface temperature. It is sufficient to take the surface of the earth as the plane $z = 0$, with the positive z-axis downward. At the earth's surface, we assume the temperature T to vary as $A \cos \omega t$. For one-dimensional flow and no sources, Equation 24.1.7 reduces to

$$\frac{\partial^2 T}{\partial z^2} = \frac{1}{\kappa} \frac{\partial T}{\partial T}$$

of which a solution is

$$T = u(z) e^{i\omega t}$$

Substituting to determine $u(z)$, and matching the boundary conditions, we find

$$T = Ae^{-kz} \cos (\omega t - kz) \qquad 24.1.9$$

where

$$k = \left(\frac{\omega}{2\kappa}\right)^{1/2}$$

The surface variations in temperature penetrate into the earth as a damped wave, with a wavelength $\lambda = 2\pi/k$, and with an amplitude attenuated by the factor $e^{-2\pi}$ for each increase in z of one wavelength. If we take a representative value of κ, 0.01 cm^2/sec, we find that $\lambda = 1$ meter for a frequency of one cycle per day, $\lambda = 20$ meters for one cycle per year, and $\lambda = 2000$ meters for one cycle in 10^4 years.

The importance of these results for the measurement of heat flow from the interior is that, below the land surface, the diurnal variation in temperature is completely negligible at a depth of a very few meters, and the seasonal variation at a few tens of meters. If, however, there have been changes in climate extending over the past few thousand years, there may well be perturbations in temperature extending to depths of hundreds of meters, and these may be expected to disturb measurements of the temperature gradient. The effect of a change in surface temperature which is not harmonic in time may be computed by performing a Fourier analysis, and applying Equation 24.1.9 to each frequency component.

In the harmonic case, maxima and minima of surface temperature propagate into the earth with a velocity given by $(2\kappa\omega)^{1/2}$. The finite velocity of propagation of thermal effects is most important, and may be further illustrated by the effect on temperature of the intrusion of a sill at some depth. Still considering one-dimensional flow, let the temperature at time $t=0$ be given by $T=f(z)$; this initial distribution of temperature represents the effect of the hot intrusion, and we wish to determine the perturbation in temperature at a distant point at some later time. For simplicity, the effect of the earth's surface will be neglected. A solution of the one-dimensional flow equation is

$$T=\frac{1}{2\sqrt{\pi\kappa t}}\,e^{-(z-z')^2/4\kappa t};$$

it may be thought of as the effect at z of the release of a quantity of heat, $2\rho c(\pi\kappa)^{1/2}$ per unit area, over the plane $z=z'$, at $t=0$. Combining solutions for all values of z', the following is also a solution:

$$T=\frac{1}{2\sqrt{\pi\kappa t}}\int_{-\infty}^{\infty} f(z')\,e^{-(z-z')^2/4\kappa t}\,dz' \qquad 24.1.10$$

and the limit of this expression, as $t\to 0$, can be found to be $T=f(z)$ as required. Equation 24.1.10 can be evaluated for given t and z, provided the function $f(z)$ is reasonably simple. In particular, let

$$T=T_0 \qquad -a<z<a, \quad t=0$$
$$T=0 \qquad \text{elsewhere}, \quad t=0$$

In other words, the plane $z=0$ is the mid-plane of a sill of thickness $2a$. Then

$$T=\tfrac{1}{2}T_0\left\{\operatorname{erf}\left(\frac{a-z}{2\sqrt{\kappa t}}\right)+\operatorname{erf}\left(\frac{a+z}{2\sqrt{\kappa t}}\right)\right\} \qquad 24.1.11$$

where

$$\operatorname{erf} x=\frac{2}{\sqrt{\pi}}\int_0^x e^{-\xi^2}\,d\xi$$

As a specific example, the thickness $2a$ may be taken as 2 km, and the initial temperature T_0 as 1000°C; for $\kappa=0.01$ cm²/sec, as before, the following increases in temperature are found:

z \ t	10^4 yrs	10^6 yrs	10^8 yrs
10 km	0°	50°	10°
100 km	0°	0°	5°

The actual temperatures would be greatly modified by the presence of the earth's surface, but the calculation does indicate that new heat sources at depth within the crust are detectable only after periods of the order of 10^6 years, and at depth in the mantle, only after times of 10^8 years.

The extension of the theory of heat conduction to more general geometries and source distributions is treated fully in Carslaw and Jaeger (1959).

24.2 HEAT FLOW MEASUREMENTS

Near the earth's surface, the vertical flux of heat q is given by

$$q = K\left(\frac{\partial T}{\partial z}\right) \tag{24.2.1}$$

where q is positive when the flow is outward, and the gradient $(\partial T/\partial z)$ is taken as positive when T increases with depth. The problem of measuring q reduces to that of measuring $\partial T/\partial z$, and the thermal conductivity K in the same location.

The determination of the gradient requires the measurement of undisturbed temperatures at several depths. For land stations, this requires access to points at least 100 m deep, and preferably much deeper. While very satisfactory measurements have been reported from short horizontal boreholes off mine workings (Misener et al., 1951), measurements are usually made by lowering the sensing instrument down vertical boreholes (Bullard, 1960; Misener and Beck, 1960). Maximum thermometers, thermocouples and thermistors have been used as sensors, but the latter appear to be the most satisfactory. Temperatures can be read at the surface to within about 0.01°C by means of a simple resistance bridge, and measurements can be repeated at short intervals of depth to give the complete temperature–depth relationship.

Virtually all boreholes are filled with fluid, so that the primary quantity measured is the fluid temperature. If this is to represent undisturbed rock temperature at the same depth, the fluid should be free of convection, of disturbances caused by the probe itself, and of the disturbance produced by drilling. Krige (1939) has indicated that convection is unlikely to be serious in holes of diameter 5 cm or less, or even in larger holes if the fluid is a viscous mud. The disturbance due to drilling is most severe if, as is usually the case, fluid was circulated during the operation; Bullard (1947) has shown that the time required for the hole to reach equilibrium depends upon the time required for the original drilling, and may be many months.

The stringency of the requirements for boreholes is such as to make satisfactory holes relatively scarce, unless they are drilled specifically for heat flow measurements. Measurements are often made in cased holes, the effect of the casing upon the gradient being negligible except near its ends.

For many years it was believed that the measurement of heat flow through the ocean floor would present insuperable difficulties, but Bullard, Maxwell, and Revelle (1956) showed that this was not so. The bottom of the ocean is, in most cases, a region of great thermal stability, so that considerations of diurnal and seasonal variations do not arise. Furthermore, the ocean floor is normally covered by a layer of soft mud, into which a probe up to 5 meters long can be made to penetrate. If the probe carries a series of thermistors, the gradient may be measured. The temperature gradient in the ocean-floor muds is such that the difference in temperature over 5 meters is about 0.25°C, and it is obvious that great care is necessary if a gradient accurate to five per cent is to be obtained. The relative uniformity of the muds, and the freedom from the disturbing effects that can occur in boreholes, appear to make this possible. In the form now generally used, the probe consists of a standard piston corer, of the type used to collect long core samples, on which the thermistors are carried on outriggers. The separation of the thermistors from the body of the probe by a few centimeters reduces the possibility of temperature disturbance by the probe itself.

The thermal conductivity K must be measured for the same interval of depth as that over which the gradient is determined. For land stations, both laboratory and *in situ* methods have been employed. In the laboratory, the usual method involves a steady-state comparison experiment, in which thin discs cut from rock samples taken from the borehole are inserted into a divided bar of brass or copper, of the same diameter as the discs (Misener and Beck, 1960). Measurements of temperature gradient across the sample and in the bar, when the latter is heated at one end, give the ratio of thermal conductivity of the sample to that of the bar material. Experimental techniques are available to correct for thermal resistance between the rock and metal, but the method suffers from the disadvantage that the determination is made on a very small sample of material (about 2 cm in diameter and 0.5 cm in thickness) whose physical properties may well have changed since it was removed from its natural site. The thermal conductivity of a non-porous rock depends on the conductivities of its constituent minerals; quartz, for example, is the best conductor among the minerals usually encountered. If the rock is porous, the conductivity is influenced by the material filling the pore spaces. For a very coarsely crystalline rock, the divided-bar method may be incapable of yielding a representative conductivity, since a single crystal of quartz can extend through the disc, and effectively short-circuit the remainder of the sample.

A method of determination of thermal conductivity *in situ* avoids these difficulties. One approach is to insert into the borehole a long probe which itself contains a linear heater, capable of developing Q units of heat per second per unit length. Remote from the ends of the probe, the heat flow is essentially radial, and the temperature at the surface of the probe, at a long time t after heating is commenced, is given by (Blackwell, 1954):

$$T = \frac{Q}{4\pi K}\left\{ \ln \frac{4\kappa t}{a^2} + A + B\,\frac{a^2}{\kappa t} + C\,\frac{a^4}{\kappa^2 t^2} + \cdots \right\} \qquad 24.2.2$$

where a is the radius of the probe and A, B, C are constants. Under suitable conditions, which may involve heating for several hours at one site (Beck, 1965), the first term of equation 24.2.2 will predominate; in this case, the thermal conductivity K is found from the slope of a plot of T against $\ln t$. There is a complication if the rock is anisotropic. The conductivity found by the simple probe technique is that for the horizontal direction, while that required for heat flux calculations is the vertical thermal conductivity. A probe method which permits the vertical conductivity of an anisotropic rock to be measured has been described by Wright and Garland (1968).

For oceanic heat flow measurements, the material usually involved is a soft mud. It is most important that its thermal conductivity be measured before drying or other physical changes have commenced. A convenient method, which may be employed on shipboard immediately after a core is recovered, makes use of a miniature thermal probe in the form of a hypodermic needle containing a heating wire and thermistor (Von Herzen and Maxwell, 1959). Application of the theory of Equation 24.2.2 shows that the logarithmic portion of the curve is obtained a very few minutes after the needle is injected into the sample and heating commenced. However, measurement of the thermal conductivity in every case is not required. It turns out that the conductivity of ocean-floor muds is rather simply related to their percentage water content (Ratcliffe, 1960), and measurement of the latter quantity is often sufficient.

Combination of the values of gradient and thermal conductivity to yield a flux usually presents little scope for adjustment in the case of oceanic measurements. For determinations on land, the depth-temperature curve may cross several formations of different thermal conductivity. If the heat flux is truly vertical, the product $K\ \partial T/\partial z$ will be constant, even though the quantities individually vary (Fig. 24.1). A convenient method of determining the best value of

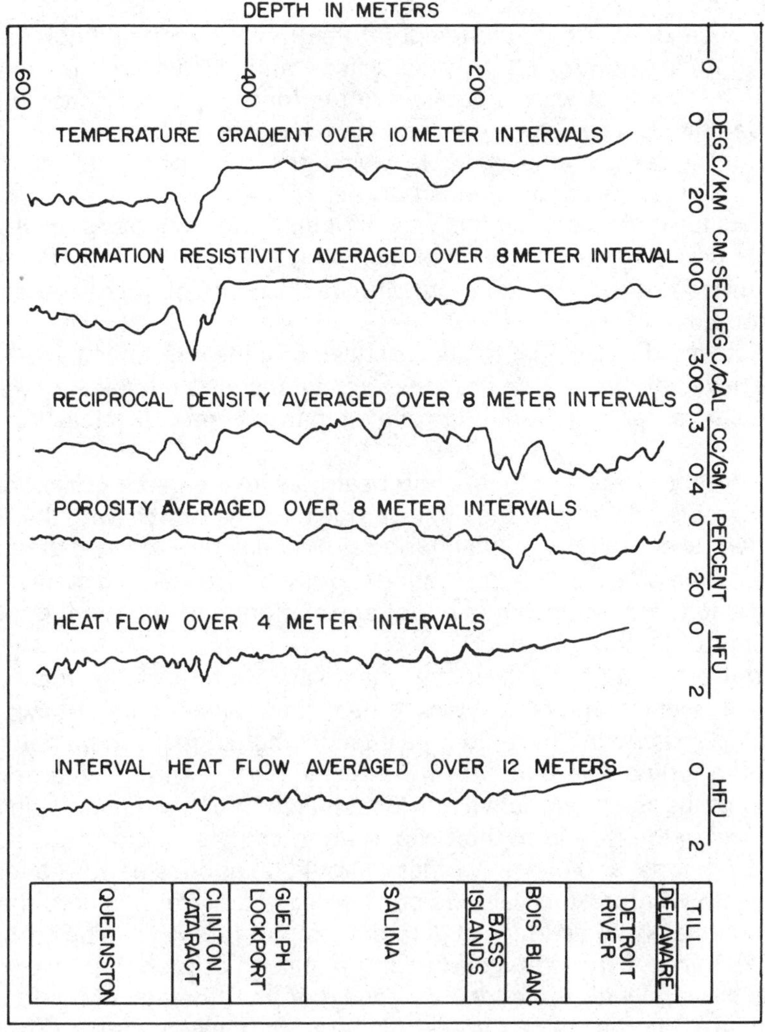

Figure 24.1 Measurements of temperature gradient and thermal resistivity, and the derived heat flow, for a borehole through a sedimentary section. Note the correlation between formation thermal resistivity (the reciprocal of conductivity), temperature gradient, and reciprocal density. Heat flow averaged over intervals of 12 m is reasonably constant with depth, below 300 m. (Courtesy of A. E. Beck)

q from measurements across a series of formations is to introduce the thermal resistance R, from the surface, defined as

$$R = \int_0^z \left(\frac{1}{K}\right) dz \qquad 24.2.3$$

For undisturbed vertical flow, the plot of T against R yields a straight line of slope q. Departures from the straight line are evidence of disturbing factors, such as circulating groundwater. The entire measurement may have to be rejected, unless a value can be determined from a portion of the vertical section in which the line is straight.

24.3 THE RESULTS OF HEAT FLOW MEASUREMENTS

When measurements of the temperature gradient were first combined with those of thermal conductivity to yield a flux—and this was not many decades ago—the chief interest was to obtain a figure for the global average outflow of heat, to determine if the earth was cooling or heating. It became apparent that the flux, for the few stations available, was of the order of 1×10^{-6} calories/sec cm^2 (about 50 mW/m^2; conversions between systems of units are discussed in Appendix E, for in this section it is very difficult to avoid quoting results in both systems of units), but the problem of determining whether or not the earth was cooling turned out to be much more difficult, because of uncertainties regarding the sources of heat.

We note incidentally that the flux of heat from the interior is a small fraction of that received at the surface from the sun, but the global total, 3.2×10^{13} watts, is at least an order of magnitude greater than the rate of release of seismic energy.

The present interest in measuring heat flow is directed much more toward the mapping of variations in the flux over the earth's surface, with the aim of inferring thermal conditions in the interior. From this point of view, the approach is similar to the study of the gravitational or magnetic fields, and it shares the characteristic that the measurement at any station may contain the influences of a number of factors.

Apart from disturbances in the immediate vicinity of the measurement, which have been mentioned above, a heat flow measurement may show the effect of topography in the region, past climatic changes, sedimentation, local variations in radioactive content of the rocks, perturbation of the gradient by variations of thermal conductivity or local injection of hot material, in addition to variations in the flux from the deep interior. Of these factors, at least the effect of topography, and possibly others, should be applied as corrections to the measured flow before the value is considered. Topography alters the flux by maintaining the earth's surface at different temperatures; it can be corrected for (Birch, 1950) in a manner analogous to the making of topographic corrections in gravity surveys, with the added complication that the age of the topography is significant. Differences in surface temperature between land and sea can be important (Lachenbruch, 1957) in the case of measurements near the coast in polar regions. Sedimentation, which can be important in measurements at sea, tends to reduce the surface flow, since heat is required to warm the deposited

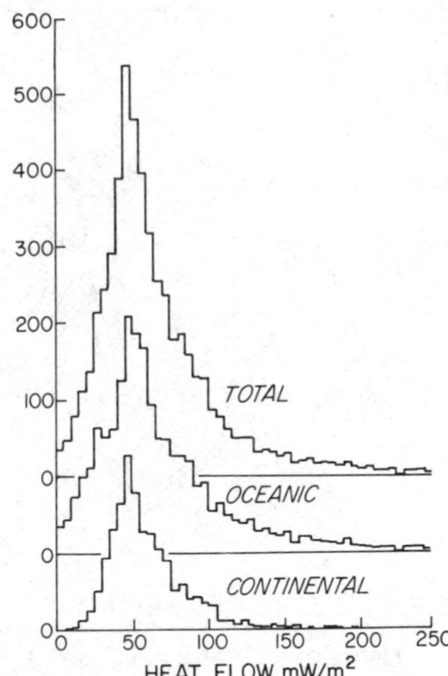

Figure 24.2 Histogram of all heat flow measurements in the catalogue of Jessop et al (1975).

material before equilibrium is restored (Jaeger, 1965). Corrections for past climatic change are theoretically possible, if the pattern of mean surface temperature during the past few thousand years can be assumed, but there is probably no substitute for measurements extending over great ranges of depth (3 or 4 km) in very deep boreholes, in which the effect of Pleistocene glaciation is minimized.

The global compilation of heat flow observations, begun by Lee and Uyeda (1965) and extended by Simmons and Horai (1968), has been published most recently by Jessop, Hobart, and Sclater (1975). Their compilation embraces 5417 measurements, of which 1699 are continental and 3718 are oceanic. Some continental areas remain very poorly covered; for example, only 20 measurements are reported for all of South America. The overall distribution of values is shown by the histograms of Figure 24.2. Immediately evident is the peak for the total curve, at about 50 mW/m^2 (1.20×10^{-6} cal/cm^2 sec), and the fact that the peaks for both oceanic and continental curves are close to this value. However, because of the longer tail toward high values on the oceanic curve, the oceanic mean is higher: 79.8 mW/m^2 as compared to 62.3 mW/m^2 for continental measurements. The average of all values for the earth is 74.3 mW/m^2.

Even approximate agreement between oceanic and continental means had not been predicted before oceanic measurements became possible. It had been expected that because of the known lower radioactivity of the oceanic crust, the mean flux through the ocean floor would be considerably lower than that through the continents. The immediate implication of the approximate equality was that a smaller flux of heat was entering the continents from below, than was crossing the base of the oceanic crust. While the agreement remains a most important geophysical observation, it must be interpreted in the light of our knowledge of ocean floor tectonics. Individual oceanic heat flow values range up to about 250 mW/m^2, and the highest values are invariably found over active ridges. These high values represent the heat escaping from the new, molten material introduced along the ridge axis, and we shall show in the next chapter that the decrease in heat flow with increasing distance from the axis is very nearly that

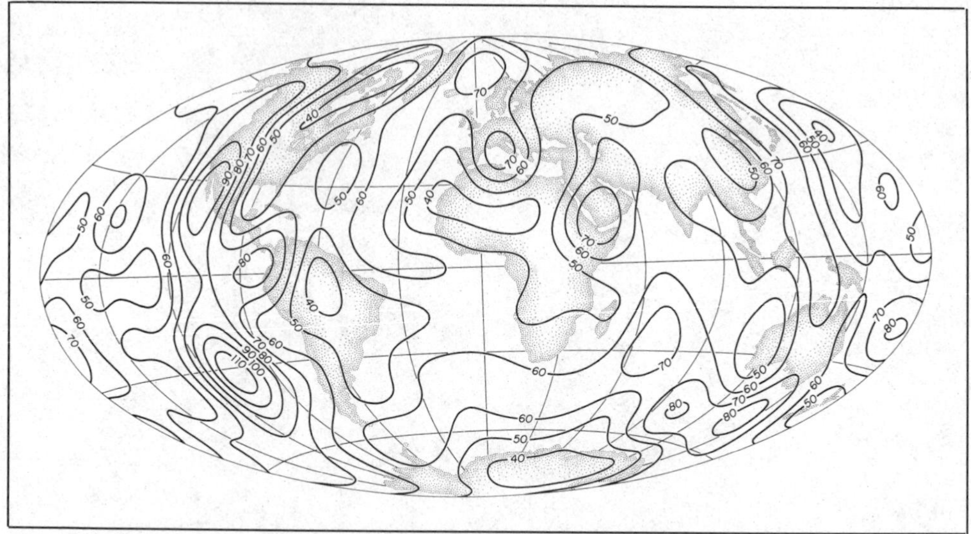

Figure 24.3 Global heat flow. Spherical harmonic representation from observations supplemented by predictor. (Courtesy of D. S. Chapman and H. N. Pollock.)

to be expected from a cooling slab. If this is the case, the heat flux at the ocean floor over young lithospheric slabs has very little to do with conduction from the interior. It is really only the mean flow over the oldest exposed ocean floor which should be compared with the continental mean. However, this is close to 50 mW/m^2 (Chapman and Pollack, 1975), so that the approximate equality with the continental average remains a valid observation. The relatively small effect of the ridges in the oceanic mean results from their small area, the few observations in axial areas, and the relatively rapid decrease in heat flow with distance from the axis (Fig. 25.3). Within the continent, the lowest values are found over old shield areas, whereas values up to 200 mW/m^2 are associated with areas of young tectonic activity and volcanism.

The attempt to construct a global map of heat flow (Lee and Uyeda, 1965) was hampered by serious gaps in the data, some of which still exist. For example, South America, northern Africa, and parts of Asia have very few measurements. The distribution of oceanic measurements is more uniform, but the oceans in high latitudes are not well covered. These gaps were particularly serious when spherical harmonic expansions were fitted to the observations, because the process of fitting placed exaggerated maxima and minima in regions of no data. Chapman and Pollack (1975) have apparently removed most of this effect, through the use of predictors to estimate the flux in unsurveyed regions. As we noted above, heat flow in the oceans decreases away from ridges and, therefore, with increasing age of oceanic crust. On the continents, the flow decreases with the age of the last orogeny, becoming lowest in the Precambrian shields. Chapman and Pollack adopted numerical relationships between flux and age for the oceans, for each of four continents (North America, Australia, Europe, Asia) and for a mean continent. Working with $5° \times 5°$ blocks of the earth's surface, they estimated the proportion of each block represented by different ages of continental or oceanic crust and, on the basis of this, assigned a mean heat flow. Twelfth-degree spherical harmonic surfaces were fitted to the results, in one case using the predicted values only and in the other (Fig. 24.3) using all measurements as well.

On this map the correlations between high heat flow and oceanic ridges and between low flow and shields are very obvious. Other correlations with major structure, or with other fields such as gravity (Fig. 11.6), are less pronounced. In fact, Chapman and Pollack confirmed an observation of Horai and Simmons (1969), that there is an important difference between the surface heat flow and gravitational fields over the earth. The amplitude of harmonics in the heat flow decreases with increasing degree, but much more slowly than in the case of gravity. In other words, higher harmonics are relatively more important, suggesting that shallow heat flow sources are making an important contribution to the flow pattern. The recognition of regional variations which might be caused by real differences in flow from the deep mantle and related to gross features of the globe, or to mantle convection, is therefore made very difficult.

Fortunately, there is a method of estimating the contribution of near-surface radioactivity to the heat flow. In Section 22.3 we introduced the heat productivity of a rock, as a property depending upon the concentration of uranium, thorium, and potassium. Let this property, designated A, be expressed in cal/cm^3 sec. Then, for rocks in the crust where the heat escapes to the surface at the same rate at which it is produced, the surface heat flow above a uniform column b cm long is simply bA cal/cm^2 sec plus whatever heat is flowing into the base of the column. It follows that if measurements of surface heat flow Q and productivity A are made over a region in which different columns extend to the same depth, a linear relation of the form

$$Q = Q_0 + bA$$

could be expected (Roy et al., 1968). In that case, the intercept Q_0 gives the flux from beneath the surface layer. In fact, not only do many continental regions yield straight-line plots [for example, the Canadian Precambrian Shield, as shown by Cermak and Jessop (1971)], but the same straight line is found for many (Fig. 24.4), as emphasized by Sclater and Francheteau (1970). The parameter b is found to be 8 km, and Q_0 is 0.8×10^{-6} cal/cm^2 sec. It is unrealistic on

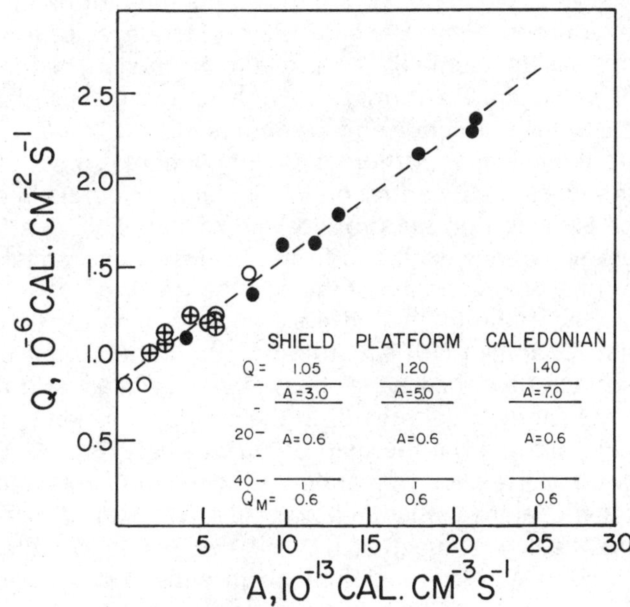

Figure 24.4 Heat flow versus heat productivity for New England (solid circles), the Central Stable region (open circles) and the Australian and Canadian shields. (open circles with crosses). Models of the distribution of productivity, for three cases consistent with the linear relationship, are shown. (After Sclater and Francheteau (1970).

geological grounds to believe that the productivity changes abruptly to zero at the same depth beneath all regions, and a continuous decrease with depth could be found to give the same surface observations (Lachenbruch, 1970).

The linear relationship between Q and A is such a fundamental observed property of heat flow that it is worth seeing just what constraints it does put on the distribution of sources. Lachenbruch showed that if we were given only the fact that the equation held over an undisturbed horizontal surface, we could not distinguish between a constant A or linear, exponential, or other variations of A with depth. But the equation holds over geological provinces of large horizontal extent, which have suffered differential erosion. The relationship must therefore remain unperturbed as material is removed from the upper surface. Suppose that the original surface was at depth $z = 0$, and that erosion in one area has lowered the surface to a depth z'. Further erosion by Δz would change the heat flow by

$$\left(\frac{dQ}{dz}\right)_{z'} \Delta z = b \left(\frac{dA}{dz}\right)_{z'} \Delta z$$

by differentiating the relationship above. Erosion would remove heat sources per unit area of strength $A_{z'}\,\Delta z$, so that

$$b \left(\frac{dA}{dz}\right)_{z'} \Delta z = -A_{z'}\,\Delta z$$

whose solution is

$$A_z = A_0\, e^{-z/b}$$

where A_0 is the value at the uneroded surface. It appears that only an exponential decrease of radioactive source concentration with depth leads to the observed fact that the linear relationship holds over broad provinces. Sclater and Francheteau (1970) adopted as a compromise the three-layered lithosphere shown in Figure 24.4, which also fits the plot of surface heat flow against surface productivity. Beneath the surface layer of 8 km, whose productivity decreases (presumably as a result of erosion) with increasing geological age (Caledonian, Platform, Shield), a uniform layer of 32 km with productivity of 0.6×10^{-13} cal/cm^3 sec contributes 0.2×10^{-6} cal/cm^2 sec to the flow. The flux from the deeper portion of the lithosphere is 0.6×10^{-6} cal/cm^2 sec. Notice that the upward concentration of the dense minerals of U and Th is not what would be expected on the grounds of gravitational separation. Radioactive minerals are apparently excluded from the closely packed lattice of the mantle, and find affinity with minerals of the continental crust.

Adopting the negative exponential variation of sources with depth, Pollack and Chapman (1977) "stripped" the contribution of crustal sources from the surface flow for several heat flow provinces, and fitted a spherical harmonic expansion, to degree 18, to the residual or mantle flow. Their resulting global map differs from the map of surface flow (Fig. 24.3) in a number of important ways. Correlation of mantle flow with continents and oceans is apparent; while the global mean mantle flow is 48 mW/m^2, the continents are characterized by a mantle flow of mean 28 mW/m^2, contrasted with the oceanic mean mantle flow of 57 mW/m^2. The difference in sub-crustal fluxes predicted above is thus

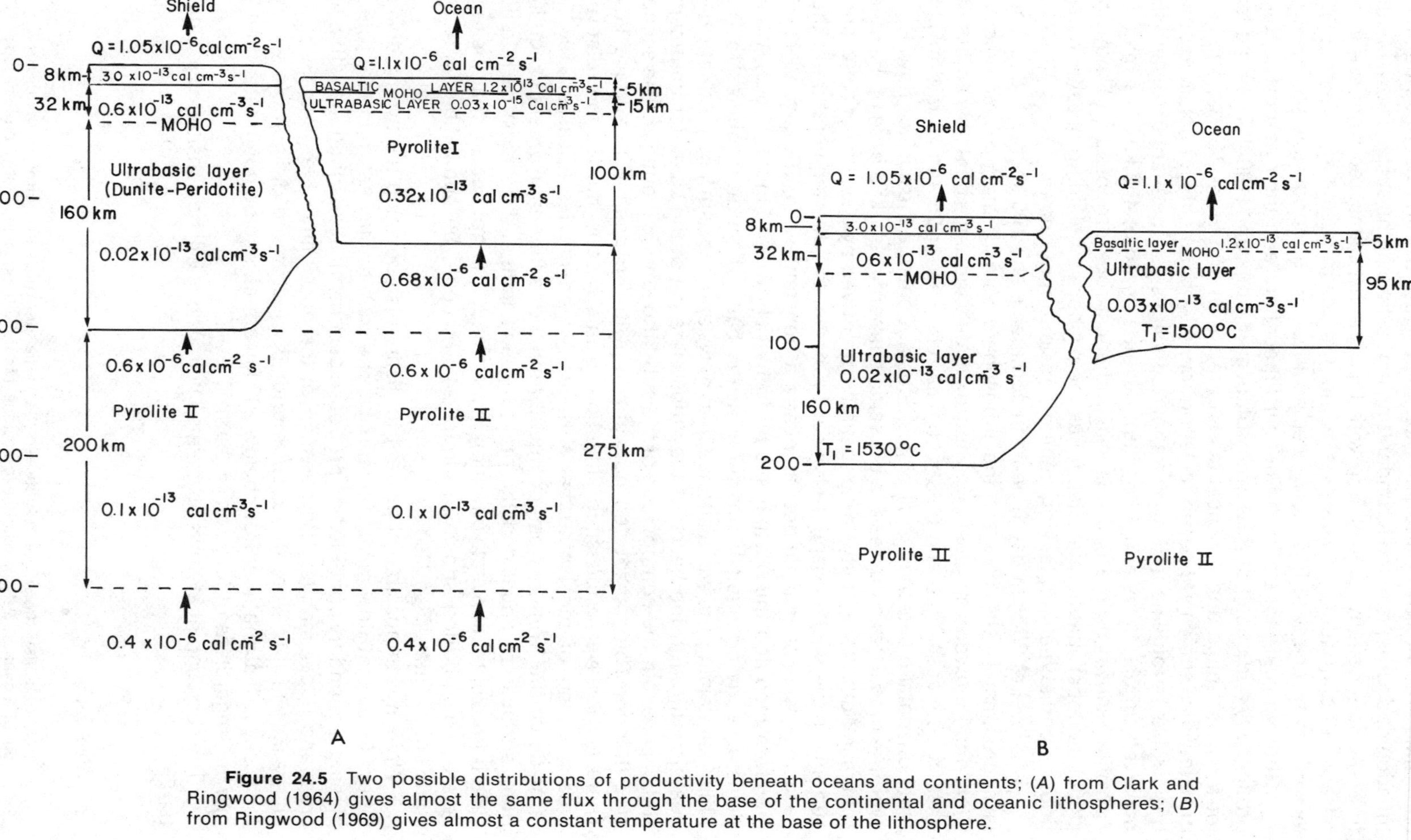

Figure 24.5 Two possible distributions of productivity beneath oceans and continents; (*A*) from Clark and Ringwood (1964) gives almost the same flux through the base of the continental and oceanic lithospheres; (*B*) from Ringwood (1969) gives almost a constant temperature at the base of the lithosphere.

confirmed. The positive correlation between heat flow and geopotential, as a function of degree, is much stronger than in the case of the surface flow. Mantle flow shows a negative correlation with topography, measured from sea level, because of continentality. However, within the ocean basins, mantle heat flow as well as surface heat flow correlates positively with topography.

Geochemical models of the deeper lithosphere and their associated heat productivity had been put forward by Clark and Ringwood (1964). Their models for a typical shield and a typical ocean are shown in Figure 24.5(*A*). The heat productivities shown are based on laboratory measurements for acidic to ultrabasic rocks, supplemented by extrapolated values for the mantle, to yield the observed surface flows and a constant flux at a depth of 400 km. The material "pyrolite" shown on the figure is a hypothetical rock proposed by Ringwood (1966) as upper mantle material. It has the property of yielding basaltic ocean floor by partial melting, leaving behind a residuum of dunite.

The level of 200 km depth beneath continents and 100 km beneath oceans is approximately the base of the lithosphere as indicated seismologically. Clark and Ringwood's model gives approximately the same heat flow across the base in both cases (0.6 and 0.68×10^{-6} cal/cm^2 sec, or 25 and 28 mW/m^2), so that horizontal displacement of plates does not change the surface flow. However, the model does not lead to the same temperature at the base of the lithosphere in both cases. If lithospheric plates are driven by a system of convection currents in the aesthenosphere beneath (Chapter 25), the temperature at their base may well be maintained at nearly a constant value. Ringwood (1969) therefore modified the oceanic model to that shown in Figure 24.5(*B*), which yields a nearly constant temperature at the base of the lithosphere (1500°C) while maintaining the same surface heat flows and near-surface productivities. Heat flow through the bases of the lithosphere is not the same for continents and oceans, but the composition of the sub-lithospheric mantle is. The section may therefore be transported horizontally, with no change in surface flow. There is, however, a possible limitation to this second model: the relatively great warping of the isothermal surfaces, which implies (Elsasser, 1967) a large departure from hydrostatic equilibrium. For a fluid in hydrostatic equilibrium, surfaces of equal density, pressure, and temperature must be parallel; even if horizontal convective motion is assumed, it is not certain that the temperature distribution shown in Figure 24.5(*B*) could be compatible with isostatic equilibrium. (See also Problem 5.)

While it is impossible at present to distinguish between the two models of Figure 24.5, heat flow measurements can be seen to play an important role in placing a boundary condition on proposed geochemical distributions within the crust and upper mantle. The question of the possibility of large scale lateral differences in flux from the deep interior is much more difficult. All we can really say is that the flow from below 200 km appears to be not greater than 25 mW/m^2, regardless of location on earth.

The very high heat flows associated with ridges will be considered in Chapter 25.

Problems

1. Establish the relations between the partial derivatives required in the derivation of Equation 24.1.3. (Hint: expressions for the changes in energy and entropy must be perfect differentials.)

2. A heat flow measurement is to be made in a continental area where the flux is normally 1.2 cal/cm^2 sec. A borehole 2.0 km deep is available for measurements, but it is known that there is a spherical mass of rock of very high thermal conductivity (0.05 c.g.s. units), whose center is 1.0 km below the base of the hole and 1.0 km to one side. The radius of this mass is 0.5 km. If the normal thermal conductivity of rocks in the region is 0.004 c.g.s. units, by how much will the gradient in the lower part of the hole be perturbed?

*3. Set up the expression for the change in temperature at the surface of a highly conducting cylindrical region, which is delivering Q calories of heat per unit length radially into the surrounding rock of thermal conductivity K. Can you show that the solution reduces to the form given in Equation 24.2.2?

4. Give order-of-magnitude figures for the energy brought to the surface annually by heat flow; for that received annually by the earth from the sun; for the annual release of energy in earthquakes; and for the energy stored in the earth's magnetic field.

5. Given only that a fluid satisfies an equation of state of the form $p = p(T, V)$, where p is pressure, T is temperature, and V is specific volume (the reciprocal of density), and that the equation of hydrostatic equilibrium is

$$V \underline{\nabla p} + \underline{g} = 0$$

where $\underline{g}$ is the acceleration due to gravity, show that surfaces of equal density, pressure, and temperature all coincide. (Hint: first take the curl of the second equation above to obtain information on the directions of the gradients of density and pressure.)

BIBLIOGRAPHY

Beck, A. E., Jaeger, J. C., and Newstead, G. N. (1956) The measurement of the thermal conductivities of rocks by observations in boreholes. *Aust. J. Phys.*, **9**, 286.

Birch, F. (1950) Flow of heat in the front range, Colorado. *Bull. Geol. Soc. Amer.*, **61**, 567.

Blackwell, J. M. (1954) A transient flow method for determination of thermal constants of insulating materials in bulk. *J. Appl. Phys.*, **25**, 137.

Bullard, E. C. (1947) The time necessary for a borehole to attain temperature equilibrium. *Mon. Not. R.A.S., Geophys. Suppl.*, **5**, 127.

Bullard, E. C., Maxwell, A. E., and Revelle, R. (1956) Heat flow through the deep sea floor. *Adv. Geophys.*, **3**, 153.

Bullard, E. C. (1960) Measurement of temperature gradient in the earth. In *Methods and techniques in geophysics,* ed. S. K. Runcorn. Interscience, New York.

Carslaw, H. S., and Jaeger, J. C. (1959) *Conduction of heat in solids* (2nd Ed.) Oxford Univ. Press, London.

Cermak, V., and Jessop, A. (1971) Heat flow and heat production in the Canadian Shield. *Tectonophysics,* **11**, 287–304.

Chapman, David S., and Pollack, Henry N. (1975) Global heatflow: a new look. *Earth Plan. Sci. Letters,* **28**, 23–32.

Clark, Sydney P. Jr., and Ringwood, A. E. (1964) Density distribution and constitution of the mantle. *Rev. Geophys.*, **2**, 35.

Elsasser, Walter M. (1967) Interpretation of heat flow equality. *J. Geophys. Res.*, **72**, 4768–4770.

Horai, K., and Simmons, G. (1969) Spherical harmonic analysis of terrestrial heat flow. *Earth Plan. Sci. Letters,* **6**, 386–394.

Hyndman, R. D., Von Herzen, R. P., Erickson, A. J., and Jolivet, J. (1976) Heat flow measurements in deep crustal holes in the Mid-Atlantic Ridge. *J. Geophys. Res.*, **81**, 4053–4060.

Jaeger, J. C. (1965) Application of the theory of heat conduction to geothermal measurements. In *Terrestrial heat flow,* ed. William H. K. Lee. American Geophysical Union, Monograph 8.

Jessop, A. M., Hobart, M. A., and Sclater, J. G. (1975) *The world heat flow data collection 1975.* Geothermal Series No. 5, Earth Physics Branch, Dept. of Energy, Mines and Resources, Ottawa.

Krige, L. J. (1939) Borehole temperatures in the Transvaal and Orange Free State. *Proc. Roy. Soc. Lond. A,* **173**, 450.

Lachenbruch, A. H. (1957) Thermal effects of the ocean on permafrost. *Bull. Geol. Soc. Amer.*, **68**, 1515.

Lachenbruch, A. H. (1970) Crustal temperature and heat production: implication of the linear heat flow relation. *J. Geophys. Res.*, **75**, 3291–3300.

Lee, William H. K., and Uyeda, Seiya (1965) Review of heat flow data. In *Terrestrial heat flow,* ed. William H. K. Lee. American Geophysical Union, Monograph 8.

Lister, C. R. B. (1972) On the thermal balance of a mid-ocean ridge. *Geophys. J.*, **26**, 515–535.

Misener, A. D., Thompson, L. G. D., and Uffen, R. J. (1951) Terrestrial heat flow in Ontario and Quebec. *Trans. Amer. Geophys. Un.*, **32**, 729.

Misener, A. D., and Beck, A. E. (1960) The measurement of heat flow over land. In *Methods and techniques in geophysics,* ed. S. K. Runcorn. Interscience, New York.

Pollack, Henry N., and Chapman, David S. (1977) Mantle heat flow. *Earth Plan. Sci. Letters*, **34**, 174–184.

Ratcliffe, E. H. (1960) The thermal conductivities of ocean sediments. *J. Geophys. Res.*, **65**, 1535.

Ringwood, A. E. (1966) The mineralogy of the upper mantle. In *Advances in earth science,* ed. P. M. Hurley, M. I. T. Press, Boston.

Ringwood, A. E. (1969) Composition and evolution of the upper mantle. In *The earth's crust and upper mantle,* ed. P. J. Hart. American Geophysical Union, Monograph 13.

Roy, Robert F., Decker, Edward R., Blackwell, David D., and Birch, Francis (1968) Heat flow in the United States. *J. Geophys. Res.*, **73**, 5207.

Sclater, John G., and Francheteau, Jean (1970) The implication of terrestrial heat flow observations on current tectonic and geochemical models of the crust and upper mantle of the earth. *Geophys. J.*, **20**, 509–542.

Simmons, Gene, and Horai, Ki-Iti (1968) Heat flow data II. *J. Geophys. Res.*, **73**, 6608.

Von Herzen, R. P., and Maxwell, A. E. (1959) The measurement of thermal conductivity of deep sea sediments by a needle probe method. *J. Geophys. Res.*, **64**, 1557.

Wright, J. A., and Garland, G. D. (1968) In situ measurement of thermal conductivity in the presence of transverse anisotropy. *J. Geophys. Res.*, **73**, 5477.

THE PRESENT TEMPERATURE OF THE EARTH 25

25.1 EXTRAPOLATION OF THE SURFACE GRADIENT

Near the earth's surface, the average geothermal gradient in the continents is about 30°C per km; this combines with an average thermal conductivity of 0.004 c.g.s. units to give the mean flow of about 1.2×10^{-6} cal/sec cm^2. It is obvious that direct extrapolation of the surface gradient to the base of the crust, at a depth of about 35 km, would lead to temperatures above the melting point of the rock. In fact, allowance must be made for heat production in the crust. The flow decreases with depth in such a way that, for uniform heat production of A cal/sec cm^3 in the crust, the change in flow is

$$-\frac{d}{dz}\left(K\frac{dT}{dz}\right)=A \qquad 25.1.1$$

For constant K, the temperature at depth z is given by

$$T=T_0+\left(\frac{q}{K}\right)z-\frac{Az^2}{2K} \qquad 25.1.2$$

where T_0 and q are the surface values of temperature and flux.

Without a detailed knowledge of A extrapolation is impossible, and other methods must be sought to estimate the temperature in the deep interior.

Methods which have been applied, and which we shall discuss in turn, include the solution of the full heat conduction equation under assumed initial conditions, estimation of the melting point temperature, and estimates based on other properties, such as the electrical conductivity.

25.2 SOLUTION OF THE CONDUCTION EQUATION

The equation of heat conduction in a sphere without lateral differences in properties is

$$\rho c\,\frac{\partial T}{\partial t}=\frac{1}{r^2}\frac{\partial}{\partial r}\left(r^2K\frac{\partial T}{\partial r}\right)+A \qquad 25.2.1$$

in which the thermal conductivity and heat productivity are taken as functions of the radius, and the contributions to A from different radioactive species have different time dependences. This equation is obtained from Equation 24.1.7 by transforming the Laplacian to spherical coordinates with the help of Equation A1, then dropping the terms in θ and λ.

When the term A is omitted, the equation describes simply the cooling of a sphere without heat sources. For example, if the initial temperature of the earth is taken to be somewhat above the melting point of rock, solution of the equation gives the time t required for the surface heat flow to be reduced to its present value. This was precisely the approach taken by Kelvin in a series of estimates (summarized in Kelvin, 1899) on the age of the earth since solidification, leading to values for that age of between 20 and 40 m.y. ("and probably much nearer 20 than 40"). Kelvin similarly estimated the age of the sun, on the basis of the total available gravitational potential energy and the present rate of radiation of energy, concluding that it was "on the whole, probable that the sun has not illuminated the earth for 100,000,000 years . . .". Both of these estimates were, of course, attacked on geological grounds, leading to a controversy that was resolved only with the discovery of radioactivity in 1896 and the recognition of its heating effect. Rutherford (1905) appears to have been the first to point out the importance of radioactivity in the histories of the earth and sun. From the known rate of heat production by radium salts in the laboratory, and assuming that all heat produced within the earth was now reaching the surface, he estimated that a distribution of radium in the amount of 2.6×10^{-13} gm/cm^3 (an order of magnitude less than the concentration in igneous crustal rock) throughout the earth would supply the observed heat flow, requiring no loss of original heat. He wrote, "the earth may have remained for very long intervals of time at a temperature not very different from that observed today . . . the presence of animal and vegetable life may be very much longer than the estimate made by Lord Kelvin . . ."

The first truly quantitative estimate of the heating effect of radioactivity within the earth was made by Strutt (1906), in a paper that led the way for many of the approaches of this and the previous chapter. Strutt measured the radium content of 28 samples of igneous rock, ranging in composition from granite to dunite, and converted the concentration to heat productivity by reference to the measured heating of a concentrated radium salt. He concluded that the mean heat productivity of typical continental rocks was 2.4×10^{-13} cal/cm^{3}sec, a value remarkably close to the present estimate, shown in Table 22.2. With a similar judicious estimate (almost a prediction, since few measurements were available) of mean heat flow as 1.75×10^{-6} cal/cm^{2}sec, he concluded that the radioactivity was confined to a layer 45 miles thick. Strutt then integrated an equation identical to Eq. 25.1.2 to give the temperature distribution within this crust, showing that the temperature would reach 1500°C at its base. Since this is approaching the melting point of rock, he argued that it was evidence of a depletion of radioactive sources in the part of the earth beneath the "crust". All of this discussion took place before the seismological identification of the crust by Mohorovičić, but Strutt, noting that Milne (1906) had tentatively proposed an abrupt change in elastic properties at a depth of approximately 30 miles, correlated his radioactive layer with this seismological evidence.

To return to Equation 25.2.1, solutions of the full equation, under simplifying assumptions, have been discussed by Jeffreys (1962), Slichter (1941), and others. If the parameters are known, and if the initial temperature and age are assumed, the present temperature at all depths can be determined by direct integration of the equation. An application of this approach was made by Jacobs and Allan (1956), who considered a number of model earths, in which K and A were constant within discrete shells, and for which a number of different initial temperature distributions were considered. Their results indicated that for any reasonable distribution of radioactivity, temperatures at depths below 500 km would today be increasing; furthermore, even if a cold origin of the earth is assumed, radioactivity in the early history of the earth would have been such that surface melting would have occurred some 3×10^9 years ago, in agreement with the fact that rock ages significantly greater than this are not found.

However, the solution of Equation 25.2.1 presents many difficulties, which have been discussed in detail by Lubimova (1967). Probably the most serious of these (apart from the assumption that conduction is the dominant mode of heat transfer) is the estimate of thermal conductivity at depth. Near the earth's surface, heat is conducted by lattice vibrations, that is, through the interactions of neighboring atoms in the crystal lattices of minerals; pressure and temperature alter the lattice conductivity in a manner that can be estimated, but at high temperatures new modes of transfer, such as radiation, may become important. The lattice, or phonon, component of conductivity is increased by pressure, but at moderate depths the dominant effect is a decrease with increasing temperature (which amplifies the effect of imperfections), and Lubimova estimates that it reaches a minimum at a depth of about 200 km.

The radiative transfer of heat through silicates appears to become important at temperatures above 200° or 300°C, with an effective thermal conductivity K_r given by

$$K_r = \frac{16}{3} \frac{n^2 \sigma T^3}{\epsilon} \qquad 25.2.2$$

where n is the index of refraction, σ is the Stefan–Boltzmann constant, T is the absolute temperature, and ϵ is the opacity, or coefficient of absorption of the material. The opacity is strongly dependent upon the wavelength of the radiation; therefore, so is the radiative conductivity.

At still higher temperatures a completely different mechanism of energy transfer apparently becomes appreciable. This involves the propagation of excited states of atoms, by a mechanism known as exciton conductivity. The resulting contribution to the thermal conductivity is given by a relation analogous to that for the electrical conductivity, or

$$K_{ex} = K_0 e^{-E/kT} \qquad 25.2.3$$

where K_0 is a constant, and E is the excitation energy. There is sufficient spectroscopic information for olivine (Clark, 1956) for E to be determined.

Combining the three mechanisms for thermal conductivity, Lubimova has estimated the total conductivity for the upper mantle to be as shown in Figure 25.1, in which estimates by Lawson and Jamieson (1958), appropriate to the lower mantle, also are plotted. Noteworthy is the sharp increase at depths greater than 700 km, and also the minimum at a depth of about 150 km. The effect of a minimum in conductivity is for the temperature to be increased just below it, and this may well be a contributing factor to the formation of the low-velocity zone.

The most detailed numerical calculations based on Equation 25.2.1 are

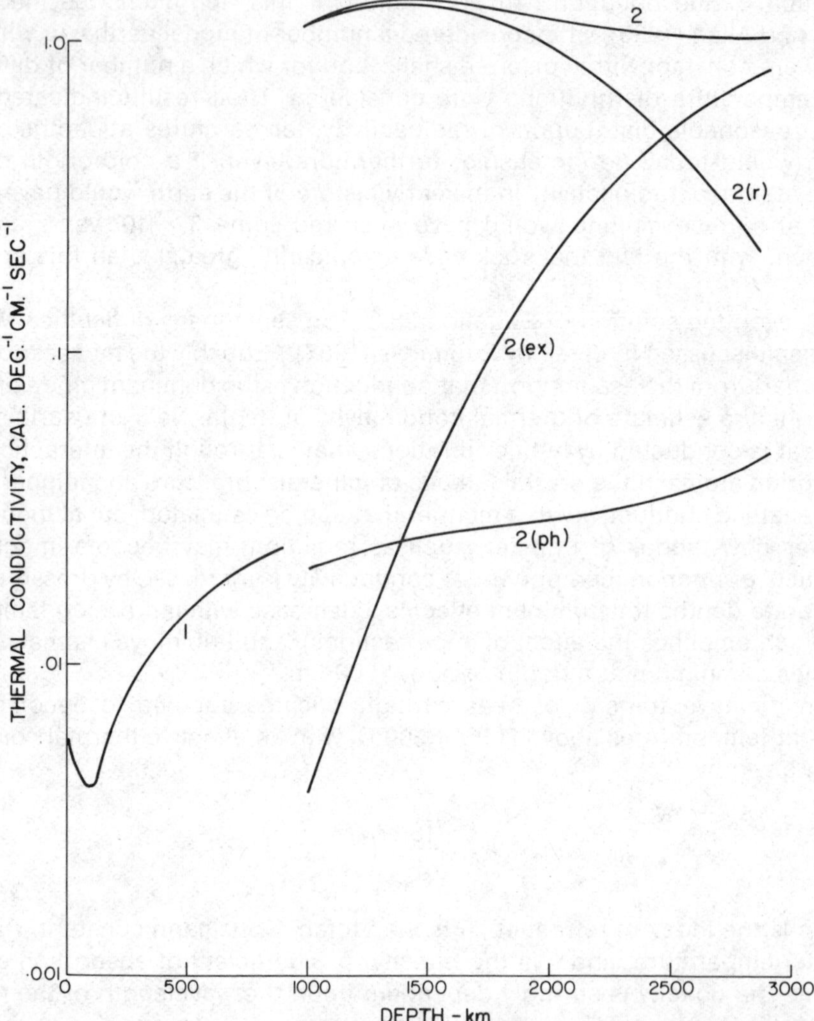

Figure 25.1 Estimates of thermal conductivity for the mantle. Curve 1 is the estimate of total conductivity by Lubimova. Curve 2 is the total conductivity of the lower mantle, according to estimates by Lawson and Jamieson of the phonon or lattice component (ph), exciton component (ex) and radiation component (r).

those of MacDonald (1959), who has produced temperature distributions for many different earth models. The actual temperatures vary between models, so that the method cannot be said to yield unequivocally the present distribution of temperature in the earth. However, some important general results follow. The first is that if the mantle is anywhere as radioactive as a chondritic meteorite, this radioactivity must be concentrated in the upper few hundred kilometers; a deeper distribution would have led to temperatures exceeding the melting point of the crust during the history of the earth. However, it must be emphasized that even this conclusion is justified only if conduction is the dominant mode of heat transfer; as we shall see, the possibility of convection greatly alters the argument. A second result, about which there is little controversy, is that at a given depth in the upper mantle, the temperature beneath the oceans is greater than that beneath the continents. This follows from the differences in crustal heat productivity, and equality of surface flows.

25.3 ESTIMATES OF THE MELTING POINT AND PHASE CHANGE TEMPERATURES

Since the mantle is solid throughout, an estimate of the melting point temperatures for mantle material provides a maximum temperature at any depth. Solid state physics shows that the melting temperature of a solid is connected to other properties, in a manner which permits it to be estimated. Many of the relationships are derived from a theory of Lindemann (1910), who proposed that melting occurred when the amplitude of atomic oscillations became sufficiently great that direct collisions between neighboring atoms could occur. Then, if r is the distance between adjacent centers, and $r(1-\rho)$ is the effective atomic diameter, the kinetic energy of an atom, at the melting temperature, is

$$L=\frac{Dr^2\rho^2}{8} \qquad 25.3.1$$

where D is the restoring force constant. But if γ is the melting temperature,

$$L=k\gamma \qquad 25.3.2$$

where k is Boltzmann's constant, and for the frequency of oscillation we may write

$$v=\frac{\sqrt{2}}{\pi\rho r}\sqrt{\frac{k\gamma}{\mu}} \qquad 25.3.3$$

where μ is the mass of an atom. In a cubic lattice, r is given by

$$N_0r^3=V \qquad 25.3.4$$

where V is the molar volume and N_0 is Avogadro's number; μ is given by M/N_0, where M is the molecular weight. If the ratio ρ can be assumed constant, Lindemann's result follows; it is

$$v=CM^{-1/2}V^{-1/3}\gamma^{1/2} \qquad 25.3.5$$

where C is a constant.

Uffen (1952) identified the frequency in Equation 25.3.5 with the Debye critical frequency, which is defined by the condition that the number of independent vibrations in a volume of substance be equal to three times the number of atoms. It is given by

$$v_D^3=\frac{9N_0}{4\pi V}\left(\frac{1}{\alpha^3}+\frac{2}{\beta^3}\right)^{-1} \qquad 25.3.6$$

where N_0 and V are as above and α and β are the velocities of P and S waves. Uffen's working equation is then

$$\frac{\gamma_1}{\gamma_0}=\left[\frac{\beta_0^3+2\alpha_0^3}{\beta_1^3+2\alpha_1^3}\right]^{2/3}\cdot\left[\frac{\alpha_1\beta_1}{\alpha_0\beta_0}\right]^2 \qquad 25.3.7$$

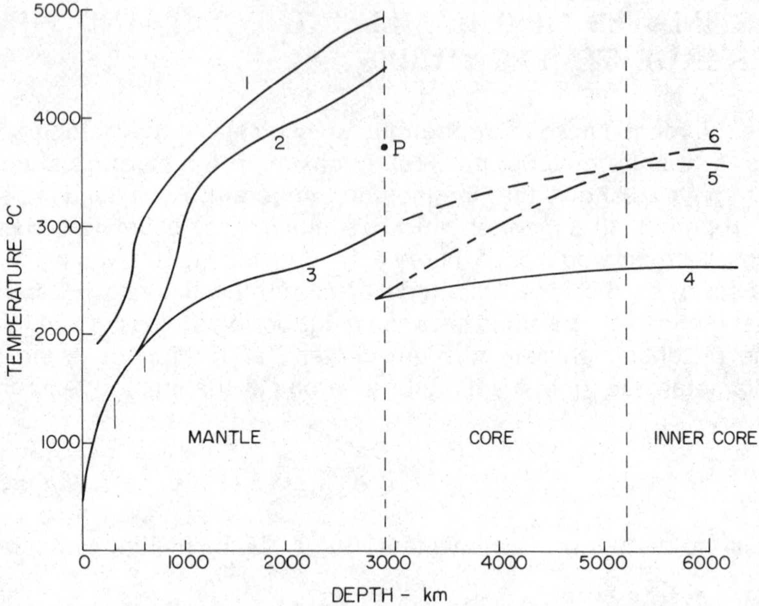

Figure 25.2 Temperature distribution curves for the earth: (1) melting point according to Uffen, (2) temperature inferred by Tozer from the electrical conductivity, (3) McKenzie modification of (2). Curve (4) is the melting point of iron extrapolated from measurements by Strong, while (6) is the required form of the core melting temperature if it is to intersect a possible actual temperature curve (5) at the boundary of the inner core. Point P is the melting point of diopside, extrapolated from measurements by Boyd and England. Vertical lines at depths near 400 and 600 km. show estimates by Anderson and Fujisawa, based on phase changes.

where the subscripts refer to two depths within the mantle. The primary assumption, of course, is that M and V remain constant through the region to which Equation 25.3.7 is applied. The great power of the method lies in the fact that α and β, as we have seen, are among the best determined parameters of the earth. The curve obtained by Uffen is shown in Figure 25.2.

The Lindemann theory of melting is also consistent with a semi-empirical equation for melting, formulated by Simon (1937) and often applied to geophysical problems. It is

$$P = A\left[\left(\frac{\gamma}{\gamma_0}\right)^c - 1\right] \qquad 25.3.8$$

where γ is the melting point temperature at pressure P, γ_0 is the melting point at atmospheric pressure, and A and c are parameters for the substance. These parameters are expressible in terms of other thermodynamic properties (Gilvarry, 1957) which may, however, themselves vary with pressure. Boyd and England (1963) have directly measured the melting points of the silicates diopside ($CaMgSi_2O_6$) and albite at pressures up to 50 kilobars, and give the values:

	γ_0(°K)	A(Kb)	c
Albite	1391	19.5	5.1
Diopside	1665	23.3	4.64

Extrapolation of their observations, by means of the Simon equation, to the pressure of 1400 kilobars at the core boundary, gives the melting temperature there as 3750°C for diopside. This is considerably below the value predicted by Uffen (Fig. 25.2).

A further empirical expression for the melting point at high pressures has been proposed by Kraut and Kennedy (1966). On the basis of the measured melting points of several substances, they suggest that there is a linear relationship between melting temperature and the isothermal compression, or fractional change in volume ($\Delta v/v$), which results from the applied pressure. The isothermal compression for the material of the earth's mantle can be obtained, as a function of depth, from shock-wave measurements, an assumed equation of state, or the seismic velocities (Chapter 26), so that the Kraut and Kennedy law could be useful in extrapolating measured melting points to high pressures. However, further measurements (Luedemann and Kennedy, 1968) on the elements lithium, potassium, sodium, and ribidium, to pressures of 80 kilobars, showed a departure from the linear relationship when $\Delta v/v$ reached 20 to 35 per cent. This degree of compression for silicates would be reached in the lower mantle, and it must be concluded that extrapolations based on the linear rate would be subject to uncertainty, just as are those which depend upon the Simon equation.

The Simon equation has also been applied to the earth's core, on the assumption that it is pure iron (Simon, 1953; Jacobs, 1953; Gilvarry, 1956). Strong (1959) measured the melting point of iron to a pressure of 96 kilobars, obtaining values of 1805°K for γ_0, 75 kilobars for A, and 8 for c. The melting point curve for pure iron extrapolated from Strong's measurements to the pressures existing in the core is also shown in Figure 25.2. It lies entirely below the melting temperatures originally estimated by Simon (which corresponded to 3600°K at the core boundary), from observations at lower pressures. Jacobs pointed out that, if the inner core is composed of solid iron, the curves of melting point and actual temperature must cross at the boundary of the inner core; in addition, the actual temperature curve must be continuous across the boundary of the outer core. We shall return to this point in the following section.

Recent theoretical extrapolations of the melting point of pure iron to high pressures have tended to yield higher temperatures (Jacobs, 1975). Thus Boschi (1974) obtained 4800°C for the mantle-core boundary, and 6600°C for the boundary of the inner core. It is difficult to escape the conclusion that an uncertainty of at least 1000°C remains in the melting point at the mantle-core boundary, but there is the additional complication that the outer core is now believed to contain alloying elements (Section 26.7). These would tend to reduce the melting point, so that curves 3–5 of Figure 25.2 remain an acceptable, but not well defined, distribution.

There is an important consequence for geomagnetism resulting from the temperature gradient in the core. To maintain convection, the actual gradient must be greater than the *adiabatic gradient.* Both of these quantities are uncertain in the core. Higgins and Kennedy (1971), on the basis of the melting relation of Kraut and Kennedy, projected that the actual gradient is sub-adiabatic. Jacobs (1971) recomputed the adiabatic gradient and showed that, on the basis of present knowledge, the actual gradient could be greater than adiabatic, so that convection is possible. While geomagnetic dynamo models can be constructed without invoking convection, a much larger class of dynamos is possible with convection.

Phase changes in the mantle are changes in the atomic structure of mantle minerals, believed to account for the seismic velocity discontinuities at depths near 380 and 620 km (for example, Figure 3.14). The probable mechanisms of reordering will be discussed in Section 26.4. The important point for the mantle temperature distribution is that the pressure and temperature conditions required for a phase change of a given mineral are known from laboratory measurements. Pressures at given depths can be estimated rather closely, so that if the mantle composition were completely known, reference points of temperature at well-determined depths would be available. The difficulty lies in assessing the relative abundances of the two olivine minerals: Fe_2SiO_4 and Mg_2SiO_4. Anderson (1967) argued for a composition that would give a temperature of about 1500°C at 365 km, and 1900°C at 620 km. Fujisawa (1968) suggested the range in temperature from 1150°C to 1530°C at a depth of 370 km. These estimates are shown on Figure 25.2; they provide reasonable controls on the temperature in the upper mantle.

25.4 ESTIMATES BASED ON THE ELECTRICAL CONDUCTIVITY

The relationship between electrical conductivity and temperature in silicates has been mentioned in connection with electromagnetic induction in the earth. Two important mechanisms of electrical conduction in olivine must be considered (Tozer, 1959). The first involves an intrinsic semiconduction process, in which electrons are excited into conduction bands; the resulting conductivity is

$$\sigma_1 = \sigma_{1_0} e^{-E_1/2kT} \qquad 25.4.1$$

where the excitation energy E_1 is of the order of 3 electron volts, and σ_{1_0} is 1 to 5 mhos/cm. The second mechanism is that of conduction by ions; the conductivity is given by

$$\sigma_2 = \sigma_{2_0} e^{-E_2/kT} \qquad 25.4.2$$

where E_2 is of the order of 3 electron volts, and σ_{2_0} can be as great as 5×10^6 mhos/cm. Ionic conduction is obviously important, as shown by the large value of σ_{2_0}, but because of the difference in exponents, the electronic mechanism is probably dominant through most of the mantle. The difference arises from the fact that both electrons and holes contribute to conduction in the semiconductor case, as opposed to the single sign of the charge carrier in ionic conduction.

The energies E_1 and E_2 are temperature- and pressure-dependent. Increasing pressure increases E_2 for olivine, as would be expected, since compression inhibits the movement of ions through the lattice, but it apparently decreases E_1. Tozer has considered these variations, and using the conductivity distribution of McDonald (Section 20.3), has estimated the temperature variation shown in Figure 25.2. However, in the years since Tozer produced this estimate, Akimoto and Fujisawa (Section 20.5) have experimentally measured an abrupt increase in the conductivity of olivine at the transition to the high-density phase spinel, and this corresponds to a lower activation energy E_1. The measurements indicate a decrease in E_1 from approximately 0.5 to 0.3 electron volts upon change of phase. A given electrical conductivity in the spinel phase therefore corresponds to a lower temperature, and McKenzie (1967) has corrected Tozer's distribution to give the revised curve, also shown in Figure 25.2. In one important way, however, the experiment of Akimoto and Fujisawa is not directly applicable to the mantle. They studied the iron olivine, Fe_2SiO_4, whereas the magnesium olivine, Mg_2SiO_4, probably is more representative of the mantle (Chapter 26). Tozer's calculations were for a magnesium-rich olivine. It will also be recalled that the distribution of electrical conductivity in the lower mantle is still uncertain, and the temperature inferred from it must share this uncertainty. However, in view of the limitations involved in applying the thermal conduction equation and the fact that melting point temperatures give only a maximum value in the mantle, the derivation of temperature from electrical conductivity appears to be the most fruitful approach.

Returning to Figure 25.2, we note the bounds placed on the actual temperature distribution curve for mantle and core by the melting point relations. The actual curve must lie below the melting point curve at the base of the mantle, but above it immediately inside the core. A curve passing through 3000°K at a depth of 2900 km (McKenzie) lies well below Uffen's estimate of melting point in the mantle, and also below extrapolation from Boyd and England. It lies above the core melting point at that depth, but could perhaps intersect the core melting curve at the boundary of the inner core, as Jacobs suggested.

An important feature of the diagram is the tendency for the melting point and actual temperature curves in the mantle to diverge with depth. This implies that mantle material is closest to its melting point in the upper mantle, a situation which could be important in connection with possible motions in the mantle.

25.5 TEMPERATURES IN LITHOSPHERIC SLABS

We have noted the high heat flow associated with oceanic ridges and the decrease in flow with increasing distance from the ridge axis. This is shown in more detail in Figure 25.3. Adopting a model of a spreading ridge with injection of hot material at the axis and a constant spreading velocity across the ridge, it is possible to show that this decrease is consistent with that to be expected in a cooling plate, and also to compute temperatures within the plate (McKenzie, 1967; LePichon et al., 1973). The two-dimensional geometry is indicated on Figure 25.3. In the steady-state case, with no local radioactive heat generation, heat is transported in the x-direction by slab motion, and flows by conduction laterally and vertically. Thus, at equilibrium

$$\rho c v \frac{\partial T}{\partial x} = K\left(\frac{\partial^2 T}{\partial x^2} + \frac{\partial^2 T}{\partial z^2}\right) \qquad 25.5.1$$

where ρ, c, v, and K are the density, specific heat, spreading velocity, and thermal conductivity of the slab. The boundary conditions are that the temperatures at the top ($z = L$) and base ($z = 0$) of the plate, and at $x = 0$, are given. In almost all cases, where the spreading velocity v is more than 1 cm/yr, the transport of heat horizontally by the movement of the plate greatly outweighs the horizontal conduction of heat. Equation 25.5.1 may be simplified to

$$\rho c v \frac{\partial T}{\partial x} = K \frac{\partial^2 T}{\partial z^2} \qquad 25.5.2$$

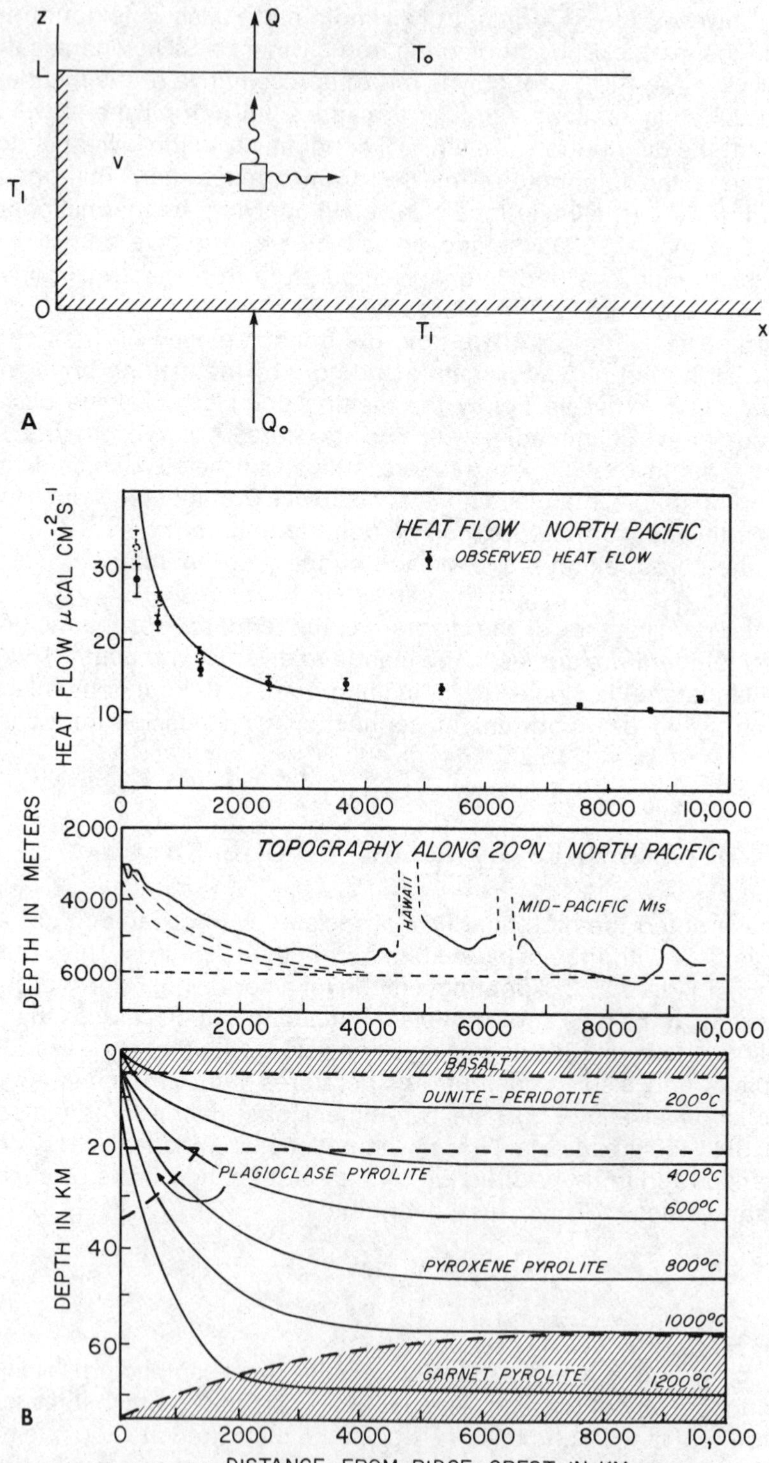

Figure 25.3 (*A*) Parameters in the modelling of a moving, cooling, plate. Heat carried by convection into a fixed volume is lost by conduction, appearing in the surface heat flux Q. (*B*) Predicted heat flow, topography and temperature distribution, compared with observations of heat flow and topography for the East Pacific Rise. On the topographic profile, the lower curve is for the simple model; the upper curve is obtained when phase changes represented in the lower section are invoked. (After Sclater and Francheteau 1970)

But, for constant spreading velocity, v, the time is $t = x/v$ (in other words, distance from the ridge and time are interchangeable as variables), and the equation becomes

$$\rho c \frac{\partial T}{\partial t} = K \frac{\partial^2 T}{\partial z^2} \qquad 25.5.3$$

This is the one-dimensional diffusion equation, which was met in Section 20.3, and the problem becomes identical to that of finding the temperature variation with *time* in a stationary slab with constant temperatures on its faces and a given initial temperature (i.e., the temperature of the injected material at $x = 0$). The solution (Carslaw and Jaeger, 1959, Chapter 3) is obtained as a series over n of terms of the form $e^{-n^2at} \sin \frac{n\pi z}{L}$, where a is determined by the thermal parameters of the slab.

The physical properties of the slab and the boundary temperatures to be used are fairly well known (Sclater and Francheteau, 1970; Sclater, 1972). The initial temperature, and the temperature at the base of the slab, is probably close to 1300°C, and thermal conductivity K is about 6.15×10^{-3} cal/cm sec °C. For the Pacific, with a spreading velocity of 5 cm/yr, Sclater and Francheteau obtained the temperature distribution and surface heat flow shown in Figure 25.3. The heat flow is a very reasonable approximation to the observed values as a function of distance from the ridge. Note that most of the excess flux due to plate cooling has disappeared at a distance of 500 km from the ridge, and that beyond this the flow is very nearly constant with distance.

Unfortunately, the uniform decrease in measured heat flow has not been confirmed by measurements made in deep holes in the basalt itself, at least for the Mid-Atlantic Ridge (Hyndman et al., 1976). Using Deep Sea Drilling Project holes up to 600 meters deep, these authors measured a flow of only 0.6×10^{-6} cal/cm^2 sec within 40 km of the median valley, and in a section of oceanic crust estimated to be only 3.5 million years old. Conventional oceanic probe type measurements made at the same sites also yielded this very low figure. The young oceanic crust must be losing heat, and the most reasonable hypothesis is that heat is carried to the sea itself by some mechanism other than conduction. Lister (1972) had already suggested that circulation of sea water in the upper part of the oceanic crust may be a major factor in determining the thermal balance of ridges. The puzzling feature in the measurements of Hyndman et al. is the uniformity in flux between holes several kilometers apart; small-scale circulation cells of sea water would be expected to produce large local variations in measured flow. The authors suggest that the cells have considerable horizontal extent, with sea water entering the crust through faults and circulating, perhaps as deep as layer 3, through tension cracks induced by cooling (Fig. 25.4). The distance from the axial valley to which such a system could extend is unknown, but it may well be of the order of 100 km. If this is the case, it means that heat flow measurements within that distance of the axis cannot be used to test the cooling slab model described above.

It has been mentioned (Section 13.4) that the ridges appear to be nearly in isostatic equilibrium. A consequence of this is that the ridge topographic profile can be estimated, given the temperature distribution and the volume coefficient of thermal expansion, by equating the masses of vertical columns of lithosphere. In the simplest case, where the base of the lithosphere remains hori-

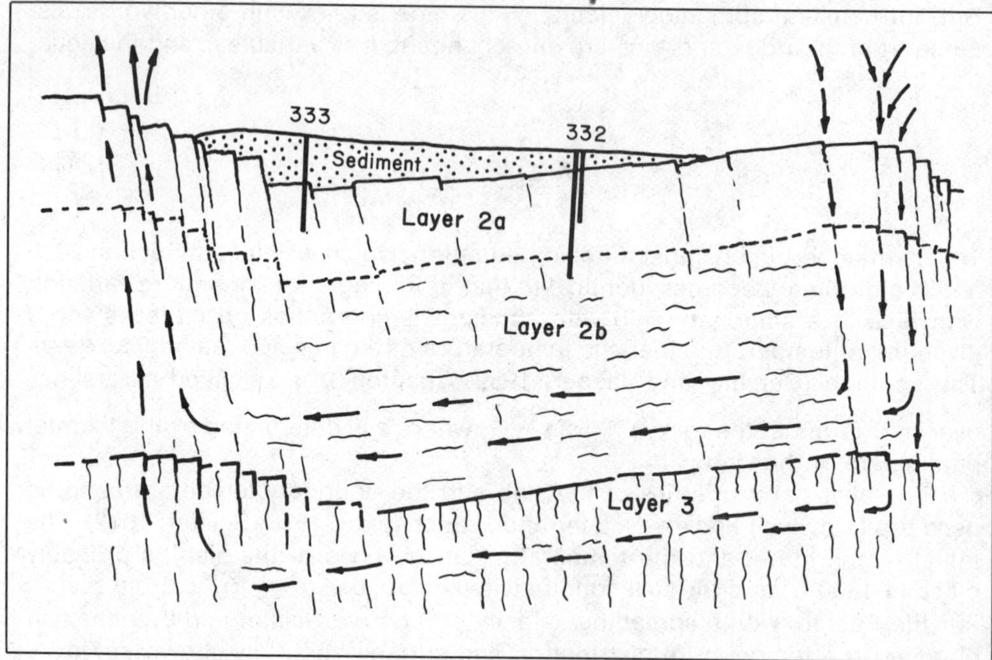

Figure 25.4 A possible model of sea water circulation in the oceanic crust near the Mid-Atlantic Ridge. From Hyndman et al. (1976), courtesy of the authors.

zontal, this is equivalent to holding constant, at different values of x, the sum

$$\rho_w dw + \int_0^L \rho(1 - \alpha T_x(z))\, dz$$

where ρ_w and ρ are the densities of sea-water and lithosphere, and α is the volume coefficient of expansion (about $3.5 \times 10^{-5}\ ^\circ C^{-1}$). The expression can be modified to provide for a non-horizontal lithosphere base. Figure 25.3 shows two topographic profiles computed by Sclater and Francheteau. The first is for a uniform expansion by the above method and the second, which gives a better fit to the observed topography, provides for changes in mineral phase with distance from the ridge, as indicated in the diagram. Phase changes involve density changes, and the topographic profile is therefore altered. While neither of the computed profiles is in complete agreement with observations, the overall conclusions regarding both heat flow and topography of oceanic ridges is that they can be explained on the basis of reasonable models of cooling, spreading plates.

For the downgoing slabs in subduction zones, we are again concerned with the transport of heat by the moving slab. In this case, however, the relatively cold slab is pushed downward into the deep mantle, with its existing temperature gradient, and the problem is to study the thermal history of the slab as it warms up. The general equation for heat flow in a moving medium is

$$\rho c \left[\frac{\partial T}{\partial t} + v \cdot \nabla T\right] = \nabla \cdot (K \nabla T) + A \qquad 25.5.4$$

where ρ, c, and v are as before, and A is the rate of heat production from all sources, per unit volume. Solutions to this equation have been obtained by Minear and Toksöz (1969). These authors assumed an initial distribution of temperature and radioactivity in the mantle as for a chondritic model (MacDonald, 1963), then considered the contributions to the term A in Equation 25.5.4 from adiabatic compression, phase changes and strain heating. For adiabatic compression the rate of energy release at a given depth is

$$\frac{dQ}{dt} = c_p \rho \frac{\partial T}{\partial t} \tag{25.5.5}$$

but

$$\frac{\partial T}{\partial t} = \frac{\partial T}{\partial z} \cdot \frac{\partial z}{\partial t} = \frac{g \alpha T}{c_p} \cdot v_z$$

since

$$\frac{g \alpha T}{c_p}$$

is the temperature gradient resulting from adiabatic compression (Section 24.1) and v_z is the vertical velocity of the material.

Thus

$$\frac{dQ}{dt} = \rho g \alpha T v_z \tag{25.5.6}$$

at each depth.

For the heat resulting from phase changes, Minear and Toksöz considered the changes plagioclase pyrolite–garnet pyrolite (60–160 km depth, 1×10^2 ergs/cm^3 yr), pyroxene–olivine + stishovite (260–360 km depth, 5×10^2 ergs/cm^3 yr) and olivine–spinel (400–500 km depth, 1×10^3 ergs/cm^3 yr). The heat released was assumed to be spread uniformly over the depth interval shown and was based upon thermodynamical calculations. For shear strain heating at the edges of the slab, the velocity distribution in the surrounding material was assumed to be of the form

$$u = -u_s e^{-y/D} \tag{25.5.7}$$

where u_s is the slab velocity, y is distance measured perpendicular to the slab, and D is the thickness of the drag zone. The rate of energy release is then

$$\frac{dE}{dt} = \eta \left(\frac{\partial u}{\partial y}\right)^2 = \eta \left(\frac{u_s}{D}\right)^2 e^{-2y/D} \tag{25.5.8}$$

The estimation of strain heating requires the assumption of the viscosity η and the thickness of the drag zone. It is probably the most uncertain of the terms contributing to A in Equation 25.5.4; furthermore, it is sensitive to the spreading rate, varying with u_s^2, whereas the heat released by adiabatic compression and phase changes in a given depth range is independent of slab

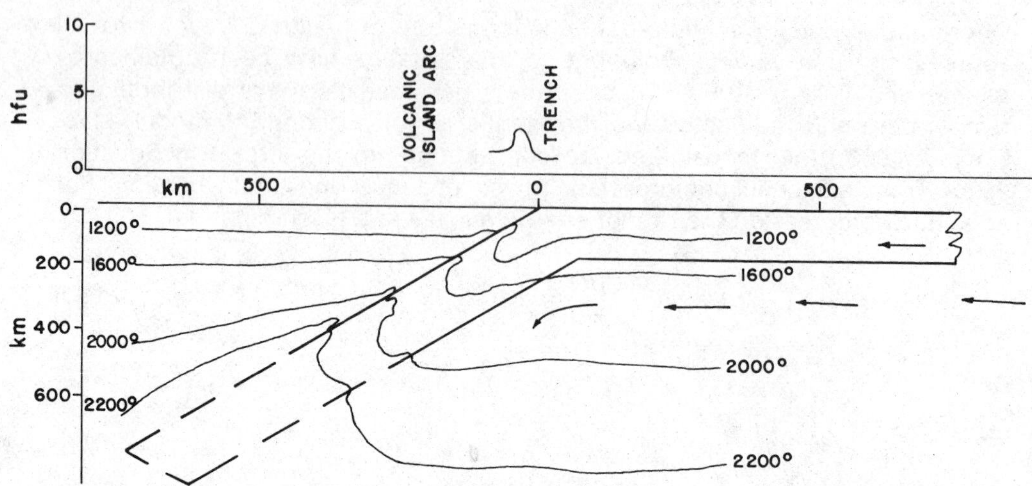

Figure 25.5 Calculated temperature in a descending slab, for a spreading velocity of 1 cm/yr, with predicted surface heat flow above. After Minear and Toksöz (1969).

velocity. On the other hand, the effect of shear strain heating is concentrated around the edges of the slab. In their calculation of temperatures, Minear and Toksöz did not employ the equation in its full form, but took account of the convective term by the method of calculation. The latter employed a finite-difference technique on a grid of points, in which temperature was computed at each grid point at an instant of time; the slab and these temperatures were moved forward, and the change in temperature in the next time interval was calculated.

The case shown in Figure 25.5 is for a spreading rate of 1 cm/yr, requiring a time of 103.7×10^6 years to produce the downwarp. It will be seen that the slab is colder than its surroundings to a depth of about 350 km, below which it is warmer. Strain heating along the edges produces the sharp upwarps of the isotherms at the margins, which should result in a very local anomaly in heat flow. This does not appear to have been observed, and it is noteworthy that no more extensive anomaly in surface heat flux is predicted over the slab. This is in conflict with observations in the northwestern Pacific, which indicate an extensive area of high flux on the continental side of trenches (Vacquier et al., 1966). If strain heating is responsible for partial melting leading to the production of volcanic magma, volcanoes would be expected in the position shown in Figure 25.5, in agreement with the location of volcanic island arcs.

In the case shown the slab would apparently lose its identity at a depth of less than 400 km. Faster spreading rates increase the strain heating, but reduce the influx of heat by conduction. With a spreading rate of 8 cm/yr the isotherms were found to be downwarped to a depth of 650 km. The zones of deepest seismic activity should therefore be associated with the fastest spreading rate. As 8 cm/yr is a very high rate, complete assimilation in all cases is to be expected at a depth of about 700 km.

25.6 THERMAL CONVECTION

The idea that the material of the mantle may be in motion, as a result of thermally generated differences in density, has been discussed for many years

(Hales, 1935; Pekeris, 1935; Griggs, 1939). More recently, it has again come into prominence, as a result of growing evidence, particularly from the ocean floors, that horizontal motions of the order of 1 cm/yr are a normal phenomenon of the earth. Accepting motions of this order at the earth's surface, it is not unreasonable to invoke similar horizontal velocities, and vertical velocities as well, in the mantle.

Convection would have at least two major impacts on the earth as a whole: first, the currents would impose stresses on the base of the lithosphere which could be the dominant agents producing the surface features of the earth, and secondly, it would completely change the thermal regime of the mantle from that inferred by a consideration of conduction alone. The latter effect is of immediate significance in connection with our present discussion, and we shall consider it first. In any convecting system, heat is carried upward by rising currents of material and is delivered at the top of the system; the cooled material sinks, to continue the cycle. A remarkable feature is the efficiency of transport of heat, as compared to conduction alone. The heat delivered, per unit area, by a rising current is $\rho c v \Delta T$, where ρc is the heat capacity, v is the vertical velocity, and ΔT is the difference in temperature between the top and bottom of the cell. For most rocks, ρc is approximately 1 cal/cm^3 °C, and for velocities of 1 cm per year the heat delivered is about 0.03 μcal/sec cm^2 for unit temperature difference. Thus differences in temperature of a few tens of degrees lead to heat transport equivalent to the outflow at the surface. Tozer (1965, 1967a, b) has pointed out that it is not sufficient to formulate a temperature distribution for the earth on the basis of conduction theories, and to consider convection as a perturbation from this. If convection exists, it must be taken into account from the beginning.

Many features of convection in relatively thin layers of fluid heated from below were illustrated in a pioneer investigation by Rayleigh (1916), and although it is now realized that any convection system in the real earth must depart from the simple Rayleigh conditions, his development is instructive. Rayleigh drew on earlier observations by Bénard (1901), that a layer of fluid, free at the upper surface and uniformly heated at the lower boundary, breaks up into a regular series of cells (which may be made visible on the upper surface by the sprinkling of powder or by other means) in which fluid rises in the center and descends along the periphery. It was found that the onset of convection is determined by the temperature gradient applied to the layer; the purpose of Rayleigh's investigation was to find an explicit relation between the critical gradient and the properties of the fluid.

We take the axis of z vertically upward, and imagine the fluid layer to be bounded by the planes $z = 0$ and $z = \zeta$. For a fluid with Newtonian viscosity η, it is well known that shearing stresses are proportional to the velocity gradients, and this condition leads to an equation, the Navier–Stokes equation, which is analogous to the equation of motion for an elastic solid (Equation 2.4.2). It is

$$\frac{\partial v_i}{\partial t} + v_j \frac{\partial v_i}{\partial x_j} = X_i - \frac{1}{\rho}\frac{\partial p}{\partial x_i} + \nu \nabla^2 v_i \qquad 25.6.1$$

where v_i and X_i are the components of velocity and body force, respectively, p is the hydrostatic pressure ($p_{11} = p_{22} = p_{33}$), and ν, the kinematic viscosity, is given by η/ρ. For body forces, we take $X_1 = X_2 = 0$ and $X_3 = -g$. The fluid is assumed to be incompressible, and the effect of temperature is considered only in connection with terms involving gravity, where its effect on $g\rho$, for a temperature disturbance θ, and undisturbed density ρ_0, becomes

$$g\rho_0 - g\rho_0\alpha\theta$$

where α is the volume coefficient of expansion. The equation for v_3 is therefore

$$\frac{\partial v_3}{\partial t} + v_j \frac{\partial v_3}{\partial x_j} = -\frac{1}{\rho}\frac{\partial p}{\partial x_3} + \gamma\theta + \nu\nabla^2 v_3 \qquad 25.6.2$$

where $\gamma = g\alpha$. We let the excess of the vertical gradient in temperature over the adiabatic gradient be β (which will be negative). Then the heat transfer by conduction and convection in the moving fluid gives for the distribution of anomalous temperature

$$\frac{d\theta}{dt} + v_3\beta = \kappa\nabla^2\theta \qquad 25.6.3$$

where κ is the diffusivity. For small velocities, Equation 25.6.1 may be simplified to

$$\frac{dv_1, v_2}{dt} = -\frac{1}{\rho}\frac{\partial p}{\partial x, \partial y} + \nu\nabla^2(v_1, v_2)$$
$$\frac{dv_3}{dt} = -\frac{1}{\rho}\frac{\partial p}{\partial z} + \gamma\theta + \nu\nabla^2 v_3 \qquad 25.6.4$$

It is now proposed that v_i, θ and p are all proportional to $e^{ilx}e^{imy}e^{nt}$, that the boundary conditions are

$$V_3 = 0 \text{ at } z = 0, \quad z = \zeta$$
$$\theta = 0 \text{ at } z = 0, \quad z = \zeta$$

and that v_3 and θ are proportional to sin (sz), where $s = q\pi/\zeta$ (q an integer).

Substitution of these forms into Equation 25.6.4 yields the equation

$$\sigma n^2 + n\sigma(\kappa\sigma + \nu\sigma) + \kappa\gamma\sigma^3 + \beta\gamma(l^2 + m^2) = 0 \qquad 25.6.5$$

where $\sigma = l^2 + m^2 + s^2$. We consider Equation 25.6.5 as an equation for n, whose value determines the time dependence of the motions. For a disturbance to become established and not be attenuated, one root for n must be real and positive; this condition is

$$\beta'\gamma(l^2 + m^2) > \kappa\gamma\sigma^3$$

where $\beta' = -\beta$ is positive.

The instability may be shown to be greatest when

$$l^2 + m^2 = \tfrac{1}{2}s^2 \qquad 25.6.6$$

which determines the horizontal scale of the most probable cells. For $s = \pi/\zeta$, the condition for a positive root n becomes

$$\beta'\gamma > \frac{27\pi^4\kappa\nu}{4\zeta^4} \tag{25.6.7}$$

which is Rayleigh's criterion for the critical temperature gradient in terms of the kinematic viscosity, thickness of the layer, and other properties. It is usual to regroup the parameters into the dimensionless Rayleigh number R, where

$$R = \frac{g\alpha\beta'\zeta^4}{\kappa\nu}$$

The above development shows that R must exceed $27\pi^4/4$ (= 658) for the boundary conditions chosen. Rayleigh did not consider these beyond the vanishing of v_3 and θ at $z = 0$ and $z = \zeta$, but Jeffreys (1926, 1928) showed that the imposition of rigid boundaries requires the vanishing also of $\partial v_3 / \partial z$ at both surfaces. Rayleigh's original solution did not provide this, and so corresponds to the improbable situation of a layer of fluid between free upper and lower boundaries. Jeffreys' solution for the critical value of R in the case of two fixed, conducting boundaries was 1709.5, and this value has often been adopted in geophysical problems. For any particular case, with a fluid of given properties, convection will take place only if the temperature gradient is such that R exceeds the critical value. After the onset of convection the velocities do not continue to increase (as might be implied by the positive value of n), but adjust themselves to carry off the heat which is not transported by conduction. It is important therefore not to mix calculations of marginal instability with those of established convection. Notice also that it is the *excess* of temperature gradient over the adiabatic gradient which enters Rayleigh's criterion. In the cooling of a fluid of low viscosity, the temperature gradient must remain close to adiabatic, since convection would rapidly reduce any excess gradient. But with lower temperature a new adiabatic gradient is defined at each point (cf. Equation 24.1.5: T itself is in the expression for the gradient). The fluid temperature therefore passes through a *family of adiabats* as the fluid cools.

An important result of the theory in this case is the relation between l, m, and s for maximum instability. Equation 25.6.6 predicts cells whose horizontal dimensions are about twice the layer thickness. We shall see that in the case of the earth it may be necessary to invoke cells of relatively greater horizontal dimension.

If we assume for the present that the mantle, or a certain depth range of it, can be treated as a viscous fluid, the calculation of the critical temperature gradient still requires the assumption of ζ, the height of the convection cells. In Chapter 26, we suggest that v falls to a minimum value, about 10^{21} poises, over a depth range of a few hundred kilometers in the upper mantle. Taking $\zeta =$ 200 km, $\kappa = 0.01$ cm²/sec, and $\alpha = 2 \times 10^{-5}$°C^{-1}, β is easily determined to be about 0.5°C/km. From the proposed temperature curves of Figure 25.2, it is seen that a temperature gradient of this order poses no difficulties. But this application of the simple Rayleigh criterion to the mantle is not justified.

The extension of the theory of convection to a realistic model of the mantle presents many complexities, as was recognized by Knopoff (1964) and Tozer (1967a, b). Most important of these, apart from the obvious modification to spherical geometry, are the facts that the treatment of the material as a viscous fluid must be justified and that, even if it is, the coefficient of viscosity will not be constant. We shall indicate in Chapter 27 that the characterization of the mantle by a coefficient of viscosity is probably justified, but that the value of viscosity varies with depth and temperature. Variation with temperature, which is expected to be of the form $e^{U/kT}$, where U is the activation energy for a creep process (Section 27.4), could have spectacular effects. The simple theory has neglected the internal generation of heat through the viscous flow itself. This process would tend to increase the temperature in zones of greatest shear, reducing the viscosity there, and further concentrating the flow in the same zones. In extreme situations "thermal runaway" could occur, in which all of the flow takes place in extremely narrow regions, leaving most of the region in a state of stagnation. The second complication is that, in addition to viscous heating, heat is probably generated within the region by radioactive decay, and the total of internally generated heat may be greater than that delivered to the base of the convecting region.

For geophysical purposes, certain characteristics of a convecting system are of great importance and any analysis should yield them. They are: the efficiency in transferring heat and, therefore, the effect on the thermal history of the earth; the *aspect ratio* of cells, or the ratio of cell width to thickness of the convecting region; the shear stress which can be delivered to a rigid plate above the cells; and, finally, the variation in gravity and heat flow across the surface above a system of cells. The experimental confirmation of convection could depend largely on the recognition of a pattern in the variation in the latter two variables, provided always that the effect of near-surface contributions can be separated. The interpretation of the observed heat flow pattern would be further complicated by the time lag required for temperature perturbations to reach the earth's surface, and the fact that lithospheric plates, where observations are made, are in motion over the aesthenosphere beneath.

Recent studies on convection have involved all three of theoretical analyses, numerical simulations, and laboratory models. An approach to a solution involving the complications noted above was given by Turcotte and Oxburgh (1967), who developed a boundary-layer model. This involved cells in which most of the flow took place in narrow cell boundaries, leaving the cell interiors isothermal and almost stagnant. In their model, the lithosphere itself is the cold, upper boundary layer. Peltier (1972) obtained a solution for a mantle with variable viscosity and with internal heat generation. His result is important because it shows that even if the mantle viscosity increases with depth, convection in the whole mantle (or in a large portion of it) is still possible. That is, the low-viscosity aesthenosphere need not decouple the lower mantle from the surface plates, but can serve to concentrate the flow. Stream lines of the flow, compressed together in the low-viscosity layer, were found by Peltier to spread out in the lower mantle, indicating that convection penetrated to a higher-viscosity region, but with greatly reduced velocity of material flow.

The problem of upper-mantle versus whole-mantle convection is intimately related to the aspect ratio of cells. If mantle convection is one, or the major, driving force in plate tectonics, then it is evident that upward flow must be associated with ridges and downward flow with subduction zones. Some cells, for example that under the Pacific plate, must be 10^4 km wide. This is greater than the entire depth of the mantle, 2900 km, so that even whole-mantle convection does not give equidimensional cells. If convection is taken to be confined to the upper mantle, perhaps even confined to the aesthenosphere or seismic low-velocity zone, the ratio of wavelength to depth of cell becomes very great. Richter (1973) has emphasized that the large-scale mantle circulation associated with ridges and subduction zones should be treated as a phenomenon discrete from, but perhaps driven by, convection of the Rayleigh-Bénard type. Richter showed that, when the two types of motion coexist, the large-scale motions tend to suppress Rayleigh-Bénard cells whose axes are parallel to spreading lines (ridges) and to replace them with cells whose axes are perpendicular to these lines (longitudinal rolls). He suggests that the longitudinal rolls may encourage the formation of lithospheric structures, for example, lines of seamounts or volcanoes, perpendicular to the ridges.

Reviews of other theoretical approaches to mantle convection have been given by Turcotte and Oxburgh (1972). Numerical simulations are illustrated by the calculations of McKenzie, Roberts, and Weiss (1974). They set up and integrated the equations for temperature and flow in cells characterized by in-

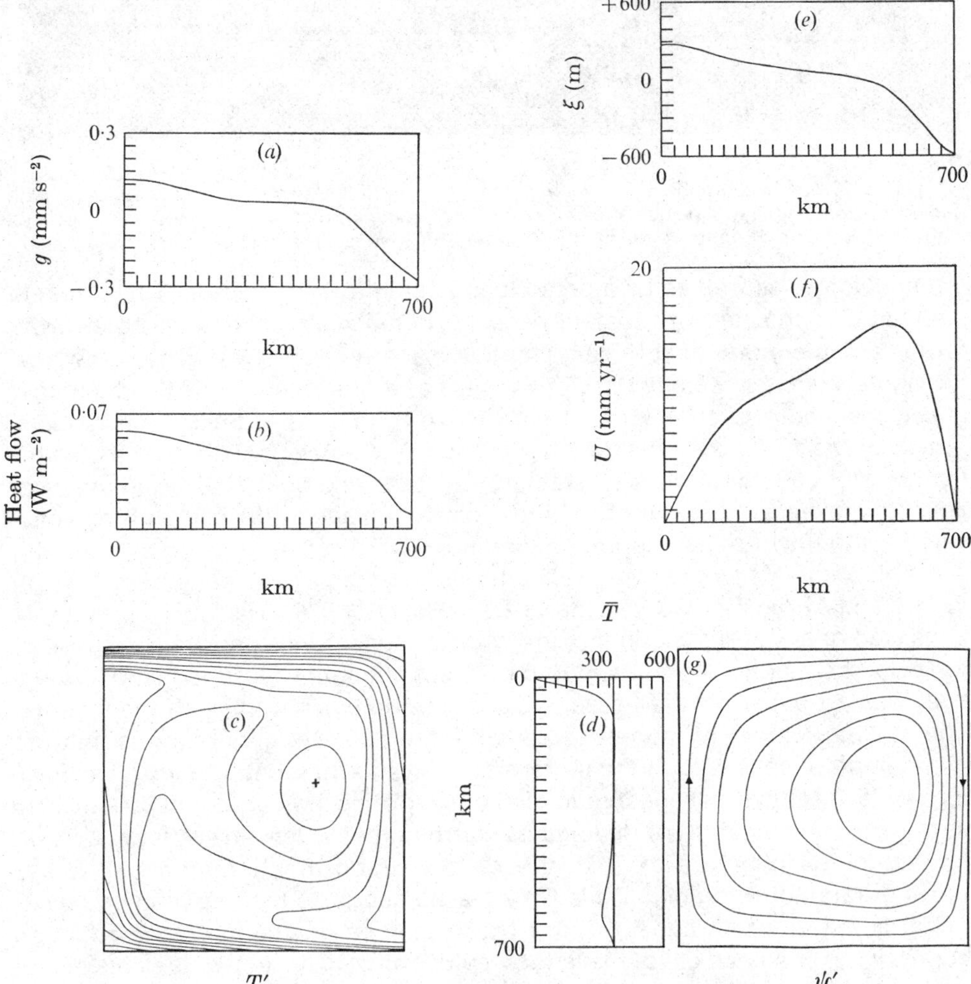

Figure 25.6 Results of a numerical simulation of convection in a cell of square cross-section with unstressed boundaries and a uniform heat flux across the lower boundary; (a) gravity, (b) heat flow, (c) temperature, (d) mean temperature averaged across the cell, (e) surface elevation, (f) horizontal velocity at the surface, (g) flow lines. Contour intervals for temperature are 50° from 0° (top of section) to 350°; then 25° to 500°. From McKenzie, Roberts and Weiss (1974); courtesy of the authors and the Journal of Fluid Mechanics; copyright Cambridge University Press.

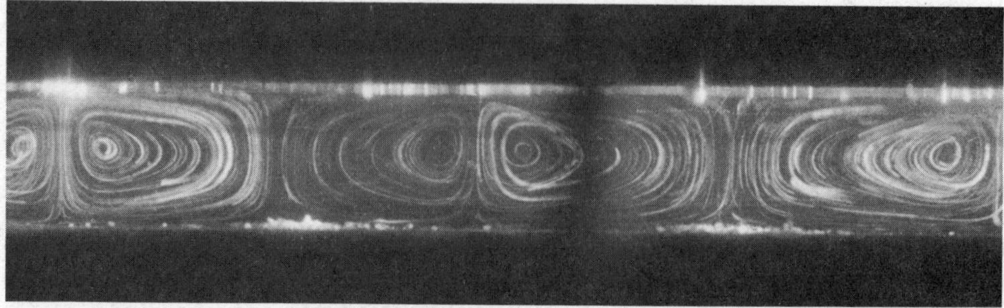

A

B

Figure 25.7 Photographs of convection cells in a liquid layer with internal heating, (*A*) with camera stationary, (*B*) with camera moving to left. Downgoing currents are narrow streams between stagnation points in *A*, dipping zones in *B*. Courtesy of S. de la Cruz Reyna.

ternal heating, as well as by a prescribed heat inflow at the base. Their models are limited to the case of constant viscosity, but the calculations are noteworthy for the completeness of surface quantities that are given: gravity, heat flow, uplift, and horizontal velocity (Fig. 25.6). The case illustrated is one in which heat is provided both internally and from below. In this, as in their other models, gravity at the surface is a maximum over the rising currents. They found that, for this case of constant viscosity, stability was obtained only for equidimensional cells; and even for these, times of the order of the age of the earth were necessary for the flow to become time-independent.

An example of observations made with a laboratory model is shown in Figure 25.6, from the work of de la Cruz-Reyna (1976). His working material consisted of a water-glycerin mixture whose viscosity was chosen to represent, to scale, the mantle. A small amount of copper sulfate was added, and internal heat was produced by the passage of an electric current through the mixture. The striking feature of model results with internal heat generation is that the wavelength of cells is found to increase, relative to their depth. Those shown in Figure 25.7(*A*) have an aspect ratio of about 3. Notice, however, that the cells are asymmetrical, the width of downgoing currents being less than that of upwelling currents. In Figure 25.7(*B*), the cells are shown photographed from a moving frame of reference. De la Cruz-Reyna suggests that the dipping downgoing currents which appear in this frame may be analoguous to subduction zones, if the observed dip of the latter results from a distortion of the convection pattern by the motion of a ridge-plate-trench system above a cell.

It cannot be said that any one of theory, numerical simulation, or laboratory experiment has yet answered all of the geophysical questions relating to mantle convection. Most geophysicists would now say that mantle convection, in some form, is possible. Few would say that surface observations of any parameter require it. Its place as a driving agent of plates, in comparison with other proposed agents, will be discussed in Chapter 27.

Problems

1. Taking Strutt's values for A and q as given in Section 25.2, and his value of K of 0.0041 c.g.s. units, calculate and plot the distribution of temperature through a uniform crust 45 miles thick, to confirm his figure for the maximum temperature. Assume that the surface temperature is 0°C, and that there is no heat flow from beneath.

2. Estimate the temperatures and pressures at the Mohorovičić discontinuity (12 km and 37 km deep below ocean and continent respectively) for the four models of Clark and Ringwood tabulated in Section 24.3. Adopt mean values for any required properties not given in the tables, and use Equation 25.1.2. Compare your results with the values plotted in Figure 26.2.

3. Consider two distributions of radioactive sources of heat in the earth. In one, the heat productivity is A cal/cm^3 sec within a surface layer 20 km thick, and zero beneath it. In the second, the productivity is $Ae^{-z/20}$ at any depth z (km). Compare the steady-state surface heat flows for the two distributions, and also the inferred temperatures at a depth of 100 km. Assume in each case that the effect of the sources is superimposed on the adiabatic gradient, and that all other sources are negligible. (The result illustrates an important principle, that for a given surface flow, the greater the depth to which radioactive material is assumed to extend, the greater are the implied temperatures.)

4. Assume that a convection cell operates within the upper mantle, between a depth of 400 km and the base of the lithosphere at a depth of 100 km. The convecting material moves at the same rate as the lithosphere that it drags, 1 cm/year. The cell extends from the Mid-Atlantic Ridge to the trench west of South America. If the difference in heat flow above rising and sinking columns does not exceed 4 cal/cm^2 sec, what is the mean difference in temperature between these columns? What fraction is this of the temperature difference over which the cell operates? How long does it take a particle to perform one cycle of the convective motion? (Take the specific heat to be 0.2 cal/gm°C.)

5. Consider the cooling of an initially fluid mantle. Where will solidification begin if the curve of melting point versus depth is (a) superadiabatic, (b) subadiabatic?

BIBLIOGRAPHY

Anderson, Don L. (1967) Phase changes in the upper mantle. *Science,* **157**, 1165–1173.

Bénard, H. (1901) Les tourbillons cellulaires dans une nappe liquide transportant de la chaleur par convection en régime permanent. *Ann. de Chimie et de Physique,* **23**, 62.

Boschi, E. (1974) Melting of iron. *Geophys. J.,* **38**, 327–334.

Boyd, F. R., and England, J. L. (1963) Effect of pressure on the melting of diopside, $CaMgSi_2O_6$, and albite, $NaAlSi_3O_8$ in the range up to 50 kilobars. *J. Geophys. Res.,* **68**, 311.

Carslaw, H. S., and Jaeger, J. C. (1959) *Conduction of heat in solids.* Oxford Univ. Press, London.

Clark, S. P. (1957) Radiative transfer in the earth's mantle. *Trans. Amer. Geophys. Un.,* **38**, 931.

de la Cruz-Reyna, Servando (1976) The thermal boundary layer and seismic focal mechanisms in mantle convection. *Tectonophysics,* **35**, 149–160.

Fujisawa, Hideyuki (1968) Temperature and discontinuities in the transition layer within the earth's mantle: geophysical application of the olivine–spinel transition in the Mg_2SiO_4–Fe_2SiO_4 system. *J. Geophys. Res.,* **73**, 3281–3294.

Gilvarry, J. J. (1956) Equation of the fusion curve. *Phys. Rev.,* **102**, 325.

Gilvarry, J. J. (1957) Temperatures in the earth's interior. *J. Atmosph. Terr. Phys.,* **10**, 84.

Griggs, D. T. (1939) A theory of mountain building. *Amer. J. Sci.,* **237**, 611.

Hales, A. L. (1935) Convection currents in the earth. *Mon. Not. R. A. S., Geophys. Suppl.,* **3**, 372.

Higgins, G. H., and Kennedy, G. C. (1971) The adiabatic gradient and the melting point gradient in the core of the Earth. *J. Geophys. Res.,* **76**, 1870–1878.

Jacobs, J. A. (1953) The earth's inner core. *Nature,* **172**, 297–298.

Jacobs, J. A. (1971) The thermal regime in the Earth's core. *Comments on Earth Science, Geophysics,* **2**, 61–68.

Jacobs, J. A., and Allan, D. W. (1954) Temperatures and heat flow within the earth. *Trans. Roy. Soc. Can.,* **48**, 33.

Jeffreys, H. (1926) The stability of a layer of fluid heated below. *Phil. Mag.*, **2**, 833.
Jeffreys, H. (1928) Some cases of instability in fluid motion. *Proc. Roy. Soc. Lond. A*, **118**, 195.
Jeffreys, H. (1962) *The earth* (4th Edition). Cambridge Univ. Press, Cambridge.
Kelvin, Lord (1899) The age of the earth as an abode fitted for life. *J. Victoria Inst.*, London, **31**, 11–35. Also in his *Mathematical Papers*, **5**, Cambridge University Press, 1911.
Knopoff, L. (1964) The convection current hypothesis. *Rev. Geophys.*, **2**, 89.
Kraut, E. A., and Kennedy, G. C. (1966) New melting law at high pressures. *Phys. Rev.*, **151**, 668.
Lawson, A. W., and Jamieson, J. C. (1958) Energy transfer in the earth's mantle. *J. Geol.*, **66**, 540.
LePichon, X., Francheteau, J., and Bonnin, J. (1973) *Plate tectonics*. Elsevier, Amsterdam.
Lindemann, F. A. (1910) The calculation of molecular vibration frequencies. *Phys. Zeit.*, **11**, 609.
Lubimova, E. A. (1967) Theory of thermal state of the earth's mantle. In *The earth's mantle*, ed. T. F. Gaskell. Academic Press, London.
Luedemann, H. D., and Kennedy, G. C. (1968) Melting curves of lithium, sodium, potassium and rubidium to 80 kilobars. *J. Geophys. Res.*, **73**, 2795.
MacDonald, Gordon J. F. (1959) Calculations on the thermal history of the earth. *J. Geophys. Res.*, **64**, 1967.
MacDonald, G. J. F. (1963) The deep structure of continents. *Rev. Geophys*, **1**, 587.
McKenzie, D. P. (1967) The viscosity of the mantle. *Geophys. J.*, **14**, 297.
McKenzie, D. P. (1967) Heat flow and gravity anomalies. *J. Geophys. Res.*, **72**, 6261.
McKenzie, D. P., Roberts, J. M., and Weiss, N. O. (1974) Convection in the earth's mantle: towards a numerical simulation. *J. Fluid Mechanics*, **62**, 465–538.
Milne, John (1906) Recent advances in seismology (Bakerian Lecture). *Proc. Roy. Soc. Lond. A*. **77**, 365–376.
Minear, J. W. and Toksöz, M. N. (1969) Thermal regime of a downgoing slab and new global tectonics. *Journ. Geophys. Res.*, **75**, 1397.
Pekeris, C. L. (1935) Thermal convection in the earth's interior. *Mon. Not. R. A. S., Geophys. Suppl.*, **3**, 343.
Peltier, W. R. (1972) Penetrative convection in the planetary mantle. *Geophys. Fluid Dyn.*, **5**, 47–88.
Rayleigh, Lord (1916) On convective currents in a horizontal layer of fluid when the higher temperature is on the underside. *Phil. Mag.*, **32**, 529.
Richter, Franck M. (1973) Convection and large-scale circulation of the mantle. *J. Geophys. Res.*, **78**, 8735–8745.
Rutherford, E. (1905) *Radio-activity* (2nd ed.) Cambridge University Press, Cambridge. 580 pp.
Sclater, J. G. (1972) New perspectives in terrestrial heat flow. In *The upper mantle*, ed. A. R. Ritsema. Elsevier, Amsterdam.
Sclater, John G., and Francheteau, Jean (1970) The implication of terrestrial heat flow observations on current tectonic and geochemical models of the crust and upper mantle of the earth. *Geophys. J.*, **20**, 509–542.
Simon, F. (1937) On the range of stability of the fluid state. *Trans. Far. Soc.*, **33**, 65.
Simon, F. E. (1953) The melting of iron at high pressures. *Nature*, **172**, 746.
Slichter, L. B. (1941) Cooling of the earth. *Bull. Geol. Soc. Amer.*, **52**, 561.
Strong, H. M. (1959) The experimental fusion curve of iron to 96,000 atmospheres. *J. Geophys. Res.*, **64**, 653.
Strutt, R. J. (1906) On the distribution of radium in the earth's crust and on the earth's internal heat. *Proc. Roy. Soc. Lond. A*, **77**, 472–485.
Tozer, D. C. (1959) The electrical properties of the earth's interior. In *Physics and chemistry of the earth*, 3, ed. L. H. Ahrens, F. Press, K. Rankama, and S. K. Runcorn. Pergamon Press, London.
Tozer, D. C. (1965) Heat transfer and convection currents. *Phil. Trans. Roy. Soc. Lond. A*, **258**, 252.
Tozer, D. C. (1967a) Toward a theory of mantle convection. In *The earth's mantle*, ed. T. F. Gaskell. Academic Press, New York.
Tozer, D. C. (1967b) Some aspects of thermal convection theory for the earth's mantle. *Geophys. J.*, **14**, 395.
Turcotte, D. L., and Oxburgh, E. R. (1967) Finite-amplitude convective cells and continental drift. *J. Fluid Mechanics*, **28**, 29–42.
Uffen, R. J. (1952) A method of estimating the melting point gradient in the earth's mantle. *Trans. Amer. Geophys. Un.*, **33**, 893.
Vacquier, V., Uyeda, S., Yasui, M., Sclater, J., Corry, C., and Watanabe, T. (1966) Studies of the thermal state of the earth: Heat flow measurements in the northwestern Pacific. *Bull. Earthquakes Res. Inst., Tokyo Univ.*, **44**, 1519.

COMPOSITION AND STATE 26

26.1 CONTINENTAL AND OCEANIC CRUST

It is appropriate here to synthesize the evidence from the individual studies based on seismic, gravimetric, magnetic, and thermal properties, as they relate to the differences between continents and oceans. In a very general way, the continental crust is marked by a greater thickness, or greater depth to the Mohorovičić discontinuity, approximately in accord with the requirements for isostatic compensation. Within the continental crust, there appears to be a gradation in properties from top to bottom, as marked by an increase in seismic velocities, and probably in density; an overall density for the crystalline portions of the continental and oceanic crusts that would provide isostatic equilibrium is not far from 2.84 gm/cm^3. The crust is subject to a great deal of regional variation. Certainly it is no longer justifiable to speak of world-wide layering, but it is a fact that the presence of horizontal refracting or reflecting boundaries within the crust is the rule rather than the exception. By contrast, the oceanic crust is thin, and until recently was thought to be relatively simple, consisting almost everywhere of two layers, probably basalt and metamorphosed basalt or gabbro, beneath a veneer of sedimentary rock. The continental crust is marked by an increased concentration of radioactive elements, and a much less regular distribution of magnetic material, as compared to the oceanic crust.

It is fair to say that the greatest increase in knowledge in recent years has been in connection with the oceanic crust. This has come about through direct observation, using submersibles, and by deep drilling as part of the Deep Sea Drilling Project. The complications in regard to magnetic character and to the thermal balance of oceanic ridges have already been mentioned. Drill cores obtained from near the Mid-Atlantic Ridge confirm the petrological complications also. In particular, the discovery of coarse-grained peridotites and gabbros within the basalts near the top of Layer 2 (Hodges and Papike, 1976) shows that the layer model is greatly oversimplified. Hodges and Papike show that these rocks must have cooled slowly, presumably within Layer 3, and suggest that they were tectonically emplaced into the basalt, only 60 meters below its top. If this phenomenon is widespread, it is remarkable that seismic observations yield such a consistent model of layering. The effect of sea water circulation, mentioned in connection with magnetic properties and heat flow, has been directly confirmed by oxygen isotope ratio measurements on samples (Muehlenbachs and Clayton, 1976). Virtually every recovered igneous rock from the oceanic crust shows the effect of isotope exchange with sea water, and it is believed that

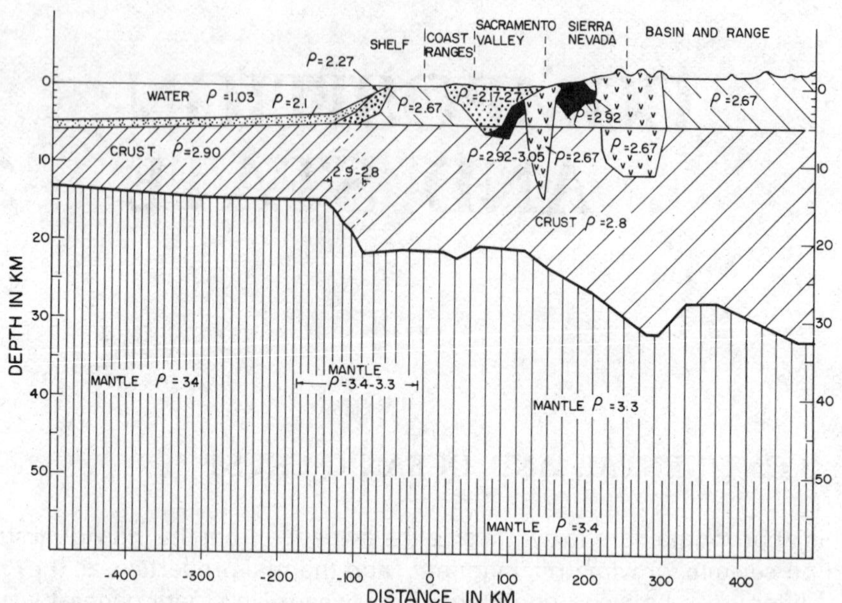

Figure 26.1 Geological section, with densities inferred from geophysical observations, across the western margin of North America, according to interpretations by Thompson and Talwani. Note the relative complexity of the continental crust and also the horizontal gradation of crust and mantle properties.

near the ridge axes, sea water circulates through all of Layer 2 and possibly Layer 3.

The boundaries between continents and oceans are evidently of two types. We have referred to those, such as the Pacific boundary of South America, which are marked by deep trenches, seismicity, and probably by a downthrusted portion of the lithosphere. The second type, exemplified by the Atlantic coast of the Americas, is without these features, and is marked chiefly by the deepening of the Mohorovičić discontinuity toward the continental side, and a gradational change in properties within the crust (Drake and Kosminskaya, 1969). A section across the Pacific margin of North America (Fig. 26.1) based on the work of Thompson and Talwani (1964) is similar to the Atlantic type in that it lacks the trench, but it has been recently deformed and is seismically active. The section indicates the irregular thickening of the crust, the lateral variation in upper mantle properties, and the gradation from a simple oceanic crust to the more complex continental crust. Intrusive masses of both basic and granitic rocks appear to be typical of the upper part of the continental crust. Gravity measurements show that the masses which crop out at the surface are unlikely to extend through the crust, but the apparent simplicity of the lower continental crust may be illusory, and the result of a lack of resolution in our measurements. The upper crust is, on the average, more intermediate than acidic in petrological character, and the lowermost material of the continental crust is, presumably, of a basaltic composition similar to that of the oceanic crust.

In many places on the continent, layered sections of rock, ranging from periodotite at the base to basalt at the top, have been recognized. These rocks, known as *ophiolites* (Dietz, 1963; Church, 1972) are believed to represent sections of ancient oceanic crust; in some cases, they may have been bodily thrust into position on top of the continental crust (Frontispiece). Apart from their great interest to geology, they have geophysical significance: first, in that measurements of gravity, electrical conductivity, or seismic velocity may assist in locating them and second, in that they may provide a convenient opportunity for physical property measurement. The latter must be treated with caution, however, because the ophiolite masses are not now under the same conditions of temperature, pressure, and water content as they were when they formed part of the oceanic crust.

26.2 THE EVIDENCE FROM VOLCANISM

Volcanoes, representing a surface expression of tectonic activity, thermal processes, and, frequently, seismicity, are an important geophysical phenomenon. Not only their distribution, but also a study of their products, can throw additional light on the nature of the interior.

We have already referred to the hypothesis that new ocean floor is created by the injection of basaltic lava along the crests of oceanic ridges, and we shall return to this in the following chapter. This activity is occasionally marked by violent outpourings, such as that which accompanied the creation of the island of Surtsey, south of Iceland. For the most part, however, the injection of this great volume of material is not directly witnessed and must be inferred from other evidence.

By contrast, there are the lines of active volcanoes which are associated with the oceanic trenches, and oceanic islands which are not on the ridges. Along the trenches, the activity often results in arcuate patterns of volcanic islands. The products of these volcanoes are predominantly basaltic, but there appears to be a systematic variation in the nature of the basalt with geographic position.

There is physical evidence that the basaltic magma poured out by volcanoes originates in the mantle rather than the crust. Eaton and Murata (1960), in a study of the volcano Kilauea, traced the foci of earthquakes which preceded an eruption, from a depth of 60 km. Furthermore, the temperature of basaltic magma is commonly about 1200°C, much higher than the temperature at the base of the oceanic, or even continental, crust. If basaltic magma is a constituent of the upper mantle, its composition must provide a clue to the nature of the mantle, and for this reason it is important to consider it in some detail.

Kuno (1967) has summarized the three main types of basalt which are observed in the Pacific region: tholeite, or SiO_2–rich basalt, high alumina basalt, and alkali-olivine basalt. The chemical composition of these types is shown in Table 26.1.

The differences in composition are not great, but they are significant, and the question is whether they can be formed from the same parent material. Kuno has argued that melting of material of appropriate composition under different pressure conditions could produce the observed differences; under this hypothesis, tholeite is formed at the shallowest depths (100 km or less) and alkali-olivine basalt at the greatest depths (300 to 400 km), in agreement with the observed association of tholeite with the shallowest seismic activity. However, melting within a single depth range, with subsequent differences in history, could produce comparable results, and it is necessary to consider more fully the source material.

TABLE 26.1 BASALT COMPOSITIONS. (AFTER KUNO)

	Tholeite (Hawaii)	High Al Basalt (Japan)	Alkali-olivine Basalt (Japan)
SiO_2	49.32	48.10	48.11
Al_2O_3	13.94	16.68	15.55
Fe_2O_3	3.03	3.88	2.99
FeO	8.53	7.75	7.19
MgO	8.44	8.89	9.31
CaO	10.30	10.48	10.43
Na_2O	2.13	2.51	2.85
K_2O	0.38	0.46	1.13
TiO_2	2.50	0.73	1.72
P_2O_5	0.26	0.15	0.56
MnO	0.16	0.54	0.16

For many years there was a tendency for geophysicists to think of the mantle as a region of great chemical simplicity. A mantle composition of olivine, $(Fe, Mg)_2SiO_4$, was widely accepted, chiefly because of the agreement with the mantle's density and elastic properties. However, if the upper mantle is to produce large amounts of basalt (with the above composition) by the local partial melting of its constituents, its chemical composition must be a good deal more complex. Ringwood (1958) proposed that the upper mantle consists of a mixture of minerals, named by him pyrolite, and defined as a substance which can produce basaltic magma through partial melting, leaving dunite-peridotite as a residuum. Partial melting of about 30 per cent of the original material, in any region, is assumed to provide the basaltic magma. Under this hypothesis, the upper mantle beneath the thick continental crust has been deprived of its basalt-producing fraction to a depth of perhaps 200 km, while under the oceans, pyrolite may extend upward almost to the Mohorovičić discontinuity. The model is therefore in agreement with the differences in heat productivity beneath continents and oceans (Section 24.3), if it is assumed that the radioactive elements are removed with the basalt fraction.

A great deal of experimental work on the partial melting of mineral assemblages similar to pyrolite, up to pressures equivalent to depths of 100 km, has been accomplished (Green and Ringwood, 1969). Melting of the individual constituents is controlled not only by pressure and temperature, but also very markedly by the amount of water present. The presence of water reduces the melting point of Mg_2SiO_4, for example, from 2000°C to 1330°C at a pressure of 20 kilobars. At the same pressure, and in the presence of water, a basaltic magma forms at temperatures less than 700°C (Lambert and Wyllie, 1969). Lambert and Wyllie propose that free water is present, in a certain depth range of the upper mantle (60 to 200 km), because of the breakdown of hydrous minerals such as hornblende. Water normally in the crystal lattice is forced out as interstitial water, at a pressure corresponding to the upper depth, but is apparently taken into the lattice again at the higher pressure. The agreement of these depths with the seismically defined low-velocity zone is remarkable.

Green and Ringwood (1969) suggest that melting of pyrolite at a depth of 35 to 70 km can yield a magma which, through subsequent fractionation at lower pressures, can provide the principal types of basalt. The quartz-rich tholeites fractionate from the parent magma at the shallowest depths (less than 15 km), and the alkali-olivine basalts at depths near the original melting. In contrast to Kuno's theory, therefore, the gradation in types of basalt is a result of subsequent history, rather than of depth of original melting. The low-pressure fractionation of the parent magma, according to the Ringwood and Green hypothesis, is in agreement with the experimental results of Yoder and Tilley (1962).

Local melting in the mantle can occur through a rise in temperature as might result from a rising convection current, a decrease in pressure, or an increase in the partial pressure of water. Decrease in pressure may result from faulting, or from the adiabatic uplift of a volume of pyrolite to lesser depths (Ringwood and Green, 1966). We have previously noted (Section 4.6) the differences in the seismic velocity profiles across the low-velocity region, between stable shields, active regions, and oceans. The relatively shallow minimum of the shields occurs in the residual sub-continental dunite, according to the suggestion of Ringwood and Green. Deeper minima of active areas may define regions of "incipient melting" where conditions are almost correct for melting, but the availability of water, for example, may be too small for large volumes of magma to form. Only in regions of active volcanism is there melting of fractions of the mantle reaching 30 per cent, and in these areas sharp minima in velocity and high attenuation of *S* waves are to be expected.

Many recent studies on volcanism are closely related to aspects of plate tectonics, and they will be referred to in Chapter 27.

26.3 DENSITY AND COMPOSITION

Density plays a role of great importance in the identification of the material of the deep interior, because it may be compared to laboratory measurements which are now available to very high pressures.

The principle of the determination of density from seismic velocities, by the method of Adams and Williamson, was outlined in Section 3.6. It was shown there that the radial derivative of density at any level was given by

$$\frac{d\rho}{dr} = -\frac{Gm\rho}{r^2\phi} \qquad 26.3.1$$

where $\phi = \alpha^2 - \frac{4}{3}\beta^2$, and m is the mass within the sphere of radius r. In practice, the application of the Adams-Williamson equation to the mantle and core of the earth, without further refinements, leads only to a first-order estimate of density. There is, first, another term to be added to the right of Equation 26.3.1 if the temperature gradient is non-adiabatic, for the equation as it stands gives the effect on density of pressure alone. Birch (1952) showed that the second term is given by $-\alpha\rho\tau$, where τ is the excess of the actual temperature gradient over the adiabatic gradient and α in this term is the volume coefficient of thermal expansion. The term becomes significant if τ is more than 1°/km, but the actual temperature gradient (Fig. 25.2) probably never exceeds the adiabatic by this amount.

The application of the method to successively refined models is due chiefly to Bullen (1965), who has described the earlier calculations. The chief difficulty is in the extension of the integration across regions of change in phase or composition, and in the assignment of the discontinuity in density at the core boundary. Bullen first established a trial density distribution for the mantle, without allowance for the rapid change in properties at the depth of about 500 km. As a test of this, he investigated the implied moment of inertia of the core alone, and found it to exceed that of a uniform sphere ($0.57\ m_c r_c^2$, as opposed to $0.40\ m_c r_c^2$). Since this would require a decrease in density with depth within the core, the trial distribution was rejected, and replaced by one with more mass in the mantle. This was accomplished by introducing a discontinuity in $d\rho/dr$ at a depth of 413 km, and assuming a quadratic distribution of density with depth down to 1000 km. The resulting density distribution thus clearly showed the rapid change in properties through the depth range of 400 to 1000 km.

The discontinuity in density at the core boundary cannot be determined directly. Bullen, on the basis of further moment of inertia calculations, suggested that the central density must lie between 12.3 and 22.3 gm/cm³, and proposed a model ("Model A") whose density at every depth was a mean of the densities of two distributions (A_i, A_{ii}) leading to those central densities.

In the course of these earlier investigations, it became apparent that there was a remarkable constancy in the modulus of incompressibility, k, and even greater constancy in dk/dp. In a region of uniform composition,

$$k = \frac{\rho\, dp}{d\rho} = \rho\phi, \tag{26.3.2}$$

and

$$\begin{aligned} \frac{dk}{dp} &= \frac{d}{dp}\,(\rho\phi) \\ &= 1 + \frac{\rho}{\phi}\frac{d\phi}{d\rho} \\ &= 1 - \frac{1}{g}\frac{d\phi}{dr} \end{aligned} \tag{26.3.3}$$

It is known directly from seismic observations that, even between the lower mantle and core, $d\phi/dr$ does not change greatly, and we have already noted that g varies little from the outer surface to the outer core. Bullen therefore constructed a model (known as "Model B") in which the chief assumption was that k and dk/dp were smoothly varying through the mantle and core. Model B, which has been very widely used as a standard of reference, places higher densities in the upper mantle than did Model A.

The central density in Model B is 18.1 gm/cm³. Birch (1963), as the result of investigations into compression in the earth (Section 26.5), suggested that the density at the pressure corresponding to the earth's center should not exceed 13 gm/cm³. The only model meeting this requirement is the distribution A_i. However, recent theoretical work on solids under great pressure (Thomsen and Anderson, 1969) suggests that the importance of an upper limit of 13 gm/cm³ may have been overestimated.

All of the earlier models, A_i, A_{ii}, and B, were based upon a figure for the moment of inertia of the earth which has since been revised downward, from $0.334\ MR^2$ to $0.3309\ MR^2$ (Section 1.3), and are therefore obsolete. In Section 3.6, reference has already been made to a later model, HB_1 (Haddon and Bullen, 1969), which was constructed to yield free oscillation frequencies in agreement with those observed. In constructing this model, Haddon and Bullen began with Model A_i (because of its lower central density), corrected it to the

TABLE 26.2 DENSITY PREDICTIONS OF EARTH MODELS

Depth, km	Model					
	A_i	A_{ii}	A	B	HB_1	JAB_2
33	3.32	3.32	3.32	3.32	3.32	3.31 (41 km)
200	3.47	3.47	3.47	3.94	3.387	3.37 (196 km)
600	4.11	4.14	4.13	4.18	4.075	3.97 (596 km)
1000	4.65	4.71	4.68	4.41	4.538	4.59 (1021 km)
2898 (mantle)	5.66	5.72	5.69	5.57	5.527 (2878 km)	5.58 (2886 km)
2898 (core)	9.7	9.1	9.4	9.74	9.927 (2878 km)	9.90 (2886 km)
5121	12.0	12.0	12.0	15.4	12.197	12.11–12.28 (5156 km)
6371	12.3	22.3	17.3	18.1	12.46	12.58

new moment of inertia, and then modified the distribution to fit the free oscillation results. Still further refinement was provided in the model, designated B_1, of Jordan and Anderson, also described in Section 8.3, and based on the best available periods of free oscillations. A comparison of the densities in gm/cm^3 at representative depths in the mantle and core is shown in Table 26.2. Notice that the Jordan-Anderson model ("JAB$_1$") includes a small discontinuity in density at the core–inner core boundary, at a depth of 5156 km.

26.4 THE COMPOSITION OF THE MANTLE: PHASE CHANGES

The possibility of minerals existing in different *phases,* represented by differences in their atomic lattices and crystalline structure, is well known in mineralogy, and was apparently first suggested for the earth by Bernal (1936). Bernal pointed out that the mineral Mg_2GeO_4 exhibited a change in phase to a denser, closer packed structure at relatively low pressures, and he proposed that olivine, Mg_2SiO_4, might do so at the pressure corresponding to a depth of about 400 km. At the time he wrote, there was believed to be a discontinuity at that depth (known as the 20° discontinuity, from the value of Δ at which the travel time curve shows abnormal behavior), which recent work on body waves would suggest is represented by two discontinuities, at depths of about 400 km and 600 km (Chapter 3).

Subsequent to this proposal, changes in phases were proposed as an explanation for the Mohorovičić discontinuity, and for the mantle-core boundary (Ramsey, 1949). Regarding the latter proposal, evidence on the density of silicates and metals from shock wave experiments has, as we shall see in Section 26.6, lent considerable support to the hypothesis that the core is dominantly iron, rather than a high-pressure form of silicate. The possibility that the Mohorovičić discontinuity is a phase change has been examined critically by Bullard and Griggs (1961). These authors point out that if a phase change is responsible for the discontinuity, the temperature-pressure curve which defines the conditions for the change must pass through the two points on a temperature-pressure plot representing the mean conditions below continents and oceans. These conditions can be estimated, and are shown on Figure 26.2, together with a hypothetical reaction curve. With increasing depth below the continents, a phase change to the denser form could occur at the point M_2, but

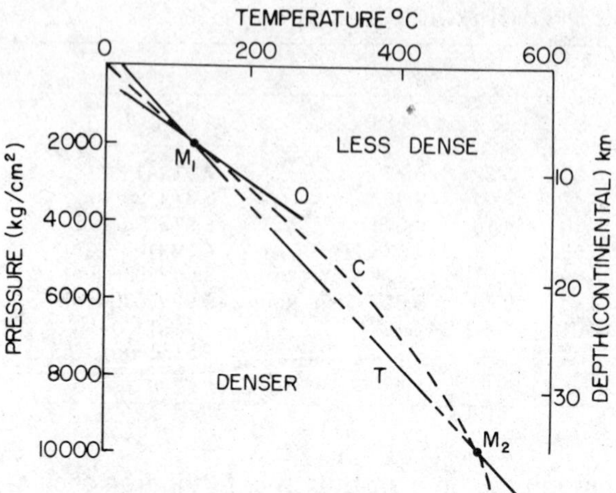

Figure 26.2 Probable distribution of temperature with depth beneath oceans (O) and continents (C), with a possible form of the phase change curve (T), which must pass through points M_1 and M_2, representing conditions at the Mohorovičić discontinuity in the two cases. (After Bullard and Griggs)

the curve for the oceans would then cross the transition curve to the less-dense phase beneath the Mohorovicic discontinuity. Bullard and Griggs conclude that the topology of the diagram (even with a curved line *T*) does not permit a single phase change to represent the discontinuity. There are other arguments that the upper mantle is of ultrabasic composition, rather than basalt in a different phase. Perhaps the strongest is the inclusion of nodules of ultrabasic composition in certain basaltic lavas. Other evidence against a global phase change at the base of the crust is the very slow rate of the reaction which had been proposed, as noted below.

In the uppermost mantle the most important phase transition is believed (Ringwood and Green, 1969) to be one involving the breakdown of plagioclase ($NaAlSi_3O_8$–$CaAl_2Si_2O_8$) in pyrolite, and the formation of garnet ($CaSiO_3$). This is an example of a phase change involving redistribution of elements between compounds, rather than a change in crystal form of a single compound. The transformation takes place over a range of pressures, and appears to be influenced by the precise chemical composition of the original pyrolite. Garnets may well appear at depths corresponding to the base of the continental crust, and when a depth of approximately 60 km is reached, garnet probably completely replaces plagioclase. The fraction of mantle material which, at lower pressures, was basalt (plagioclase plus pyroxene) is known in this condition as eclogite. Ahrens and Schubert (1975) have examined in detail the rate of the gabbro-eclogite reaction and have shown that it is greatly dependent upon the presence of water and minor elements, as much as upon temperature and pressure. In some situations, the change proceeds so slowly that the eclogite phase can hardly be present in significant amounts, even over periods of geological time. Locally, as in the downgoing plates of subduction areas, the reaction probably does take place rapidly enough to be an important factor. Of greatest interest in the mantle is the rapid change in properties between depths of 400 and 1000 km. The change in crystal structure of olivine, proposed by Bernal, does appear to mark the onset of this region. At low pressures olivine has an orthorhombic crystal structure (Fig. 26.3), while at the critical pressure there is a rearrangement, to give the closest packing of the large oxygen atoms. The resulting cubic form is known as spinel structure. Because of the closer packing, an increase in density of as much as 10 per cent may result. The transformation of the olivine fayolite (Fe_2SiO_4) has already been mentioned in connection with the electrical conductivity measurements of Akimoto and Fujisawa (Section 20.5). However, as suggested by density measurements, the olivine of the mantle is probably closer to forsterite (Mg_2SiO_4), and the transformation of this mineral has only recently been observed (Ringwood and Major, 1970; Akimoto and Syono, 1970; Kawai et al., 1970). For composition between Fe_2SiO_4 and $(Mg_{0.8}Fe_{0.2})_2SiO_4$, a direct transformation to spineal structure, with increase in density of 10 per cent, was observed. For composition between $(Mg_{0.8}Fe_{0.2})_2SiO_4$ and Mg_2SiO_4, the transformation was to a distorted orthorhombic spinel, with density increase of 8 per cent. The latter transformation was observed (Ringwood and Major, 1969) at a pressure of 130 kilobars at 1000°C, conditions that would be met in the mantle at a depth of about 400 km.

The seismic discontinuity at a depth of about 600 km is believed to coincide with a less understood phase, known as "post-spinel" (Anderson, 1967). At still greater depths in the transition region, both olivine (post-spinel phase) and pyroxene ($MgSiO_3$) probably break down into other compounds. The pressures required to produce these changes in silicates have not in every case been attained in the laboratory, but the reactions may be inferred from the lower-pressure behavior of their germanate analogues (Ringwood and Major, 1967). Pyroxene may be transformed into olivine through the reaction

$$2MgSiO_3 \rightarrow Mg_2SiO_4 + SiO_2$$

The final form of Mg_2SiO_4 (spinel) is not yet known. Olivine may break down into other simpler compounds, for example through the reaction

$$Mg_2SiO_4 \rightarrow 2MgO \text{ (periclase)} + SiO_2$$

or it may suffer a further change in structure. In any case, it is probable that all changes in phase are accomplished above the depth of 1000 km, that is, the base of the region of rapid change in properties.

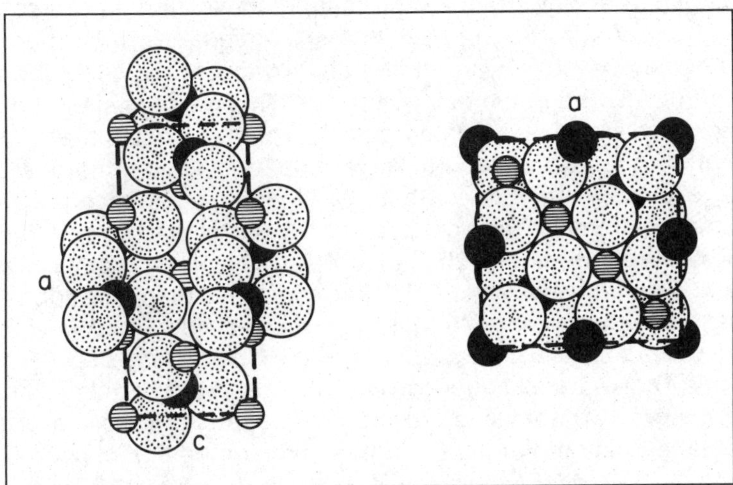

Figure 26.3 Left, section of a crystal of olivine, looking along the **b**-crystallographic axis. Large stippled atoms are oxygen, horizontally striped ones are silicon, black atoms are magnesium. Right, the high-pressure or spinel phase of the same mineral. Notice the decrease of voids resulting from the close packing of oxygen atoms.

It will be noticed, in the two reactions above, that free SiO_2 is produced. Quartz itself suffers a change in phase, first to coesite (Coes, 1953) and then to stishovite (Stishov and Popova, 1961). The latter phase occurs at a pressure of about 130 kilobars, so that in the region considered, below 400 km, any SiO_2 formed as a separate compound would occur as stishovite.

Phase changes in the mantle have a number of important geophysical consequences (Schubert and Turcotte, 1971). We have already noted, in Section 25.3, that they can probably be identified with the minor seismic discontinuities at depths near 400 km and 600 km, and can therefore provide controls on mantle temperature. Much more complex is the relationship of a phase change boundary to mantle convection. Let us realize, first, that if the seismic discontinuities represented changes in composition, they would be strong evidence against mantle convection, simply because convection would have broken up the interfaces. But if the boundaries are phase changes, through which convection proceeds, what is the effect of the change on the convection? The pressure required to produce a higher-density phase increases with temperature, as suggested by the hypothetical curve of Figure 26.2; in other words, colder material changes phase at a shallower depth. Suppose that a downgoing convection current is bringing material toward a phase change boundary. The material is cooler than stagnant mantle material at the same depth, and will tend to change to the dense form above the normal position of the boundary. Becoming denser, its downward motion will tend to increase the convection velocity. However, the phase changes are known to be exothermic, so that heat is released, and the material is warmed upon changing. This would tend to restore the boundary to its original position, lowering the density and reducing the tendency for instability. The balance between the opposing factors is critically dependent upon the values of the heats of phase change and the gradient of the phase-change curve. In extreme situations, phase changes may be so stabilizing as to inhibit convection, or they may permit it in a region in which it would be otherwise attentuated by viscosity. The consensus at present (Schubert and Turcotte, 1971; Peltier, 1972) appears to be that the net effect is much smaller than these extremes, but that it is impossible to say whether the phase changes are, to a degree, assisting or inhibiting mantle convection. In any case, if convection is proceeding through phase change boundaries, it is to be expected that the boundaries will be warped slightly upward at the sites of downgoing currents and downward at the rising currents. This warping might be sufficient to be detected seismically, in which case it would provide extremely valuable independent information on the phase change–convection interplay.

We have already mentioned that the downgoing lithospheric slabs are probably sites of dramatic temperature contrast with their surroundings. Their relatively low temperature (Fig. 25.5) means that the olivine-spinel change must take place at a much shallower depth than in normal mantle. Adopting a slope for the phase-change curve of 40 bars/°K, Schubert and Turcotte estimate that the change occurs as shallow as 200 km in the downgoing slab. This increase in density may have a consequence on plate motion; it certainly is a factor in producing the gravity anomaly observed over subduction areas and it, and the gabbro-eclogite change mentioned above, explain the complex density structure invoked in the model shown in Figure 13.8.

If a change in phase were widespread at the base of the crust, it would have important consequences for isostasy and vertical crustal movements. Loading of an area by sedimentation could promote the formation of the denser phase, thus assisting in downward movement and encouraging further loading (cf.

Problem 2). But, as discussed above, the evidence at present seems to be against a widespread phase change at the base of the crust, and the accepted changes at depths of 400 km and greater appear to be too deep to correlate directly with individual sedimentary basins and vertical crustal movements.

26.5 EQUATIONS OF STATE

A quantitative method of investigating the uniformity of regions of the mantle is to set up an equation of state, connecting pressure and density at a given temperature, and to determine whether the same parameters apply over the entire region. Equations of state are normally written for isothermal conditions, although the properties determined both from the seismic velocities and from the thermal state of the mantle are very nearly adiabatic. The first detailed investigation of state in the mantle by this method was that of Birch (1952), who adopted the equation

$$p = \tfrac{3}{2}k_0\left[\left(\frac{\rho}{\rho_0}\right)^{7/3} - \left(\frac{\rho}{\rho_0}\right)^{5/3}\right] \qquad 26.5.1$$

where p is pressure, k_0 is the isothermal incompressibility at atmospheric pressure, and ρ is density. The form of the equation has theoretical justification (Gilvarry, 1957) and is derivable from a theory of finite strain developed by Murnaghan (1937). In this theory, terms in the second order of the elastic strains, e_{ij}, are retained, and the ratio of density ρ, under compression, to its unstrained value ρ_0 is given by

$$\frac{\rho}{\rho_0} = (1 + 2f)^{3/2} \qquad 26.5.2$$

where f is the positive quantity known as compression. For the maximum compressions met in the mantle, simple relations follow for p, k, and $\partial k/\partial p$:

$$\left.\begin{aligned} p &= \tfrac{3}{2}k_0 f(1 + 2f)^{5/2} \\ k_T &= k_0(1 + 2f)^{5/2}(1 + 7f) \\ \left(\frac{\partial k}{\partial p}\right)_T &= \frac{12 + 49f}{3(1 + 7f)} \end{aligned}\right\} \qquad 26.5.3$$

where the subscript T denotes isothermal values.

We have already referred to the fact that one of the best determined parameters in the mantle is the ratio

$$\phi = \frac{k_s}{\rho} = \alpha^2 - \tfrac{4}{3}\beta^2 \qquad 26.5.4$$

where κ_s is the adiabatic incompressibility. To apply the theory of finite compression, it is necessary to replace k_s by k_T through the relation

$$k_s = k_T(1 + \alpha\gamma T) \qquad 26.5.5$$

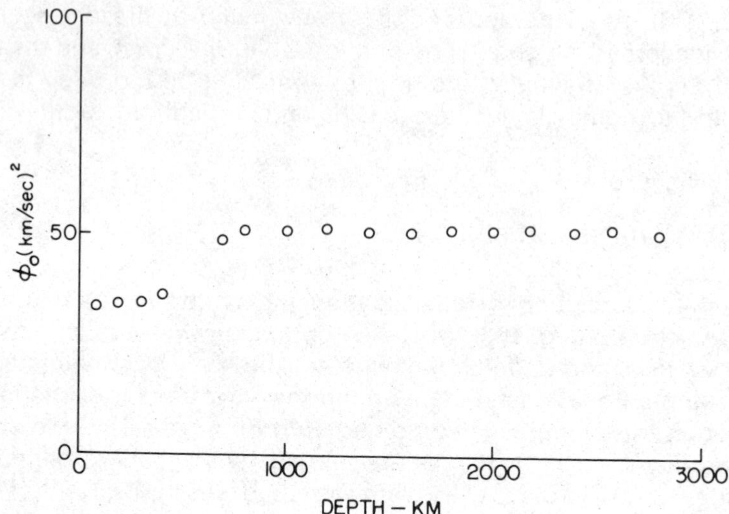

Figure 26.4 The quantity ϕ (bulk modulus/density) for the mantle, reduced to atmospheric pressure. (After Birch)

where α is here the coefficient of thermal expansion and γ is a parameter of the material known as Gruneisen's ratio. The variation of ϕ with radius, for hydrostatic conditions, was then found by Birch to be

$$1 - g^{-1}\frac{d\phi}{dr} = \left(\frac{\partial k_T}{\partial P}\right)_T - 5T\alpha\gamma - 2\tau\alpha\frac{\phi}{g} \qquad 26.5.6$$

where τ is the difference between the actual and adiabatic temperature gradients (Section 24.1). The right-hand side of Equation 26.5.6 was shown by Birch to be dominated by the first term, which is given in terms of the strain by Equation 26.5.3. It was therefore possible, without a detailed assumption regarding the distribution of temperature, to obtain the strain from the observed values of ϕ, and from the strain, to obtain values of ϕ_0, the equivalent ϕ at zero pressure. The result of this determination of ϕ_0, as shown in Figure 26.4, is striking evidence for the uniformity of the mantle, except in the depth region from 400 to 700 km.

Anderson (1967) has investigated an equation of state of the form

$$\rho = A\overline{M}\phi^n \qquad 26.5.7$$

where A is a constant, $\overline{M}$ is the mean atomic weight, and n is a constant between $\frac{1}{4}$ and $\frac{1}{3}$. On the basis of seismically determined values of ϕ for the mantle and core, mean atomic weights of somewhat greater than 20, and somewhat greater than 50, respectively, are indicated for the two values of M. For comparison, M for olivine is 22.0 and for pure iron it is 55.85, so that once again there is evidence of alloying of the core with lighter elements.

26.6 SHOCK-WAVE MEASUREMENTS OF PROPERTIES AT HIGH PRESSURES

The possibility of the identification of materials within the earth has greatly increased with the availability of laboratory measurements at pressures greater than 1 megabar. This is several times greater than the maximum pressure attainable in a static experiment, and has been attained through the use of shock-wave techniques. The observations are usually presented as an equation of state for the material under study, but because this equation of state differs from those usually encountered, a brief description of the procedure is necessary (Rice, McQueen, and Walsh, 1958). A strong shock is generated in the material, usually by a specially shaped charge of high explosive, in such a way that the resulting wave-front of the shock waves is a plane. This shock front is marked by a discontinuous increase in density, to very high values behind the front. It moves through the material at a velocity dependent upon the strength of the charge, but greater than both the normal sound velocity in the material and the particle velocity in the shocked material. Across the front, mass, momentum, and energy are conserved and these conditions lead to the three basic equations:

$$\left.\begin{aligned} \rho_0 U_s &= \rho(U_s - U_p) \\ P_1 - P_0 &= \rho_0 U_s U_p \\ \left[E - E_0 - \frac{U_p^2}{2}\right]\rho_0 U_s &= P_0 U_p \end{aligned}\right\} \qquad 26.6.1$$

where U_s and U_p are shock and particle velocities, ρ, P, and E are density, pressure, and internal energy respectively, and the subscript 0 refers to the unshocked state.

Equations 26.6.1 can be combined to give

$$\left.\begin{aligned} U_s &= V_0\left[\frac{P - P_0}{V_0 - V}\right]^{1/2} \\ U_p &= [(P - P_0)(V_0 - V)]^{1/2} \end{aligned}\right\} \qquad 26.6.2$$

where V is the specific volume. The measurement of U_s and U_p thus suffices to determine a relationship between P and V, and if measurements are made with shocks of various strengths, a curve of V against P is obtained. The shock wave velocity measurements are made by photographically recording the instant of arrival of the shock at the surface of the sample, where it is in contact with a gas which becomes luminous when shocked. Particle velocities can be determined by recording the subsequent motion of the free surface, but the availability of standards of known behavior now permits the use of a comparison method in which only the shock value is required (McQueen, Marsh, and Fritz, 1967).

The relation between P and V obtained through Equations 26.6.2 is an equation of state. It is neither adiabatic nor isothermal, but is defined by the three conditions expressed in Equations 26.6.1. Such an equation of state is known as the *Hugoniot.* Provided thermodynamic equilibrium is obtained in the sample, and this has usually been assumed in the measurements which have been reported, it is possible to calculate the temperature at each point, and therefore to derive other equations of state from the Hugoniot. Conversely, a Hugoniot equation of state may be calculated for the earth, if its temperature distribution is assumed.

To obtain a relation involving temperature, we write the standard thermodynamic equation

$$dQ = dE + p\,dV$$

where dQ is the quantity of heat added during a small change, and evaluate dE for the Hugoniot conditions from the third equation of 26.6.1. This gives

$$dQ = \frac{[(V_0 - V)\,dP + (P - P_0)\,dV]}{2} \qquad 26.6.3$$

However, a second standard thermodynamic equation is

$$dQ = C_V\,dT + C_V T\left(\frac{\partial P}{\partial E}\right)\,dV \qquad 26.6.4$$

where C_V is the specific heat at constant volume. Equation 26.6.4 is often written in terms of Gruneisen's parameter γ, where

$$\gamma = V\left(\frac{\partial P}{\partial E}\right)_V$$

giving

$$dQ = C_V\,dT + C_V T\left(\frac{\gamma}{V}\right)\,dV \qquad 26.6.5$$

Equating dQ between Equations 26.6.3 and 26.6.5 gives

$$dT = \frac{(V_0 - V)\,dP}{2C_V} + \left[\frac{P - P_0}{2C_V} - T\left(\frac{\gamma}{V}\right)\right]\,dV \qquad 26.6.6$$

Equation 26.6.6 can be evaluated numerically along the Hugoniot, provided C_V and γ are known, to give the successive changes in temperature from the starting temperature T_0.

Geophysically, the results of greatest interest have been those on metallic elements, particularly iron (McQueen and Marsh, 1960; Altschuler et al., 1960), and on a series of rocks, including dunite (McQueen, Marsh, and Fritz, 1967). Observations on the density of iron to a pressure corresponding to the depth of the core boundary were obtained both by McQueen and Marsh and by Altschuler et al. The indicated density is 11 gm/cm^3. Measurements by Altschuler et al. to a pressure corresponding to the earth's center gave a density of iron of 13 gm/cm^3. If we compare the densities in the Haddon-Bullen model (Section 26.3) with these, we see a suggestion of alloying by lighter elements, particularly in the outer core.

All of the rocks investigated exhibited a change in phase, as evidenced by a break in the Hugoniot curves. For the two dunites studied, the phase change appeared to begin at a pressure of 0.45 megabars, which would correspond to a depth of about 1000 km in the earth. This is at the base of the region of rapid change in properties in the mantle, and the pressure is also greater than that reported from static measurements of phase change (Section 26.4). It is possible that during the onset of a phase change, equilibrium is not attained in the shock-wave experiments.

One of the dunites (from Mooihoek Mine in Transvaal) consisted largely of an iron-rich olivine, giving the rocks a mean atomic weight of 25.1. The other

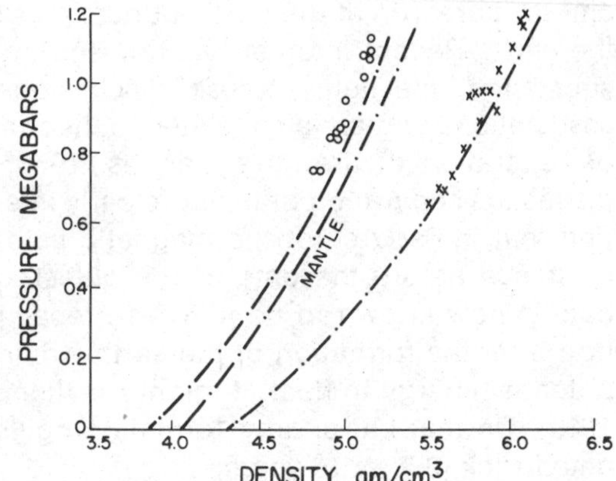

Figure 26.5 Observed Hugoniots for olivines: magnesium rich (circles) and iron rich (crosses), with the computed Hugoniot for the mantle. The observed curves have been shifted to pass through the computed zero-pressure density of the high-pressure phase. (After McQueen, Marsh, and Fritz)

(Twin Sisters Peaks, Washington) contained magnesium-rich olivine, and had a mean atomic weight of 21.2. The pressure-density relations for the dense phases of the two rocks, at pressures up to 1.2 megabars (corresponding to a depth of 2600 km), straddled the computed Hugoniot for the mantle (Fig. 26.5). McQueen, Marsh, and Fritz suggest that an olivine rock with a mean atomic weight of 21.7 (i.e., close to the magnesium-rich sample) would satisfy the pressure-density requirement for the lower mantle.

26.7 THE CORE AND THE CORE-MANTLE INTERFACE

Properties of the core have been mentioned in connection with seismology, earth tides, geomagnetism, and the temperature of the earth, but it is important to distinguish between those properties which are reasonably well established and those which are inferred from compatibility with some hypothesis. For example, the pressure distribution is well known, because it depends largely on the density distribution through the whole earth rather than on the composition of the core itself. The fluidity of the outer core and the solidity of the inner core have been well established by free oscillation observations, and the density within the core is reasonably well known. The multiply-reflected core phases confirm the lateral homogeneity of the outer core and the sharpness of the core-mantle boundary. By contrast, the electrical conductivity can be inferred only within orders of magnitude for compatibility with geomagnetic dynamo models. Geomagnetism would also favor a relatively low viscosity for the outer core and a state of convective motion. The latter in turn implies an energy source and a temperature gradient steeper than the adiabatic gradient. If the outer and inner cores have the same composition, the inner core boundary is at the melting point temperature of core material, as proposed by Jacobs (1953). Very briefly, these represent the established and inferred physical properties of the core.

The composition has long been accepted to be dominantly a nickel-iron alloy. Undoubtedly, the composition of iron meteorites was influential in suggesting this association, and the fact that iron meteorites have not come from the core of a planet of earth size (Section 1.5) has not caused the association to be abandoned. Shock-wave measurements of density at core pressures suggest that, to the first order, iron is a reasonable candidate, but that it is probably alloyed with lighter elements. Approximate values for density would be 11 gm/-

cm^3 for pure iron at the core boundary, as opposed to 9.9 gm/cm^3 observed for the earth. Recent interest has focused on the probable nature of the alloying substance. Inevitably, these discussions depend heavily on geochemical associations and are also related to theories on the origin of the earth and mode of separation of the core (Jacobs, 1975; Brett, 1976). Geophysical evidence places the separation of the core early in earth history, because on the assumption that the source of the magnetic field is in the core, it must have been in existence before the date of the oldest known magnetization in a rock. This date is now known to be early in Precambrian time. Also, since separation of iron after the formation of the earth would result in a very large conversion of potential energy to heat, it might result in the melting of the entire crust (Wise, 1969). On that basis, core formation could be expected to predate the oldest dated rock (3.7×10^9 years).

Elements that have been proposed as minor constituents of the core include carbon, silicon, hydrogen, sulfur, magnesium, oxygen, and potassium, or various combinations of these. Of these, only magnesium can be excluded (Brett, 1976), but the evidence for most of the others is very indirect. For example, sulfur has been proposed because of a depletion of sulfur in crustal rocks, as compared to its cosmic abundance, but this depletion could also result from the volatile nature of sulfur and its loss during the formation of the earth. On the other hand, the known affinity of sulfur with iron, as indicated by the abundance of FeS, suggests that a core of Fe-Ni-FeS is not unreasonable. Potassium was suggested as a core constituent by Lewis (1971) and Hall and Murthy (1971) on the basis of its affinity to sulfur. Noting that potassium-rich sulfides occur in meteorites, they proposed that compounds of the form K_2S exist in the core. The great interest in the possibility of potassium as a core constituent arises from the fact that K^{40} is radioactive, as discussed in Chapter 22. A source of energy from the driving of convection, as part of the geomagnetic dynamo, would thus be available. The testing of the hypothesis will require further chemical measurements on the partitioning of potassium between silicates (representing the mantle) and metallic sulfides, at high temperatures and pressures.

If the outer core is alloyed with lighter elements, the inner core need not be. Seismological evidence is consistent with a density increase at the inner core boundary, although a more critical test of this would be available if the Slichter mode of oscillation of the inner core (Section 14.3) could be observed. If the outer core is looked upon as a eutectic mixture, the inner core could be the frozen-out end member: i.e., pure metal. This point of view is rather different from that which places the outer boundary of the inner core at the melting temperature of a single substance.

An important question is the mechanism of coupling of the fluid outer core to the base of the solid mantle. Evidence of this coupling must come from the analysis of motions, at the earth's surface, which are excited within the core. For example, variations in the length of the day (Chapter 1) of decade duration are believed to arise from mass redistributions in the fluid core, and to be transferred to the mantle by the core-mantle coupling.

The mechanisms of coupling which have been proposed include the nonspherical shape of the core boundary, the viscosity of the fluid core material, electromagnetic coupling between the earth's magnetic field and the conducting lower mantle, and coupling by "topography" on the core-mantle interface. If the evidence of the westward drift of the secular change of the earth's magnetic field is accepted (Section 18.5), the mantle must be assumed to be rotating more rapidly about the earth's axis than the outermost layer of the core. Viscous coupling, however, would require the mantle to rotate more slowly, since the decelerating tidal couples act chiefly upon it. Electromagnetic coupling (which would result from the phenomenon that a body of high electrical conductivity does not move relative to a magnetic field) would be compatible with the westward drift, provided it is assumed that the main field to which the mantle is coupled is generated at some depth within the core. The outer layer of the core then rotates more slowly than either the bulk of the core or the mantle. Electromagnetic coupling is generally accepted to be sufficient to transmit changes in the length of the day, but it has been shown to be incapable of transmitting the Chandler wobble to the mantle if the wobble were generated in the core (Rochester and Smylie, 1965).

The hypothesis of topography on the core-mantle interface has been discussed by Hide and Horai (1968); evidence for it, based on gravity anomalies, was discussed in Section 13.3. Departures of the boundary from its regular form could produce regions in the fluid core which are at rest relative to the mantle. Such regions, observed in other cases of fluid-solid interfaces, are known as Taylor columns. It is difficult to reconcile the westward drift with the presence of Taylor columns, unless the magnitude of drift is a regional, rather than a global, parameter (Section 18.5). Indeed, a further analysis of the westward drift in terms of regions may be the best method of investigating this hypothesis.

Problems

1. Verify Equations 26.6.1 for the conservation of mass, momentum, and energy across a shock front.

2. An effect of phase changes in the earth that could be of considerable importance is that related to subsidence during sedimentation, if the Mohorovičić discontinuity represents a phase change. Reconsider the calculations of Problem 3, Chapter 13, under the assumption that, as the crust subsides, it suffers a phase change with a density increase of 0.3 gm/cm^3 as it reaches the normal depth of the Mohorovičić discontinuity.

BIBLIOGRAPHY

Ahrens, Thomas J., and Schubert, Gerald (1975) Gabbro-eclogite reaction rate and its geophysical significance. *Rev. Geophys. Space Phys.,* **13**, 383–400.

Akimoto, Syun-Iti, and Syono, Y. (1970) High-pressure decomposition of the system $FeSiO_3$-$MgSiO_3$. *Phys. Earth Plan. Int.,* **3**, 186.

Altschuler, L. V., Kormer, S. B., Brazhnik, M. I., Vladimirov, L. A., Speranskaya, M. P., and Funtikov, A. I. (1960) The isentropic compressibility of aluminum, copper, lead and iron at high pressures. *Soviet Physics, JETP,* **11**, 766.

Anderson, D. L. (1967) Phase changes in the upper mantle. *Science,* **157**, 1165–1173.

Bernal, J.D. (1936) Discussion in *The Observatory,* **59**, 268.

Birch, Francis (1952) Elasticity and constitution of the earth's interior. *J. Geophys. Res.,* **57**, 227.

Birch, Francis (1963) Geophysical applications of high-pressure research. In *Solids under pressure,* ed. W. Paul and D. M. Warschauer. McGraw-Hill, New York.

Brett, R. (1976) The current status of speculations on the composition of the core of the earth. *Rev. Geophys. Space Phys.,* **14**, 375–383.

Bullard, E. C., and Griggs, D. T. (1961) The nature of the Mohorovičić discontinuity. *Geophys. J.,* **6**, 118.

Bullen, K. E. (1965) *Introduction to the theory of seismology* (3rd Edition). Cambridge Univ. Press, Cambridge.

Church, W. R. (1972) Ophiolite: its definition, origin as oceanic crust and mode of emplacement in orogenic belts, with special reference to the Appalachians. In *The ancient oceanic lithosphere,* ed. E. Irving. Publications of the Earth Physics Branch, **42**, Department of Energy, Mines and Resources, Ottawa.

Coes, L. (1953) A new dense crystalline silica. *Science,* **118**, 131.

Dietz, R. S. (1963) Alpine serpentines as oceanic rind fragments. *Geol. Soc. Amer. Bull.,* **74**, 947–952.

Drake, C. L., and Kominskaya, I. P. (1969) The transition from continental to oceanic crust. *Tectonophysics,* **7**, 363.

Eaton, J. P., and Murata, K. J. (1960) How volcanoes grow. *Science,* **132**, 925.

Gilvarry, J. J. (1957) Temperatures in the earth's interior. *J. Atmosph. Terr. Phys.,* **10**, 84.

Green, D. H., and Ringwood, A. E. (1969) The origin of basalt magmas. In *The earth's crust and upper mantle,* ed. P. J. Hart, Amer. Geophys. Un., Monograph 13.

Haddon, R. A. W., and Bullen, K. E. (1969) An earth model incorporating free earth oscillation data. *Phys. Earth Plan. Int.,* **2**, 35.

Hall, H. T., and Murthy, V. R. (1971) The early chemical history of the earth: Some critical elemental fractionations. *Earth Plan Sci. Letters,* **11**, 239–244.

Hide, R., and Horai, K. (1968) On the topography of the core-mantle interface. *Phys. Earth Plan. Int.,* **1**, 305.

Hodges, Floyd N., and Papike, J. J. (1976) DSDP Site 334: Magmatic cumulates from oceanic Layer 3. *J. Geophys. Res.,* **81**, 4135–4151.

Jacobs, J. A. (1953) The Earth's inner core. *Nature,* **172**, 297–298.

Jacobs, J. A. (1975) *The earth's core.* Academic Press, London.

Kawai, N., Endo, S., and Ito, K. (1970) Split-sphere high pressure vessel and phase equilibrium relation in the system Mg_2SiO_4-Fe_2SiO_4. *Phys. Earth Plan. Int.,* **3**, 182.

Kuno, H. (1967) Volcanological and petrological evidences regarding the nature of the upper mantle. In *The earth's mantle,* ed. T. F. Gaskell, Academic Press, London.

Lambert, I. B., and Wyllie, P. J. (1970) Melting in the deep crust and upper mantle and the nature of the low velocity layer. *Phys. Earth Plan. Int.,* **3**, 316.

Lewis, J. S. (1971) Consequences of the presence of sulfur in the core of the earth. *Earth Plan Sci. Letters,* **11**, 130–134.

McQueen, R. G., and Marsh, S. P. (1960) Equation of state for nineteen metallic elements from shock-wave measurements to two megabars. *J. Appl. Phys.,* **31**, 1253.

McQueen, R. G., Marsh, S. P., and Fritz, J. N. (1967) Hugoniot equation of state of twelve rocks. *J. Geophys. Res.,* **72**, 4999.

Muehlenbachs, K., and Clayton, R. N. (1976) Oxygen isotope composition of the oceanic crust and its bearing on seawater. *J. Geophys. Res.,* **81**, 4365–4369.

Murnaghan, F. D. (1937) Finite deformation of an elastic solid. *Amer. J. Math.,* **59**, 235.

Peltier, W. R. (1972) Penetrative convection in the planetary mantle. *Geophys. Fluid Dyn.,* **5**, 47–88.

Ramsey, W. H. (1949) On the nature of the earth's core. *Mon. Not. R. A. S., Geophys. Suppl.,* **5**, 409.

Rice, M. H., McQueen, R. G., and Walsh, J. M. (1958) Compression of solids by strong shock waves. In *Solid state physics,* Vol. 6, Academic Press, New York.

Ringwood, A. E. (1958) Constitution of the mantle, 3. *Geochim. et Cosmochim. Acta,* **15**, 195.

Ringwood, A. E., and Green, D. H. (1966) An experimental investigation of the gabbro-eclogite transformation and some geophysical consequences. *Tectonophysics,* **3**, 383.

Ringwood, A. E., and Green, D. H. (1969) Phase transitions. In *The earth's crust and upper mantle,* ed. P. J. Hart. Amer. Geophys. Un., Monograph 13.

Ringwood, A. E., and Major, Alan (1967) Some high-pressure transformations of geophysical significance. *Earth Plan. Sci. Letters,* **2**, 106.

Ringwood, A. E., and Major, A. (1970) The system Mg_2SiO_4-Fe_2SiO_4 at high pressures and temperatures. *Phys. Earth Plan. Int.,* **3**, 89.

Rochester, M. G., and Smylie, D. E. (1965) Geomagnetic core-mantle coupling and the Chandler wobble. *Geophys. J.,* **10**, 289.

Schubert, G., and Turcotte, D. L. (1971) Phase changes and mantle convection. *J. Geophys. Res.,* **76**, 1424–1432.

Stishov, S. M., and Popova, S. V. (1961) New dense polymorphic modification of silica. *Geokhimiya,* **1961**, 837.

Thompson, G. A., and Talwani, M. (1964) Crustal structure from Pacific Basin to central Nevada. *J. Geophys. Res.,* **69**, 4813.

Thomsen, Leon, and Anderson, Orson L. (1969) On the high-temperature equation of state of solids. *J. Geophys. Res.,* **74**, 981.

Wise, Donald U. (1969) Origin of the moon from the earth: some new mechanisms and comparisons. *J. Geophys. Res.,* **74**, 6034.

Yoder, H. S., and Tilley, C. E. (1962) Origin of basalt: an experimental study of natural and synthetic rock systems. *J. Petrol.,* **3**, 342.

GEODYNAMICS 27

27.1 MOVEMENTS: PAST, RECENT, AND PRESENT

The record of geology is one of relative motions of the outer part of the earth, although, until recently, vertical motions were recognized much more widely than horizontal displacements. For example, the deposition of several kilometers of sedimentary rock represents a downward displacement of the crust by a comparable amount. This displacement is evidence of a downward force exerted on the crust, in addition to the weight of the sediments; otherwise, isostatic compensation would limit the maximum possible accumulation to a few kilometers. Conversely, there is ample evidence for the recurring uplift of regions through geological time. Horizontal displacements were recognized through the folding of sedimentary strata and through strike-slip faults, but the evidence for horizontal displacements of major size is more indirect. We have seen the contributions of paleomagnetism, and the magnetic character of the ocean floors, to the idea that the latter are spreading, with consequent motion of the continents. Further evidence, albeit inconclusive, that continents have moved is provided by the matching of coastlines across oceans to obtain a fit. As is well known, this approach was employed by Wegener (1924) but it has been placed on a quantitative basis by Bullard, Everett, and Smith (1965), who (Fig. 27.1) determined the fit of the Americas against Europe and Africa. The procedure was to select a depth contour (at 500 fathoms) to define the continental margins, and to minimize the areas of misfit between the contours on either side of the Atlantic Ocean. For the actual computation, use was made of a theorem of Euler, which states that any displacement over a sphere can be reduced to a rotation about a vertical line through some pole. The position of this pole was initially assumed, and differences in "longitude" (ϕ_n, measured relative to the pole) of N pairs of points on opposite coasts at the same "latitude" from the pole were computed. For a rotation of the Americas by an angle ϕ_0 about the pole, the mean square misfit was defined to be $\frac{1}{N}\sum^{N}(\phi_n - \phi_0)^2$; for a given pole this is a minimum for

$$\phi_0 = \frac{1}{N}\sum^{N}\phi_n$$

A computer program was designed to hunt over small changes in the coordinates of the assumed pole, until the mean square misfit, computed in each case with the best value of ϕ_0, was a true minimum. The result was indeed striking, as is the fit of the southern continents (Fig. 27.2) obtained by Smith and Hallam (1970) through application of the same method. These continental fits,

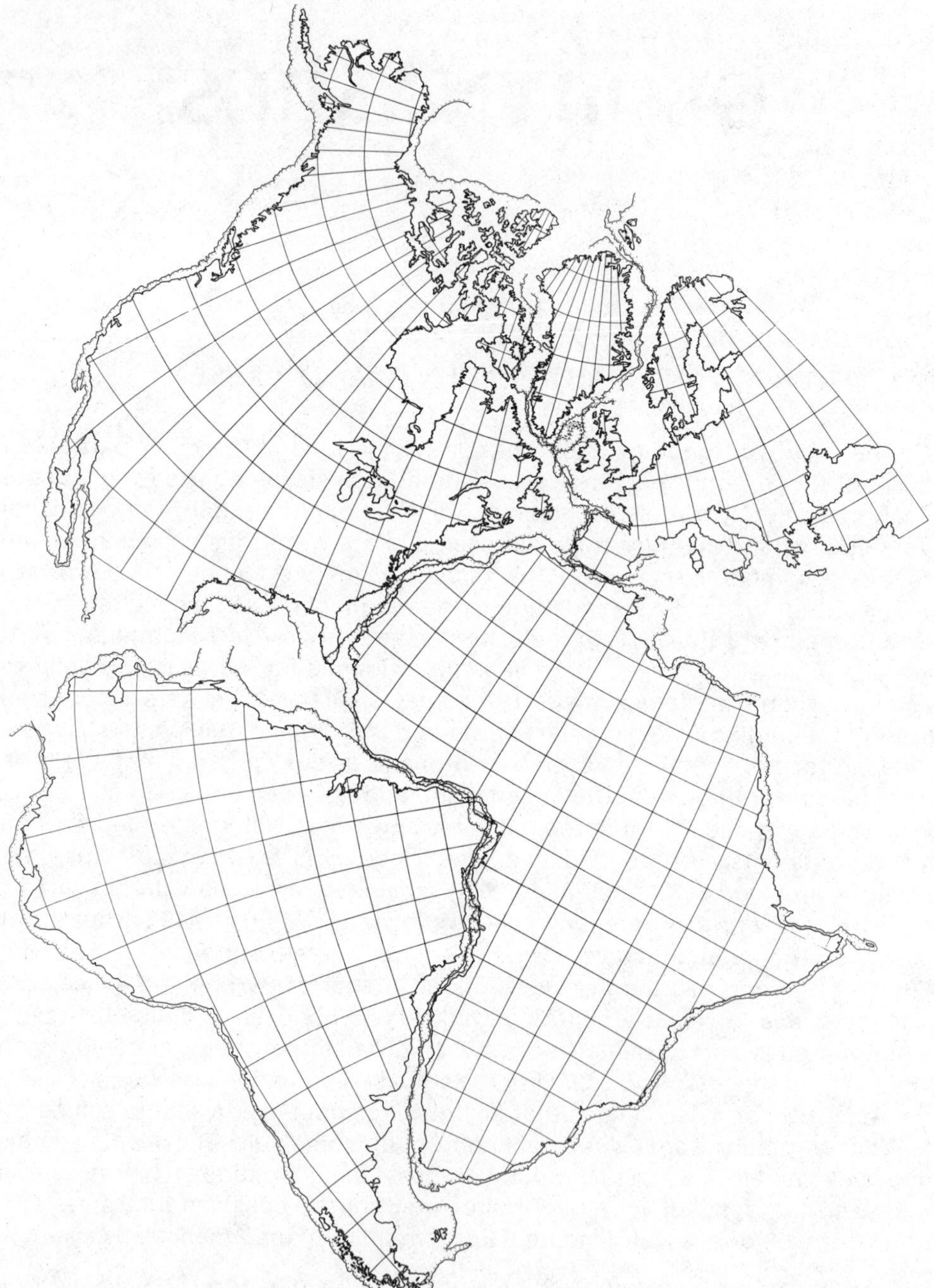

Figure 27.1 The fit of the Americas against Europe and Africa, at the 500 fathom depth contour, according to Bullard, Everett, and Smith. Dots are on the continental side of the contour. (Courtesy of E. C. Bullard)

determined by a purely geometrical procedure, are independent of the detailed assumptions of plate tectonics. But the fact that good fits can be obtained by pure rotations helps to confirm the rigid behavior of the outer part of the earth. In addition, there is impressive agreement with paleomagnetism. In Figure 27.2 are shown the areas covered by the magnetic pole positions corresponding to these continents after reassembly, for different geological periods, as obtained by McElhinny and Embleton (1974). In other words, polar positions obtained for individual continents were rotated, according to the principles of Section 21.3, by the same amounts as were the continents in obtaining the fit. The resulting pole positions for different ages fall into the remarkably small areas shown on Figure 27.2. From the two youngest groups shown, one can infer the motion of the assembled mass across the south pole. However, it would be unjustified and even erroneous to assume that the reassembled polar regions dating from earlier Palaeozoic and Precambrian times imply that the assembled continents remained as one from those times. In a detailed reconsideration of Palaeozoic pole positions from all major continents, Irving (1977) concluded that there were important relative displacements between the middle Carboniferous (325 m.y. ago) and the late Triassic (200 m.y. ago). These involve chiefly the motion of a combined North America-Europe from a position northwest of South America into the position shown in Fig. 27.1. Relative motions between the "southern continents" were much smaller (although there is evidence of independent movement for Australia), hence their assembled poles, in general, fall into the tight group shown.

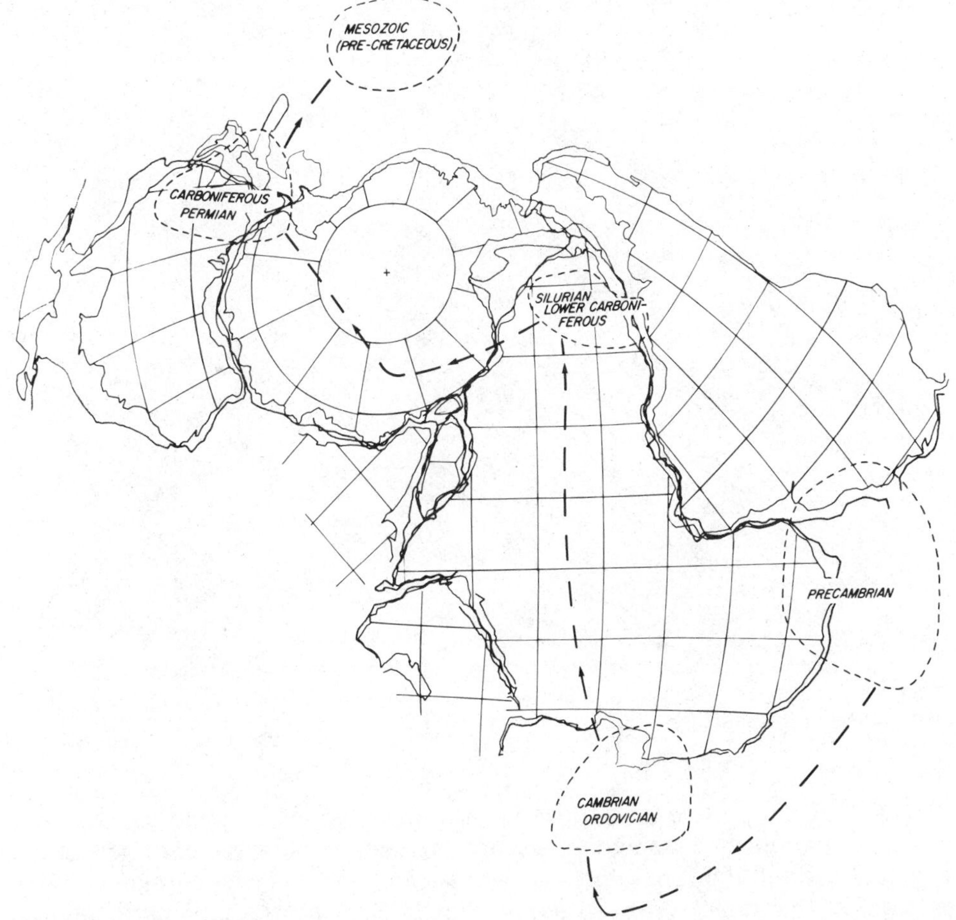

Figure 27.2 The fit of the southern continents according to Smith and Hallam. Superimposed in dotted lines are the areas occupied by magnetic poles, determined from these continents, after re-assembly. Original diagram courtesy of A. Gilbert Smith; paleomagnetic results from McElhinny and Embleton (1974).

Figure 27.3 Contours of the uplift of land in Fennoscandia, based on repeated precise leveling. (Courtesy of E. Kääriäinen)

The problem of detecting present-day displacements of the crust is essentially one of geodesy. Once again, vertical movements are easier to detect and measure, through the repetition of precise-level nets, and the change of mean sea level at tide gauges. For example, Figure 27.3 (from Kääriäinen, 1966) illustrates the change in elevation of a portion of Fennoscandia, based on relevelling between 1892 and 1962. There have been successful applications of local triangulation across active faults (Fig. 27.4) to show the accumulation of strain (Burford, 1966), and these provide a most useful extension to direct measurements of offsets for the study of the mechanics of faulting. However, the use of the new tools of geodesy to determine the present rate of relative displacement over long distances is in its infancy. For intervisible points, the geodimeter (which employs an inverse velocity-of-light experiment with velocity adopted and distance to be measured) provides positions accurate to centimeters in lines up to several kilometers in length. If adjacent blocks of the crust are moving relatively at the rate of a few centimeters per year, there is the possibility of detecting this through repeated geodimeter surveys. Geodimeter networks have already been established between such blocks as Greenland and Ellesmere Island, although these are probably now on the same plate. Intraplate boundaries which have been instrumented include the Gulf of California and the Mid-Atlantic Ridge axis in Iceland. Another area of great interest, where lines of sight between separate plates would be available, is the Red Sea–Gulf of Aden intersection. For the measurements across Iceland, Decker et al. (1971) initially reported a widening of 6 to 7 cm between 1967 and 1970. Later measurements (Schäfer, 1975) showed that the situation was complex, with movement taking place on several narrow grabens within the general rift zone. Across some of these, compression took place in some years, resulting in no net extension of the zone. One major graben, near Thingvellir, showed a net extension of 4 cm between 1967 and 1973. Allowing for several grabens in the rift area, there is probably no difficulty in reconciling the observations with the 1 to 2 cm per year suggested by sea-floor spreading, but the motion is by no means uniform in time. For intraplate measurements across very much longer distances, the greatest hope at present lies in lunar ranging and very long baseline interferometry. In the first of these (Bender and Silverberg, 1975), trilateration is

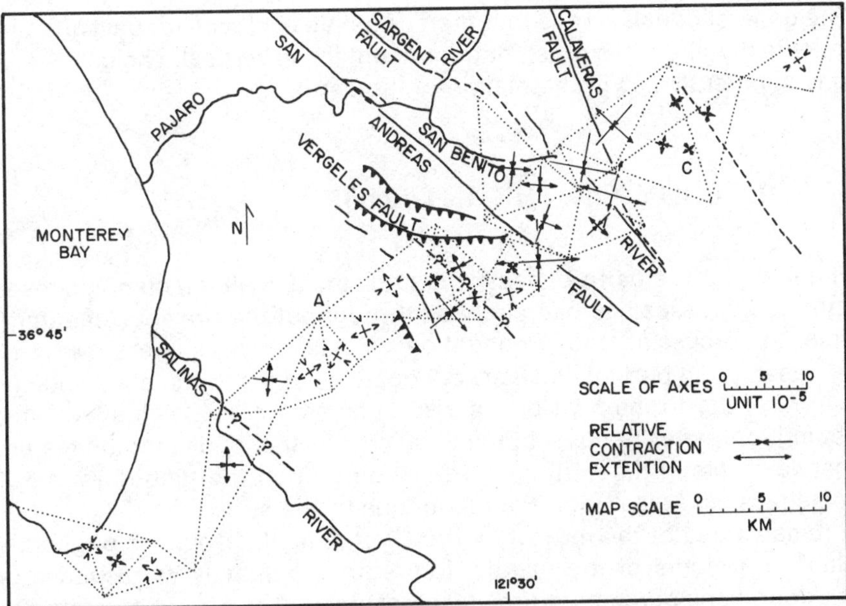

Figure 27.4 Strains across the San Andreas fault, as shown by the computed extension or compression of the crust, based on repeated networks (1932–1951) of triangulation. The triangulation is indicated by fine lines. Note the evidence for crustal compression in a direction perpendicular to the San Andreas fault as well as slip along the fault. (Courtesy of R. O. Burford)

carried out between laser sites on two plates on earth, and a reflector installed on the moon. Knowledge of the moon's orbit permits the calculation of the distance between points on earth when the distances to the moon are observed, by the timing of laser pulses to the reflector and back. Bender and Silverberg report that an observatory on Hawaii has achieved an accuracy of 2 to 3 cm in the lunar distance and that the relative motion between the Pacific and American plates will be observable to 1 cm/yr within a short time. Very long baseline interferometry (Coates et al., 1975) employs two receivers for stellar radio sources, these receivers also being located on different plates. Records of a pulsating source in a given direction show a very large, narrow correlation peak when one record is delayed, in processing, by an amount corresponding to the time of travel of the wavefront from one receiver to the other. It is not necessary that the processing be done in real time; the digital records made at separate radio telescopes may be processed later. From the known direction to the source, and the wavefront travel time, the absolute distance between sites can be calculated to an accuracy currently estimated to be between 5 and 8 cm. Part of the uncertainty arises from the variation in the velocity of radio waves with atmospheric conditions, and its reduction will depend upon an improved knowledge of these. In any case, direct confirmation of relative plate motion should be possible with repeat observations after a few years.

Between the time scale of geology and that of human experience, there are motions extending over periods of 10^3 to 10^4 years. The most striking of these are the vertical displacements apparently related to a recovery of the crust following the removal of Pleistocene ice sheets. We have noted the evidence of gravity anomalies for depressed areas; these regions are usually also marked with raised beaches which, if they contain organic material, can be dated by the carbon-14 method to provide an accurate record of the uplift. The combination of past and present motions in these regions permits the determination of a viscosity for the upper mantle.

The general conclusion is that there is ample evidence for past and present movements of the earth's crust, both horizontal and vertical. The cause of these motions apparently lies in the mantle of the earth.

27.2 RHEOLOGY OF THE INTERIOR

The fact that the earth beneath the crust must exhibit a complex response to impressed stresses has been implied throughout the previous chapters. For example, in discussing the propagation of elastic waves, we showed that a theory based upon perfect elasticity adequately predicted the observed effects, while in referring to the post-glacial rise of areas, we noted an effect which is continuing long after removal of the ice. The explanation is that the material of the mantle reacts in different ways, depending upon the time duration of the applied stress, and also upon the magnitude of the stress.

It remains one of the most important, yet difficult, problems of geophysics to establish a model of the mantle, to predict adequately its response under varying conditions of temperature and pressure, and as a function of frequency. Without such a model it is impossible to conclude whether or not such motions as convection are possible.

The various approaches which have been used to determine the behavior

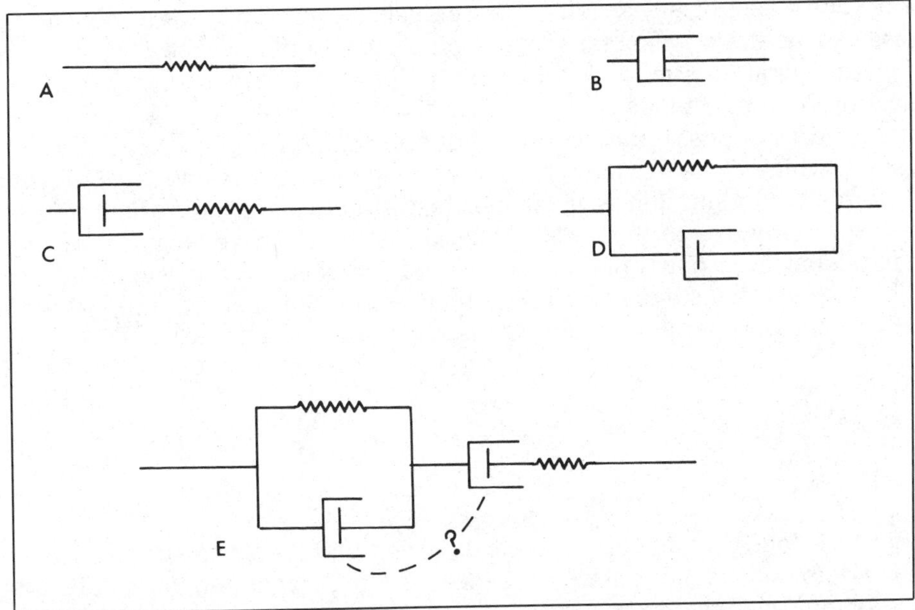

Figure 27.5 Representations of behavior: (A) elastic body; (B) Newtonian fluid; (C) Maxwell substance; (D) Kelvin or Voigt substance; and (E) a combination of the latter two.

include the extrapolation from laboratory measurements, the study of departure from perfect elasticity at the high-frequency end of the spectrum, and the measurement of response at the longest possible periods, to determine an effective viscosity. We shall look into each of these, but it is desirable first to introduce certain standard models for the behavior of material. The application of these to earth problems has been summarized by Jaeger (1956).

A perfectly elastic substance obeys Hooke's Law, and can be represented as a spring (Fig. 27.5(a)), a force constant K, responding instantaneously to a stress p through the relation

$$p = Ke \qquad 27.2.1$$

where e is the strain.

A Newtonian fluid with coefficient of viscosity η responds through the relation

$$p = \eta \dot{e} \qquad 27.2.2$$

where e is the time rate of change of strain; the fluid can be represented by a dashpot (Fig. 27.5(b)). To explain the elastic and viscous properties of materials, combinations of springs and dashpots have been proposed. It may also be appropriate to introduce a critical yield stress, which can be represented by a frictional contact capable of providing a resistive force up to some maximum value.

The Maxwell substance (Fig. 27.5(c)) consists of a spring and dashpot in series; application of a constant stress produces an immediate elastic strain, followed by a strain which increases linearly with time, without limit. If the spring and dashpot are arranged in parallel (Fig. 27.5(d)), the application of stress causes strain to increase exponentially, depending on the viscosity, to an asymptotic limit determined by the spring. Such an arrangement is known as a Kelvin or Voigt substance.

At short periods, the Kelvin or Voigt substance displays a departure from elasticity which is greater than that observed in the passage of seismic waves. It was formerly thought that because the Maxwell substance has no limit to strain, it is inappropriate in earth models. This view is changing, and Maxwell or visco-elastic behavior is now often accepted as a representation of earth behavior. In the Maxwell substance, the total strain at any time can be represented as partly elastic and partly viscous. That is,

$$e = e_E + e_V$$

$$\dot{e} = \dot{e}_E + \dot{e}_V = \frac{\dot{p}}{E} + \frac{p}{\eta} \qquad 27.2.3$$

where E is Young's modulus and it is assumed that the same stress p contributes to both deformations. If we write the operator D for time-differentiation, the stress-strain relationship becomes

$$De = \left(\frac{D}{E} + \frac{1}{\eta}\right) p$$

or

$$p = \frac{E\eta}{D\eta + E} \cdot De \qquad 27.2.4$$

If the body were purely elastic, p would be simply Ee, for longitudinal deformation. The solution for many elastic problems may therefore be carried over to the Maxwell substance, simply by replacing E with the operator $E\eta D/(D\eta + E)$.

If the material is given a constant strain e_0 at time $t = 0$, it is not difficult to show that the stress decreases as the dashpot yields, according to

$$p = Ee_0 e^{-Et/\eta}$$

(where the unsubscripted e is the base of the natural logarithms). The stress falls to $1/e$ of the initial value in a time η/E known as the Maxwell relaxation time.

A more general arrangement is also shown in Figure 27.5(e). Here, there is a Maxwell element in series with a Kelvin-Voigt element, with, in general, different viscosities in the two dashpots. Departures from elasticity at short periods are controlled by the lower dashpot, while the response at long periods is governed by the greater viscosity of the upper one. Such an arrangement comes closer to explaining the observed features of the response of the earth. It will be recognized, of course, that the springs and dashpots are diagrammatic conveniences only; their interpretation in terms of the atomic, ionic, crystal, and larger scale properties of the material is a challenging problem of solid-state physics. Furthermore, Orowan (1967) has pointed out that in a composite material, such as the mantle, it is necessary to invoke the general arrangement of Figure 27.5(e) for each molecular constituent, and that different constituents may be activated under different conditions of temperature and pressure.

27.3 CREEP: OBSERVATIONS AND THEORY

Many measurements of the elastic properties, and breaking stresses, of rocks under different conditions of temperature and pressure have been made. The results have been summarized in Clark (1966) and reference has already been made, for example, to the use of measured values of K/ρ as a diagnostic tool in seismic interpretation, and to the crushing strength of igneous rock in connection with the stresses induced by surface loads. We are here concerned rather with the measured time variation of stress and strain, as a guide to non-elastic behavior. Griggs (1939) and more recently Lomnitz (1956) have studied the effect of stress on samples, for duration times up to one week. For gabbro at room temperature, under torsion, Lomnitz obtained the relation

$$e = \frac{p}{\mu}\left[1 + q \ln\left(1 + at\right)\right] \tag{27.3.1}$$

where e is strain, p stress, μ the shear modulus and q and a constants. Numerical values observed were 4.0×10^{11} dynes/cm^2 for μ, 0.003 for q, and 1000 sec^{-1} for t. The strain therefore increases as the logarithm of time, and the specimen is said to undergo *logarithmic creep*. The strain rate decreases markedly with time after the onset of stress, and the consequence of this, if it were applied to the earth, would be very great. In fact, Jeffreys (1962) has shown that, in such a material, sustained thermal convection would not be possible.

The phenomenon of logarithmic creep is well known in engineering, and it is attributed to the motion of dislocations within crystals. At low temperatures, these dislocations can move only until they reach a barrier, and as they are used up, the strain rate decreases. However, a most important fact is that, as the temperature is raised, other effects become important. Dislocations may pass barriers, the barriers may move, or, independent of dislocations, atoms may migrate down the stress gradient. The latter process, known as diffusion or *Herring-Nabarro creep,* has been investigated in detail (Herring, 1950). It leads to a strain rate which is proportional to stress, so that the material behaves as a Newtonian fluid, with an effective viscosity η (corresponding to the upper dash-pot of the generalized arrangement in Figure 27.5(e)). This effective viscosity is expressible in terms of temperature and properties of the material, but since the latter involve such parameters as grain radius, which are unknown for the earth, the calculation is uncertain. Gordon (1967) and McKenzie (1967) obtained the estimates of the radial variation in viscosity shown in Figure 27.6, in which the large values in the lower mantle are notable features. More recent work in post-glacial uplift (Section 27.4) suggests that the viscosity of the mantle does not increase with depth to the extent originally predicted, and support for this view is given by a recalculation based on creep theory (Sammis et al., 1977). The activation energy required for creep to occur enters the expression for viscosity, and this energy contains a term dependent upon the product of pressure and activation volume. In the original estimates of viscosity, the increase in pressure with depth dominates the calculations; Sammis et al. suggest that the activation volume decreases with depth in such a way as to greatly reduce the net change in this term. We can conclude that there is at present no serious contradiction between theory and experiment on the viscosity variation in the mantle, and, as a corollary, no obstacle to whole-mantle convection. The recog-

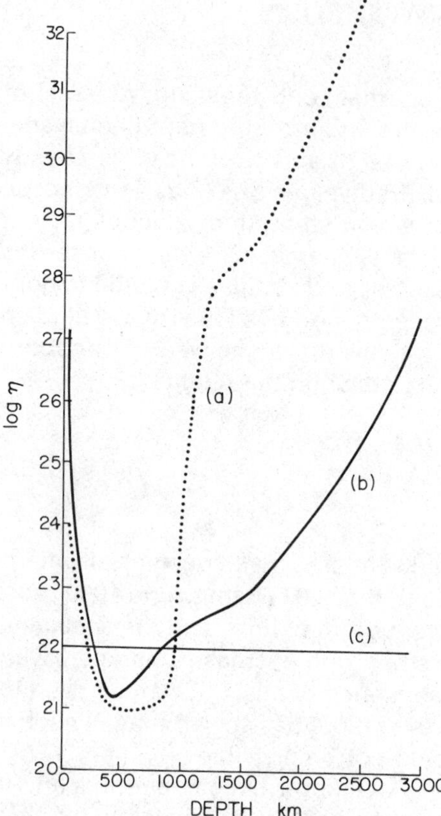

Figure 27.6 The viscosity of the mantle, according to Gordon (a) and McKenzie (b), based on creep theory, and the preferred model (c) of Peltier and Andrews, based on post-glacial uplift. The ordinate is the logarithm of viscosity in poises.

nition of the importance of high temperature creep has had a tremendous influence on geophysical thinking.

However, a word of caution is in order. Laboratory measurements on minerals typical of the mantle suggest that the rheology is more complicated than an effective viscosity relationship would imply. For example, Kohlstedt and Goetze (1974) measured creep in single crystals of olivine, at temperatures up to 1650°C and stresses up to 1500 bars. They found that the stress-strain relationship could not be adequately represented by a simple function. If one sought to fit a relation of the form *strain rate* = *stress difference*^{n}, a value of n close to 3 would be indicated by their results.

27.4 OBSERVATIONS AT SHORT AND LONG PERIODS

At seismic periods, including the periods of the free oscillations, departures from ideal elasticity can be expressed in terms of the reciprocal of the dimensionless *quality factory* Q (Knopoff, 1964). For an oscillatory process in which some losses occur, Q is defined through the equation

$$\frac{2\pi}{Q} = \frac{\Delta E}{E} \qquad 27.4.1$$

where ΔE is the energy dissipated per cycle and E is the peak energy stored,

ΔE and E referring to the same volume of material. In a homogeneous material, there are other definitions of Q which are equivalent to that of Equation 27.4.1. These include its relation to an attenuation factor, α, for the amplitude of a traveling wave with distance, by means of the equation

$$Q = \frac{\omega}{2c\alpha} \qquad 27.4.2$$

where c is the phase velocity and ω the angular frequency. The factor α is determined from the decrease of amplitude with distance x in the direction of propagation, by $e^{-\alpha x}$. At a fixed point in space, amplitude decreases by a factor $e^{-\gamma t}$, and Q can be obtained through

$$Q = \frac{\omega}{2\gamma} \qquad 27.4.3$$

Finally, if a system is driven in forced oscillations, the width $\Delta\omega$ of a peak response at a resonant frequency ω is related to Q by

$$Q = \frac{\omega}{\Delta\omega} \qquad 27.4.4$$

Q has been measured for rocks in the laboratory by a number of techniques, applying one or another of the above relations. A remarkable fact is its relative insensitivity to frequency, over ranges from cycles per second to megacycles per second. The simple Maxwell and Kelvin-Voigt models described in Section 27.2 do not, in fact, predict such a constancy, and this in itself is evidence of more complicated behavior.

For measurements on the actual earth, use has been made of body waves, surface waves, and the free oscillations. The comparison of amplitudes of body waves at different stations is not completely satisfactory, because of differences in calibration and local conditions, but body waves may be studied at a single station if multiple reflections are observed (Anderson and Kovach, 1964). For the core, Q for P-waves is determined from the multiply-reflected phases P_mKP (Chapter 3). Surface waves are conveniently studied if a wave train passes a single station several times (Anderson et al, 1965). Q for free oscillations can be obtained from line-width measurements, or from successive calculations of the power spectrum during a series of oscillations. Dratler et al (1971) have described the latter technique, showing spectral analyses made up to 80 hours following a deep focus earthquake of magnitude 7, approximately, in Columbia in 1970.

The interpretation of the observed values of Q in the earth is complicated because of the possible existence of overlapping causes of variation (Anderson, 1967; Kanamori, 1970). Ideally, we should like to have, for different regions of the earth, the behaviour of Q with frequency, for comparison with laboratory measurements and for the testing of models of material behaviour. But travelling waves or free oscillation modes of different periods sample to different depths, so that the spectrum of Q purely for one region is not obtained. In general Q is greater for P-waves than for S-waves, and for spheroidal than for torsional oscillations, suggesting that attenuation is strongest in shear deformations. Values of between 100 and 200 are typical for spheroidal oscillations sampling the upper mantle, with Q increasing to approximately 300 for the longest period fundamental modes (Roult, 1975, Jobert and Roult, 1976). The most striking increase of Q is with radial overtone number for the spheroidal modes, Dratler et al (1971) reporting values as high as 1000 for overtones such as ${}_4S_0$ and ${}_5S_0$. These overtones have shorter periods than the corresponding fundamental, so that Q in this case decreases with period. But the higher overtones are affected more by material at great depth in the earth. The tentative conclusion is that Q increases with depth in the mantle, from values as low as 100 in the upper mantle, but probably not by more than a factor of three. Within the upper mantle, it may very well show lateral variations related to tectonic regions (Fig. 20.13). For the core, Q for shear waves is, of course, zero, but for P-waves or spheroidal oscillations, it is presumably in excess of 1000, thus contributing to the very high Q observed in radial overtones. The determination of the variation of Q with frequency in a given region of the earth must await further study.

By contrast, it is possible to estimate Q for the entire earth at a few periods through the technique of spectral analysis of some tidal or rotational distortion, and the utilization of equation 27.4.4. Much of the effort in this direction has been directed toward the estimate of Q for the Chandler wobble described in Section 1.3 (Currie, 1974; Graber, 1976). Currie analysed 74 years of polar-motion data obtained through the international services, while Graber selected 15 years of observations, then reanalysed these in overlapping three-year segments. Graber's value of Q is 600 at the mean Chandler period of 430.8 solar days, an order of magnitude higher than Currie's. His reanalysis suggested that the Chandler motion shows abrupt shifts in peak period, up to 30 days from the mean, presumably related to phase shifts in the exciting mechanism, with the consequence that a spectrum based on a very long sequence of observations yields too broad a peak, and too low a value of Q.

The earliest estimates of mantle viscosity at long periods came from observations of the rate of postglacial vertical movement of the crust, treating the mantle as a half space of uniform viscosity. The equation used in these studies can be derived without difficulty, for the stress relationships in a viscous fluid are very similar to those in an elastic solid. Shearing stresses in the fluid are equal to the rate of change of shear strain multiplied by the coefficient of viscosity of the fluid, designated η. If we assume that all strain rates are linear functions of all stress components, an analysis identical to that of Section 2.3 for the elastic case gives us, for an isotropic fluid,

$$p_{ij} = \lambda'\theta\delta_{ij} + 2\eta\dot{e}_{ij} \tag{27.4.5}$$

where $\dot{e} = \partial e/\partial t$, and λ' is a constant corresponding to Lamé's parameter λ in the solid. For an incompressible fluid, $\theta = 0$. Thus, the stress-strain rate relationships are of the form

$$\left.\begin{aligned} p_{11} &= 2\eta\dot{e}_{11} = 2\eta\frac{\partial\dot{u}_1}{\partial x_1} \\ p_{22} &= 2\eta\dot{e}_{22} = 2\eta\frac{\partial\dot{u}_2}{\partial x_2} \\ p_{12} &= 2\eta\dot{e}_{12} = \eta\left(\frac{\partial\dot{u}_2}{\partial x_1} + \frac{\partial\dot{u}_1}{\partial x_2}\right) \end{aligned}\right\} \tag{27.4.6}$$

The equations of motion, as before, are

$$\rho \frac{\partial^2 u_i}{\partial t^2} = \frac{\partial p_{ij}}{\partial x_j} \qquad 27.4.7$$

when body forces are neglected. When the motion is so slow that acceleration can be neglected, the left side of Equation 27.4.7 is zero; when the right side is expressed in terms of the strain rates, through Equations 27.4.6, we obtain

$$\nabla^2 \dot{u}_i = 0 \qquad 27.4.8$$

as the equation governing slow fluid motion.

We shall investigate a two-dimensional case, in which $\partial/\partial x_2 = u_2 = 0$ and x_3 is taken vertically downward into the earth. It follows from Equation 27.4.8 that

$$\dot{u}_1 = -\frac{\partial \psi}{\partial x_3}; \qquad \dot{u}_3 = +\frac{\partial \psi}{\partial x_1}$$

where ψ, known as the stream function, satisfies

$$\frac{\partial^4 \psi}{\partial x_1^4} + 2\frac{\partial^4 \psi}{\partial x_1^2 \partial x_3^2} + \frac{\partial^4 \psi}{\partial x_3^4} = 0 \qquad 27.4.9$$

Treating the mantle as the uniform half space $x_3 > 0$, it is easily found by substitution that a solution of Equation 27.4.9 is

$$\psi = Ae^{-kx_3} \sin kx_1 \qquad 27.4.10$$

where A and K are constants, and the fluid velocities are

$$\dot{u}_1 = Ake^{-kx_3} \sin kx_1$$
$$\dot{u}_3 = Ake^{-kx_3} \cos kx_1 \qquad 27.4.11$$

The vertical normal stress is, from Equation 27.4.6,

$$p_{33} = 2\eta k^2 Ae^{-kx_3} \cos kx_1 \qquad 27.4.12$$

Suppose now that the load results from the distortion of the surface of the half space. Thus, the plane $x_3 = 0$ is warped into

$$x_3 = h \cos kx_1$$

The load applied (effectively at $x_3 = 0$) is 27.4.13

$$p_{33} = \rho gh \cos kx_1$$

where ρ is the density of the fluid. Comparison of Equations 27.4.12 and 27.4.13 gives the constant A as $\rho gh/2\eta k^2$.

Finally, the vertical velocity at the surface is obtained:

$$\dot{u}_{3(x_3=0)} = \frac{dh}{dt} = \rho g h / 2\eta k \qquad 27.4.14$$

In the application of Equation 27.4.14 to determine the viscosity, it is necessary to know *h, dh/dt,* and *k.* If a deglaciated region, such as Fennoscandia or northern Canada, is undergoing uplift, *dh/dt* can be measured by the methods already mentioned: the repeating of lines of precise level, or the recognition and dating of uplifted beaches. The quantity *h* is the distance the region must still rise to reach isostatic equilibrium; it may be estimated from the isostatic gravity anomaly, as discussed in Chapter 13. The parameter *k* is determined by the wavelength of the distortion; a single value can be estimated, or, preferably, the uplift pattern can be decomposed into harmonic components. Notice that in the simple application, no account is taken of the strength of the lithosphere, the argument being that, for a broad load, this can be neglected in a first approximation.

Early applications of the method were the studies on the uplift of Fennoscandia by Haskell (1935) and Niskanen (1939). Niskanen's value of η for the mantle was 1.9×10^{22} poises. Crittenden (1963, 1967) applied the same principles to the uplift of a postglacial lake, Lake Bonneville in Utah, whose uplift following removal of the water is extremely well recorded by raised beaches. Crittenden's value for η was 1.0×10^{21} poises. The mean dimension of Lake Bonneville is 200 km, one tenth that of Fennoscandia, and it could be argued that, in sampling a shallower portion of the mantle, evidence was obtained of very low viscosity in the upper mantle aesthenosphere. But there is reason to believe that the Lake Bonneville area is itself anomalous.

The obvious possible extensions of the above approach are to treat the viscosity as variable (McConnell, 1968) and to take account of the spherical geometry of the earth (O'Connell, 1971; Cathles, 1975). A very complete recent analysis of the North American uplift data has been carried out by Peltier and Andrews (1976), using a solution obtained by Peltier (1974) for the response of a Maxwell earth to loads of any prescribed distribution. For the assumed Laurentide ice load shown in Figure 27.7, Peltier and Andrews compared uplift data, principally from raised beaches, at sites from Baffin Island to Florida, with the predictions of two models of mantle viscosity. In the first model, η increases from 10^{22} poises to 10^{24} poises at a depth of 1000 km, while in the second it is constant at 10^{22} poises. A selection of their comparisons is shown in Figure 27.7. Most sites give much closer agreement between the observed and predicted uplifts when the constant viscosity model is used. Peltier and Andrews emphasize that their conclusions are preliminary, partly because the strength of the lithosphere has been neglected and partly because the complete solution will involve inversion of the observations to obtain the best viscosity distribution (Peltier and Andrews, 1976). But at least it can be said that there is no evidence from postglacial uplift for a dramatic increase in viscosity in the lower mantle. This observation supports the view of Goldreich and Toomre (1969), discussed in Chapter 13, that the figure of the earth need no longer be taken as evidence of lower mantle strength. The important conclusion for tectonics is that the conditions for mantle convection (Section 25.6), particularly whole-mantle convection, become much less restrictive.

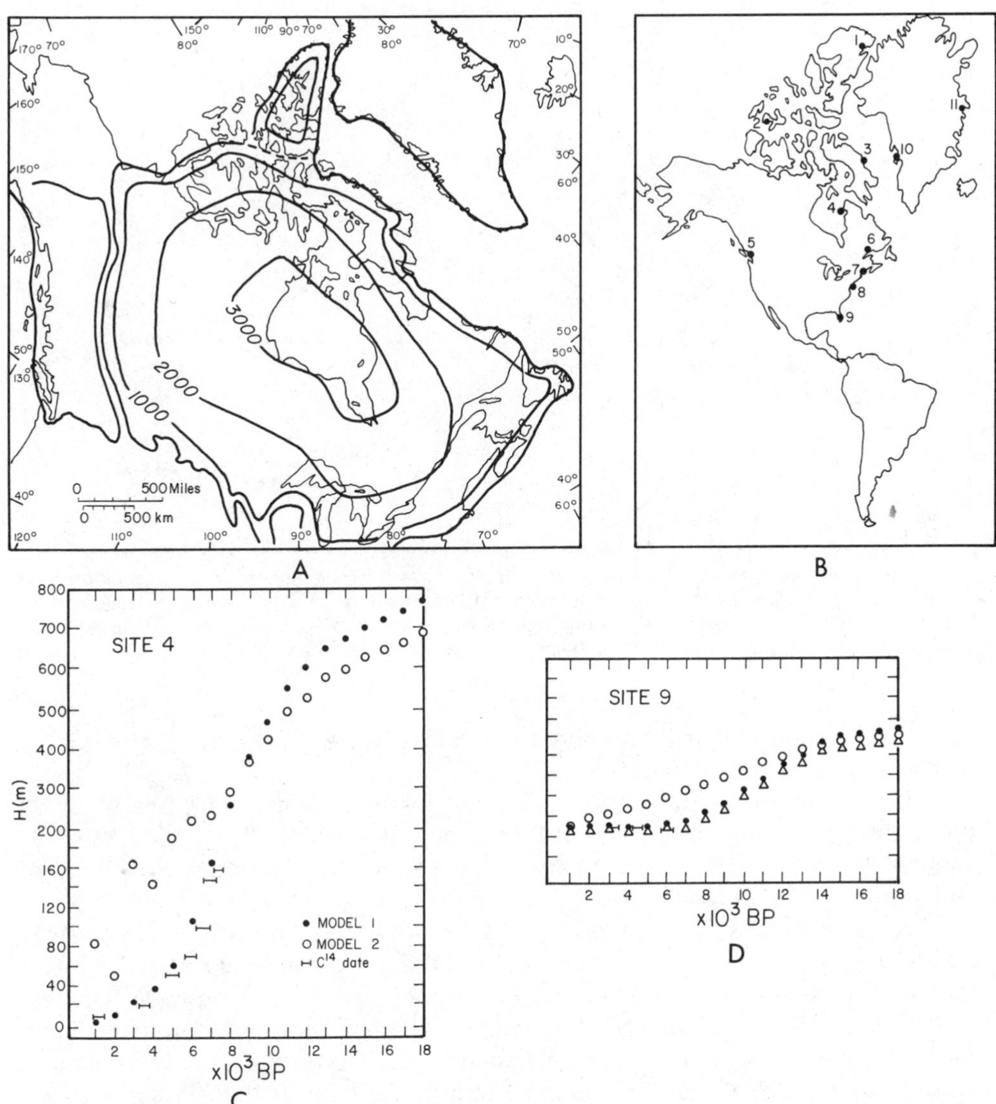

Figure 27.7 Observed and computed vertical movement (*C*) and (*D*) of the earth's surface at the sites 4 and 9, shown in (*B*). Computed values are based on the assumed ice load at 18,000 years before present, shown in meters of thickness in (*A*), for a mantle of constant viscosity (solid circles) and with viscosity increase at depth. The results represented by triangles for site 9 contain an additional correction for increased ocean loading as a result of ice melting. The latter effect is seen to be small. After Peltier and Andrews (1976).

27.5 FLEXURE OF THE LITHOSPHERE

So far the only indication of the thickness of the lithosphere we have noted is the depth of the seismic low velocity zone, it being assumed that the lithosphere extends from the earth's surface to the top of this zone. It is extremely desirable to have an independent estimate of lithospheric thickness, because the response at the long times associated with geology could be quite different from that at seismic frequencies. This independent check is furnished by the

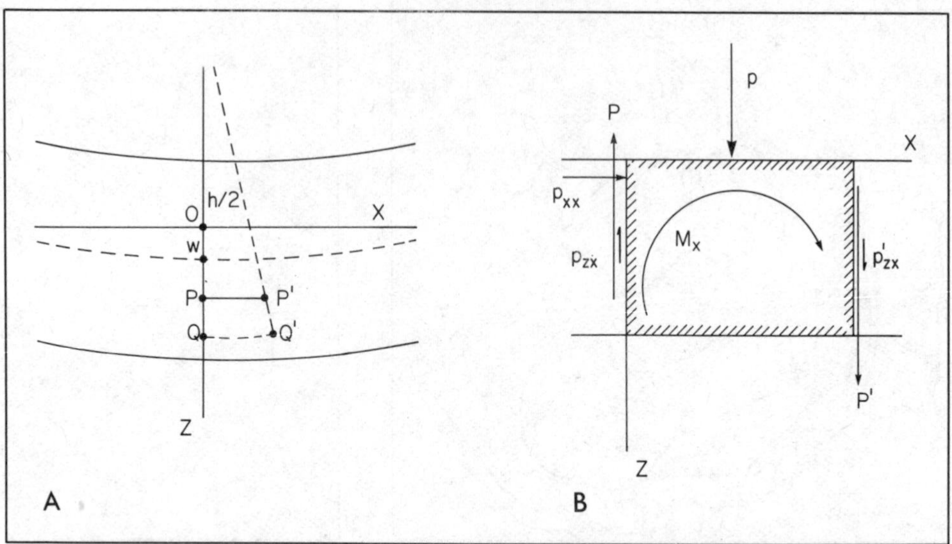

Figure 27.8 (a) Flexure of a thin plate. For downward displacements w < h, an element at PP′ is strained to QQ′, in proportion to its z-coordinate position. (b) Equilibrium of an element of the plate. The bending moment M_x results from the integrated effect of p_{xx}; the surface load p is supported by variations in the shear forces P which are the integrated effect of p_{zx}.

direct observation of the bending or flexure of the lithosphere, around the margins of superposed loads.

The treatment of the lithosphere as a floating, elastic plate has received the attention of many authors, but the pioneer writing of Gunn (1943, 1947) is especially notable. The basic theory of the bending of floating plates is found very conveniently in Nadai (1963).

We take (Figure 27.8) a plate of small thickness h, which lies, before bending, with its midplane coinciding with the x-y plane. It is a familiar result that, upon bending by a load added to the upper surface leading to vertical displacements w, the mid-surface of the plate is neither compressed nor stretched. At coordinate z within the plate, a fiber parallel to the x-axis is increased in length by $(R+z)/R$, where R is the local radius of curvature in the x-z plane; $1/R$ is equal to $\partial^2 w/\partial x^2$. The elastic strains are thus

$$e_{xx} = -z\,\frac{\partial^2 w}{\partial x^2}; \qquad e_{yy} = -z\,\frac{\partial^2 w}{\partial y^2}; \qquad e_{xy} = -2z\,\frac{\partial^2 w}{\partial x \partial y} \qquad 27.5.1$$

The stresses can be found from the general relationship (Equation 2.3.3)

$$p_{ij} = \lambda\theta\delta_{ij} + 2\mu e_{ij}$$

under the condition that p_{zz} be zero. For if p_{zz} is not very small compared to the horizontal stress components, the plate will crumple instead of bend. Setting $p_{zz} = 0$, e_{zz} may be expressed in terms of e_{xx} and e_{yy}; also, it is convenient, for the normal stresses, to use the elastic constants E (Young's modulus) and σ (Poisson's ratio) instead of λ and μ (using Equations 2.3.9). These substitutions lead without difficulty to

$$p_{xx} = \frac{E}{1-\sigma^2}(e_{xx} + \sigma e_{yy}) = -\frac{Ez}{1-\sigma^2}\left[\frac{\partial^2 w}{\partial x^2} + \frac{\sigma \partial^2 w}{\partial y^2}\right]$$

$$p_{yy} = \frac{E}{1-\sigma^2}(e_{yy} + \sigma e_{zz}) = -\frac{Ez}{1-\sigma^2}\left[\frac{\partial^2 w}{\partial y^2} + \frac{\sigma \partial^2 w}{\partial x^2}\right] \qquad 27.5.2$$

$$p_{xy} = \mu e_{xy} = -2\mu z \frac{\partial^2 w}{\partial x \partial y}$$

If we isolate a section of unit width in the x-z plane (Fig. 27.8), we find the section subject to a bending moment

$$M_x = \int_{-h/2}^{h/2} p_{xx}\, z\, dz = -N\left(\frac{\partial^2 w}{\partial x^2} + \frac{\sigma \partial^2 w}{\partial y^2}\right)$$

Similarly, there is a bending moment

$$M_y = \int_{-h/2}^{h/2} p_{yy}\, z\, dz = -N\left(\frac{\partial^2 w}{\partial y^2} + \frac{\sigma \partial^2 w}{\partial x^2}\right) \qquad 27.5.3$$

and a twisting moment

$$M_{xy} = \int_{-h/2}^{h/2} p_{xy}\, z\, dz = -(1-\sigma)\frac{N\partial^2 w}{\partial x \partial y}$$

where $N = Eh^3/12(1-\sigma^2)$ is the *flexural rigidity* of the plate; it has the dimensions of moment and is conveniently measured in Newton-meters. We shall see presently that it is the fundamental lithospheric parameter determined by observations. The isolated section shown in Figure 27.8 is subject also to vertical forces, resulting from the shear components p_{xz} and p_{yz} which support the load. These forces can be written

$$P = \int_{-h/2}^{h/2} p_{xz}\, dz \qquad Q = \int_{-h/2}^{h/2} p_{yz}\, dz \qquad 27.5.4$$

For the static equilibrium of the section, the forces and moments must be related through

$$P = \frac{\partial M_x}{\partial x} + \frac{\partial M_{xy}}{\partial y}; \qquad Q = \frac{\partial M_y}{\partial y} + \frac{\partial M_{xy}}{\partial x}$$

$$\frac{\partial P}{\partial x} + \frac{\partial Q}{\partial y} + p = 0 \qquad 27.5.5$$

where $p(x,y)$ is the imposed load per unit area.

Substituting for M_x, M_y, M_{xy} in Equation 27.5.5, from Equation 27.5.3, we find

$$P = -\frac{N\partial}{\partial x}\left(\frac{\partial^2 w}{\partial x^2} + \frac{\partial^2 w}{\partial y^2}\right); \qquad Q = -\frac{N\partial}{\partial y}\left(\frac{\partial^2 w}{\partial x^2} + \frac{\partial^2 w}{\partial y^2}\right) \qquad 27.5.6$$

or

$$\frac{\partial^4 w}{\partial x^4}+\frac{2\partial^2 w}{\partial x^2 y^2}+\frac{\partial^4 w}{\partial y^4}=\frac{p}{N} \qquad 27.5.7$$

Equation 27.5.7 is the fundamental relation for the study of plate bending. If the plate, of density ρ, floats in a denser liquid, of density ρ_m, a buoyant force $(\rho_m-\rho)gw$ arises whenever the plate is deflected. The relationship in this case becomes

$$\frac{\partial^4 w}{\partial x^4}+\frac{2\partial^4 w}{\partial x^2 \partial y^2}+\frac{\partial^4 w}{\partial y^4}=\frac{p-(\rho_m-\rho)gw}{N} \qquad 27.5.8$$

For an extensive plate, whose edges are far removed from the location of interest, the equation can be solved most easily if the load p is harmonic in x and y, or if it can be resolved into harmonic components. Nadai, for example, has given solutions for floating plates bent by concentrated point or line loads. Beyond the edge of the load, where $p=0$, the solution for w contains terms of the form

$$e^{-n\,x/\alpha}\cos\,(nx/\alpha)$$

where n is determined by the harmonic component of the load, and

$$\alpha^4=4N/(\rho_m-\rho)g$$

The quantity α (with the dimensions of length) is called the flexural parameter of the plate and its environment; it is closely related to N, but involves the densities also. The form of the plate, beyond the edge of the load, is evidently composed of damped harmonic waves of wavelength $2\pi\alpha/n$.

Extensive examinations of lithospheric flexure under varying tectonic conditions have been made by Walcott (1970a, b). An example of the approach, in the case of Hawaii, is shown in Figure 27.9. The Hawaiian ridge is taken as a two-dimensional line load, on a plate having one free edge along the axis of the ridge. The forms of the top surface of a bent plate, for the given load, are shown on the figure for values of α of 100 km and 150 km, where they may be compared with the bathymetric profile (particularly the crest of the arch) and the form of the Mohorovičić discontinuity. Walcott concluded that, for Hawaii, α lies between 100 and 150 km. Upon the assumption of reasonable densities ρ and ρ_m, the corresponding flexural rigidity is obtained. Walcott (1970a) showed that the apparent flexural rigidities for eight major structures, ranging in age from 10^3 to 5×10^8 years, lay between 6×10^{24} and 5×10^{22} Newton-meters. With the exception of the relatively young Lake Bonneville uplift in Utah (Crittenden, 1963), which gives the smallest value of N, the apparent values of N decrease with the age of the feature. Lake Bonneville is believed to represent an anomalous region of the continental lithosphere, with very low flexural strength. For the other cases, Walcott suggests that the decrease of the apparent flexural rigidity with age is due to the fact that the lithosphere itself is not perfectly elastic, but is a Maxwell body (Section 27.2), in which the flexures have increased with time, as the lithosphere sinks into the substratum. The equivalent Maxwell decay time indicated by the variation of apparent N with age is 10^5 years. The important

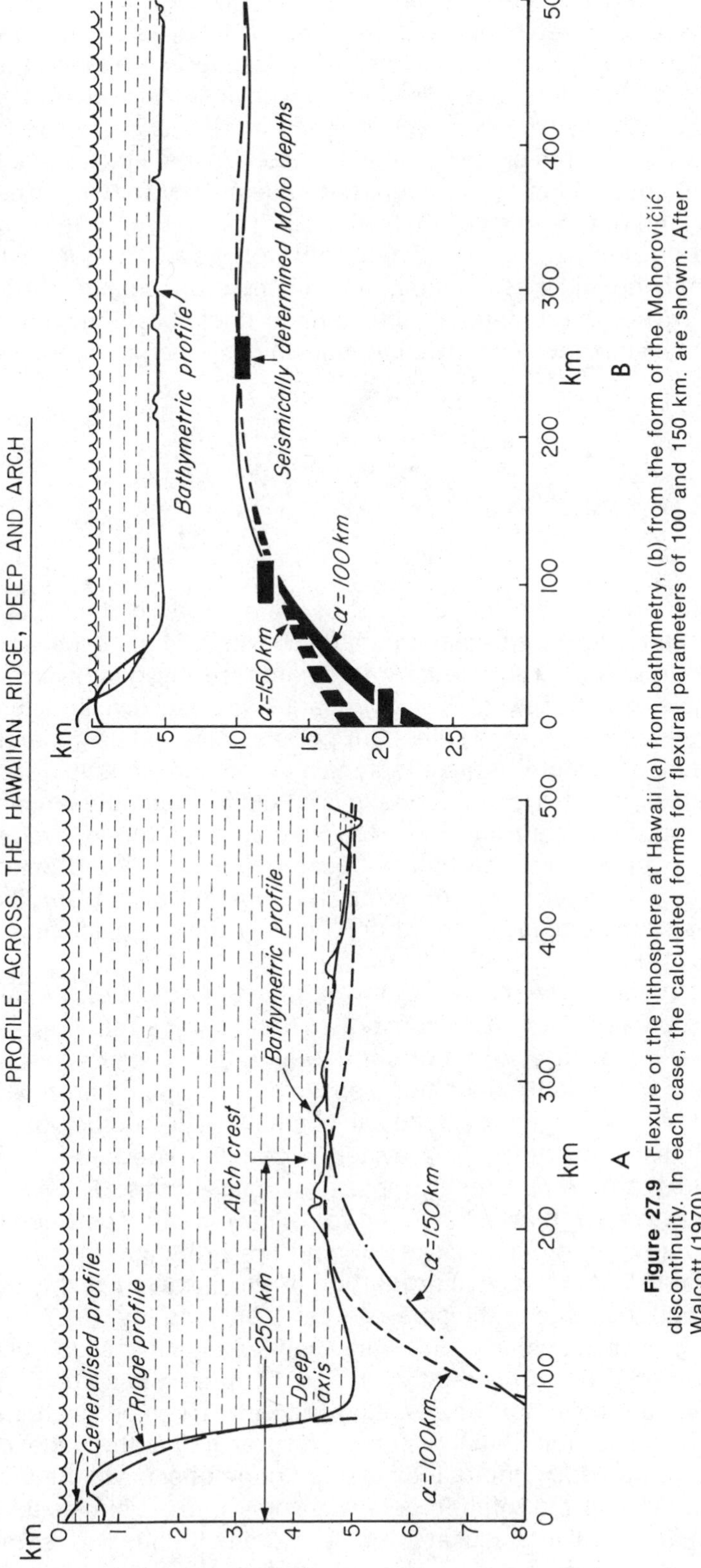

Figure 27.9 Flexure of the lithosphere at Hawaii (a) from bathymetry, (b) from the form of the Mohorovičić discontinuity. In each case, the calculated forms for flexural parameters of 100 and 150 km. are shown. After Walcott (1970).

point is that, while a Maxwell-type body, rather than a purely elastic body, is definitely indicated for the lithosphere, the decay time is so long that very appreciable loads can be supported for times of the order of 10^8 years.

To pass from values of N to values of lithospheric thickness requires the adoption of E and σ, but these lie within very small limits. Walcott took $\sigma = 1/2$ and E between 5×10^{10} and 1×10^{11} Newtons/m^2. For $N = 6 \times 10^{24}$ Newton-m, typical of young features on continental lithosphere, the lithospheric thickness is between 90 and 130 km, with a preferred value of about 110 km. The values of N for Hawaii and island arcs are somewhat less than those for features of comparable age on continental lithosphere, but not greatly so. Walcott suggests that 75 km may be taken as the minimum thickness of oceanic lithosphere. We have thus arrived at estimates of lithospheric thickness, independent of the depth of the seismic low-velocity zone, but which, in general, agree with those depths.

27.6 PLATE TECTONICS II

The basic principles of plate tectonics were introduced in Chapter 1, and these have been referred to again in the discussions of almost every aspect of the solid earth. It is appropriate here to summarize the geophysical evidence concerning plates themselves, their boundaries, their relative and absolute motion, and the possible driving forces. Our point of view will be general or global in nature. The application of plate tectonics to the interpretation of the history of specific areas is outside our scope; it is a field in which there are already an enormous number of contributions. (An example, from a geophysical point of view, is the discussion of the Indian Ocean by McKenzie and Sclater (1971); from a more geological standpoint, the development of the Cordillera by Atwater (1970), or the emplacement of ophiolites in the Appalachians, by Dewey and Bird (1971).)

The first point concerns the rigidity of lithospheric plates. While in a general way the concentration of seismicity along narrow lines supports plate rigidity, a kinematic test is also important. It was noted that the 12-plate global model gave a very much better fit to observed tectonics than the original 6-plate model, and in regions of special complexity, a number of very minor plates are indicated. What is the evidence that the supposed major plates, America, Eurasia and Africa, for example, are single entities? The best test is the consistency of the rotation poles found for relative motion between two plates separated by a ridge. If the plates are in rigid-body rotation, the spreading rate across the ridge will increase in proportion to sin θ, where θ is the arc distance from a point on the ridge to the pole (Morgan, 1968). As Figure 27.10 indicates, the spreading rates indicated for the Mid-Atlantic Ridge by ocean-floor magnetic anomalies conform to this relationship. (Of course, the spreading rates are used to locate the pole, but the resulting fit to the assumed relationship gives confidence in the assumptions). Secondly, if spreading at the ridge is normal to ridge sections, transform faults indicated by ridge offsets will be tangential to circles drawn around the rotation pole. Conversely, great circles taken normal to transform faults will intersect at this pole. Morgan found, also for the Atlantic, that these great circles did indeed intersect near the pole obtained from the

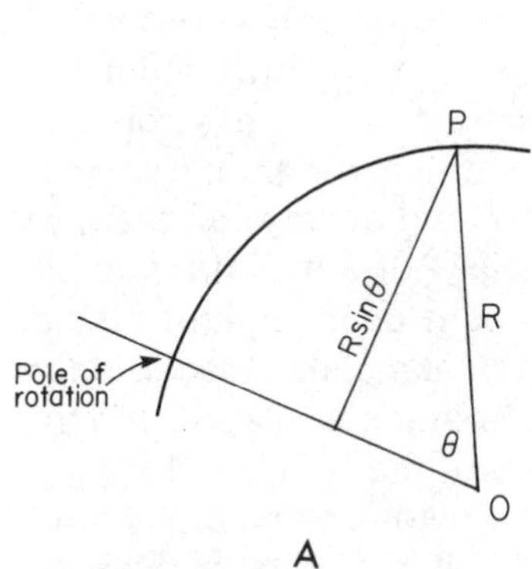

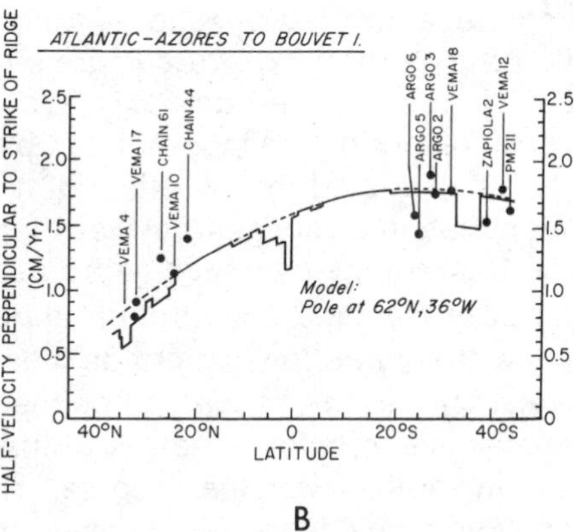

Figure 27.10 (a) Spreading rate at a ridge varies as sin θ, where θ is angular distance from the rotation pole. (b) observed spreading rates for the Atlantic, compared to the sine function. After Morgan (1968).

spreading rates. There is good evidence, therefore, that the major plates behave kinematically as rigid blocks (McKenzie 1972a).

It has been mentioned that complex areas require the presence of minor plates. An outstanding example is the eastern Mediterranean, where McKenzie (1972b) showed that the seismicity could best be explained by invoking the presence of 15 minor plates or plate slices (Figure 27.11), separated by the three possible types of boundary, and in a most complicated pattern of relative movement. A somewhat different point of view would be that in these regions plate rigidity no longer holds and the major plates become crumpled.

Geophysical evidence on the nature of the boundaries: ridges, transform faults, and trenches has become reasonably complete, but not without raising

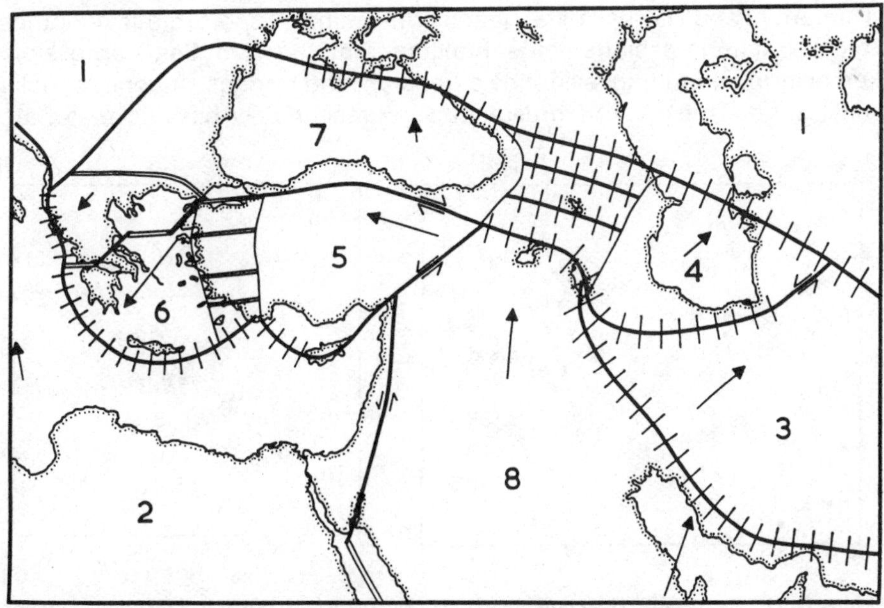

Figure 27.11 The complex of minor plates in the area of the eastern Mediterranean. Plate boundaries are spreading (double lines), converging (ticked lines) or transform. Relative plate motions are indicated. From D. McKenzie (1972b); courtesy of the author and the *Geophysical Journal*.

some unanswered questions. Seismicity on the ridges (Sykes, 1967) confirms that the earthquake sources are shallow and have directions of maximum tension normal to the ridge axis. Where ridges are offset by transform faults, the sense of motion is that predicted by the spreading directions, which is opposite to that expected for the offset of a static feature by a strike-slip fault. We have seen that, in a very general way, bathymetry, heat flow, and gravity suggest that the topography of ridges can be explained on the basis of a moving, cooling mass in isostatic equilibrium, although the most recent deep drilling results show that heat flow cannot be measured by conventional means close to ridge axes. The reason appears to be the great influence of water percolation and the convective transfer of heat, within the central portion of the ridges.

In another way, the deep-sea drilling project has provided some of the most convincing evidence for ocean-floor spreading. A critical test of the hypothesis is that the ages of the ocean floor increase with distance from spreading axes, in accord with the magnetic polarity ages. Unfortunately, the ocean-floor basaltic rocks, which have now been obtained by drilling through the sedimentary cover, do not lend themselves to radiometric dating. Their youth would restrict the methods to potassium-argon, but problems of low potassium content and excess argon pose many problems. However, the sedimentary cover lying on the ocean floor has been dated palaeontologically (Maxwell et al., 1970), with the result, for the Atlantic Ocean, shown in Figure 27.12. The accord between the sediment age and the age of the underlying basalt, inferred from the magnetic pattern, is remarkably good.

The ridge system itself can be studied not only in the deep oceans, but in those places where ridges run into continents: the western coast of North America, the east African zone of rift valleys, and the Red Sea (Fig. 1.1). It would appear that rift valleys (Irvine, 1966) represent the first stage in the development of a new spreading locus, and that the Red Sea illustrates a later stage. The cross-section of Figure 27.13, based on the work of Drake and Girdler (1964) and Tramontini and Davies (1969), lends convincing support to this hypothesis. Seismic refraction measurements indicate that the Red Sea depression is marked by normal faulting, evidence of tension, and by dense magnetic rocks in the axial zone. These rocks presumably represent new basaltic material in-

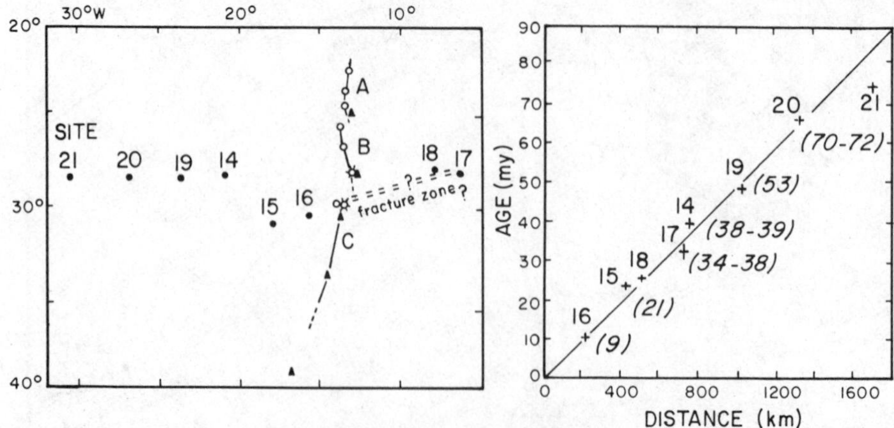

Figure 27.12 Right: age of sedimentary rocks overlying the oceanic crust as a function of distance from the ridge. Figures in parentheses are magnetic ages, in millions of years, of the basement. Drilling sites are shown at the left. After Maxwell et al, *Science* **168**. Courtesy of the authors and the American Association for the Advancement of Science; original copyright 1970, A.A.A.S.

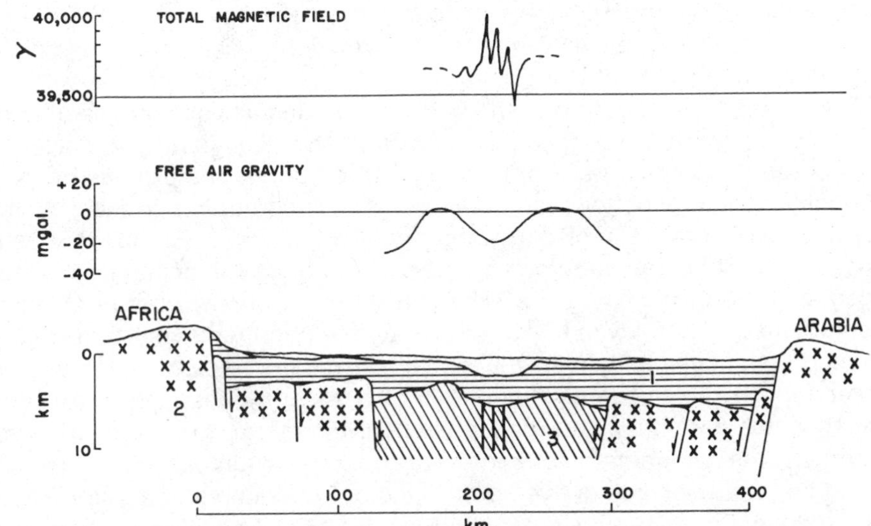

Figure 27.13 Cross-section of the Red Sea, based on the work of Drake and Girdler (1964) and Tramontini and Davies (1969). Rocks are (1) unconsolidated and consolidated sedimentary rock, (2) continental crust, (3) basic rocks with magnetic dike-like sections. The relatively small free-air gravity anomaly indicates approximate isostatic compensation.

jected into the crust, across a width of 100 km; their excess density compensates for the thickness of low-density sedimentary rock in the trough. The magnetic anomalies decrease in amplitude very rapidly with distance from the axis, even within the presumed area of basic rock. Tramontini and Davies suggest that because of a very slow earlier spreading rate, perhaps as low as 0.5 cm/yr, strips of alternate magnetization in the Red Sea floor, away from the axis, are so thin that their anomaly pattern becomes vanishingly small. The topography on both sides of the rift valleys and the Red Sea is high, suggesting that uplift precedes spreading. This would be expected over a new rising convection current. If rifting proceeds without interruption, the result may be a new ocean with axial ridge, bounded by two continental blocks whose margins are of the Atlantic North American type. That is, they lack oceanic trenches and subduction zones.

The recently discovered anisotropy of upper mantle velocity near the ridges at sea (Section 7.4) may well be related to the movements associated with plate tectonics. It will be recalled that the direction of maximum *P*-velocity was found to be perpendicular to the ridge. Francis (1969) has pointed out that the olivine crystal itself is highly anisotropic, with a *P*-velocity of 9.87 km/sec in the **a** crystallographic direction, as against a velocity of 7.73 km/sec in the **b** direction. He proposes that gliding of olivine crystals as mantle material spreads out from the ridge tends to produce an alignment of **a** axes perpendicular to the ridge. The process must take place very close to the axis, after which the crystal directions become frozen into the rigid material of the lithosphere. Ave'Lallement and Carter (1970) agree that a preferred orientation of olivine crystals is related to mantle flow, but on the basis of laboratory studies of recrystallization under stress, they propose that this orientation results from recrystallization in and beneath the lithospheric plate. They show that whether the slab is dragged from beneath, or pushed apart at the ridge, the stress field near its base would promote recrystallization with the required orientation. There are thus alternative explanations for the mechanism producing anisotropy, but as they are both related to plate flow the measurement of anisotropy promises to be an important tool for the study of tectonics.

The detailed investigation of the structure of the thin veneer of sedimentary material which lies on the oceanic crust in most places is also important. Information on this veneer has come largely from continuous profiles of vertical incidence seismic reflections, as obtained on oceanographic cruises, and supplemented with core samples (Ewing and Ewing, 1967; Ewing, Houtz, and Ewing, 1969). The very young sedimentary material on the ocean floors can be divided into two broad categories: that carried out from the continents by currents of density instability in the waters along the continental margins, and that produced in the oceans themselves from the remains of biological material. The coarser material from the continents, known as turbidites from the turbidity currents which carry them, are deposited near the margins of the deep oceans. In contrast, the biological sediments are found over almost the entire ocean floor and, in many places in the South Pacific, for example, reach thicknesses of over 200 meters (Ewing, Houtz, and Ewing, 1969). Alteration of some of this material to layers of dense chert provides reflecting horizons within the sedimentary section. The range of accumulation varies with latitude, and indeed the present axis of maximum accumulation does not correspond to the predicted latitude of maximum production. Nevertheless, the mean rate of accumulation is estimated to be 2 cm in 1000 years. If ocean-floor spreading progressed at a uniform rate in time, the sediment thickness would increase linearly with distance from the mid-ocean ridges. However, the striking feature of the sediment isopach maps is the relative constancy of thickness away from the ridges, and the abrupt thinning within 200 km of them. This is attributed to the rapid creation of ocean floor within the last 10 million years, or the time required for 200 m of sediment to accumulate, with the suggestion that spreading was slower in the preceding period. This evidence on the variability of the rate of spreading is important, for the counting of magnetic reversals does not provide an absolute time scale in the past, unless the magnetic epochs are idependently dated. As we have seen, only the most recent reversals have been dated geochronologically. Further support for rigid-plate motions is also provided by the studies of ocean-floor sediments, because the sedimentary layers are found to be lying flat, except in the vicinity of ridges, fracture zones, and trenches. In the latter regions they are crumpled, but the lack of distortion elsewhere is testimony to the uniform motion of the individual lithospheric plates.

At the trenches or subduction zones, many lines of geophysical evidence support the presence of cold descending slabs. These include, as we have seen, the seismicity, particularly the presence of deep-focus earthquakes, the variations of seismic velocity and "*Q*," gravity, and electrical conductivity. A point on which there is not yet agreement is the planar nature of the seismic Benioff zone (Fig. 6.6). The earthquake foci appear to define the upper surface of a slab, rather than both surfaces. Theoretical studies on the warming of slabs by conduction show that the slabs will remain cooler than their surroundings for the time required for their complete emplacement to depths of the order of 700 km, at appropriate spreading rates. The motion within subduction zones is inherently more complicated than at ridges, where spreading normal to the ridge axis appears to be the rule. Subduction zones are randomly oriented relative to the spreading direction on the side being consumed and because of the curved form of the zones in plan, the angle between the trench axis and the motion varies along this axis. The curved plan of subduction zones is probably the natural result of the intersection of a dipping slab with the earth's surface; variations in radius of the arcs should relate to the near-surface dip of the slab, but this is not a well-defined quantity. The observed curvature in some cases requires a near-surface dip that is much smaller than that defined at greater depth by the seismic Benioff zone (LePichon, Francheteau, and Bonnin, 1972).

The near-surface structure of the trench associated with subduction zones has been revealed in a spectacular way by reflection seismic profiling (Fig. 27.14). The bathymetric trench, in some cases at least, does not itself represent the boundary between two plates. Sedimentary rocks carried in the plate being consumed persist to the continental side of the trench, where they are crumpled against the second plate. The observed trench is a rather superficial feature.

Theoretical studies of the temperature distribution in descending slabs show the importance of strain heating along the slab margins (Fig. 25.5). This heating is believed to be of great consequence in the formation of volcanic island arcs which form above the descending slab (Hasebe, Fujii, and Uyeda, 1970). Lithospheric slabs presumably consist of both crust and mantle; in the case of a descending oceanic slab, the oceanic crust carried on the top contains rocks which have, in general, lower melting points than mantle minerals. The melting of this crust as a result of strain heating, and its migration to the surface, could contribute to the volcanism of island arcs (cf. Figure 20.13).

Detailed interpretations of volcanic island arcs have been carried out by many workers (Dickinson and Hatherton, 1967; Dickinson, 1970; Dewey and Bird, 1970; Takeuchi and Uyeda, 1965). But the situation on the underridden plate is often complex. Karig (1970, 1971) has emphasized the presence of multiple arcs, separated by basins. Within these systems, there is evidence of tension, which could be taken as evidence of the development of local spreading

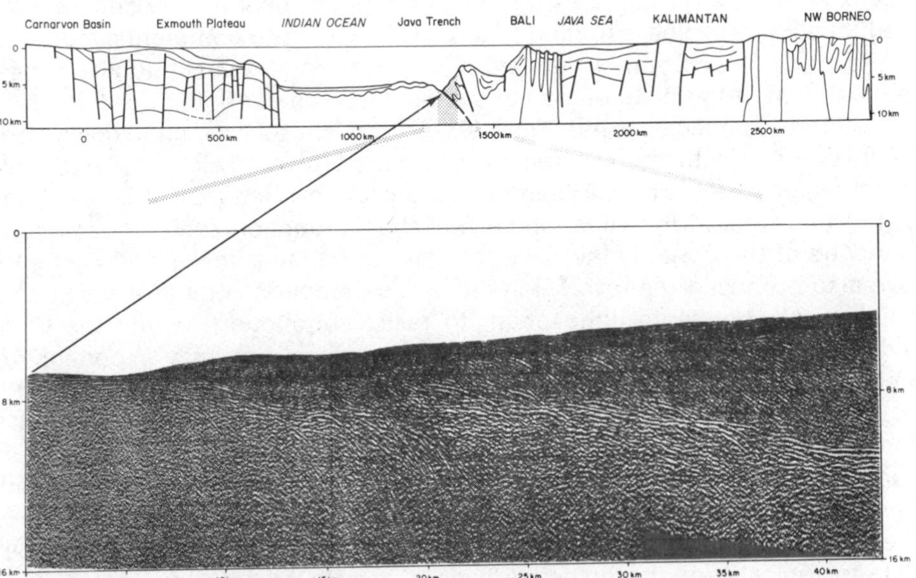

Figure 27.14 Vertical reflection seismic profile across the Java Trench. The surface of the oceanic basement dips from 7 km. at the Trench to about 12 km. at the right. Deformed sedimentary rocks, still pertaining to the oceanic plate, make up the landward slope of the trench. From R. H. Beck and P. Lehner (1974); courtesy of the authors.

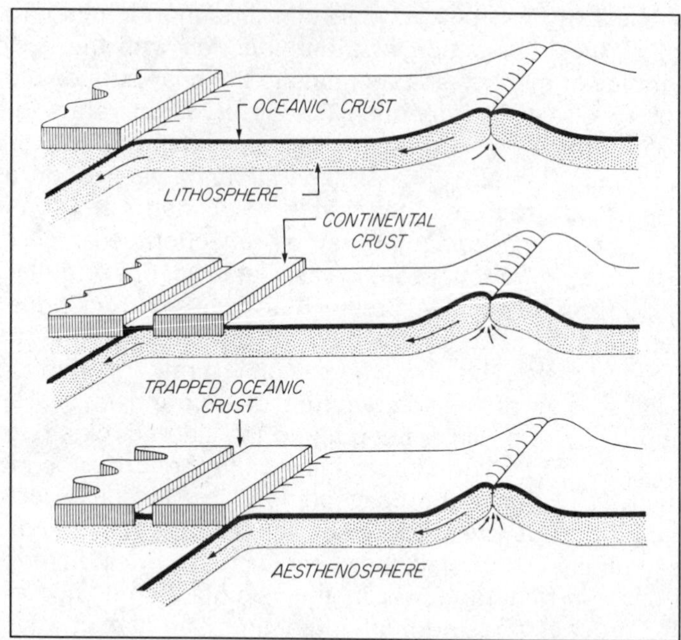

Figure 27.15 Possible mechanism of continental collision, in which oceanic crust becomes entrapped. Note the migration of the subduction zone. After Irving (1972).

axes. Karig suggests, however, that the tensile failure results from a pattern of block faulting and rotation induced by the primary underthrusting.

We have not yet referred to the role of continents in plate tectonics. Certainly this role is more passive than in Wegener's original concept of continental drift, but continents do represent thick rafts of relatively low-density rocks within a lithosphere that is probably thicker than the average oceanic lithosphere. McKenzie (1969) has argued that continent-bearing lithosphere resists subduction. In a typical situation (Fig. 27.15), leading to continental margins of the Pacific type of South America, oceanic lithosphere is subducted beneath the continent. If the plate being consumed carries a continent itself, the situation is more complicated. As Figure 27.15 suggests, a continental collision will result, possibly with the entrapment of oceanic crust. Thick, elevated continental regions such as the Tibetan plateau probably had their origin in a collision of this type. In the situation shown, the subduction zone migrates to the new edge of the continental block, but McKenzie, and Dewey and Bird (1970) have also proposed the reversal in dip of a subduction zone as a result of the tendency of continental lithosphere to resist subduction. Within the underridden continent, the region above the descending slab may experience the effects of increased temperature and heat flow noted in connection with island arc formation, leading to continental volcanoes and Andean-type deformation. Observations of seismicity, seismic velocity, gravity, electrical conductivity, and magnetization direction may all contribute to the recognition of ancient oceanic crust or collision zones ("sutures") within present-day continental blocks. These studies may lead to an understanding of the role of the current type of plate tectonics through geological history. It must be recognized that, as yet, there is very little evidence of the type of global tectonics which existed through the long history of the earth from the solidification of the crust to 200 million years ago.

27.7 ABSOLUTE PLATE MOTIONS AND THE DRIVING FORCE

The most familiar methods of studying plate motions: ocean-floor magnetic stripes, motion on transform faults, and palaeomagnetic observations, lead to relative motions between plates. There is great interest in attempting to find a frame of reference, fixed relative to the earth as a whole, within which to describe the motion more absolutely. It is true, as we saw in Chapter 21, that if all inter-plate relative motions and some palaeomagnetic pole positions are known, a component of "true polar wander" can be isolated. In this sense the motion of plates relative to the rotation poles of the earth is known. But the method, at present, has been applied only to net displacements over rather long periods of time and does not lead to an instantaneous, present-day, measure of absolute plate velocity. The situation of course, is very complicated kinematically; not only are all plates of the lithosphere in relative motion with each other, but plate boundaries can be migrating with respect to the plates on either side and the lithosphere as a whole conceals the body of the earth beneath from our direction observations. In this connection, the suggestion of Wilson (1963) that there may be surface manifestations of volcanism from sources beneath the lithosphere assumes great importance. Wilson's original suggestion related to the origin of the Hawaiian Islands and was based upon the apparent age sequence, that the oldest islands were furthest "downstream" in the spreading direction from the ridge axis. He proposed that a source, anchored within the deeper mantle, was able to eject lava to the surface through the overriding plate. If the source is indeed at rest, the length of the island chain and its ages give the absolute motion of the plate (Fig. 27.16). This original suggestion was extended by Wilson (1965a, b) to include lateral ocean-floor ridges as manifestations of active points on the principal mid-ocean ridges, and the concept of using these points as a global reference frame was developed by Morgan (1972). Morgan determined the motion of the major plates relative to the frame, as shown in Figure 27.17. Noteworthy is the considerable variation in plate motion from plate to plate.

Whether or not the frame is truly fixed to the deep mantle depends on the nature of the reference points, which have become known as *hot spots,* or the surface manifestations of *mantle plumes.* There may be a considerable difference between hot spots on a ridge, which represent unusually active short sections of the ridge, and those beneath the central region of a large plate, such as that inferred for the origin of the Hawaiian chain. As Figure 27.17 indicates, most of the hot spots used by Morgan are on plate boundaries and are marked

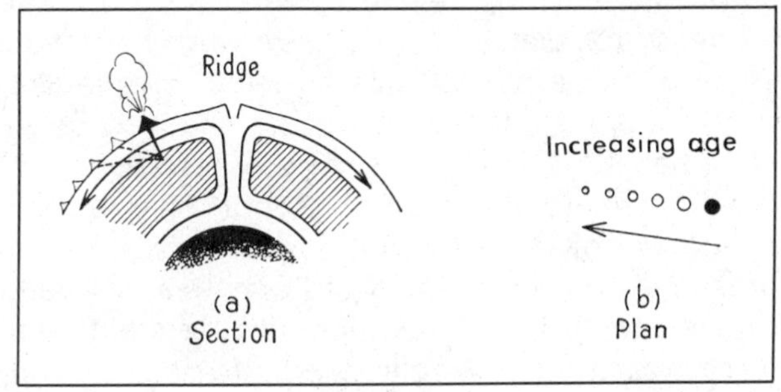

Figure 27.16 Wilson's original diagram illustrating the mode of formation of the Hawaiian island chain, by plate motion over a stationary hot spot. The plume is here shown as originating in the stagnant center of a convection cell. Courtesy of J. T. Wilson and the Canadian Journal of Physics.

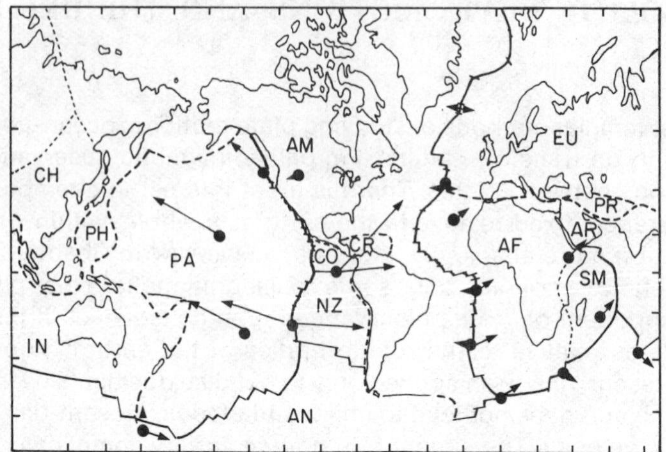

Figure 27.17 Motion of the major plates, shown by arrows, over the frame of hot spots, shown by black circles. After Morgan (1972).

by active volcanism. A larger number of hot spots has been proposed by Wilson (1973); in many cases these are marked by persistent uplift through long geological periods, or former volcanism, rather than current activity.

The first geophysical evidence in support of the suggestion that plumes extend to great depth in the mantle was presented by Kanasewich et al (1973). They observed that earthquake waves generated in the southwest Pacific and observed at arrays in western Canada penetrated to near the core-mantle boundary under Hawaii. The array observations permitted a sensitive measure of the velocity at the deepest point of the ray and from these, the authors concluded that the lowermost mantle immediately beneath Hawaii was a region of anomalously high seismic wave velocity. Increased temperature alone in a rising plume would result in a lowering of velocity, so that these observations would have to be interpreted in terms of compositional differences, possibly resulting from material dragged upward from the core.

The total geophysical evidence for the existence of mantle plumes remains very sparse. One suggestion from thermal studies is the possibility of convective instability resulting from the strain heating-temperature-viscosity interrelationship (Section 25.6), but this has not been developed for the extreme aspect ratio demanded by a small-area plume extending through the mantle.

From the earliest days of plate tectonics speculations have been made concerning the driving force for plate motion. In general, two types of forces have been proposed: those applied at plate boundaries and those applied across the base of the lithosphere over an entire plate. The first includes the tendency for plates to slide downward from ridge axes (Orowan, 1964; Hales, 1969) or to be pushed apart by the injection of new material. At subduction zones, there is a tendency for the relatively dense down-going slab to pull the plate behind it. The second type of force is that exerted by convection beneath the lithosphere; it is a shear proportional to the local velocity gradient and the coefficient of viscosity. None of these forces can be ruled out on the basis of present knowledge. In fact, Forsyth and Uyeda (1975) show eight possible forces acting on a plate, including, in addition to those mentioned above, the forces resulting from collisions and a possible suction toward subduction zones.

However, they argue that mantle convection is unlikely to be a significant driving force, partly because of the unfavorable aspect ratio of cells and partly because of decoupling in the aesthenosphere, and they include a mantle drag force, rather than a mantle driving force. This is an important *a priori* decision, as it influences their conclusions, discussed below. They then correlate the absolute plate velocity, using the hot spot reference frame, with plate area, continental area, length of ridge boundary, length of transform boundary, and length of trench boundary. The most striking correlation is with the length of trench boundary, from which the authors suggest that forces exerted at subduction zones are of dominant importance. There is some evidence of negative correlation between area of continent carried by a plate and plate velocity, suggesting that the thickened lithosphere beneath continents acts as a sea anchor to oppose motion. Proceeding to a quantitative evaluation, Forsyth and Uyeda make the assumption that all plates are in dynamic equilibrium, based on the vanishingly small accelerations. This requires that the net torque on each plate, about any axis, is zero. They express these torques, for 12 plates and three axes per plate, in terms of the 8 proposed types of force, integrated over the plate boundaries or plate areas as the case may be. This leads to 36 equations and 8 unknowns, on the assumption that the same intensity coefficients for each force type applied to each plate. Their results show that two opposing forces, slab pull and slab resistance, are an order of magnitude greater than the others, but the difference between these, at a velocity of 1 cm/yr, is of the same order as other forces, with slab pull being the greater. Slab pull is the effect of the dense slab, mentioned above. Slab resistance is the viscous force exerted by the surrounding mantle at the boundaries of the descending slab. It is proportional to plate velocity. At a terminal velocity of 7 cm/yr, slab resistance becomes equal to slab pull, leaving the other forces in control of the plate. The authors conclude that, on their model, forces exerted at plate edges (as opposed to those exerted on plate lower boundaries) can explain the observed pattern of plate motions, but they admit that convection in the aesthenosphere cannot be ruled out as a controlling influence. An important point concerning slab forces is that some other type of force must have initiated plate movement, since slab forces obviously come into being only after the formation of subduction zones.

The further understanding of the driving mechanism, and the related question of the role of plate tectonics before 200 million years ago, remain among the most challenging problems of geophysics for the future.

Problems

1. Set up the specific relations for strain as a function of the time after the application of a stress p to the Maxwell and Kelvin substances respectively (Section 27.2).

2. An elastic sphere of material with linear coefficient of thermal expansion α cools from the outside, so that the decrease in temperature t is a function of radius r. Show that the fractional elastic shortening of an arc at any radius within the sphere is given by

$$-\alpha t + \frac{1}{r^3}\int_0^r 3r^2 \alpha t \, dr$$

Hence show that in a region $0 < r < r_1$ where t is nearly constant, the strain vanishes; in the region $r_1 < r < r_2$ where the sphere is cooling most rapidly there is extension, while for $r > r_2$, if the sphere is already cool, there is compression.

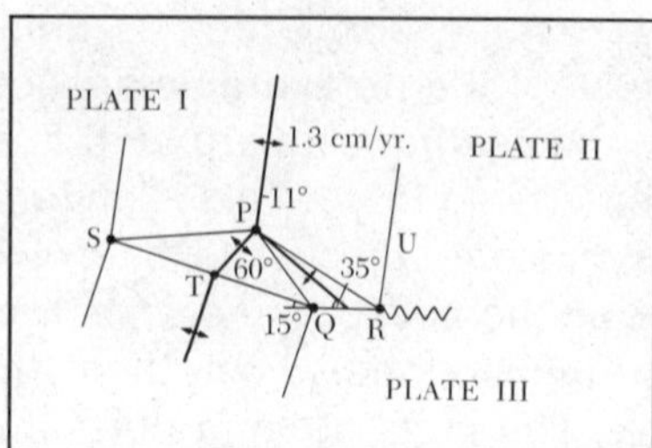

Figure 27.A

3. Deduce the relative positions of the ocean-floor structures shown in Figure 1.9, after an interval of spreading has taken place.

4. In the vicinity of the Azores, the ocean-floor tectonics appear to be as shown in Figure 27.A, with abrupt changes in the direction of the Mid-Atlantic Ridge at points *P* and *T*, and a secondary ridge running southeastward from *P* for 500 km. Given the single spreading rate shown, and the knowledge that the crust within the triangles *PSQ and PQR* was formed in the same time interval as crust across the main ridge (*SQ, SU*), determine the other 3 spreading rates and the relative motions of Plates I, II and III. (The segment *QR* is east-west; north of *P* the main ridge trends 11° east of north.)

5. One explanation for the arcuate shape of the globe of trenches and island arcs is that they represent the intersection of a dipping plane with the spherical surface of the earth (Lake, 1931). Determine the relation between the dip of the plane and the radius of the arc. Seismological evidence suggests that the deeper part of the down-going lithospheric slab under South America, for example, dips at approximately 45°. By measuring on a map the radii of associated arcs, determine whether this is consistent with your relation. If it is not, can you offer an explanation which would permit the retention of Lake's suggestion?

6. In the text, the departure from spherical symmetry of the earth has been emphasized by maps of the globe, or parts of it, showing the variation of earthquake frequency, P_n velocity, geoid height, main magnetic field, secular change, heat flow, and motion between rigid plates. Arrange these parameters in a matrix and give arguments why you would expect a positive correlation, a negative correlation, or no correlation between each possible pair.

BIBLIOGRAPHY

Alley, C. O., and Bender, P. L. (1967) Information obtainable from laser range measurements to a lunar corner reflector. In *Continental drift, secular motion of the pole, and rotation of the earth,* ed. W. Markowitz, and B. Guinot. Int. Astron. Union, Symposium, **32**, 86.

Anderson, D. L. (1967) Latest information from seismic observations. In *The earth's mantle,* ed. T. F. Gaskell. Academic Press, London.

Anderson, D. L., and Kovach, R. L. (1964) Attenuation in the mantle and rigidity of the core from multiply reflected core phases. *Proc. Nat. Acad. Sci. U.S.*, **51**, 168.

Anderson, D. L., Ben-Menahem, A., and Archambeau, C. B. (1965) Attenuation of seismic energy in the upper mantle. *J. Geophys. Res.*, **70**, 1441.

Atwater, T. (1970) Implications of plate tectonics for the Cenozoic tectonic evolution of western North America. *Bull. Geol. Soc. Amer.*, **81**, 3513.

Ave'Lallement, H. G., and Carter, N. L. (1970) Syntectonic recrystallization of olivine and modes of flow in the upper mantle. *Bull. Geol. Soc. Amer.*, **81**, 2203.

Beck, R. H. and Lehner, P. (1974) Oceans, new frontier in exploration. *Bull. Amer. Assoc. Pet. Geol.* **58**, 376–395.

Beloussov, V. V. (1962) *Basic problems in geotectonics.* McGraw-Hill, New York.

Bender, P. L., and Silverberg, E. C. (1975) Present tectonic-plate motions from lunar ranging. In *Recent crustal movements,* ed. N. Pavoni and R. Green. Elsevier, Amsterdam (also *Tectonophysics,* **29**).

Bodvarsson, G., and Walker, G. P. L. (1964) Crustal drift in Iceland. *Geophys. J.*, **8**, 285.

Bullard, E. C., Everett, J. E., and Smith, A. G. (1965) The fit of the continents around the Atlantic. *Phil. Trans. Roy. Soc. London A,* **258**, 41.

Burford, R. O. (1966) Strain analysis across the San Andreas fault and coast ranges of California. In *Proceedings of the Second International Symposium on Recent Crustal Movements,* ed. V. Auer, and T. J. Kukkamaki. Helsinki.

Cathles, L. M. (1975) *The viscosity of the earth's mantle.* Princeton University Press, New Jersey.

Clark, S. P. (1966) Handbook of physical constants. Geol. Soc. Amer., Memoir **97**.

Coates, R. J., Clark, T. A., Counselman, C. C. III, Shapiro, I. I., Hinteregger, H. F., Rogers, A. E., and Whitney, A. R. (1975) Very long baseline interferometry for centimeter accuracy geodetic measurements. In *Recent crustal movements,* ed. N. Pavoni and R. Green. Elsevier, Amsterdam (also *Tectonophysics,* **29**).

Crittenden, Max. D., Jr. (1963) Effective viscosity of the earth derived from isostatic loading of Pleistocene Lake Bonneville. *J. Geophys. Res.,* **68**, 5517.

Crittenden, Max. D., Jr. (1967) Viscosity and finite strength of the mantle as determined from water and ice loads. *Geophys. J.,* **14**, 261.

Currie, Robert G. (1974). Period and Q_w of the Chandler wobble. *Geophys. J.,* **38**, 179–185.

Decker, R. W., Einarsson, P., and Mohr, P. A. (1971) Rifting in Iceland: new geodetic data. *Science,* **173**, 530.

Dewey, J. F., and Bird, J. M. (1970) Mountain belts and the new global tectonics. *J. Geophys. Res.,* **75**, 2615.

Dewey, J. F., and Bird, J. M. (1971) Origin and emplacement of the ophiolite suite: Appalachian ophiolites in Newfoundland. *J. Geophys. Res.,* **76**, 3179.

Dicke, R. H. (1969) Average acceleration of the earth's rotation. In *The earth-moon system,* ed. B. G. Marsden and A. G. W. Cameron. Plenum Press, New York.

Dicke, R. H. (1969) Average acceleration of the earth's rotation and the viscosity of the deep mantle. *J. Geophys. Res.,* **74**, 5895.

Dickinson, W. R. (1970) Relations of andesites, granites, and derivative sandstones to arc-trench tectonics. *Rev. Geophys. Space Phys.,* **8**, 813.

Dickinson, W. R., and Hatherton, T. (1967) Andesitic volcanism and seismicity around the Pacific. *Science,* **157**, 801.

Drake, C. L., and Girdler, R. W. (1964) A geophysical survey of the Red Sea. *Geophys. J.,* **8**, 473.

Dratler, J., Farrell, W. E., Block, B., and Gilbert, F. (1971) High-Q overtone modes of the earth. *Geophys. J.* **23**, 399–410.

Ewing, J., and Ewing, M. (1967) Sediment distribution of the mid-ocean ridges with respect to spreading of the sea floor. *Science,* **156**, 1590.

Ewing, M., Houtz, R., and Ewing, J. (1969) South Pacific sediment distribution. *J. Geophys. Res.,* **74**, 2477.

Fischer, A. G., Heezen, B. C., Boyce, R. E., Bukry, D., Douglas, R. G., Garrison, R. E., Kling, S. A., Krasheninnikor, V., Lisitzin, A. P., and Pimm, A. C. (1970) Geological history of the western North Pacific. *Science,* **168**, 1210.

Forsyth, Donald, and Uyeda, Seiya (1975) On the relative importance of the driving forces of plate motion. *Geophys. J.,* **43**, 163.

Francis, T. J. G. (1969) Generation of seismic anisotropy in the upper mantle along the mid-oceanic ridges. *Nature,* **221**, 162.

Goldreich, P., and Toomre, A. (1969) Some remarks on polar wandering. *J. Geophys. Res.,* **74**, 2555.

Gordon, Robert B. (1967) Thermally activated processes in the earth: creep and seismic attenuation. *Geophys. J.,* **14**, 33.

Graber, Michael A. (1976) Polar motion spectra based upon Doppler, IPMS and BIH data. *Geophys. J.* **46**, 75–85.

Griggs, D. (1939) A theory of mountain building. *Amer. J. Sci.,* **237**, 611.

Gunn, R. (1943) A quantitative evaluation of the influence of the lithosphere on anomalies of gravity. *J. Franklin Inst.,* **236**, 47.

Gunn, R. (1947) Quantitative aspects of juxtaposed ocean deeps, mountain chains, and volcanic ranges. *Geophysics,* **12**, 238.

Hales, A. L. (1969) Gravitational sliding and continental drift. *Earth Plan. Sci. Letters,* **6**, 31.

Hasebe, K., Fujii, N., and Uyeda, S. (1970) Thermal processes under island arcs. *Tectonophysics,* **10**, 335.

Haskell, Norman A. (1935) The motion of a viscous fluid under a surface load. *Physics,* **6**, 265.

Haskell, Norman A. (1937) The viscosity of the asthenosphere. *Amer. J. Sci.,* **33**, (5), 22.

Herring, C. (1950) Diffusional viscosity of a polycrystalline solid. *J. App. Phys.,* **21**, 437.

Irvine, T. N., Ed. (1966) *The world rift system.* Geol. Surv. Canada, Paper **66–14**.

Irving, E., Editor (1972) The ancient oceanic lithosphere. *Pub. Earth Physics Branch,* Dept. of Energy, Mines and Resources, Ottawa, **42**, No. 3.
Irving, E. (1977) Drift of the major continental blocks since the Devonian. *Nature,* **270**, 304–309.
Isacks, B., Oliver, Jack, and Sykes, Lynn R. (1968) Seismology and the new global tectonics. *J. Geophys. Res.,* **73**, 5855.
Isacks, Bryan, and Molnar, Peter (1969) Mantle earthquake mechanisms and the sinking of the lithosphere. *Nature,* **223**, 1121.
Jaeger, J. C. (1956) *Elasticity, fracture, and flow.* Methuen and Co., London.
Jeffreys, H. (1962) *The earth* (4th ed.) Cambridge University Press, Cambridge.
Jobert, N. and Roult, G. (1976) Periods and damping of free oscillations observed in France after sixteen earthquakes. *Geophys. J.* **45**, 155–176.
Kääriäinen, E. (1966) Land uplift in Finland as computed with the aid of precise levellings. In *Proceedings of the Second International Symposium on Recent Crustal Movements,* ed. V. Auer and T. J. Kukkamäki. Helsinki.
Kanamori, H. (1970) Velocity and Q of mantle waves. *Phys. Earth Plan. Int.,* **2**, 259–275.
Kanasewich, Ernest R., Ellis, Robert M., Chapman, Chris. H. and Gutowski, Paul R. (1973) Seismic array evidence of a core boundary source for the Hawaiian linear volcanic chain. *J. Geophys. Res.,* **78**, 1361.
Karig, D. E. (1970) Ridges and basins of the Tonga-Kermadec island arc system. *J. Geophys. Res.,* **75**, 239.
Karig, D. E. (1971) Structural history of the Mariana Island arc system. *Bull. Geol. Soc. Amer.,* **82**, 323.
Knopoff, L. (1964) *Q. Rev. Geophys.,* **2**, 625.
Kohlstedt, D. L., and Goetze, C. (1974) Low-stress high-temperature creep in olivine single crystals. *J. Geophys. Res.,* **79**, 2045.
Krause, D. C., and Watkins, N. D. (1970) North Atlantic crustal genesis in the vicinity of the Azores. *Geophys. J.,* **19**, 261.
Lake, P. (1931) Mountain and island arcs. *Geol. Mag.,* **68**, 34.
LePichon, Xavier (1968) Sea-floor spreading and continental drift. *J. Geophys. Res.,* **73**, 3661.
LePichon, Xavier, Franchetau, Jean, and Bonnin, Jean (1972) *Plate tectonics.* Elsevier, Amsterdam.
Lomnitz, C. (1956) Creep measurements in igneous rocks. *J. Geol.,* **64**, 473.
MacDonald, G. J. F. (1963) The deep structure of continents. *Rev. Geophys.* **1,** 587.
Maxwell, Arthur E., Von Herzen, Richard P., Jinghwa, Hsu K., Andrews, James E., Saito. Tsunemasa, Percival, Stephen F., Jr., Milow, E. Dean, and Boyce, Robert E. (1970) Deep sea drilling in the South Atlantic. *Science,* **168**, 1047.
McConnell, R. K., Jr. (1968) Viscosity of the mantle from relaxation time spectra of isostatic adjustment. *J. Geophys. Res.,* **73**, 7089.
McElhinny, M. W., and Embleton, B. J. J. (1974) Australian palaeomagnetism and the phanerozoic plate tectonics of eastern Gondwanaland. *Tectonophysics,* **22**, 1.
McKenzie, D. P. (1967) The viscosity of the mantle. *Geophys. J.,* **14**, 297.
McKenzie, D. P. (1967) Some remarks on heat flow and gravity anomalies. *J. Geophys. Res.,* **72**, 6261.
McKenzie, D. P. (1969) Speculations on the consequences and causes of plate motions. *Geophys. J.,* **18**, 1.
McKenzie, D. P. (1972a) Plate tectonics. In *The nature of the solid earth,* ed. Eugene C. Robertson. McGraw-Hill, New York.
McKenzie, Dan (1972b) Active tectonics of the Mediterranean region. *Geophys. J.* **30**, 109–185.
McKenzie, D. P., and Morgan, W. J. (1969) Evaluation of triple junctions. *Nature,* **224,** 125.
McKenzie, D. P., and Sclater, J. G. (1971) The evolution of the Indian Ocean since the Late Cretaceous. *Geophys. J.,* **24**, 437.
Menard, H. W., and Morgan, W. J. (1968) Changes in direction of sea-floor spreading. *Nature,* **219**, 463.
Morgan, W. J. (1968) Rises, trenches, great faults, and crustal blocks. *J. Geophys. Res.,* **73**, 1959.
Morgan, W. (1972) Plate motions and deep mantle convection. In *Studies in earth and space sciences,* Geol. Soc. Amer. Mem. **132** (Harry H. Hess Volume).
Nadai, A. (1963) *Theory of flow and fracture of solids.* McGraw-Hill, New York.
Niskanen, E. (1939) On the upheaval of land in Fennoscandia. *Ann. Acad. Sci. Fennicae, AIII,* **53**, 1.
O'Connell, R. J. (1971) Pleistocene glaciation and the viscosity of the lower mantle. *Geophys. J.,* **23**, 299.
Oliver, J., and Isacks, B. (1967) Deep earthquake zones, anomalous structures in the upper mantle, and the lithosphere. *J. Geophys. Res.,* **72**, 4259.

Orowan, E. (1964) Continental drift and the origin of mountains. *Science,* **146**, 1003.
Orowan, E. (1967) Seismic damping and creep in the mantle. *Geophys. J.,* **14**, 191.
Peltier, W. R. (1974) The impulse response of a Maxwell earth. *Rev. Geophys. Space Phys.,* **12**, 649.
Peltier, W. R., and Andrews, J. T. (1976) Glacial-isostatic adjustment—I. The forward problem. *Geophys. J.,* **46**, 605.
Roult, G. (1975) Attenuation of seismic waves of very low frequency. *Phys. Earth Plan. Int.,* **10**, 159–166.
Sammis, Charles G., Smith, John C., Schubert, Gerald, and Yuen, David A. (1977) Viscosity-depth profile of the earth's mantle: effects of polymorphic phase transitions. *J. Geophys. Res.,* **82**, 3747.
Schäfer, Karlheinz (1975) Horizontal and vertical crustal movements in Iceland. In *Recent crustal movements,* ed. N. Pavoni and R. Green. Elsevier, Amsterdam (also *Tectonophysics,* **29**).
Scheidegger, A. E. (1958) *Principles of geodynamics.* Springer, Berlin.
Smith, A. Gilbert, and Hallam, A. (1970) The fit of the southern continents. *Nature,* **225**, 139.
Sykes, L. R. (1967) Mechanism of earthquakes and nature of faulting on the mid-oceanic ridges. *J. Geophys. Res.,* **72**, 2131.
Takeuchi, H., and Uyeda, S. (1965) A possibility of present-day regional metamorphism. *Tectonophysics,* **2**, 59.
Tramontini, C., and Davies, D. (1969) A seismic refraction survey in the Red Sea. *Geophys. J.,* **17**, 225.
Walcott, R. I. (1970a) Flexural rigidity, thickness and viscosity of the lithosphere. *J. Geophys. Res.,* **75**, 3941.
Walcott, R. I. (1970b) Flexure of the lithosphere at Hawaii. *Tectonophysics,* **9**, 435.
Wegener, A. (1924) *The origin of continents and oceans.* E. P. Dutton and Co., New York.
Wilson, J. T. (1963) A possible origin for the Hawaiian islands. *Can. J. Phys.,* **41**, 863.
Wilson, J. T. (1965a) A new class of faults and their bearing on continental drift. *Nature,* **207**, 343.
Wilson, J. T. (1965b) Evidence from ocean islands suggesting movement in the earth. *Phil. Trans. Roy. Soc. Lond.,* **258**, 145.
Wilson, J. T. (1965c) Submarine fracture sones, aseismic ridges and the International Council of Scientific Unions line: Proposed western margin of the East Pacific Ridge. *Nature,* **207**, 907.
Wilson, J. T., (1966) Did the Atlantic close and then re-open? *Nature,* **211**, 676.
Wilson, J. T. (1973) Mantle plumes and plate motion. *Tectonophysics,* **19**, 149.

APPENDIX A: SPHERICAL HARMONICS

An appreciation of the important properties of spherical harmonics is virtually essential in working with the earth's gravitational and magnetic fields, and with the free oscillations. We will show first how they arise in connection with problems involving potential fields or the wave equation, then examine such applications as expressing a measured quantity as a sum of spherical harmonics.

A.1 LAPLACE'S EQUATION AND THE WAVE EQUATION IN SPHERICAL COORDINATES

With spherical polar coordinates as in Figure A.1, an elementary cube has edges dr, $rd\theta$, and $r \sin\theta \, d\lambda$. The components of a force field which is derivable from a scalar potential U are given by the spatial gradient of U in the directions of increasing r, θ, and λ; they are therefore $\partial U/\partial r$, $(1/r)\,(\partial U/\partial\theta)$, and $[1/(r \sin\theta)]\,(\partial U/\partial\lambda)$. The fundamental equation of potential theory, Laplace's equation, is essentially a statement that the net flux of a force field across a closed surface surrounding no attracting matter is zero. If we write the expression for the flux (i.e., force times area), entering the cube across the three surfaces meeting at P, then, by taking partial derivatives of the force components, write down the flux leaving the other three faces of the cube (the procedure is similar to that employed in Section 2.4 for the stress), and equate the net sum to zero, we obtain

$$\frac{1}{r^2}\frac{\partial}{\partial r}\left(r^2\frac{\partial U}{\partial r}\right)+\frac{1}{r^2 \sin\theta}\frac{\partial}{\partial\theta}\left(\sin\theta\frac{\partial U}{\partial\theta}\right)+\frac{1}{r^2\sin^2\theta}\frac{\partial^2 U}{\partial\lambda^2}=0 \qquad \text{A.1}$$

This is Laplace's equation in spherical coordinates, corresponding to the Cartesian form

$$\frac{\partial^2 U}{\partial x^2}+\frac{\partial^2 U}{\partial y^2}+\frac{\partial^2 U}{\partial z^2}=0; \qquad \text{or} \qquad \nabla^2 U=0$$

In dynamical problems, the corresponding wave equation is

$$(\nabla^2+h^2)u=0$$

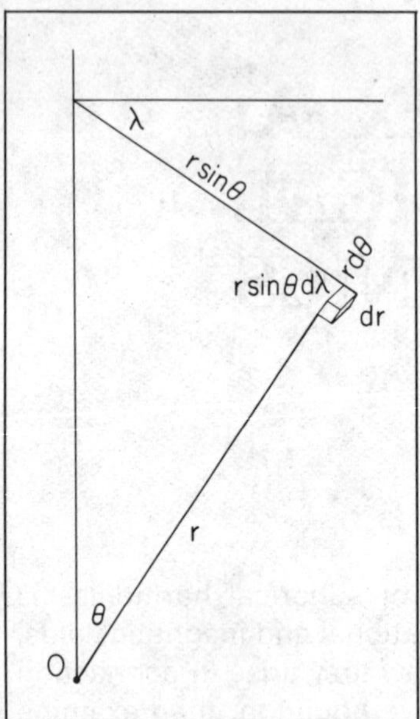

Figure A.1 Definition of a volume element in spherical polar coordinates.

It transforms to

$$\frac{1}{r^2}\frac{\partial}{\partial r}\left(r^2\frac{\partial u}{\partial r}\right)+\frac{1}{r^2\sin\theta}\frac{\partial}{\partial\theta}\left(\sin\theta\frac{\partial u}{\partial\theta}\right)+\frac{1}{r^2\sin^2\theta}\frac{\partial^2 u}{\partial\lambda^2}+h^2u=0$$

In either case, the procedure in obtaining solutions is to separate variables, writing, for example,

$$U=R(r)\,P(\theta)\,Q(\lambda)$$

To demonstrate the method, we take a simple form for $R(r)$:

$$R(r)=r^l$$

where l is an integer. Then, substituting in A.1, the equation to be satisfied is

$$l(l+1)+\frac{1}{\sin\theta\,P(\theta)}\frac{\partial}{\partial\theta}\left(\sin\theta\frac{\partial U}{\partial\theta}\right)+\frac{1}{\sin^2\theta\,Q(\lambda)}\frac{\partial^2 Q(\lambda)}{\partial\lambda^2}=0 \qquad \text{A.2}$$

Only the third term contains λ; therefore

$$\frac{1}{Q(\lambda)}\frac{\partial^2 Q(\lambda)}{\partial\lambda^2}=\text{constant, or}$$

$$Q(\lambda)=A\cos m\lambda+B\sin m\lambda$$

where m is an integer and A and B are constants.

Substituting in A.2, we find that $P(\theta)$ must satisfy

$$\frac{1}{\sin\theta}\frac{d}{d\theta}\left(\sin\theta\frac{dP(\theta)}{d\theta}\right)+\left[l(l+1)-\frac{m^2}{\sin^2\theta}\right]P(\theta)=0$$

This is known as Legendre's associated equation, Legendre having carried out pioneering studies on the theory of attraction. The solutions for $P(\theta)$ are polynomials in $\cos\theta$, involving the integers l and m. For $m \leqslant l$, the general term is written

$$P_l^m(\cos\theta)=\frac{(1-\cos^2\theta)^{m/2}}{2^l l!}\frac{d^{l+m}(\cos^2\theta-1)^l}{d(\cos\theta)^{l+m}}$$

In this case, the solution for U is

$$U=r^l P_l^m(\cos\theta)\,[A\cos m\lambda+B\sin m\lambda] \qquad \text{A.3}$$

It is not difficult to show that a second solution is

$$U=r^{-(l+1)}P_l^m(\cos\theta)\,[A\cos m\lambda+B\sin m\lambda] \qquad \text{A.4}$$

The expressions on the right of equations A.3 and A.4 are known as solid spherical harmonics of degree l and order m.

In the case of the wave equation, a more general form of $R(r)$ is required. It is found that $R(r)$ must satisfy

$$r^2\frac{d^2R}{dr^2}+2r\frac{dR}{dr}+[h^2r^2-l(l+1)]\,R=0$$

known as Ricatti's equation. The general solution is

$$R_l(\mathrm{r})=r^l\left(\frac{1}{r}\frac{d}{dr}\right)^l\left[\frac{C_l\sin hr+(D_l\cos hr)}{r}\right] \qquad \text{A.5}$$

where C_l and D_l are constants.

In a particular problem, the choice of solutions A.3 or A.4, or of terms in C_l or D_l in A.5, depends on boundary conditions. A solution which is to remain finite at $r=0$, for example, requires the solution A.3, or $D_l=0$ in A.5. Conversely, the solution A.4, which approaches zero as $r\to\infty$, gives the potential outside a surface which completely encloses the attraction matter.

That portion of the solutions which is a function of θ and λ, that is

$$P_l^m(\cos\theta)\,[A\cos m\lambda+B\sin m\lambda]$$

is known as a surface spherical harmonic. Because any quantity defined over a spherical surface can be expressed as a sum of surface spherical harmonics, it is extremely important in geophysics to be able to visualize their behavior in terms of latitude and longitude. In the first place, if $m=0$, there is no dependence on longitude, and the dependence in latitude is given by the Legendre polynomials. These functions, also known as zonal harmonics, were introduced in Section 11.2, in the series expansion of $1/r$; they have the property of vanishing on l parallels of latitude, the solution changing sign at these parallels

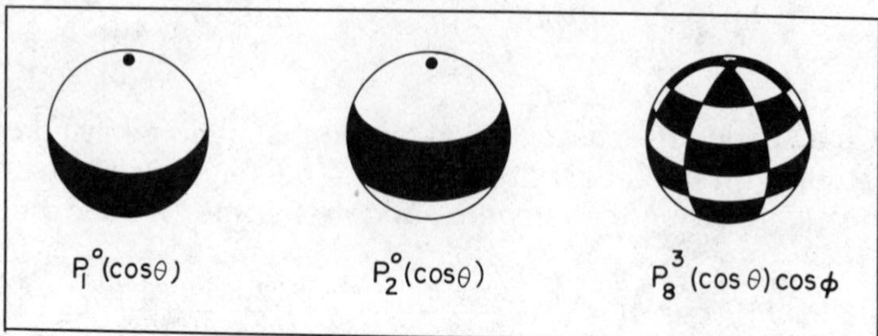

Figure A.2 Examples of surface spherical harmonics. P_1^0 (cos θ) and P_2^0 (cos θ) are dominant terms in the earth's magnetic and gravitational fields respectively. P_8^3 (cos θ cos ϕ) shows a tesseral harmonic.

(Figure A.2). If $0 < m < l$, the surface spherical harmonic vanishes on $2m$ meridians of longitude and on $(l - m)$ parallels of latitude. The surface of the sphere is broken up into regions, known as tesserae, between which the solution changes sign. Increasing values of m and l correspond to successively shorter wavelengths on the surface of the sphere. For $m = l$, the solution does not change sign with latitude, the surface of the sphere being divided into sectors bounded by meridians of longitude (cf. Figure 11.4).

The first few values of Legendre polynomials and associated polynomials are:

$$P_1\ (\cos\theta) = \cos\theta \qquad P_2^1\ (\cos\theta) = 3\sin\theta\cos\theta$$

$$P_1^1\ (\cos\theta) = \sin\theta \qquad P_2^2\ (\cos\theta) = 3\sin^2\theta$$

$$P_2\ (\cos\theta) = \frac{1}{2}\,(3\cos^2\theta - 1)$$

A.2 THE ORTHOGONALITY OF SPHERICAL HARMONICS

Two sets of functions are orthogonal in a given interval if the definite integral of the product of two members over that interval vanishes, except in the case of two identical members. This property, which is shared by spherical harmonics, is of great importance in the determination of coefficients to represent a given field.

We examine therefore the integral

$$\int_{-1}^{1} P_l^m(\mu)\ P_{l'}^m\ (\mu)\ d\mu \qquad \begin{array}{l} \mu = \cos\theta \\ l \neq l' \end{array}$$

The equations satisfied by the Legendre associated polynomials are

$$\frac{d}{d\mu}\left[(1-\mu^2)\ \frac{dP_l^m\ (\mu)}{d\mu}\right] + \left[l(l+1) - \frac{m}{1-\mu^2}\right] P_l^m\ (\mu) = 0$$

and

$$\frac{d}{d\mu}\left[(1-\mu^2)\frac{dP_{l'}^m(\mu)}{d\mu}\right]+\left[l'(l'+1)-\frac{m}{1-\mu^2}\right]P_{l'}^m(\mu)=0$$

Multiply the first equation by $P_{l'}^m(\mu)$, the second by $P_l^m(\mu)$, subtract, and integrate. Then

$$(l'-l)\ (l'+l+1)\int_{-1}^{1} P_l^m(\mu)\ P_{l'}^m(\mu)\ d\mu$$

$$=\int_{-1}^{1}\left[P_{l'}^m(\mu)\frac{d}{d\mu}\left\{(1-\mu^2)\frac{dP_l^m(\mu)}{d\mu}\right\}-P_l^m(\mu)\frac{d}{d\mu}\left\{(1-\mu^2)\frac{dP_{l'}^m}{d\mu}\right\}\right]d\mu$$

$$=\int_{-1}^{1}\frac{d}{d\mu}\left[(1-\mu^2)\left\{(P_{l'}^m(\mu)\frac{d}{d\mu}P_l^m(\mu)-P_l^m(\mu)\frac{d}{d\mu}P_{l'}^m(\mu)\right\}\right]d\mu$$

$$=0$$

Therefore, unless $l=l'$,

$$\int_{-1}^{1} P_l^m(\mu)\ P_{l'}^m(\mu)\ d\mu=0$$

It follows that the surface integrals, over θ and λ, of the product of two surface spherical harmonics of the same order vanish if $l \neq l'$. Because of the orthogonality of the elementary trigonometric functions, the surface integral of the product of two spherical harmonics of the same degree, but of different order, also vanishes. In other words,

$$\int_0^{2\pi}\int_{-1}^{1}\left[\begin{Bmatrix}\cos\\ \sin\end{Bmatrix}(m_1\lambda)\ P_l^{m_1}(\mu)\begin{Bmatrix}\cos\\ \sin\end{Bmatrix}(m_2\lambda)\ P_l^{m_2}(\mu)\right]d\lambda\, d\mu=0$$

unless $m_1=m_2$.

The above analysis does not determine the value of

$$\int_{-1}^{1}\left[P_l^m(\mu)\right]^2 d\mu$$

which is most easily found by using a relation between $P_l^m(\mu)$ and $P_l^0(\mu)$:

$$P_l^m(\mu)=(\mu^2-1)^{m/2}\frac{d^mP_l^0(\mu)}{d\mu^m}$$

Squaring, integrating, and using recurrence relations on the right, it is found that

$$\int_{-1}^{1}\left[p_l^m(\mu)\right]^2 d\mu=(-1)^l\frac{(l+m)!}{(l-m)!}\int_{-1}^{1}\left[P_l^0(\mu)\right]^2 d\mu$$

To investigate

$$\int_{-1}^{1}\left[P_l^0(\mu)\right]^2 d\mu$$

we use the expression

$$P_l^0(l)=\frac{1}{2^l\, l!}\frac{d^l(\mu^2-1)^l}{d\mu^l}$$

$$\therefore \qquad \int_{-1}^{1}\left[P_l^0(\mu)\right]^2 d\mu=\frac{1}{2^{2l}(l!)^2}\int_{-1}^{1}\left[\frac{d^l(\mu^2-1)^l}{d\mu^l}\right]^2 d\mu$$

Integration by parts l times on the right gives

$$\frac{(2l)!}{2^{2l}(l!)^2}\int_{-1}^{1}(1-\mu^2)^l\, d\mu$$

The substitution $\mu=(2x-1)$ reduces the integral to a standard form:

$$\int_0^1 x^l(1-x)^l\, dx=\frac{(l!)}{(2l+1)(2l)!}$$

Finally

$$\int_{-1}^{1}\left[P_l^0(\mu)\right]^2 d\mu=\frac{2}{(2l+1)}$$

A.3 EXPANSION OF A FUNCTION DEFINED OVER A SPHERE IN TERMS OF SPHERICAL HARMONICS

Let the function $f(\theta,\lambda)$ be written

$$f(\theta,\lambda)=\sum_{l=0}^{\infty}\left\{A_lP_l(\mu)+\sum_{m=1}^{l}(A_l^m\cos m\lambda+B_l^m\sin m\lambda)\,P_l^m(\mu)\right\}$$

Multiply the equation by $P_l(\mu)$ and integrate over the sphere. Then

$$\int_0^{2\pi}\int_{-1}^{1}P_l(\mu)\,f(\theta,\lambda)\,d\theta\,d\lambda=\frac{2A_l}{2l+1}\int_0^{2\pi}d\lambda=\frac{4\pi A_l}{2l+1}$$

since all other terms on the right vanish by orthogonality.

$$\therefore \qquad A_l=\frac{(2l+1)}{4\pi}\int_0^{2\pi}\int_{-1}^{1}f(\theta,\lambda)\,P_l(\mu)\,d\mu\,d\lambda$$

To determine A_l^m multiply by $\cos m\lambda P_l^m(\mu)$ and integrate. Then

$$\int_0^{2\pi}\int_{-1}^{1} P_l^m(\mu)\cos m\lambda\, f(\theta,\lambda)\,d\lambda\,d\theta = \frac{(l+m)\,!^2}{(l+m)\,!\,(2l+1)} A_l^m \int_0^{2\pi} \cos^2 m\lambda\, d\lambda$$

as before, and all other terms on the right vanish.

$$\therefore \qquad A_l^m = \frac{(2l+1)\,(l-m)\,!}{2\pi\,(l+m)\,!}\int_0^{2\pi}\int_{-1}^{1} f(\theta,\lambda)\,P_l^m(\mu)\cos m\lambda\,d\lambda\,d\mu$$

and similarly for B_l^m, with $\sin m\lambda$ replacing $\cos m\lambda$.

The expressions for the coefficients are normally evaluated by computer, using values of the function $f(\theta, \lambda)$ assigned to small elements of the spherical surface.

A.4 NORMALIZATION OF SPHERICAL HARMONICS

The mean value of the squares of the surface harmonics corresponding to $P_l^m(\mu)$ over a sphere are of very different magnitude. Using the results from pages 439–440, the mean square is:

$$\frac{1}{4\pi}\int_{-1}^{1}\int_0^{2\pi}\left[P_l^m(\mu)\cos m\lambda\right]^2 d\mu d\lambda = \frac{(l+m)\,!}{2(2l+1)(l-m)\,!}$$

For a given l, the variation in the value of the mean square with m is very rapid. As an example, the ratio for $P_4^1(\mu)$ and $P_4^4(\mu)$ is $\frac{5\,!}{3\,!}:\frac{8\,!}{1} = 1:2016$. A consequence is that the numerical coefficients (A_l, A_l^m, B_l^m) do not by themselves realistically convey the relative importance of various terms. For this reason, various workers have defined related functions to replace the Legendre associated polynomials, in such a way that the mean square values over the sphere are more nearly normal.

Schmidt (1889), in an analysis of the earth's magnetic field, introduced partially normalized harmonics, given by $P_l(\mu)$ when $m = 0$ and

$$\frac{2(l-m)\,!}{(l+m)\,!} P_l^m(\mu) \qquad m > 0$$

The Schmidt functions are defined so that the mean square value of the corresponding surface spherical harmonic over the sphere, for any m, is $\frac{1}{2l+1}$, the same as for $P_l(\mu)$. They have been extensively used in geomagnetism.

A function defined as

$$2(2l+1)^{1/2}\frac{(l-m)\,!}{(l+m)\,!} P_l^m(\mu)$$

is known as a fully normalized harmonic. The mean square value of the corresponding surface spherical harmonic over a sphere, for all l and all m, is unity.

It is most important, when considering the numerical coefficients in a specific case, to know whether unnormalized, partly normalized, or completely normalized polynomials have been used. In the case of the earth's gravitational field, the symbol J_l^m is used for the coefficients relating to unnormalized functions and C_l^m, and S_l^m for those related to normalized ones.

A.5 POTENTIAL OF A SPHERICAL SHELL OF DENSITY EXPRESSED IN SPHERICAL HARMONICS

Let the surface density σ be expanded as a series of surface spherical harmonics:

$$\sigma(\theta, \lambda) = \sum_{l=0}^{\infty} Z_l(\theta, \lambda)$$

where $Z_l(\theta, \lambda)$ is given by

$$Z_l(\theta, \lambda) = C_l P_l(\mu) + \sum_{m=1}^{l} \left\{ A_l^m \cos m\lambda + B_l^m \sin m\lambda \right\} P_l^m(\mu)$$

At an external point on the polar axis (Fig. A.3)

$$U = G \int_0^{2\pi} \int_{-1}^{1} \frac{\sigma a^2 \, d\mu \, d\lambda}{R}$$

The term $1/R$ can be expanded, as in Section 11.2, as

$$\frac{1}{R} = \frac{1}{r} P_0(\mu) + \frac{a}{r} P_1(\mu) + \ldots$$

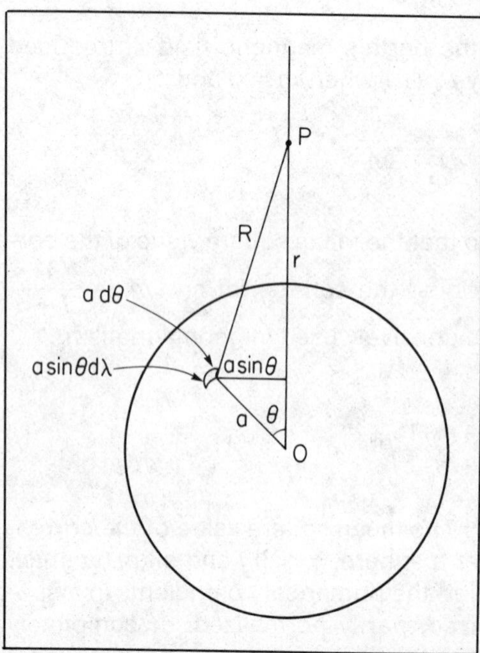

Figure A.3 Attraction of a spherical shell.

so that

$$U = G \sum_{l=0}^{\infty} \frac{a^{l+2}}{r^{l+1}} \int_0^{2\pi} \int_{-1}^{1} \sigma\, P_l\,(\mu)\; d\mu\; d\lambda$$

When the series for σ is inserted and the integration carried out term by term, a general term of the series involves

$$\int_0^{2\pi} \int_{-1}^{1} \left[C_l P_l\,(\mu) + \sum_{m=1}^{l} \left\{ A_l^m \cos m\lambda + B_l^m \sin m\lambda \right\} P_l^m\,(\mu) \right] P_l\,(\mu)\; d\lambda\; d\mu$$

By orthogonality, this equals $\frac{4\pi}{2l+1}$ C_l. But, since $P_l^m\,(\mu)$ vanishes at $\mu = 1$ for all m, C_l is the value of $Z_l\,(\theta, \lambda)$ at the polar axis. Therefore

$$U_{(\text{axis})} = 4\pi\; G \sum_{l=0}^{\infty} \frac{1}{2l+1} \frac{a^{l+2}}{r^{l+1}} Z_{l(\text{axis})}$$

It follows from the rules for transforming spherical harmonics to other axes that we may write, in general:

$$U(\theta, \lambda) = 4\pi G \sum_{l=0}^{\infty} \frac{1}{2l+1} \frac{a^{l+2}}{r^{l+1}} Z_l\,(\theta, \lambda)$$

It must follow that each coefficient in the expansion for the potential is equal to the corresponding coefficient in the expansion for the density multiplied by

$$\frac{4\pi G}{2\,l+1} \cdot \frac{a^{l+2}}{r^{l+1}}$$

For an internal point

$$U(\theta, \lambda) = 4\pi G \sum_{l=0}^{\infty} \frac{1}{2l+1} \frac{r^l}{a^{l-1}} Z_l\,(\theta, \lambda)$$

The expression for the potential at an external point is used in Section 12.3 in connection with the direct interpretation of gravity anomalies.

A.6 OTHER FORMS FOR P_l^m (cos θ)

There are many equivalent expressions for the Legendre polynomials, as references to specialized works (MacRobert, 1947; Hobson, 1931) will show. Some expressions are valid only for large values of l, and are known as *asymptotic expansions.* One of these has already been mentioned (Section 8.2) in connection with free oscillations. The form of it may be established as follows:

The equation satisfied by P_l^m (cos θ) is

$$\frac{d^2 P_l^m\,(\cos\theta)}{d^2\theta} + \cot\theta \frac{dP_l^m\,(\cos\theta)}{d\theta} - m^2 \operatorname{cosec}^2\theta\; P_l^m\,(\cos\theta) + l(l+1)\; P_l^m\,(\cos\theta) = 0$$

Introduce a new variable z defined by

$$P_l^m\ (\cos\theta) = (\sin^{-1/2}\theta)\ z$$

Then the equation for z is readily found to be

$$\frac{d^2z}{d\theta^2} + \left[(l+\tfrac{1}{2})^2 - (m^2 - \tfrac{1}{4})\ \mathrm{cosec}^2\,\theta\right] z = 0$$

When l is large, and m is zero or small, this reduces to

$$\frac{d^2z}{d\theta^2} + (l+\tfrac{1}{2})^2\ z = 0$$

from which

$$z = e^{\pm(l+\frac{1}{2})\,i\,\theta}$$

Under these conditions

$$P_l^m\ (\cos\theta) \propto \sin^{-\frac{1}{2}}\theta\, e^{\pm(l+\frac{1}{2})\,i\,\theta}$$

This is the relationship that allows a standing wave defined by a spherical harmonic of degree l to be associated with travelling wave of length determined by $(l+\frac{1}{2})$.

The determination of the constants of proportionality requires a more detailed investigation of the properties of $P_l^m\ (\cos\theta)$. The result is

$$P_l^m\ (\cos\theta) = \frac{l\,!}{(l-m)\,!}\left(\frac{2}{\pi(l+\frac{1}{2})\ \sin\theta}\right)^{1/2} \cos\left[(l+\tfrac{1}{2})\ \theta - \frac{m\pi}{2} - \frac{\pi}{4}\right]$$

and the error in this approximation decreases with increasing l as $l^{-3/2}$.

REFERENCES

Hobson, E.W. (1931) *The theory of spherical and ellipsoidal harmonics.* Cambridge University Press, Cambridge.

MacRobert, T.M. (1947) *Spherical harmonics* (2nd ed.). Methuen, London.

Schmidt, A. (1889) Mathematische Entwickelungen zier allgemeine Theorie des Erdmagnetismus. *Archiv. d. deutsche Seewarte* **12**, 3.

APPENDIX B: ROTATIONAL DYNAMICS

Qualitative aspects of the earth's rotation were described in Section 1. It is desirable to trace the development in more quantitative terms, in order to show how such quantities as the moment of inertia of the earth are determined. In the most straightforward analysis, the earth is taken as a rigid body, subject to a disturbing torque due to the attraction of the sun alone. This torque arises from the non-spherical shape of the earth and it is first necessary to express it in terms of the principal moments of inertia of the earth. This is most conveniently done by considering the potential energy of the sun-earth system.

We begin with an expression for the potential of the earth at an external point, in a manner very similar to that employed in Section 11.2. Using Figure 11.1, but letting the external point P be in the x-axis, we have, as before

$$U = G\int \frac{dm}{\rho} = \frac{G}{r'}\int dm\left(1 - \frac{2r'x - r^2}{r'^2}\right)^{-1/2} \qquad \text{B.1}$$

since $\rho^2 = r'^2 + r^2 - 2r'x$, and the element of mass is at (x, y, z).

To terms of order $\frac{1}{r'^2}$

$$U = \frac{G}{r'}\int dm\left(1 + \frac{x}{r'} + \frac{3x^2 - r^2}{2r'^2}\right) \qquad \text{B.2}$$

The first term integrates to give the total mass, while the second vanishes upon integration, since the origin is at the mass center of the earth. The third can be expressed in terms of principal moments of inertia, since

$$\int r^2 dm = \tfrac{1}{2}(A + B + C)$$

and

$$\int x^2 dm = (r^2 - y^2 - z^2)\, dm = \tfrac{1}{2}\,(A + B + C) - I \qquad \text{B.3}$$

where I is the moment of inertia about the line OP.

$$U = G\left[\frac{M}{r'} + \frac{A + B + C - 3I}{2r'^3}\right] \qquad \text{B.4}$$

This expression is known as McCullagh's formula.

To obtain the potential energy of the sun-earth system, we must write the expression above for the earth potential at every point of the sun, multiply by an element of mass of the sun, and sum over the sun. An identical procedure to that used above permits the latter summation to be expressed in terms of the principal moments of inertia of the sun. The potential energy of the system is then given by the symmetrical expression

$$V = G\left[\frac{MM'}{R} + \frac{M(A' + B' + C' - 3I')}{2R^3} + \frac{M'(A + B + C - 3I)}{2R^3}\right] \qquad \text{B.5}$$

where primed quantities refer to the sun, I and I' are moments of inertia about the line joining the sun-earth centers, and R is the length of this line.

This result is general. Turning to the specific problem of the earth's rotation, we take axes as shown in Figure B.1 (A). The origin O is the mass center of the earth; axes OX, OY, and OZ have fixed directions in space, with the plane XY coinciding with the plane of the sun's orbit. (The choice of left-handed axes in this case makes it somewhat easier to employ astronomical conventions). Axes OA, OB, and OC are moving axes, with OC the axis of figure of the earth and OA and OB in the equatorial plane. OA is chosen so that A, C, and Z are in the same plane. The moving axes rotate, to the extent that the direction of the earth's axis of rotation varies in space.

We are assuming that the earth has symmetry about OC, so that the principal moments A and B are equal. The attraction of the sun on the equatorial bulge produces torques T_1 and T_2 about OA and OB, but not about OC. The equations of motion are obtained by equating the rates of change of angular momentum to the torques. Reference to any standard work on dynamics shows that these are:

$$\left.\begin{aligned} A\frac{d\omega_1}{dt} - A\omega_2\omega_3' + C\omega_3\omega_2' &= T_1 \\ A\frac{d\omega_2}{dt} - C\omega_3\omega_1' + A\omega_1\omega_3' &= T_2 \end{aligned}\right\} \qquad \text{B.6}$$

$$C\frac{d\omega_3}{dt} = 0$$

in which ω_1, ω_2, and ω_3 are the angular velocities of the earth relative to the moving axes, and ω_1', ω_2', and ω_3' are the instantaneous angular velocities of the moving axes about themselves. Because of the fixing of A and B in the equatorial plane, $\omega_1' = \omega_1$ and $\omega_2' = \omega_2$. The angular velocity ω_3' is of the same order, since if $\omega_1 = \omega_2 = 0$, the direction of the axis of rotation OC is fixed, and $\omega_3' = 0$. Neglecting the squares and products of small quantities, we have

$$\left.\begin{aligned} A\frac{d\omega_1}{dt} + C\omega_3\omega_2 &= T_1 \\ A\frac{d\omega_2}{dt} - C\omega_3\omega_1 &= T_2 \end{aligned}\right\} \qquad \text{B.7}$$

in which ω_3 equals a constant, as shown by the third equation of B.6.

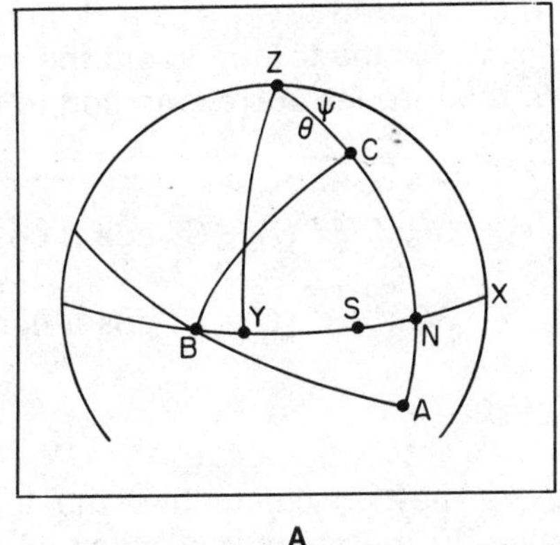

A

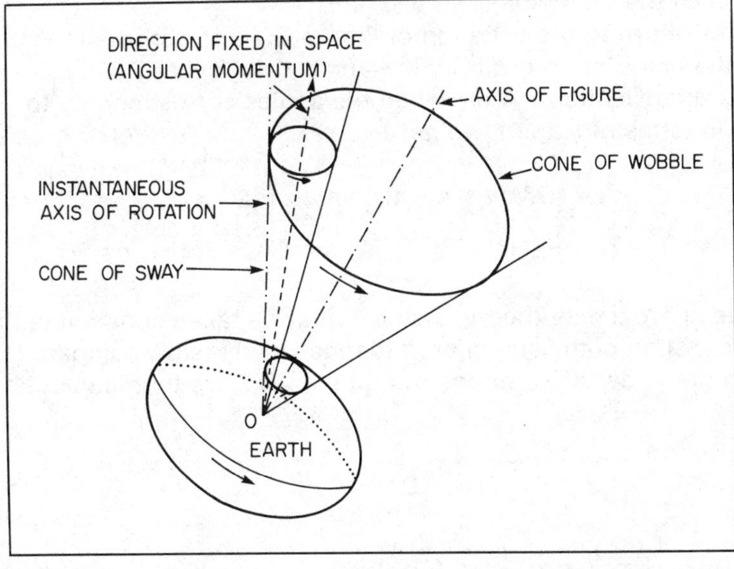

B

Figure B.1 (A) Geometry of the earth's precession. XYZ are fixed points on the celestial sphere with XY taken as the plane of orbit on the sun, S. Point C is the pole of rotation, which moves around Z at the angle θ.

(B) Motion of the earth's axis of rotation, relative to the axis of figure and to a direction fixed in space.

The torques are now to be obtained from the expression for the potential energy of the sun-earth system, Equation B.5. Any component of torque is given by $dV/d\phi$ where ϕ is the angle between the plane containing the axis concerned and the sun-earth line, and the equatorial plane.

In Equation B.5, only I varies appreciably as ϕ varies; therefore any component is given by

$$T = -\frac{3}{2}\frac{GM'}{R^3}\left(\frac{dI}{d\phi}\right) \qquad \text{B.8}$$

If α, β, and γ are the direction angles of the earth-sun line relative to the principal axes of inertia of the earth, a theorem in moments of inertia gives

$$I = A\cos^2\alpha + B\cos\ \beta + C\cos^2\gamma$$

and since $\cos\gamma = \sin\beta\sin\phi$, and $\cos\alpha = \sin\beta\cos\phi$, $dI/d\phi$ is of the form $-2(A-C)\sin^2\beta\sin\phi\cos\phi$ (for the torque about the axis corresponding to B). In the case where $A = B$, the torques corresponding to the three principal axes in the earth are

$$\left.\begin{aligned} T_A &= -3GM'/R^3\,(A-C)\cos\beta\cos\gamma \\ T_B &= -3GM'/R^3\,(C-A)\cos\alpha\cos\gamma \end{aligned}\right\} \qquad \text{B.9}$$

$$T_C = 0$$

Also, since $A = B$, we may, in calculating the torques, replace the two hypothetical principal axes in the equatorial plane with the axes OA and OB through which the earth spins.

We now return to the situation of Figure B.1 and assume the sun to be located, in the plane of its orbit, by the angle l, such that $BS = \pi - l$ (this is an astronomical convention). Then, when the angles corresponding to α, β, and γ are found in terms of θ and l, we have

$$\left.\begin{aligned} T_1 &= \tfrac{3}{2}GM'/R^3(A-C)\sin\theta\sin 2l \\ T_2 &= -\tfrac{3}{2}GM'/R^3(C-A)\sin\theta\cos\theta(1-\cos 2l) \end{aligned}\right\} \qquad \text{B.10}$$

For the approximate theory, T_1 and T_2 may be taken constant in time, since the relative motion of the sun in orbit (defined by l) is slow compared to the rate of rotation of the earth. Solutions of Equation B.7 are then immediately found to be

$$\omega_1 = -\frac{T_2}{C\omega_3} \qquad \omega_2 = \frac{T_1}{C\omega_3} \qquad \text{B.11}$$

We may visualize the motion of the earth more readily in terms of the change with time of the angles ψ and θ, given by

$$\omega_1 = \sin\theta\, d\psi/dt, \qquad \omega_2 = d\theta/dt \qquad \text{B.12}$$

or

$$\left.\begin{aligned} \frac{d\theta}{dt} &= -3\,GM'/2\omega_3\,R^3 \cdot \frac{(C-A)}{C}\sin\theta\sin 2l \\ \frac{d\psi}{dt} &= -3GM'/2\omega_3 R^3 \cdot \frac{(C-A)}{C}\cos\theta(1-\cos 2l) \end{aligned}\right\} \qquad \text{B.13}$$

The expressions in Equations B.13 have a well-known interpretation, for the term in $d\psi/dt$ which is independent of l represents a uniform *precession* about the fixed axis OZ, at a mean angular distance θ. The terms in l infer a small elliptical motion of the axis of rotation, which is the forced nutation (in the present case, that part of the forced nutation excited by the sun).

Astronomical observations have fixed the mean value of θ at approximately 23.5° and the period of precession at 26,000 years. When the simplified theory above is extended to allow for the effect of the moon, the value $(C-A)/C$, known as the precessional constant H, can be determined in terms of observed quantities. The accepted value is

$$H = \frac{C - A}{C} = 0.00327293 \pm 0.00000075 \qquad \text{B.14}$$

We note that the precession does not yield C and A individually, but in Chapter 11 we derived a second relation between A and C, which is involved in the figure of the earth and in the expression for the earth's potential field. The determination of the latter has been greatly improved by satellite observations, and the simultaneous solution of two expressions for A and C yields

$$C = 0.3309\ MR_e^2 \qquad \text{B.15}$$

where R_e is the equatorial radius.

For a uniform sphere, the moment of inertia about a diameter is $0.400\ MR^2$, so that the value of C provides further proof of a concentration of mass near the earth's center.

To return to Equations B.7, it will be noticed that the complementary function, representing free motion, has not yet been included. This involves adding the solution of

$$\left.\begin{aligned} A\ d\omega_1/dt + C\omega_3\omega_2 &= 0 \\ A\ D\omega_2/dt - C\omega_3\omega_1 &= 0, \end{aligned}\right\} \qquad \text{B.16}$$

which is easily found to be

$$\omega_1 = K_1 \sin\left(\frac{C\omega_3}{A}t + K_2\right) \quad \omega_2 = -K_1 \cos\left(\frac{C\omega_3}{A}t + K_2\right) \qquad \text{B.17}$$

where K_1 and K_2 are constants. The motion is a *"wobble,"* corresponding to that observed when a disc is spun about an axis making a small angle with the axis of figure.

The addition of the small components ω_1 and ω_2 to ω_3 to form the instantaneous vector angular velocity (Fig. B.1(B)) causes the latter to describe a cone about OC, which is the axis of figure. (It will be recalled that ω_1, ω_2, and ω_3 are all relative to the moving axes, of which OC is constrained to be the axis of figure.) The period required for the axis to describe the cone is $C/(C-A)$ days, or about 305 days. The resulting motion is known as the wobble of the earth on its rotation axis. The components ω_1 and ω_2 also contribute to motion relative to directions fixed in space, but because the angular velocities are about instantaneous axes which are themselves in motion, and because the relative motions have the symmetrical form shown by Equation B.16 there is a cancellation effect. The instantaneous axis of rotation does describe a cone about a direction fixed in space, but the ratio of the angle of this cone to the cone of wobble is only $(C-A)/C$, or about 1/305. Since we are here considering free motion, with the external torques set equal to zero, the "direction fixed in space" is the direction of the angular momentum vector, whose direction is essentially fixed for periods of a year or so. The period with which the instantaneous axis of rotation describes this very small cone is about one sidereal day, and the resulting motion is known as *sway.* Observationally, it produces the same effect as precession (Figure 1.5), altering the co-declination of a star.

All of the above treatment is for a rigid earth. In the case of the real earth, several modifications to the theory have already been noted. First, the period of the free wobble is increased to about 14 months. Second, there are possible effects of the liquid core on a number of aspects of rotational dynamics, including wobble and the earth's obliquity (Rochester, 1975, 1976; Rochester et al., 1974). Third, redistribution of mass within the earth, for example by tectonic activity, could be expected to cause the axis of greatest inertia to move relative to the body of the earth. This would result in "true polar wander," but as we saw in Chapter 21, the palaeomagnetic evidence is that this has been small, at least in the past 150 million years. If polar wander of greater amplitude took place at any earlier geological time, it could have contributed to lithospheric fracturing, since the rigid lithosphere would have to adjust to a new axis of figure. Thus, tectonics and the history of the earth's rotation may be closely linked. Finally, the rotation speed, ω_3, is not constant, but suffers relatively short-period variations due to mass redistributions and shows a long-term decrease due to tidal friction (Chapter 14).

REFERENCES

Rochester, M.G. (1975) Chandler wobble and viscosity in the Earth's core, *Nature* **255**, 655; **257,** 828.

Rochester, M.G. (1976) The secular decrease of obliquity due to dissipative core-mantle coupling, *Geophys. J.* **46**, 109.

Rochester, M.G., Jensen, O.G. and Smylie, D.E. (1974) A search for the Earth's nearly diurnal free wobble, *Geophys. J.* **38**, 349.

APPENDIX C: SEISMOLOGICAL TOPICS

C.1 BOUNDARIES

In Chapter 3, the directions of new waves formed at a boundary were predicted on the basis of Huygens' principle, and the partition of amplitude was determined only for the case of a normally-incident P wave. The more complete treatment which follows confirms the directions and shows how amplitude may be determined in a more general case.

We take the x_1-x_2 plane to be the boundary between two media of different elastic properties and allow a *P* wave in the plane x_1-x_3 to be incident on it. (cf. Figure 3.2; the axis x_1 extends to the right and the axis x_3 downward from the point of incidence. A principle we may employ is that no particle motion perpendicular to the motion in the incident wave can be produced. Therefore, while *S* waves will be produced upon refraction and reflection, they must be polarized with particle motion in the x_1 - x_3 plane. We will further assume that no parameters vary in the direction x_2, perpendicular to that plane.

The procedure is to write expressions for the displacement and stress related to each of the five waves and to satisfy the continuity of these quantities across the boundary. Discontinuity in stress would imply infinite acceleration somewhere, while discontinuity in displacement would imply a "non-welded" boundary. Notice that, in general, both types of waves contribute to the displacement in any direction. For this reason, it is convenient to express the displacement in terms of two new functions, ϕ and $\boldsymbol{\psi}$, such that

$$\boldsymbol{u} = \nabla\phi + \nabla \times \boldsymbol{\psi}$$

that is, C.1.1

$$u_1 = \frac{\partial\phi}{\partial x_1} + \frac{\partial\psi_3}{\partial x_2} - \frac{\partial\psi_2}{\partial x_3}$$

and so forth. We will continue to assume that no particle motion is generated in the direction x_2, normal to the plane of the figure, and that no properties vary with x_2.

The functions ϕ and $\boldsymbol{\psi}$ are known, respectively, as scalar and vector potentials of the displacement.

Certain properties of these functions follow immediately. By substituting Equations C.1.1 in the expression for the dilatation θ, we have

$$\theta = \nabla^2 \phi \tag{C.1.2}$$

and by substitution in the equation of motion 2.4.4 we find

$$\nabla^2 \phi = \frac{1}{\alpha^2} \frac{\partial^2 \phi}{\partial t^2}$$
$$\nabla^2 \psi_i = \frac{1}{\beta^2} \frac{\partial^2 \psi_i}{\partial t^2} \tag{C.1.3}$$

In other words, disturbances in the potentials ϕ and $\boldsymbol{\psi}$ are propagated with velocities appropriate to P and S waves, respectively.

In our specific case, we try for solutions of the form

$$\phi = f(x_3) e^{ik(ct - x_1)}, \qquad \psi = g(x_3) e^{ik(ct - x_1)} \tag{C.1.4}$$

in which ψ may be considered a scalar, since only ψ_2 is required. (Why?)

Such solutions represent disturbances which travel along the boundary with an apparent velocity c. It may be questioned why the same exponent, $ik(ct - x_1)$, is chosen for ϕ and ψ; we shall presently write the displacements and stress in terms of these quantities, and we should find that the boundary conditions could not be satisfied for all points and all times if the exponents were not identical. Substitution in the wave Equation C.1.3 gives, for example,

$$\phi = A_1 \exp ik\left(ct + x_3\sqrt{\frac{c^2}{\alpha^2} - 1} - x_1\right)$$
$$+ A_2 \exp ik\left(ct - x_3\sqrt{\frac{c^2}{\alpha^2} - 1} - x_1\right) \tag{C.1.5}$$

The apparent velocity c and the true P wave velocity α must be related (Figs. 3.2 and C.1) by

$$\left(\frac{c^2}{\alpha^2} - 1\right)^{1/2} = \tan e = \tan e_0 \tag{C.1.6}$$

Similarly, the expression for ψ will involve a term

$$\left(\frac{c^2}{\beta^2} - 1\right)^{1/2} = \tan f \tag{C.1.7}$$

We may therefore write for the first medium

$$\phi = A_1 \exp\,[ik(ct - x_1 + ax_3)] + A_2 \exp\,[ik(ct - x_1 - ax_3)]$$
$$\psi = B_1 \exp\,[ik(ct - x_1 + bx_3)] + B_2 \exp\,[ik(ct - x_1 - bx_3)] \tag{C.1.8}$$

and for the second medium

$$\phi' = A' \exp\,[ik(ct - x_1 + a'x_3)]$$
$$\psi' = B' \exp\,[ik(ct - x_1 + b'x_3)] \tag{C.1.9}$$

where

$$a = \tan e$$

$$b = \tan f$$

$$a' = \tan e'$$

$$b' = \tan f'$$

The difference in form of Equations C.1.8 and C.1.9 expresses the fact that, in the second medium, there is propagation only away from the boundary in the x_3 direction; also, in our particular case, for incident P waves, constant B_1 is zero.

For the quantity c to be identical in all exponents, we must have

$$c = \frac{\alpha}{\cos e_0} = \frac{\alpha}{\cos e} = \frac{\beta}{\cos f} = \frac{\alpha'}{\cos e'} = \frac{\beta'}{\cos f'} \qquad \text{C.1.10}$$

We note first that the apparent velocity along the boundary, c, is greater than any of the true wave velocities, except when $e_0 = 1$. Secondly, Equation C.1.10 is equivalent to the generalized Snell's law, Equation 3.2.2, but the more rigorous procedure shows that the new wave directions are determined by the conditions on stress and displacement at the boundary.

To satisfy the boundary conditions on displacement and stress, it is necessary to write equations for u_1 and u_3, and for p_{31} and p_{33}, on each side of the boundary. The first pair of these follows by substitution of C.1.8 and C.1.9 into C.1.1. The second pair is obtained from

$$p_{13} = \mu\left(\frac{\partial u_3}{\partial x_1} + \frac{\partial u_1}{\partial x_3}\right) = \mu\left(\frac{2\partial^2\phi}{\partial x_1 \partial x_3} + \frac{\partial^2\psi}{\partial x_1^2} - \frac{\partial^2\psi}{\partial x_3^2}\right) \qquad \text{C.1.11}$$

$$p_{33} = \lambda\theta + 2\mu\frac{\partial u_3}{\partial x_3} = \lambda\nabla^2\phi + 2\mu\left(\frac{\partial^2\phi}{\partial x_3^2} + \frac{\partial^2\psi}{\partial x_1 \partial x_3}\right)$$

Writing the corresponding equations for media 1 and 2, and equating at $x_3 = 0$, leads immediately to

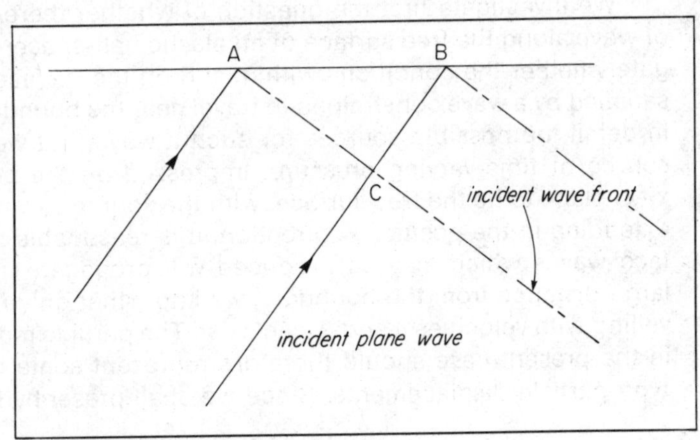

Figure C.1 Apparent velocity along a boundary: in the time the incident wavefront advances from C to B, the boundary disturbance travels from A to B.

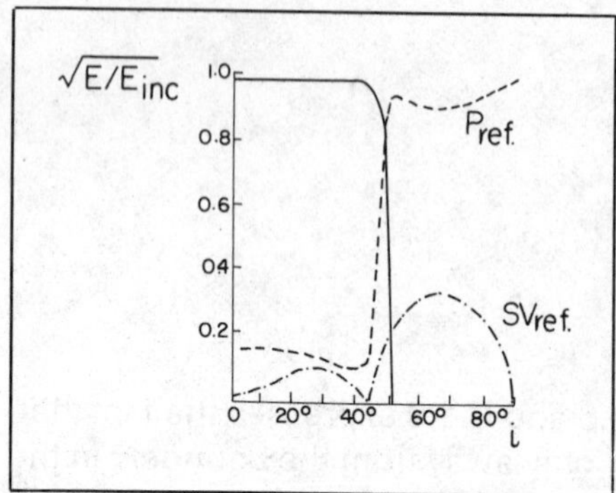

Figure C.2 Variation with angle of incidence of the strength of the transmitted wave (solid curve), reflected *P*, and reflected *S*, for a *P* wave incident against a faster medium ($\alpha_2/\alpha_1 = \beta_2/\beta_1 = 1.286$; $\rho_2/\rho_1 = 1.103$). Ordinates are plotted as the square root of the energy ratio. (After Gutenberg)

$$
\begin{aligned}
A_1 + A_2 - bB_2 &= A' + b'B' \\
a(A_1 - A_2) - B_2 &= a'A' - B' \\
\rho\beta^2[-(b^2-1)(A_1+A_2) - 2bB_2] &= \rho'\beta'^2[-(b'^2-1)A' + 2b'B'] \\
\rho\beta^2[2a(A_1 - A_2) + (b^2-1)B_2] &= \rho'\beta'^2[2a'A' + (b'^2-1)B']
\end{aligned}
\qquad \text{C.1.12}
$$

These equations permit determination of the amplitudes of the reflected and refracted waves in terms of A_1, the amplitude of the incident *P* wave. It is necessary, evidently, to adopt values of the velocities on both sides of the boundary and then to perform the calculation for different values of *a*, which is determined by the angle of incidence of the incoming wave. Solutions of equations of this type were given by Knott (1899) and by many later workers (e.g., Gutenberg, 1912; Ergin, 1952). We show an example for the case of an incident *P* wave for one particular combination of velocities in Figure C.3. We notice the small amplitude (the square root of the energy ratio, which is plotted, is equivalent to amplitude ratio) for normal incidence. Application of Equation 3.2.6 to the property ratios shown in the caption of Figure C.3 gives a normal incidence reflection coefficient of 0.17, in agreement with the diagram. The significance for experimental work is that reflections are much easier to observe at critical incidence than at normal incidence.

C.2 RAYLEIGH AND LOVE WAVES

We investigate first the question of whether there can exist a special type of wave along the free surface of an elastic half-space. The analysis will investigate whether the conditions within, and on the surface of, the medium can be satisfied by a wave constrained to travel near the boundary. We will not consider in detail the possible sources for such a wave, but we could imagine a linear source of time-varying pressure, impressed on the boundary. If we take the x_1-x_2 plane to be the free surface, with the source parallel to x_2, and the medium extending in the positive x_3 direction, it is reasonable to suppose that any surface waves which may be produced will propagate in the $\pm$ directions. At a large distance from the boundary, we know that only *P* and *S* type waves, travelling with velocities α and β, can exist. The particle motions near the boundary in the present case should therefore represent some combination of *P* and *S* type particle displacements. Since we shall presently have to establish conditions on the stresses at the boundary, it is a considerable simplification to express the displacements in a way that will lead to scalar, rather than vector, equations. This can be done in the same way as in Section C.1, writing

$$u_i = \nabla\phi + \nabla x \boldsymbol{\psi}$$

where ϕ and $\boldsymbol{\psi}$ are the scalar and vector potentials of displacement. It was also pointed out in Section C.1 that direct substitution confirmed that

$$\phi = \nabla^2\phi$$

and

$$\nabla^2\phi = \frac{1}{\alpha^2}\frac{\partial^2\phi}{\partial t^2}$$

C.2.1

$$\nabla^2\psi_i = \frac{1}{\beta^2}\frac{\partial^2\psi_i}{\partial t^2}$$

Now in the case we are considering, no quantities vary with x_2; therefore $\partial/\partial x_2 \equiv 0$. Also, it is only in the x_1-x_3 plane that P and S type particle displacements interact, and we shall have to express only u_1 and u_3 in terms of the potentials. Under these conditions, only ψ_2 is required and we proceed with purely scalar equations for ϕ and ψ. Solutions of equations C.2.1 which represent travelling harmonic waves in the x_1 direction with amplitude decreasing in the x_3 direction, are of the form

$$\phi = A \exp\left[ik\,(ct \pm ax_3 - x_1)\right]$$

where A and a are real and imaginary constants respectively, and k and c are wave number and velocity.

For Equation C.2.1 to be satisfied, we find

$$\phi = A \exp\left[ik(ct \pm \sqrt{c^2/\alpha^2 - 1}\; x_3 - x_1)\right] \qquad \text{C.2.2}$$
$$\psi = B \exp\left[ik(ct \pm \sqrt{c^2/\beta^2 - 1}\; x_3 - x_1)\right]$$

provided $\sqrt{c^2/\alpha^2 - 1}$ and $\sqrt{c^2/\beta^2 - 1}$ are imaginary quantities and the proper sign is chosen for the coefficient of x_3. Obviously, therefore

$$c < \beta < \alpha$$

in contrast to the nature of c, as an apparent velocity greater than the body wave velocities, found in the analysis of reflection and refraction.

The boundary conditions are the vanishing of normal and shear stress at $x_3 = 0$. We have

$$p_{33} = \lambda\nabla^2\phi + 2\mu\left(\frac{\partial^2\phi}{\partial x_3^2} + \frac{\partial^2\psi}{\partial x_1 \partial x_3}\right) = 0$$

C.2.3

$$p_{31} = \mu\left(2\frac{\partial^2\phi}{\partial x_1 \partial x_3} + \frac{\partial^2\psi}{\partial x_1^2} - \frac{\partial^2\psi}{\partial x_3^2}\right) = 0$$

At this point we may consider the possibility of displacements u_2 being propagated as a surface wave, although it is unlikely that the source proposed would generate this type of displacement. In any case, u_2 would result from S-type waves only and rather than express it in terms of potentials, we could seek a solution of the form

$$u_2 = C \exp\left[ik(ct \pm \sqrt{c^2/\beta^2 - 1}\, x_3) - x_1\right]$$

Note that only one sign of the $\pm$ pair is to be chosen, so that amplitude decreases with increasing x_3. Among the stress components which must vanish everywhere on the plane $x_3 = 0$ are p_{12} and p_{32} where

$$p_{12} = \mu\left(\frac{\partial u_1}{\partial x_2} + \frac{\partial u_2}{\partial x_1}\right)\ ;\ p_{32} = \mu\left(\frac{\partial u_3}{\partial x_2} + \frac{\partial u_2}{\partial x_3}\right)$$

But $\partial/\partial x_2 = 0$; therefore u_2 can only be a constant, which is inconsistent with its being propagated as a wave. We may therefore proceed, confident that all particle motion is in the x_1-x_3 plane. Substituting for ϕ and ψ from Equation C.2.2, and putting $x_3 = 0$, gives

$$\begin{aligned} (2 - c^2/\beta^2)\, A \pm 2\sqrt{c^2/\beta^2 - 1}\, B = 0 \\ \mp 2\sqrt{c^2/\alpha^2 - 1}\, A + (2 - c^2/\beta^2)\, B = 0 \end{aligned} \qquad \text{C.2.4}$$

If A and B are not to vanish

$$(2 - c^2/\beta^2)^2 = 4(1 - c^2/\alpha^2)^{1/2}(1 - c^2/\beta^2)^{1/2} \qquad \text{C.2.5}$$

or

$$c^2/\beta^2\left[\frac{c^6}{\beta^6} - 8c^4/\beta^4 + c^2\left(\frac{24}{\beta^2} - \frac{16}{\alpha^2}\right) - 16(1 - \beta^2/\alpha^2)\right] = 0$$

One value of c, such that $0 < c < \beta$, can always be found, for which the bracketted term in Equation C.2.5 vanishes. We may illustrate a special, but representative case, by taking $\lambda = \mu$ and $\alpha = \sqrt{3}\,\beta$ (a relationship which is not only convenient, but is approximated in the earth). Then

$$c^6/\beta^6 - 8c^4/\beta^4 + \frac{56}{3}\, c^2/\beta^2 - \frac{32}{3} = 0 \qquad \text{C.2.6}$$

which has roots

$$c^2/\beta^2 = 4,\quad 2 + 2/\sqrt{3},\quad 2 - 2/\sqrt{3}$$

of which only the last is acceptable.

$$\therefore \qquad c = 0.9194\beta \qquad \text{C.2.7}$$

The above development, proposed by Rayleigh (1885), shows the possibility of existence of a wave along a free surface. We note immediately one characteristic: c is independent of k, hence of wavelength or frequency. The nature of the particle motion can be shown by substituting Equation C.2.7 into C.2.2 and operating to obtain u_1 and u_3. It is found that

$$u_1 = C(e^{-0.8475kx_3} - 0.5773e^{-0.3933kx_3}) \sin k(ct - x_1)$$
$$u_3 = C(-0.8475e^{-0.8475kx_3} + 1.4679e^{-0.3933kx_3}) \cos k(ct - x_1) \qquad \text{C.2.8}$$

where C is a constant.

For small values of x_3, a particle describes an ellipse in the x_1-x_3 plane, the ellipse having its major axis vertical, and the motion being retrograde (i.e., particle motion opposed to propagation direction at the top of the ellipse).

We noted above that in the case of a uniform half-space, transverse motion could not be propagated. Love (1911) investigated the possibility of the transmission of a horizontally polarized shear wave through a surface layer. To follow his development, we again take the plane of propagation to be the x_1-x_3 plane, but assume that only the displacement u_2 is propagated. We let the x_2-x_3 plane be the base of the surface layer, take the free surface to be at $x_3 = +H$ and let primed quantities refer to the layer.

In the layer, we seek a solution of the form

$$u_2' = C' \exp\,[ik(-\sqrt{c^2/\beta'^2 - 1}\,x_3 + x_1 - ct)]$$
$$+ F' \exp\,[ik(\sqrt{c^2/\beta'^2 - 1}\,x_3 + x_1 - ct)] \qquad \text{C.2.9}$$

and in the substratum, of the form

$$u_2 = C \exp\,[ik(-\sqrt{c^2/\beta'^2 - 1}\,x_3 + x_1 - ct)] \qquad \text{C.2.10}$$

Equations C.2.9 and C.2.10 are set up so that u_2' and u_2 satisfy the appropriate wave equations for shear waves, and in such a way that u_2 vanishes with x_3 in the substratum, but u_2' is periodic in x_3 in the layer. For this to be the case, $\sqrt{c^2/\beta^2 - 1}$ must be imaginary.

The boundary conditions are the vanishing of stress at the free surface, and the continuity of displacement and stress at $x_3 = 0$. These require

$$C' \exp\,(-ik\sqrt{c^2/\beta'^2 - 1}\,H) = F' \exp\,(ik\sqrt{c^2/\beta'^2 - 1}\,H)$$
$$C = C' + F' \qquad \text{C.2.11}$$
$$\mu\sqrt{c^2/\beta^2 - 1}\,C = \mu'\sqrt{c^2/\beta'^2 - 1}\,(C' - F')$$

The first of Equations C.2.11 indicates the important difference from the case of free Rayleigh waves. Because the wave in the layer consists of two components, with amplitudes C' and F', travelling in opposite senses in the x_3-direction, it is possible to have the shear stress vanish at the free surface without requiring the displacement to vanish.

Eliminating C, C', and F', we find

$$\mu i\sqrt{c^2/\beta^2 - 1} + \mu'\sqrt{c^2/\beta'^2 - 1}\tan\,(k\sqrt{c^2/\beta'^2 - 1}\,H) = 0$$

or, in terms of real quantities,

$$\tan\,(k\sqrt{c^2/\beta'^2-1}\,H) = \mu\sqrt{1-c^2/\beta^2}/\,\mu'\sqrt{c^2/\beta'^2-1} \qquad \text{C.2.12}$$

Equation C.2.12 determines values of the velocity c which are compatible with the assumed conditions. For all restrictions on real and imaginary quantities, we must have

$$\beta' < c < \beta$$

C.3 THE HEAD WAVE

The theoretical treatment, as developed by Stoneley (1924), Cagniard (1939), Heelan (1953), Ewing, Jardetzky, and Press (1957) and Cerveny and Ravindra (1971), is somewhat involved, but an outline of it will show the methods that are applied. We consider, not a plane wave, but a spherical wave front propagating from a point source of compressional waves located a distance h above the boundary between two media (Fig. C.3). Initially, we assume that the source emits a single angular frequency ω, and we omit the time factor $e^{i\omega t}$ in the equations that follow. The problem as stated obviously has cylindrical symmetry about an axis through the source; we take the axis of z through this point, normal to the interface (the plane $z=0$), and we express all quantities in terms of cylindrical coordinates (r, z). Compressional waves spread out from the source, and some of these strike the interface; upon refraction, S waves will be generated, but since the incident particle motion is entirely in vertical planes, these waves will be polarized in vertical planes also. Particle motion is com-

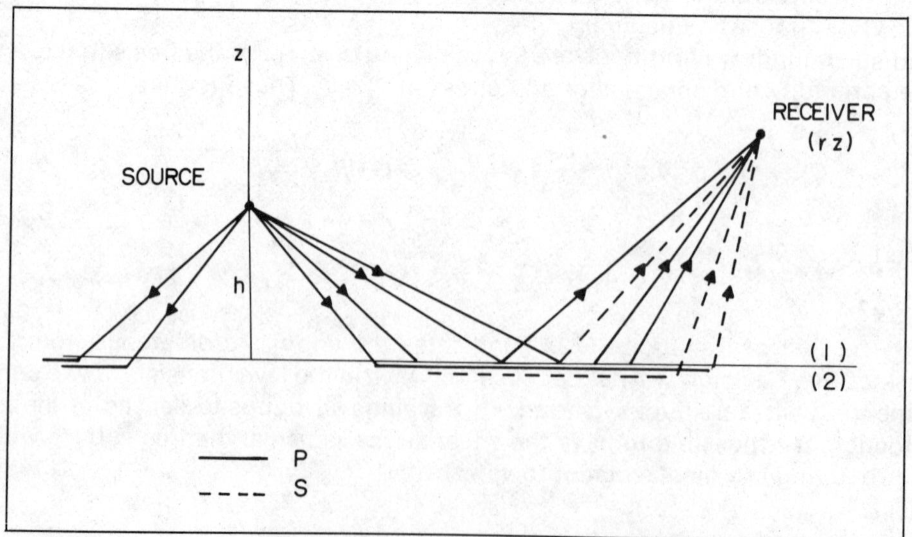

Figure C.3 Paths by which energy may travel from source to receiver, in the vicinity of an interface. Medium 2 has the higher seismic velocities. The points to observe are that spherical rather than plane waves are incident on the boundary, and energy may arrive by paths involving conversion from P to S waves upon either reflection or refraction.

pletely described in terms of components (q, w), in the r and z directions. These displacements can be derived from scalar potentials ϕ and ψ, through the relations

$$\left.\begin{aligned} q &= \partial\phi/\partial r + \partial^2\psi/\partial r\,\partial z \\ w &= \partial\phi/\partial z - \partial^2\psi/\partial r^2 - \frac{1}{r}\,\partial\psi/\partial r \end{aligned}\right\} \qquad \text{C.3.1}$$

and the potentials satisfy

$$\left.\begin{aligned} \nabla^2\phi &= 1/\alpha^2\ \partial^2\phi/\partial t^2 \\ \nabla^2\psi &= 1/\beta^2\ \partial^2\psi/\partial t^2 \end{aligned}\right\} \qquad \text{C.3.2}$$

However, the operator on the left side is now the Laplacian in cylindrical coordinates, which is given by

$$\nabla^2 = \partial^2/\partial r^2 + 1/r\ \partial/\partial r + \partial^2/\partial z^2 \qquad \text{C.3.3}$$

We seek solutions for the potentials, ϕ and ψ, at the location of an observer (r, z).

The equation to be satisfied by ϕ, for example, is then

$$\partial^2\phi/\partial r^2 + \frac{1}{r}\,\partial\phi/\partial r + \partial^2\phi/\partial z^2 + \omega^2/\alpha^2(\phi) = 0 \qquad \text{C.3.4}$$

We proceed in the standard way to separate variables, and seek a solution of the form $\phi = R(r)\ Z(z)$. In this case,

$$\frac{1}{R}\,\partial^2R/\partial r^2 + \partial R/\partial r + \frac{1}{Z}\,\partial^2Z/\partial z^2 + \omega^2/\alpha^2 = 0 \qquad \text{C.3.5}$$

and since the last two terms on the left are independent of r, we must have

$$1/Z\ \partial^2Z/\partial z^2 + \omega^2/\alpha^2 = \text{constant} = k^2 \text{ say.} \qquad \text{C.3.6}$$

$$\therefore\quad Z = e^{-(\omega^2/\alpha^2 - k^2)z} = e^{-\nu_1 z}$$

The function R then satisfies

$$\partial^2R/\partial r^2 + \frac{1}{r}\,\partial R/\partial r + kR = 0 \qquad \text{C.3.7}$$

which is Bessel's equation of zero order. The solution is a polynomial in (kr), known as the Bessel function, also of zero order, $J_0(kr)$.

The complete solution for ϕ is obtained by superimposing the results for all positive values of k, so that its form is

$$\phi = \int_0^\infty P(k)\,J_0(rk)e^{-\nu_1 z}\,dk \qquad \text{C.3.8}$$

where $P(k)$ is a coefficient to be determined by boundary conditions.

In our particular problem a slight modification is required for ϕ in the first medium because of the location of the source. There is a contribution to ϕ through a direct wave from source to observer, and this contribution must be symmetrical about the plane $z = h$. It is convenient to write, therefore,

$$\phi_1 = \int_0^\infty P(k) J_0(kr) e^{-\nu_1|z-h|}\, dk + \int_0^\infty Q_1(k) J_0(kr) e^{-\nu_1(z-h)} dk$$

and similarly

$$\psi_1 = \int_0^\infty S_1(k) J_0(kr) e^{-\nu_1'(z-h)}\, dk \qquad \text{C.3.9}$$

in the first medium, and

$$\begin{aligned} \phi_2 &= \int_0^\infty Q_2(k) J_0(kr) e^{\nu_2(z-h)}\, dk \\ \psi_2 &= \int_0^\infty S_2(k) J_0(kr) e^{\nu_2'(z-h)}\, dk \end{aligned} \qquad \text{C.3.10}$$

in the second medium. In these equations,

$$\nu_1 = (\omega^2/\alpha_1^2 - k^2)^{1/2}, \text{ as before,} \quad \text{and} \quad \nu_1' = (\omega^2/\beta_1^2 - k^2)^{1/2},$$

$$\nu_2 = (\omega^2/\alpha_2^2 - k_2)^{1/2}, \quad \nu_2' = (\omega^2/\beta_2^2 - k^2)^{1/2}$$

Also, investigation of the direct pulse shows that $P(k)$ must be given by k/ν_1. The boundary conditions are the continuity of displacement and of normal and shear stress at the interface. The displacements are obtained from Equation C.3.1 and the stresses from Equation C.3.2 (converted to cylindrical coordinates).

The conditions are:

$$\begin{aligned} \frac{\partial\phi_1}{\partial r} + \frac{\partial^2\psi_1}{\partial z\,\partial r} &= \frac{\partial\phi^2}{\partial r} + \frac{\partial^2\psi_2}{\partial z\,\partial r} \\ \frac{\partial\phi_1}{\partial z} - \frac{\partial^2\psi_1}{\partial r^2} - \frac{1}{r}\frac{\partial\psi_1}{\partial r} &= \frac{\partial\phi_2}{\partial z} - \frac{\partial^2\psi_1}{\partial r^2} - \frac{1}{r}\frac{\partial\psi_2}{\partial r} \\ \lambda_1\nabla^2\phi_1 + 2\mu_1\frac{\partial\omega_1}{\partial z} &= \lambda_2\nabla^2\phi_2 + 2\mu_2\frac{\partial\omega_2}{\partial z} \\ \mu_1\left(\frac{\partial q_1}{\partial z} + \frac{\partial\omega_1}{\partial r}\right) &= \mu_2\left(\frac{\partial q_2}{\partial z} + \frac{\partial\omega_2}{\partial r}\right) \end{aligned} \qquad \text{C.3.11}$$

When the expressions for ϕ and ψ from Equations C.3.9 and C.3.10 are substituted into Equations C.3.11, four equations for Q_1, S_1, Q_2, and S_2 are obtained. These solutions, and the time factor $e^{i\omega t}$, are then substituted into the expressions for ϕ and ω, and the integration over k is performed. In practice, the methods of complex integration are always employed, k being replaced by a complex variable. The integrands are found to contain one pole (zero of the denominator), and terms of the form $e^{-\nu_1(z-h)}$. The latter lead to "branch points" when any of ν_1, ν_1', ν_2, or ν_2' vanish (as each must, for some values of k). Contributions to the integrals come from the pole and from the cuts made, in the complex plane, around the branch points. Each contribution represents the existence of energy propagated from the source to the receiver, with ϕ_1 and ψ_1 yielding respectively the total *P*-type and *S*-type motions received by the observer. However, the above analysis, in which a continuous source is assumed, does not directly indicate the travel times of these waves. To do so, it is necessary to consider an applied pulse, which can be represented by a combination of values of ω. When the expressions for the potentials at a distant point are combined to give the effect of a source pulse, the times at which energy arrives can be determined. The final result is that an observer in the first medium can receive energy via a reflection, or via the critically refracted path, as predicted by the simple Huygens' principle analysis. In addition, he may receive energy by means of the converted waves, *PSS* or *PPS* where the letters indicate the wave type for each of the three legs of the head-wave path (Fig. C.3).

The conclusion, therefore, is that the critically refracted arrival is confirmed as a real event, in the case of incident spherical waves. It will be recognized that the surface defined by the first leg of the path is a cone extending from the source to the interface; also, the wave front associated with the refracted portion defines a spreading frustrum of a cone, and the waves have been known as "conical waves." Strangely enough, the contributions from the branch point cuts in the integral are almost second order effects, as compared to that of the pole, and it is somewhat paradoxical that the refraction event is often so prominent on seismic records. The contribution of the pole is none other than the Stonely wave, mentioned in Section 4.3; because it is confined to the interface, it is rarely observed in practice.

C.4 FREE OSCILLATIONS OF A UNIFORM ELASTIC SPHERE

The treatment by Lamb (1882) follows most directly from our previous discussions of the equations of elasticity (Section 2.2), although the problem was studied by Poisson more than 50 years earlier. The procedure is to find the most general solution of the equation of motions, subject to the boundary condition that the stresses vanish at the surface of the sphere. It will be necessary to develop the solution of the wave equation in spherical coordinates as we proceed; the development is somewhat lengthy, but the solutions will be required later in connection with the magnetic and gravitational fields of the earth.

The starting point is the equation of motion, in the absence of body forces

$$\rho \frac{\partial^2 \mathbf{u}_i}{\partial t^2} = (\lambda + \mu)\, \nabla\Theta + \mu\nabla^2 \mathbf{u}_i \qquad \text{C.4.1}$$

The neglect of body forces, hence of compression, may appear strange in a sphere the size of the earth; in fact, the specialized treatment yields natural frequencies which are only slightly modified when the effect of gravitation is included. If all displacements vary with time as e^{ipt}, we have

$$(\lambda + \mu)\, \nabla\Theta + \mu\nabla^2 \mathbf{u}_i + \rho p^2 \mathbf{u}_i = 0 \qquad \text{C.4.2}$$

By differentiating Equation C.4.2 with respect to x_i, and adding the components, we obtain, as before,

$$(\nabla^2 + h^2)\,\Theta = 0 \qquad \text{C.4.3}$$

where $h^2 = p^2\rho/(\lambda + 2\mu)$; in other words, Θ satisfies the wave equation for longitudinal waves. A fact which we have not used is that an equation for the displacements may be obtained from Θ; from Equation C.4.2 we have

$$(\nabla^2 + k^2)u_i = (1 - k^2/h^2)\ \nabla\Theta_i \qquad \text{C.4.4}$$

where

$$k^2 = p^2\rho/\mu$$

The procedure to determine the possible displacements is then to find solutions of Equations C.4.3 for Θ, and obtain u_i from Equation C.4.4. However, the most general solution of Equation C.4.4 includes also the complementary function, or solution of

$$(\nabla^2 + k^2)\,\mathbf{u}_i = 0 \qquad \text{C.4.5}$$

In the solution of both C.4.3 and C.4.5, functions must be found which are appropriate to the spherical boundary conditions. Equation C.4.3 involves the scalar quantity Θ, whereas Equation C.4.5 is the vector wave equation for the displacement $\mathbf{u}_i$. We investigate first the solution of the scalar equation. Spherical coordinates (r, θ, ϕ) are taken, with origin at the center of the sphere; in this system, r is radial distance, θ is colatitude, and ϕ is longitude, measured from some reference meridian plane. When the Laplacian operator is expressed in these coordinates, the wave equation becomes:

$$\frac{1}{r^2}\frac{\partial}{\partial r}\left(r^2\frac{\partial\Theta}{\partial r}\right) + \frac{1}{r^2\sin\theta}\frac{\partial}{\partial\theta}\left(\sin\theta\frac{\partial\Theta}{\partial\theta}\right) + \frac{1}{r^2\sin^2\theta}\frac{\partial^2\Theta}{\partial\phi^2} + h^2\theta = 0 \qquad \text{C.4.6}$$

If a solution is sought, in the form $\Theta = R(r)P(\theta)\,Q\,(\phi)$, it is found that the functions of the separated variables must satisfy

$$\left.\begin{aligned} r^2\frac{d^2R}{dr^2} + 2\frac{rdR}{dr} + (h^2r^2 - l(l+1))R &= 0 \\ \frac{1}{\sin\theta}\frac{d}{d\theta}\left(\sin\theta\frac{dP}{d\theta}\right) + (l(l+1) - m^2/\sin^2\theta)P &= 0 \\ d^2\Theta/d\phi^2 + m^2Q &= 0 \end{aligned}\right\} \qquad \text{C.4.7}$$

The second two equations are the same as those met in Appendix A, for the solution of Laplace's equation. It is evident, therefore, that the pattern of displacements over the surface of the sphere will be given by a sum of spherical harmonic functions.

The general solution of the equation satisfied by $R(r)$, known as Riccati's equation, is well-known to be

$$R_l(r) = r^l \left(\frac{1}{r}\frac{d}{dr}\right)^l \frac{A_l \sin hr + B_l \cos hr}{r} \qquad \text{C.4.8}$$

For solutions which remain finite at $r = 0$, it is necessary to choose $B_l = 0$.

We may now express the dilatation Θ in our particular problem in the compact form

$$\Theta = \sum_l \omega_l \psi_l(hr) \qquad \text{C.4.9}$$

where

$$\omega_l = r^l \sum_m P_l^m (\cos\theta)\,(A_l^m \sin m\phi + B_l^m \cos m\phi)$$

is known as a spherical solid harmonic of degree l, and

$$\psi_l(hr) = \left(\frac{1}{hr}\frac{d}{d(hr)}\right)^l \frac{\sin (hr)}{hr} \qquad \text{C.4.10}$$

Before we investigate the displacements or stresses related to this solution for Θ we shall return to the complementary equation, C.4 5 The solution of it presents some important differences to that employed above, because of the vector nature of $\boldsymbol{u}_i$, and it will be found that two approaches have been used. The more elegant is to transform the vector equation into spherical coordinates; if this is done, it is found that the Laplacian operation, ∇^2, is no longer represented by its application to each component of the vector. We shall follow Lamb in maintaining the displacements $\boldsymbol{u}_i$ in Cartesian coordinates, and in seeking solutions for each component in terms of spherical functions. By analogy with our solution for Θ, we must have

$$\boldsymbol{u}_i = U_{il}\psi_l(kr) \qquad \text{C.4.11}$$

where U_{il} is a solid spherical harmonic, and $\psi_l(kr)$ is the function defined above for $\psi_l(hr)$. It will be recalled that the condition for the existence of the complementary solution is the vanishing of the right side of Equation C.4.4, and therefore of Θ. Hence,

$$\frac{\partial u_i}{\partial x_i} = 0 \qquad \text{C.4.12}$$

and there are obvious constraints upon the choice of the functions U_{il}. Equation C.4.11 is satisfied if

$$U_{1l} = x_2 \frac{\partial \chi_l}{\partial x_3} - x_3 \frac{\partial \chi_l}{\partial x_2} \qquad \text{C.4.13}$$

and similarly for U_{2l} and U_{3l}, where χ_l itself is a spherical solid harmonic of degree l. The substitution of Equation C.4.13 into C.4.11 therefore gives one possible solution for the displacements corresponding to the complementary function; but a characteristic of the vector wave equation, C.4.5, is that for any vector satisfying it, the curl of that vector is also a solution. The 1-component of $\nabla \times u_i$ is

$$(\nabla \times u_i)_1 = k\psi_l'(kr)\left(l\frac{x_1}{r}\chi_l - r\frac{\partial \chi_l}{\partial x_1}\right) - l(l+1)\psi_l(kr)\ \partial\psi_l/\partial x_1 \qquad \text{C.4.14}$$

where the prime indicates $d/d(kr)$. The properties of χ_l and ψ_l allow this to be rewritten as

$$(\nabla \times u_i)_1 = \frac{l+1}{2l+1}\psi_{l-1}(kr)\frac{\partial \chi_l}{\partial x_1} - \frac{l}{2l+1}\psi_{l+1}(kr)k^2r^{2l+3}\ \partial/\partial x_1\left(\frac{\chi_l}{r^{2l+1}}\right) \qquad \text{C.4.15}$$

which is therefore also a possible solution for the displacement. In fact, the solid spherical harmonic χ_l of Equation C.4.15 is not necessarily the same as the χ_l of Equation C.4.13; we shall denote the former by Φ_l, and obtain the final expression for u_1:

$$u_1 = -\frac{1}{h^2}\frac{\partial \Theta}{\partial x_1} + \sum\left[\psi_l(kr)\left(x_2\frac{\partial \chi_l}{\partial x_3} - x_3\frac{\partial \chi_l}{\partial x_2} + \frac{\partial \Phi_{l+1}}{\partial x_1}\right)\right.$$
$$\left. - \frac{l+1}{l+2}\psi_{l+2}(kr)k^2r^{2l+5}\frac{\partial}{\partial x_1}\left(\frac{\Phi_{l+1}}{r^{2l+3}}\right)\right] \qquad \text{C.4.16}$$

in which the first term is the particular integral, and the summation is the complementary function.

To apply the boundary condition of the vanishing of stress at the surface of the sphere, $r = a$, we must express the components of stress in terms of the displacements. It is here that the use of Cartesian coordinates could introduce an inconvenience. However, since all stresses, both normal and shear, must vanish at the surface, we may introduce stress components of the form p_{1r}, which give the force in the 1-direction, exerted across a unit area of the surface $r =$ constant. The component p_{1r}, for example, is given by

$$p_{1r} = \frac{x_1}{r}\left(\lambda\Theta + 2\mu\frac{\partial u_1}{\partial x_1}\right) + \frac{x_2}{r}\mu\left(\frac{\partial u_2}{\partial x_1} + \frac{\partial u_1}{\partial x_2}\right) + \frac{x_3}{r}\mu\left(\frac{\partial u_1}{\partial x_3} + \frac{\partial u_3}{\partial x_1}\right) \qquad \text{C.4.17}$$

which is obtained without difficulty from Equation 2.3.3. When the expressions for u corresponding to Equation C.4.16 are substituted into C.4.17, the condition for the vanishing of stress is found to be given by three equations of which the one for p_{1r} is

$$p_l\left(x_2\frac{\partial \chi_l}{\partial x_3} - x_3\frac{\partial \chi_l}{\partial x_2}\right) + a_l\frac{\partial \omega_l}{\partial x_1} + b_lr^{2l+3}\frac{\partial}{\partial x_1}\left(\frac{\omega_l}{r^{2l+1}}\right) + c_l\frac{\partial \Phi_l}{\partial x_1}$$
$$+ d_lr^{2l+3}\frac{\partial}{\partial x_1}\left(\frac{\Phi_l}{r^{2l+1}}\right) = 0 \qquad \text{C.4.18}$$

where p_l, a_l, c_l and d_l are constants given by

$$p_l = (l-1)\psi_l(ka) + ka\psi_l'(ka)$$

$$a_l = \frac{1}{(2l+1)h^2}\left\{k^2a^2\psi_l(ha) + 2(l-1)\,\psi_{l-1}(ha)\right.$$
$$b_l = -\frac{1}{2l+1}\left\{\frac{k^2}{h^2}\,\psi_l(ha) + \frac{2(l+2)}{ha}\,\psi_l'(ha)\right. \qquad \text{C.4.19}$$
$$c_l = k^2a^2\psi_l(ka) + 2(l-1)\psi_{l-1}(ka)$$
$$d_l = k^2\frac{l}{l+1}\left\{\psi_l(ka) + \frac{2(l+2)}{ka}\,\psi'(ka)\right\}$$

The condition may be expressed in terms of two simpler equations, by first operating on the three equations corresponding to C.4.18 with $\partial/\partial x_i$ and adding to give

$$b_l\omega_l + d_l\Phi_l = 0 \qquad \text{C.4.20}$$

or by multiplying by x_i and adding, to give

$$a_l\omega_l + c_l\Phi_l = 0 \qquad \text{C.4.21}$$

It follows also that there are three conditions of the form

$$p_l\left(x_2\frac{\partial\chi_l}{\partial x_3} - \frac{x_3\,\partial\chi_l}{\partial x_2}\right) = 0 \qquad \text{C.4.22}$$

We are finally in a position to see how the boundary condition can be satisfied. One possibility is that the spherical solid harmonics ω_l and Φ_l are identically zero, which requires, from Equation C.4.18, that p_l also vanish. This leads to what are known as vibrations of the first class. The dilatation is everywhere zero (since $\omega_l = 0$), and there is no radial component of displacement. In modern terminology, these vibrations are known as torsional oscillations. Before we investigate their properties further, we note that a second way of satisfying Equation C.4.18 is to have $\chi_l = 0$. If the harmonics ω_l and Φ_l are not to vanish, Equations C.4.20 and C.4.21 require that

$$a_l\,d_l - b_lc_l = 0 \qquad \text{C.4.23}$$

The corresponding vibrations are known as those of the second class, or coupled oscillations. They involve both radial displacements and torsional motions.

Of particular interest are the natural frequencies for the different types of oscillations. For torsional oscillations, we must have

$$p_l = (l-1)\psi_l(ka) + ka\,\psi_l'(ka) = 0 \qquad \text{C.4.24}$$

where $\psi_l(ka)$ is of the form given by Equation C.4.10. The precise function depends on l; when this is given, k, and from it the frequency, are determined by Equation C.4.24.

The displacements are given by the surviving terms of Equation C.4.16; for torsional oscillations these are of the form

$$\psi_l(kr)\left(x_2\frac{\partial\chi_l}{\partial x_3} - x_3\frac{\partial\chi_l}{\partial x_2}\right)$$

and these vanish for

$$\psi_l(kr) = 0 \qquad \text{C.4.25}$$

We should, therefore, expect to find values of r corresponding to zero displacement within the sphere.

Since the displacements depend upon the derivatives of χ_l, there is no solution corresponding to $l = 0$. For $l = 1$, χ_l changes sign at the equator, but its gradient does not, and the displacement has no node over the surface. In fact, concentric shells within the sphere execute small rigid-body rotations about the polar axis (Fig. 8.2). Putting $l = 1$ in Equation C.4.10 gives

$$\tan ka = 3ka/(3 - k^2a^2)$$

of which the lowest roots are

$$\frac{ka}{\pi} = 1.8346,\ 2.8950,\ 3.9225 \cdots$$

The number π/ka is the ratio of the period to the time required for an S wave to travel across the diameter of the sphere. For a uniform sphere of the diameter of the earth, with $\beta = 6$ km/sec, the value of 1.8346 corresponds to a period of 19 minutes. Nodal surfaces in radius are given by

$$\tan kr = kr$$

of which the lowest roots are

$$\frac{kr}{\pi} = 1.4303,\ 2.4590,\ 3.4709 \cdots$$

The oscillation with the longest period has a single nodal surface at $r = 0.7796a$; oscillations corresponding to higher values of ka/π have increasing numbers of nodal surfaces. We refer to the first as the fundamental, and to the latter oscillations as overtones. The analysis of this special case, therefore, has demonstrated that, for a particular harmonic representation over the sphere (i.e., value of l), there is a series of possible torsional oscillations, with increasing frequency and increasing number of radial nodal surfaces. For $l = 2$, there is a nodal line for displacement around the equator, and the two hemispheres oscillate in anti-phase.

The coupled oscillations, or vibrations of the second class, are less straightforward, since the determination of natural frequency, corresponding to the condition given in Equation C.4.23, requires that the ratio h/k be given. Usually, results are quoted for the condition $\lambda = \mu$ (i.e., $k^2/h^2 = 3$) which we have already noted to be a reasonable approximation in the actual earth. For $l = 0$, the sphere retains its form, but suffers a periodic variation in radius. In this case,

$$b_0 = 0$$

or

$$\psi_0(ha) + (4/k^2a^2)ha\psi_0'(ha) = 0$$

which implies

$$\frac{\tan ha}{ha}=\frac{1}{1-\frac{1}{4}(k^2/a^2)}$$

The lowest roots are

$$\frac{ha}{\pi}=0.8160,\ 1.9285,\ 2.9359\ \cdots$$

where the number π/ha is now the ratio of the period to the time required for a P wave to traverse the diameter of the sphere. For a uniform sphere of the diameter of the earth, with $\alpha=10$ km/sec and $\lambda=\mu$, the greatest period is about 16 minutes. As in the case of torsional oscillations, there is a series of overtones for this single value of l.

Coupled oscillations with $l=1$ would require the application of an external force, and are not permitted under our assumptions. The case $l=2$, $m=0$, however, is of considerable importance. There are two nodal parallels of latitude, dividing the surface into three zones. As the sphere oscillates, it is distorted alternately into an oblate and a prolate spheroid. This particular oscillation has been referred to as the football mode (Fig. 8.2).

A word of caution is in order concerning the representation of the free modes by spherical harmonics. In the development above, nothing has been said about the source of the oscillations, and the pole of the harmonics, in Figure 8.2 for example, has been taken at the north pole of the sphere. In a real case the symmetry of the pattern will be determined by the location of the earthquake source; oscillations in the football mode, as one example, would be set up with the extremities of the football at the epicentre and the antipodal point.

APPENDIX D: GEOMAGNETIC TOPICS

D.1 ELECTROMAGNETIC INDUCTION IN A SPHERE

The purpose of this section is to outline the procedure established by Lahiri and Price (1939) in more detail than was possible in Chapter 20. It will be recalled that the observational material consists of the ratio of magnetic field at the earth's surface due to external sources, to that due to internal currents, for a series of known periods, each related to a known spherical harmonic distribution over the earth's surface. The problem is to place limits on the radial distribution of electrical conductivity within the earth. In the classical approach, which is described here, the indirect problem only is discussed. That is, earth models of given conductivity are assumed and the ratios of externally-to-internally produced fields are to be calculated. As noted in Chapter 20, the recent approach is to invert the observations to obtain the conductivity, but this is a second and more difficult step.

We begin again with Maxwell's equations (Section 15.2) and eliminate H to obtain

$$\nabla \times \nabla \times \boldsymbol{E} = -4\pi\sigma(\rho)\,\dot{\boldsymbol{E}} \qquad \text{D.1.1}$$

Here, σ is taken to be a function of the dimensionless radial parameter ρ $(\rho \times r/a)$, and the permeability is assumed to be unity. If $\boldsymbol{E}$ is written as

$$\boldsymbol{E} = -\dot{\boldsymbol{A}} \qquad \text{D.1.2}$$

it is found that new vector $\boldsymbol{A}$ satisfies

$$\boldsymbol{A} = \boldsymbol{r} \times \nabla u \qquad \text{D.1.3}$$

where the scalar function u is a solution of

$$\nabla^2 u = 4\pi\sigma(\rho)\dot{u} \qquad \text{D.1.4}$$

Equation D.1.4 is the diffusion equation, whose solution in the case of spherical symmetry is similar to that of the wave equation which we met in connection with the free oscillations of the earth (Section 8.2). In particular, u may be written as a sum of surface spherical harmonics (we take, for simplicity, the reference sphere for these to be the conducting sphere of radius a; this is not necessary in general):

$$u = a \, \Sigma \, f_l(t, \rho) \, S_l \qquad \text{D.1.5}$$

where, by methods analogous to those of Appendix A, it is found that $f_l(t, \rho)$ satisfies

$$\frac{\partial}{\partial \rho}\left(\rho^2 \frac{\partial f_l}{\partial \rho}\right) = \left\{l(l+1) + 4\pi a^2 \rho^2 \sigma(\rho) \frac{\partial}{\partial t}\right\} f_l(t, \rho) \qquad \text{D.1.6}$$

The lth term in the sum for the auxiliary vector $\boldsymbol{A}$ inside the sphere is then

$$\boldsymbol{A}_l = a[\boldsymbol{r} \times \nabla (f_l) \, S_l] = a[\boldsymbol{r} \times \nabla S_l] f_l(t, \rho) \qquad \text{D.1.7}$$

Both electromagnetic field vectors are derivable from $\boldsymbol{A}$; $\boldsymbol{E}$ from Equation D.1.2. and $\boldsymbol{H}$ from

$$\boldsymbol{H} = \nabla \times \boldsymbol{A} \qquad \text{D.1.8}$$

But $\boldsymbol{H}$, on or outside the sphere, is also derivable from the scalar potential U, which is taken to be expressed as a series of spherical harmonics, as in Appendix A, in the form:

$$U = a\Sigma \, \{e_l(t) \, \rho^l + i_l(t) \, \rho^{-l-1}\} \, S_l \qquad \text{D.1.9}$$

where the coefficients e_l and i_l give the contributions of external and internal sources to each harmonic term.

Since $\nabla \times \underline{A} = -\nabla U$, Equation D.1.9 is equivalent to writing

$$\boldsymbol{A}_l = a(\boldsymbol{r} \times \nabla S_l) \left\{\frac{\rho^l}{l+1} e_l(t) + \frac{\rho^{-l-1}}{l} i_l(t)\right\} \qquad \text{D.1.10}$$

for $\boldsymbol{A}$ outside the sphere.

The boundary conditions to be satisfied are the continuity, at the surface of the sphere, of the tangential components of $\underline{E}$ and $\underline{H}$. Deriving these from Equations D.1.7 and D.1.10 for regions inside and outside the sphere, and equating at $r = a$, gives

$$\left.\begin{aligned} f_l(t, \rho) &= \frac{1}{l+1} e_l(t) - \frac{1}{l} i_l(t) \\ f_l(t, \rho) + \frac{\partial}{\partial \rho} f_l(t, \rho) &= e_l(t) + i_l(t) \end{aligned}\right\}_{\rho=1} \qquad \text{D.1.11}$$

Recalling that the terms $e(t)$ are related to the primary source, and $f(t, \rho)$ and $i(t)$ to the internally induced currents and the secondary fields due to them, let us write the latter as the result of operations on $e(t)$:

$$\begin{aligned} i(t) &= I(i\alpha) e(t) \\ f(t, \rho) &= F(i\alpha, \rho) e(t) \end{aligned} \qquad \text{D.1.12}$$

where the functions I and F are obtained by solving Equations D.1.11, subject to D.1.6. All field quantities are taken to vary as $e^{i\alpha t}$. For a given distribution of conductivity and given frequency of the field, Equation D.1.6 determines $F(i\alpha, \rho)$ except for a coefficient independent of ρ. We can therefore write

$$F(i\alpha, \rho) = C(i\alpha) R(i\alpha, \rho) \quad \text{D.1.13}$$

where $R(i\alpha, \rho)$ is known. Then, from D.1.11,

$$\left.\begin{aligned} C(i\alpha) &= \frac{2l+1}{l+1} \frac{1}{(l+1) R + R'} \\ I(i\alpha) &= \frac{l}{l+1} \left\{ 1 - \frac{(2l+1) R}{(l+1) R + R'} \right. \end{aligned}\right\} \quad \text{D.1.14}$$

where R and R' are the values of $R(i\alpha, \rho)$ and $\frac{d}{d\rho} R(i\alpha, \rho)$ at $\rho = 1$. Thus, when a conductivity distribution $\sigma(\rho)$ has been assumed, $R(i\alpha, \rho)$ can be determined from Equation D.1.6, and from it the operators $I(i\alpha)$ and $C(i\alpha)$ can be found. The key to the application of the approach is that the spherical harmonic analysis of a periodic field variation, by the method of Gauss, yields the amplitude and phase of $i(t)$ relative to $e(t)$. The modulus and argument of the complex operator $I(i\alpha)$ are therefore known quantities. What is required is to calculate $I(i\alpha)$ for different earth models, and to compare the results with the value given by observations.

To illustrate the method of solution in the simplest case, let us consider a uniform sphere of conductivity σ, immersed in a uniform magnetic field H_z. If the z-axis is taken as the polar axis of spherical harmonics, the potential over the surface of the sphere, due to the external field, is given by the first degree zonal harmonic, P_1^0. Equation D.1.6 then reduces to, writing R in place of $f(\rho)$,

$$\rho^2 \frac{\partial^2 R}{\partial \rho^2} + 2\rho \frac{\partial R}{\partial \rho} - (2 + k^2\rho^2) R = 0 \quad \text{D.1.15}$$

where

$$k^2 = 4\pi i\alpha\sigma$$

This equation can be easily converted to Bessel's equation of order $(1 + \frac{1}{2})$ through the substitutions

$$R = \rho^{-1/2} w \qquad v = k\rho$$

for then,

$$v^2 \frac{d^2w}{dv^2} + 2 \frac{dw}{dv} - [(1 + \tfrac{1}{2})^2 + v^2] w = 0 \quad \text{D.1.16}$$

The general solution is given in terms of the modified Bessel functions $I_{l+1/2}(k\rho)$ and $K_{1+1/2}(k\rho)$ as

$$R = \rho^{-1/2} w = \rho^{-1/2}[C_1 I_{1+1/2}(k\rho) + C_2 K_{1+1/2}(k\rho)] \quad \text{D.1.17}$$

or, from properties of Bessel functions and our conditions,

$$R = \frac{3\rho}{k^2\rho^2}\left(\cosh k\rho - \frac{\sinh k\rho}{k\rho}\right) \qquad \text{D.1.18}$$

Equation D.1.14 then gives, in terms of a,

$$I(i\alpha) = -\frac{3}{2}a^3\left[\frac{1}{3} + \frac{1}{k^2a^2} - \frac{\cosh ka}{ka \sinh ka}\right] \qquad \text{D.1.19}$$

The potential over the surface of the sphere due to the internally induced currents is also distributed as a first-degree spherical harmonic. It is therefore equivalent to a central magnetic dipole, oriented in the direction of the applied field, of moment μ, where

$$\mu = -\frac{3}{2}H_za^3\left[\frac{1}{3} + \frac{1}{k^2a^2} - \frac{\cosh ka}{ka \sinh ka}\right] \qquad \text{D.1.20}$$

The term in square brackets is a complex number, whose real and imaginary parts have been computed as a function of ka by Wait (1951). As ka increases (it is impossible to separate the effects of radius, conductivity and frequency) the imaginary component passes through a maximum, while the real component continues to increase. For very large values of ka, the effective dipole is 180° out of phase with H_z and its magnitude is the same as that predicted in Equation 18.4 1 for the rotating conducting sphere.

The chief interest in the case of a uniform sphere in a uniform magnetic field is in connection with the detection of local variations in conductivity (Section 20.4). In the study of the earth as a whole, it is necessary to provide for a radial variation in conductivity, and to consider higher harmonics of the inducing field. For the conductivity to decrease with increasing radius, Lahiri and Price took the distribution

$$\sigma = \sigma_0\rho^{-m} \qquad \text{D.1.21}$$

When this is substituted in Equation D.1.6, the equation for R is found to reduce to Bessel's differential equation; $I(i\alpha)$ and $C(i\alpha)$ are expressible as sums of Bessel functions. Their numerical analysis was based chiefly on the third spherical harmonic of the daily variation. To match the observed ratio e/i of 2.2, and phase lag of 18°, it was found that a value of m as high as 30 would be required.

D.2 ELECTRODYNAMIC CONSIDERATIONS IN THE MAGNETOSPHERE

To investigate some properties of the neutral, conducting plasma which proceeds from the sun, mentioned in Section 19.4, we start from the same equation met in Section 17.5:

$$4\pi\sigma\dot{\boldsymbol{H}} = 4\pi\sigma\nabla \times (\boldsymbol{v} \times \boldsymbol{H}) - \nabla \times \nabla \times \boldsymbol{H} \qquad \text{D.2.1}$$

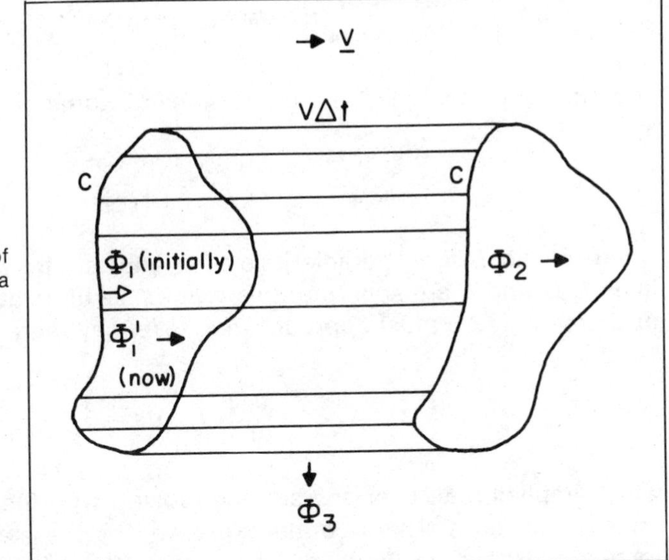

Figure D.1 A contour of integration *C* moving with a highly conducting fluid.

which holds at every point of the plasma. Consider a surface *S*, bounded by a curve *C*, initially fixed in space in the plasma. Integrating the above equation over *S* and applying Stokes' formula to express all quantities on the right in terms of integrals around *C*, we have

$$\begin{aligned} \frac{\partial}{\partial t}\int_s \boldsymbol{H}\cdot d\boldsymbol{s} &= \oint (\boldsymbol{v}\times\boldsymbol{H})\cdot d\boldsymbol{c} - \frac{1}{4\pi\sigma}\oint \nabla\times\boldsymbol{H}\cdot d\boldsymbol{c} \\ &= \oint \boldsymbol{H}\cdot(d\boldsymbol{c}\times\boldsymbol{v}) - \frac{1}{4\pi\sigma}\oint \nabla\times\boldsymbol{H}\cdot d\boldsymbol{c} \end{aligned} \qquad \text{D.2.2}$$

We now let the curve *C* move with the fluid and understand *d/dt* to infer differentiation with respect to the moving fluid. In a time Δt, a surface is swept out as shown in Figure D.1, where the magnetic flux Φ across each portion of it is indicated.

$$\frac{d}{dt}\int_s \boldsymbol{H}\cdot d\boldsymbol{s} = \frac{\Phi_2-\Phi_1}{\Delta t} \qquad \text{D.2.3}$$

but since

$$-\Phi_1' + \Phi_2 + \Phi_3 = 0$$

$$\begin{aligned} \frac{d}{dt}\int_s \underline{\boldsymbol{H}}\cdot \underline{d\boldsymbol{s}} &= \frac{\Phi_1'-\Phi_1}{\Delta t} - \frac{\Phi_3}{\Delta t} \\ &= \frac{\partial}{\partial t}\int_s \underline{\boldsymbol{H}}\cdot \underline{d\boldsymbol{s}} - \frac{\Phi_3}{\Delta t} \end{aligned} \qquad \text{D.2.4}$$

However, $-\Phi_3/\Delta t$ is given by the first integral on the right of Equation D.2.2, so we have

$$\frac{d}{dt}\int_s \mathbf{H}\cdot d\mathbf{s} = -\frac{1}{4\pi\sigma}\oint \nabla\times\mathbf{H}\cdot d\mathbf{c} \qquad \text{D.2.5}$$

The ratio of magnitudes of the two sides of Equation D.2.5 is given by

$$\frac{\text{R.S.}}{\text{L.S.}} = \frac{[T]}{\sigma[L^2]} \qquad \text{D.2.6}$$

where T and L are characteristic time and length, respectively. If the quantities T, L, and σ are such that this ratio is small compared to unity, both sides of Equation D.2.5 must approach zero. We may therefore take

$$\frac{d}{dt}\int_s \mathbf{H}\cdot d\mathbf{s} = 0 \qquad \text{D.2.7}$$

which implies that, over any surface moving with the fluid, the magnetic flux remains constant. This is often expressed by the statement that the lines of force are "frozen" to the fluid. If a plasma with no field impinges on a magnetic field in space, the lines of the latter field are frozen out of the fluid.

The (old) concept of tubes of magnetic force is useful in dealing with plasmas. We visualize a tube as being generated by all lines of force which pass through a given curve. The flux along any tube remains constant, and since the density of energy in a magnetic field is given by

$$u = \frac{H^2}{8\pi} \qquad \text{D.2.8}$$

the total energy in a tube is

$$U = \frac{\Phi^2}{8\pi}\int \frac{dl}{A} \qquad \text{D.2.9}$$

where A is the cross-sectional area and the integration is along the tube.

For a given length and volume, the energy is least if A is constant along the volume; a narrow neck tends to increase the energy. For a given volume and area, the energy is proportional to the square of the length. Tubes of force may therefore be thought of as elastic filaments in tension, exerting a pressure $(H^2/4\pi)$, perpendicular to the boundaries of the tubes.

The final property of a plasma which we require is its ability to sustain hydromagnetic waves. The existence of such waves was shown by Alfven (1950), initially for the simple case of plane waves in an incompressible fluid permeated by a uniform field H_0. Let axes be chosen so that z is parallel to H_0; we investigate solutions of Maxwell's equations which depend upon z and time t only. The total field is given by

$$\mathbf{H} = \mathbf{H}_0 + \mathbf{h}$$

where $\mathbf{h}$ results from the current induced by the mave motion. The plane wave condition requires the current density $\boldsymbol{J}$ to have no z components; we may take J as J_x and write

$$J_x = -\frac{1}{4\pi}\frac{\partial h_y}{\partial z} \qquad \text{D.2.10}$$

$$h_x = 0$$

The equation of motion of the fluid, neglecting body forces, is

$$\rho\frac{d\mathbf{v}}{dt} = \mathbf{J} \times \mathbf{B} - \nabla p \qquad \text{D.2.11}$$

where ρ is density and p the pressure.

Again, the plane wave condition (∇p parallel to z) and the incompressibility of the fluid ($\nabla \cdot \mathbf{v} = 0$) lead to a single component of velocity, v_y. Equations D.2.10 and D.2.11 yield

$$\frac{\partial v_y}{\partial t} = \frac{H_0}{4\pi\rho}\frac{\partial h_y}{\partial z} \qquad \text{D.2.12}$$

This may be written in terms of h_y only by introducing the electric field

$$E_x = \frac{J_x}{\sigma} - v_y H_0 \qquad \text{D.2.13}$$

and noting that

$$\frac{\partial h_y}{\partial t} = \frac{\partial E_x}{\partial z}, \qquad \text{D.2.14}$$

for then

$$\frac{\partial^2 h_y}{\partial t^2} = H_0\frac{\partial^2 v_y}{\partial t\,\partial z} - \frac{1}{\sigma}\frac{\partial^2 J_x}{\partial t\,\partial z} \qquad \text{D.2.15}$$

The equation of motion becomes

$$\frac{\partial^2 h_y}{\partial t^2} = \frac{H_0^2}{4\pi\rho}\frac{\partial^2 h_y}{\partial z^2} + \frac{1}{4\pi\sigma}\frac{\partial^3 h_y}{\partial z^2\,\partial t} \qquad \text{D.2.16}$$

If the conductivity is sufficiently high, the second term on the right of Equation D.2.16 may be neglected. The resulting equation represents plane waves propagating with the Alfven velocity

$$v = \frac{H_0}{\sqrt{4\pi\rho}} \qquad \text{D.2.17}$$

along the lines of force. The waves represent the propagation of the small distortional field h_y; they may be thought of as oscillations of the lines of force of the total field. Since the lines are locked to the fluid, the fluid is set into oscillatory motion perpendicular to the lines of H_0. In the more general case of compressible fluid, the waves are found to be dispersive.

We turn next to a review of the motion of a single charged particle in a magnetic field, as this is the basis for the trapping of solar particles in the earth's field.

A particle of mass m and charge e (assumed positive) injected into a uniform field $\boldsymbol{H}$ spirals around a line of force at a constant pitch angle. Its equation of motion is

$$m \frac{d\boldsymbol{v}}{dt} = e(\boldsymbol{v} \times \boldsymbol{H}) \qquad \text{D.2.18}$$

If we let the particle velocity v be composed of components $v_{\|}$ and $v_{\perp}$, respectively parallel and perpendicular to the field, $v_{\|}$ remains constant. The motion represented by $v_{\perp}$ has constant speed, the particle rotating about the line of force with angular velocity ω where

$$\omega = \frac{eH}{m} \qquad \text{D.2.19}$$

and at radius

$$a = v_{\perp} \frac{m}{eH} \qquad \text{D.2.20}$$

The circular motion of the charged particle is equivalent to a magnetic dipole of strength μ, where

$$\underline{\mu} = \frac{(-\frac{1}{2}mv_{\perp}^2)\ \underline{H}}{H^2} \qquad \text{D.2.21}$$

If the field is non-uniform, two new effects arise, which are of the greatest consequence to the theory of magnetic disturbances. The equation of motion in this case is complicated, and is usually solved by a series of approximations. In the case of a slowly varying field, we may take

$$\boldsymbol{H} = \boldsymbol{H}_0 + \alpha \boldsymbol{r} \qquad \text{D.2.22}$$

where $\boldsymbol{H}_0$ is the field in the z-direction at a local origin, $\boldsymbol{r}$ is the position vector of a particle and α is the tensor $\partial H_i / \partial x_j$. For a first approximation the particle's position at time t is assumed to be given by the results from the uniform field case, so that

$$x = v_{\perp}/\omega \sin \omega t$$

$$y = v_{\perp}/\omega \cos \omega t$$

$$z = v_{\|} t$$

These values are used to determine $\boldsymbol{r}$, and the resulting $\boldsymbol{H}$ is substituted in the equation of motion 19.4.18. The solution indicates that there is an acceleration

$$\frac{dv_{11}}{dt} = \frac{-v_\perp^2}{2H}\frac{\partial H}{\partial z} \qquad \text{D.2.23}$$

and a drift velocity

$$v_D = \frac{1}{\omega}\frac{(\frac{1}{2}v_\perp^2 + v_{||}^2)}{H^2}(\boldsymbol{H} \times \nabla H) \qquad \text{D.2.24}$$

where ∇H is the vector formed by applying the grad operation to the scalar magnitude field of $\boldsymbol{H}$. These results show that a particle no longer spirals at a constant pitch angle. As Equation D.2.23 indicates, if the field is increasing in the direction of advance, $v_{||}$ is reduced, and the forward motion is stopped. In fact, if the energy of a particle remains constant, its magnetic moment μ can be shown to be constant. If θ is the angle between the trajectory and the line of force, we have

$$v_\perp = v \sin \theta$$

so that

$$\mu = \frac{mv^2 \sin^2 \theta}{2H} = \text{constant} \qquad \text{D.2.25}$$

Therefore

$$\frac{\sin^2 \theta}{H} = \text{constant}$$

and the particle will advance only to a point where θ becomes 90°. At such a point, the particle is said to strike a magnetic mirror, after which it spirals in the opposite sense along the line of force. The second effect, the drift given by Equation D.2.24, is perpendicular to the field lines and to the gradient of the scalar magnitude of field. For a dipole field, such as the earth's main field, ∇H is essentially in the radial direction, so that the drift for a particle above the equator is longitudinal. The sense of drift is opposite for positively and negatively charged particles.

REFERENCE

Wait, J. R. (1951) A conducting sphere in a time varying magnetic field. *Geophysics* **16**, 666.

APPENDIX E: SI UNITS

Many journals now insist upon SI units, but as much of the geophysical literature is written in other systems, it has not been considered necessary to use the SI system throughout this book. Conversion factors for the most important quantities found in geophysics are given here and illustrated in the end papers.

The SI system has been described in detail by Markowitz (1973). Basically, it is a meter-kilogram-second system, with approved derived units, and with standard prefixes corresponding to powers of 10 increasing by multiples of 3 (or −3). Electrical quantities are derived from current, defined through the ampere in terms of the force between parallel conductors, while magnetic fields are described fundamentally by the flux density, expressed in teslas.

Conversions of mechanical quantities (e.g., mass, length, density, moment of inertia) expressed in c.g.s. units should present no difficulty.

Pressure is expressed in Pascals, with 1 Pa equal to 1 Newton/m^2. The kilobar, so often used in geophysics, is 10^8 Pa.

Magnetic field is conveniently given in nanoteslas (1×10^{-9} tesla):

$$1 \text{ gamma} = 1 \text{ nanotesla}$$

The ohm meter is the SI unit of resistivity, while its reciprocal, the unit of conductivity, is the siemen per meter.

For heat flow, the unit is milliwatts/m^2:

$$1 \times 10^{-6} \text{ cal/cm}^2\text{sec (hfu)} = 41.9 \text{ mW/m}^2,$$

while for heat productivity:

$$1 \times 10^{-13} \text{ cal/cm}^3\text{sec} = 0.419\ \mu\text{W/m}^3.$$

REFERENCE

Markowitz, W. (1973) SI, the international system of units. *Geophys. Surv.* **1**, 217.

APPENDIX F: INTERNATIONAL COOPERATION, PROJECTS, AND EXCHANGE OF DATA IN GEOPHYSICS

Geophysics owes much to international cooperation in the adoption of standards, the processing of data, and the exchange of original data between countries. The international (but non-governmental) body responsible for the physics of the earth is the International Union of Geodesy and Geophysics (IUGG), one of the scientific unions federated in the International Council of Scientific Unions (ICSU). Within IUGG there are seven associations, of which four are directly concerned with the solid earth: Geodesy; Seismology and Physics of the Earth's Interior; Geomagnetism and Aeronomy; and Volcanology and the Chemistry of the Earth's Interior. Resolutions of IUGG or of the associations, passed at General Assemblies, have established, for example, scales of magnitudes for earthquakes (Section 6.2), the International Gravity Formula (Section 11.4) and the International Geomagnetic Reference Field (Section 17.3).

The reduction and processing of geophysical data includes the study of variations in the earth's rate of rotation, conducted by the Bureau International de l'Heure, and the tracing of the polar motion, conducted by the International Polar Motion Service. Indices of geomagnetic activity are produced and published by the Permanent Service for Geomagnetic Indices. Gravity maps are published by the International Gravity Bureau. Information about earth tides is assembled and published by the International Center for Earth Tides. The work of all of these Permanent Services is supervised by IUGG, in some cases in cooperation with other scientific unions. Located as they are in host institutes around the world, the Services can be reached through the permanent office of ICSU (51 Bd. de Montmorency, Paris, France).

The storage and exchange of unreduced data (e.g., copies or microfilms of seismograms and magnetograms) is the responsibility of the World Data Centers, which are also organized under ICSU. World Data Center *A* is located in Washington, D.C. (c/o National Academy of Sciences, 2101 Constitution Ave.), while World Data Center *B* is in Moscow, USSR. Data centers publish catalogues of data in storage, which may usually be obtained upon payment of the cost of reproduction. Unfortunately, not all measurements of geophysical interest are filed with the Centers; gravity measurements, for example, are considered in some countries to be classified information.

The international scientific unions organized the International Geophysical Year in 1957–58 (the project which led to the establishment of the World Data Centers), and the Upper Mantle Project in 1963–1970. One of the outstanding achievements of the latter was a series of symposia on Theory and Computers in Geophysics, still continuing under the title Symposia on Mathematical Geophysics. The reader who wishes to pursue further such topics as the applications of computers to the inversion of geophysical observations, model calculations, statistical analysis in geophysics, and time-series analysis, cannot do better than to refer to the published papers of these symposia. The references are:

Symposium	*Collected Papers Published in:*
1 Moscow-Leningrad, 1964	*Reviews of Geophysics* **3**, No. 1 (Special Issue), 1965.
2 Rehovoth, 1965	*Geophysical Journal, Roy. Ast. Soc.* **11**, Nos. 1–2 (Special Issue), 1966.
3 Cambridge, England, 1966	*Geophysical Journal, Roy. Ast. Soc.* **13**, Nos. 1–3 (Special Issue), 1967.
4 Trieste, 1967	*Supplemento al Nuovo Cimento* **6**, No. 1 (Special Issue), 1968.
5 Tokyo-Kyoto, 1968	*Journal of Physics of the Earth* 16 (Special Issue), 1968.
6 Copenhagen, 1969	*Geophysical Journal, Roy. Ast. Soc.* **21**, Nos. 3–4 (Special Issue), 1970.
7 Stockholm, 1970	*Geophysical Journal, Roy. Ast. Soc.* **25**, Nos. 1–3 (Special Issue), 1971.
8 Moscow, 1971	–
9 Banff, 1972	*Geophysical Journal, Roy. Ast. Soc.* **35**, Nos. 1–3 (Special Issue), 1973.
10 Cambridge, 1974	*Geophysical Journal, Roy. Ast. Soc.* **42**, No. 2, 1975.
11 Seeheim/Odenwald (F.R.G.), 1976	*Journal of Geophysics* **43**, Nos. 1–2 (Special Issue), 1977.

AUTHOR INDEX

Page references in *italics* refer to bibliographic entries.

SUBJECT INDEX